Birkhäuser

More information about this series at http://www.springer.com/series/4927

Aviation Meteorology: Observations and Models

Edited by
Ismail Gultepe
Wayne F. Feltz

Previously published in *Pure and Applied Geophysics* (PAGEOPH), Volume 176, No. 5, 2019

Birkhäuser

Editors
Ismail Gultepe
Meteorological Research Division
Environment and Climate Change Canada (ECCC)
Toronto, ON, Canada

The Faculty of Engineering and Applied Science
Ontario Technical University (UOIT)
and Automotive Center of Excellence (ACE)
Oshawa, ON, Canada

Wayne F. Feltz
Institute for Meteorological Satellite Studies (CIMSS)
University of Wisconsin-Madison
Madison, WI, USA

ISSN 2504-3625
Pageoph Topical Volumes

Cover illustration: Taken by the editor I. Gultepe during an ice fog project took place nearby Yellowknife International Airport during January 2011, NWT, Canada

This book is published under the imprint Birkhäuser, www.birkhauser-science.com by the registered company Springer Nature Switzerland AG
The registered company address is: Gewerbestrasse 11, 6330 Cham, Switzerland

Contents

Pure Appl. Geophys. 176 (2019), 1863–1867
© 2019 Crown
https://doi.org/10.1007/s00024-019-02188-2

Pure and Applied Geophysics

Aviation Meteorology: Observations and Models. Introduction

Ismail Gultepe[1,2] and Wayne F. Feltz[3]

This special issue of *Pure and Applied Geophysics* contains an introduction plus 14 papers related to *Aviation Meteorology*. The papers cover important topics such as wind, gust, turbulence, visibility related to fog and precipitation, convection and lightning, icing, and high ice water content and engine icing. These research areas are important to develop applications for aviation meteorology, and in particular, aviation operations. The results discussed in this special issue are generated by research efforts conducted internationally, and include several review papers. Review papers on aviation meteorology (Gultepe et al. 2019a), gravity waves and convectively-induced turbulence (Sharman and Trier 2019), lidars (Thobois et al. 2019), and geostationary satellites (Ellrod and Pryor 2019), as well as supersite observations (Gultepe et al. 2019b) provided extensive details for operational applications. Some of these papers were included in various international conferences, including the EGU (European Geophysical Union) aviation meteorology sessions, that took place in Vienna, Austria, during 2016–2018.

Aviation meteorology covers many interrelated subjects and engineering topics. These include ground-based operations that are related to adverse effect of frost, ice, snow, and visibility on aircraft performance and in-flight icing and turbulence. The effects of these on aircraft are decreased thrust and lift, increased drag, increased stall speed, and modified handling characteristics (TC AIM 2019; Bragg et al. 2000). TC AIM report suggested that the type of frost, ice or snow, that can accumulate on an aircraft, while on the ground, is a key factor in determining the type of de-icing/anti-icing procedures. Figure 1 shows an icing event that occurred during an aircraft icing project, which took place in Eastern Canada (a) (Isaac et al. 2005), and a severe frost condition that occurred during an ice fog project (b) (Gultepe et al. 2015). At the ground stops for aircraft, wet snow, freezing drizzle, or rain, with the ambient temperature around 0 °C or even in a temperature range between 8° and 14 °C, can be critical for aviation operations (TC AIM 2019).

Wind shear and turbulence conditions can also be critical for aircraft flight conditions, as clear air turbulence (CAT) can be reported anywhere during a flight path. The CAT events are usually related to frontal weather systems and jet cores called jet streaks (Fig. 2), with a sketch adapted from TC AIM (2019), showing the possible areas of CAT conditions at the higher level of atmosphere (Fig. 2a). Figure 2b shows the airflow streamlines behind an aircraft taking off, indicating vortex core and vortex flow directions. Based on these issues, the current special issue will present the recent works to advance the aviation meteorology science and operations.

For this volume we focused on scientific and operational issues that affect flight conditions, such as aircraft climb, and drag and lift forces. The review paper by Gultepe et al. (2019a) clearly shows that wind, turbulence (including gust), and visibility are more often a cause of aircraft flight disruptions, rather than other factors, such as icing conditions studied in Chuang et al. (2019), convection, and pollution-dust, which also play important roles for aviation operations. Therefore, more research is needed to

[1] Meteorological Research Division (MRD), Environment and Climate Change Canada (ECCC), Toronto, ON M3H 5T4, Canada. E-mail: ismail.gultepe@canada.ca; Ismail.gultepe@uoit.ca

[2] The Faculty of Engineering and Applied Science, Ontario Technical University (OTU), Oshawa, ON L1G 0C5, Canada.

[3] Space Science and Engineering Center/Cooperative Institute for Meteorological Satellite Studies (SSEC/CIMSS) UW, Madison, WI, USA. E-mail: wayne.feltz@ssec.wisc.edu

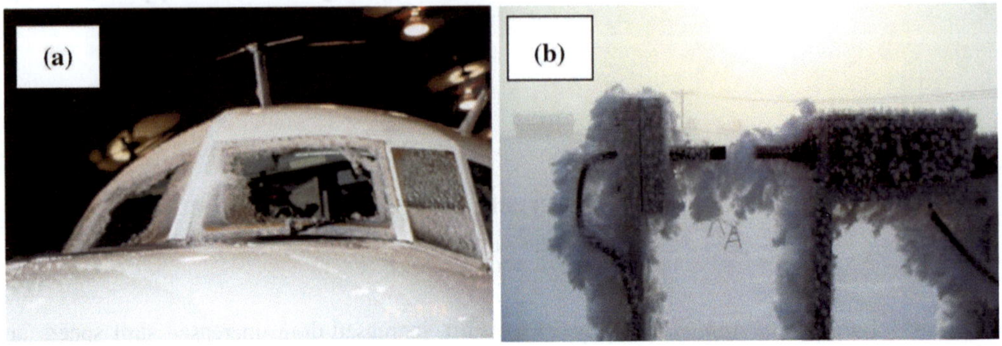

Figure 1
Shows icing occurred on NRC Convair during AIRS II field project (Isaac et al. 2005) (**a**) and frost happened during FRAM ice fog project (Gultepe et al. 2015, which took place in Barrow, Alaska, USA in April of 2008 (**b**). Pictures were taken by I. Gultepe

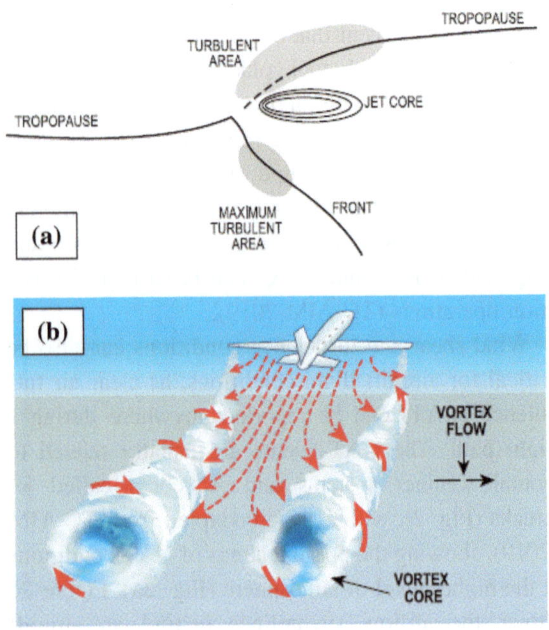

Figure 2
Possible areas of turbulence and CAT related to a jet streak, front, and tropopause (**a**), and a wake vortex conceptual design (**b**) (Adapted from TC AIM 2019)

characterize aircraft flying conditions and ground based operations and environments.

Aviation meteorology needs new instrumental platforms and analysis techniques to improve safety conditions for flight environments. These platforms need to be developed as a part of integrated systems, such as ground based in situ observations, and remote sensing platforms, that include both ground and space based operational satellites. Although presently unmanned aerial vehicle (UAV) and unmanned aerial systems (UAS) platforms have not been used for operations, they are becoming very popular for atmospheric research, especially for dangerous conditions such as hurricanes, convective storms, and icing environments (Griffin and Velden 2019), as well as NWP (numerical weather prediction) simulations and initializations. The UAVs can carry various miniature sensors for weather research and observations (Gultepe et al. 2019b). In this topical special issue, the importance of UAVs for future atmospheric research has also been emphasized for high level and low-level flight conditions and weather research.

Ground based in situ observations of various aviation meteorology related parameters can be obtained for extreme weather conditions within climate and wind tunnel laboratories (Gultepe et al. 2019b). Two of these, including the University of Ontario Institute of Technology (UOIT) climatic wind tunnel (CWT) and NASA Glen wind tunnel, have been used for simulating Arctic and icing environments, respectively (Gultepe et al. 2019a). Climatic wind tunnel tests are also important for simulating aircraft aerodynamic conditions that had been summarized by many others (Dodson 2005; Tomek et al. 2006; Bell 2002; Ross 2011).

Weather forecasts have been improved over short and long-time periods, using various analysis systems, such as NWP models (Chachere and Pu 2019), nowcasting-based systems, integrated methods, and statistical-based methods such as neural network and artificial intelligence (AI) systems (Kneringer et al. 2019; Zheng et al. 2018). This topical issue provides the current status and challenges of these systems. While none of these systems alone can provide accurate predictions of meteorological parameters over short or long-time periods, integrating these methods can be used more confidently to improve aircraft physical and managerial flying conditions, including conditions on the ground (Kutty et al. 2019; França et al. 2019). Several papers focusing on these issues are included in this issue.

The accuracy of forecasting and nowcasting aviation weather using numerical models and remote sensing techniques still remains difficult, mainly because of the lack of observations for physical and dynamical parameterizations as a part of numerical models (Weston et al. 2019). Presently, typical numerical forecast models lack sufficient time and space resolutions and appropriate physical parameterizations to represent aircraft scales (10–100 m) in estimating aerodynamic and microphysical processes. Simple approaches, e.g., using only temperature and relative humidity with respect to water (RH_w) to calculate icing rates, can lead to uncertainties of more than 30% in many conditions.

This topical special issue points out that improvements of in situ observations and microphysical parameterizations are required to reduce aviation related accidents. Pu et al. (2019) stated the importance of microphysical parameterizations for convective systems and NWP sensitivity for various parameterizations. However, detailed 3D models do not represent certain scales and use large scale parameterizations for predictions; furthermore, these 3D models are currently not practical to be used operationally because of processing time of numerical algorithms based on smaller time and space scales. The capability of these models needs to be improved mainly for wind speed and direction, gust, visibility and ceiling, convection, and icing conditions. For example, Gultepe et al. (2019a) stated that on average, 60 people die because of fog-related accidents every year in Canada, which is comparable to impacts resulting from other severe weather events, and these numbers can be much higher based on international considerations. In fact, the worst accident in aviation history occurred due to fog and low visibility impacting aircraft take off processes on March 27, 1977. Two Boeing 747 passenger airliners collided on the runway of Tenerife Airport on the Spanish island of Tenerife, one of the Canary Islands. The crash claimed a total of 583 fatalities (Smith 2017). As is the case in most weather-related aviation accidents, several conditions combined leading to the accident occurrence, however certainly dense fog played a critical role. The two aircraft would have been aware of one another's presence in fair conditions, but because of dense fog, neither aircraft could see one another until it was too late.

Global climate change issues are also critical for aviation operations. For example, temperature changes of a few °C can impact aircraft aerodynamical conditions such as lift and drag, as well as climb rates (Storer et al. 2019). Regional air pollution issues, including volcanic ash and dust conditions, also affect flying conditions; but aircraft contrail formation due to exhaust gases during taking off or landing can also affect the weather and climate (Gultepe et al. 2014; Heymsfield et al. 1998, 2010; Schumann 2002, 2005). Schumann and Heymsfield (2017) stated that key parameters controlling contrail properties, besides aircraft and fuel properties, ambient pressure, temperature, and humidity, can be related to the number of ice particles per flight distance surviving the wake vortex phase, the contrail depth and particle sedimentation, wind shear, turbulence, and vertical motions controlling contrail dispersion. Contrail effect on climate change depends also on the ratio of shortwave to longwave radiative forcing (RF) that contributes to surface warming. Figure 3 shows a contrail formation during an aircraft take off at the Yellow Knife International Airport, NWT, Canada (Fig. 3a, b). Figure 3c shows a contrail occurred due to a condensation process related to an aircraft jet exhaust (Schumann and Heymsfield 2017). However, we are just starting to understand these conditions and develop new methods to better evaluate contrail impact on the climate change, and develop, mitigation steps to reduce risk-related aircraft accidents and aircraft-related pollution.

Wind and turbulence activities factor into the aircraft landings and takeoffs conditions (Sharman and Trier 2019). In reality, wind measurements at airports lack the needed resolution over the horizontal and vertical scales. Therefore, improvements in wind measurements can only be done by using state of the art

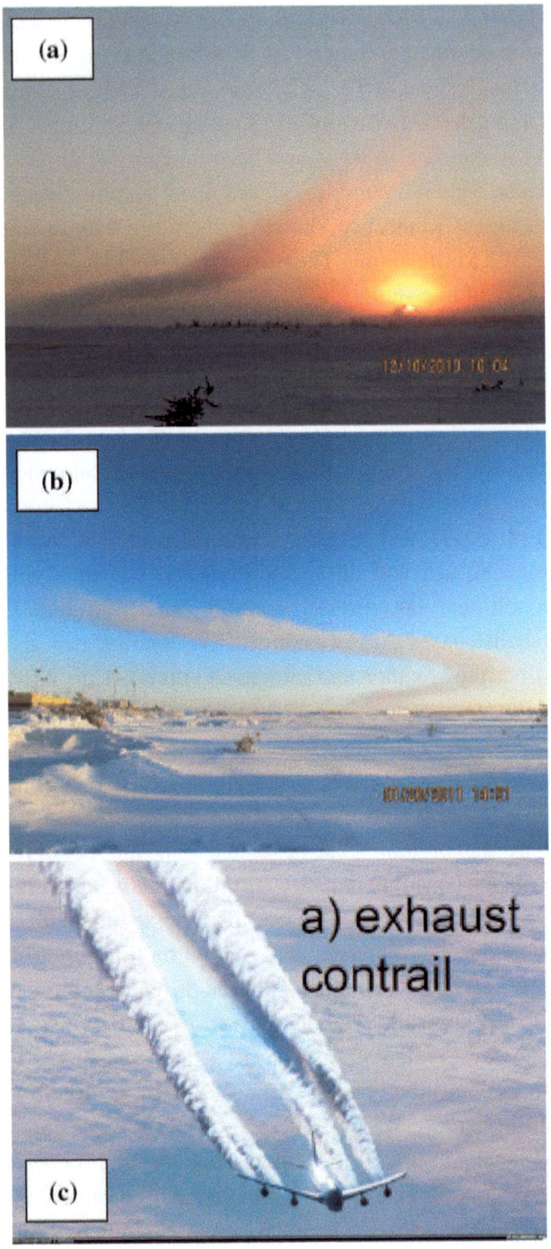

Figure 3
a is for a contrail formation after an aircraft takes off at 1004 local standard time (LST) 10 December 2010; **b** another contrail on 1451 LST 20 January 2011 over Yellowknife International Airport (adapted from Gultepe et al. 2014), and **c** exhaust contrail at the upper atmosphere (adapted from Schumann and Heymsfield 2017). Permission by American Meteorological Society

observational platforms such as lidars and UAVs (Thobois et al. 2019; Gultepe et al. 2019b). Doppler lidars (Thobois et al. 2019) adaptable for airport use can provide wind speed, directional shear, gust, and turbulence conditions over short time and space scales at airports and during landings and takeoffs. As a result, these new platforms can be used to more accurately define safety conditions for aviation applications.

While this publication broadly covers the whole of aviation meteorology, less information on Polar flight conditions is noted, although major issues related to cold temperature and snow conditions, as well as ice fog, are discussed extensively. To simulate Polar conditions, climate wind tunnels (e.g., UOIT CWT, Gultepe et al. 2019b) can be used to test aircraft parts and pieces, but these also need to be validated using real flight environments. In reality, freezing fog and ice fog occur often in cold climates, particularly in the high Arctic and Antarctic regions, and can create adverse weather conditions for aviation and marine applications, that need to be addressed seriously as air traffic and marine transportation increases over the Arctic regions.

Acknowledgements

Guest editors wish to thank the scientific contributors to this topical issue and we also gratefully acknowledge funding support for Editor I. Gultepe by Environment and Climate Change Canada (ECCC) and the Canadian National Search and Rescue (SAR) Secretariat. In addition, our sincere thanks go to Drs. Renata Dmowska and Alexander Rabinovich, Editors in Chief for topical and regular issues at *Pure and Applied Geophysics*, for inviting our working group to prepare this topical issue, handling the review process and providing invaluable advice. Special thanks are also given to all contributors, who submitted their valuable work to this topical issue as well as to all reviewers.

Publisher's Note Springer Nature remains neutral with regard to jurisdictional claims in published maps and institutional affiliations.

References

Bell, J. (2002). Comparison of crew exploration vehicle (CEV) tests in two supersonic wind tunnels. In *AIAA 2007-1008, 45th AIAA aerospace sciences meeting and exhibit*, Reno, NV, January 2007.

Bragg, M., Hutchison, T., Merret, J., Oltman, R., & Pokhariyal, D. (2000). Effect of ice accretion on aircraft flight dynamics. In *38th AIAA aerospace sciences meeting & exhibit*. Urbana, Illinois 61801.

Chachere, C. N., & Pu, Z. (2019). Numerical simulations of an inversion fog event in the salt lake valley during the MATERHORN-Fog field campaign. *Pure and Applied Geophysics*. https://doi.org/10.1007/s00024-018-1770-8.

Chuang, H. Y., Mao, Y., & Zhou, B. (2019). R2O transition of NCAR's icing and turbulence algorithms into NCEP's operations. *Pure and Applied Geophysics*. https://doi.org/10.1007/s00024-018-1975-x.

Dodson, M. G. (2005). An historical and applied aerodynamic study of the Wright Brothers' wind tunnel test program and application to successful manned flight. In *US Naval Academy Technical Report*. USNA-334. Accessed 03 Nov 2009.

Ellrod, G. P., & Pryor, K. (2019). Applications of geostationary satellite data to aviation. *Pure and Applied Geophysics*. https://doi.org/10.1007/s00024-018-1821-1.

França, G. B., do Carmo, L. F. R., de Almeida, M. V., & Neto, F. L. A. (2019). Fog at the Guarulhos International Airport from 1951 to 2015. *Pure and Applied Geophysics*. https://doi.org/10.1007/s00024-018-1781-5.

Griffin, S. M., & Velden, C. S. (2019). Hazard avoidance products for convectively-induced turbulence in support of high-altitude global hawk aircraft missions. *Pure and Applied Geophysics*. https://doi.org/10.1007/s00024-018-1772-6.

Gultepe, I., Agelin-Chaab, M., Komar, J., Elfstrom, G., Boudala, F., & Zhou, B. (2019b). A meteorological supersite for aviation and cold weather applications. *Pure and Applied Geophysics*. https://doi.org/10.1007/s00024-018-1880-3.

Gultepe, I., Kuhn, T., Pavolonis, M., Calvert, C., Gurka, J., Isaac, G. A., et al. (2014). Ice fog in Arctic during FRAM-IF project: Aviation and nowcasting applications. *Bulletin of American Meteorological Society, 95,* 211–226.

Gultepe, I., Sharman, R., Williams, P. D., Zhou, B., Ellrod, G., Minnis, P., Trier, S., Griffin, S., Yum, S. S., Gharabaghi, B., Feltz, W., Temini, M., Pu, Z., Storer, L. N., Kneringer, P., Weston, M. J., Chuang, H. Y., Thobois, L., Dimri, A. P., Dietz, S. J., Gutenberg, M., Almeida, M. V., & Neto, F. L. A. (2019a). A review of high impact weather for aviation meteorology. *Pure and Applied Geophysics*. https://doi.org/10.1007/s00024-019-02168-6.

Gultepe, I., Zhou, B., Milbrandt, J., Bott, A., Li, Y., Heymsfield, A. J., et al. (2015). A review on ice fog measurements and modeling. *Atmospheric Research, 151,* 2–19.

Heymsfield, A. J., Baumgardner, D., DeMott, P., Forster, P., Gierens, K., & Kärcher, B. (2010). Contrail microphysics. *Bulletin of American Meteorological Society, 91,* 465–472. https://doi.org/10.1175/2009BAMS2839.1.

Heymsfield, A. J., Lawson, R. P., & Sachse, G. W. (1998). Growth of ice crystals in a precipitating contrail. *Geophysical Research Letters, 25,* 1335–1338. https://doi.org/10.1029/98GL00189.

Isaac, G. A., Ayers, J. K., Bailey, M., Bissonnette, L., Bernstein, B. C., Cober, S. G., Driedger, N., Evans, W. F. J., Fabry, F., Glazer, A., Gultepe, I., Hallett, J., Hudak, D., Korolev, A. V., Marcotte, D., Minnis, P., Murray, J., Nguyen, L., Ratvasky, T. P., Reehorst, A., Reid, J., Rodriguez, P., Schneider, T., Sheppard, B. E., Strapp, J. W., & Wolde, M. (2005). First results from the alliance icing research study (AIRS-II). In *43rd AIAA aerospace sciences meeting and exhibit* 10–13 January 2005, Reno, Nevada, AIAA 2005-252.

Kneringer, P., Dietz, S. J., Mayr, G. J., & Zeileis, A. (2019). Probabilistic nowcasting of low-visibility procedure states at Vienna International Airport during cold season. *Pure and Applied Geophysics*. https://doi.org/10.1007/s00024-018-1863-4.

Kutty, S. G., Agnihotri, G., Dimri, A. P., & Gultepe, I. (2019). Fog occurrence and associated meteorological factors over Kempegowda International Airport, India. *Pure and Applied Geophysics*. https://doi.org/10.1007/s00024-018-1882-1.

Pu, Z., Lin, C., Dong, X., & Krueger, S. K. (2019). Sensitivity of numerical simulations of a mesoscale convective system to ice hydrometeors in bulk microphysical parameterization. *Pure and Applied Geophysics*. https://doi.org/10.1007/s00024-018-1787-z.

Ross, J. C. (2011). Aerodynamic and aeroacoustic wind tunnel testing of the Orion spacecraft. In *AIAA 2011-0014395. Applied Aerodynamics*; 15 June 2011; Honolulu, HI; United States.

Schumann, U. (2002). Contrail cirrus. In D. K. Lynch, et al. (Eds.), *Cirrus* (pp. 231–255). Oxford: Oxford University Press.

Schumann, U. (2005). Formation, properties and climate effects of contrails. *Comptes Rendus Physique, 6,* 549–565. https://doi.org/10.1016/j.crhy.2005.05.002.

Schumann, U., & Heymsfield, A. J. (2017). On the life cycle of individual contrails and contrail cirrus. *Meteorological Monographs, 58,* 3.1–3.24. https://doi.org/10.1175/AMSMONOGRAPHS-D-16-0005.1.

Sharman, R. D., & Trier, S. B. (2019). Influences of gravity waves on convectively-induced turbulence (CIT): A review. *Pure and Applied Geophysics*. https://doi.org/10.1007/s00024-018-1849-2.

Smith, P. (2017). The true story behind the deadliest air disaster of all time. The telegraph. https://www.telegraph.co.uk/travel/comment/tenerife-airport-disaster/. Accessed 5 Mar 2019.

Storer, L. N., Williams, P. D., & Gill, P. G. (2019). Aviation turbulence: dynamics, forecasting, and response to climate change. *Pure and Applied Geophysics*. https://doi.org/10.1007/s00024-018-1822-0.

TC AIM. (2019). *Aeronautical information manual*. TP 14371E (2019-1). TC. ISSN#1715-7390.

Thobois, L., Cariou, J. P., & Gultepe, I. (2019). Review of lidar based applications for aviation weather. *Pure and Applied Geophysics*. https://doi.org/10.1007/s00024-018-2058-8.

Tomek, D. M., Sewall, W. G., Mason, S. E., & Szchur, W. A. (2006). Next generation of high-speed dynamic stability wind tunnel testing. In: *AIAA 2006-3148, 25th AIAA Aerodynamic Measurement Technology and Ground Testing Conference*, San Francisco, CA, June 2006.

Weston, M., Chaouch, N., Valappil, V., Temimi, M., Ek, M., & Zheng, W. (2019). Assessment of the sensitivity to the thermal roughness length in Noah and Noah-MP land surface model using WRF in an arid region. *Pure and Applied Geophysics*. https://doi.org/10.1007/s00024-018-1901-2.

Zheng, N., Luo, M., Zou, X., Qiu, X., Lu, J., Han, J., Wang, S., Wei, Y., Zhang, S., & Yao, H. (2018). A novel method for the recognition of air visibility level based on the optimal binary tree support vector machine. *Atmosphere, 9,* 481. https://doi.org/10.3390/atmos9120481.

(Published online May 2, 2019)

Pure Appl. Geophys. 176 (2019), 1869–1921

https://doi.org/10.1007/s00024-019-02168-6

Pure and Applied Geophysics

A Review of High Impact Weather for Aviation Meteorology

Ismail Gultepe,[1,2] R. Sharman,[3] Paul D. Williams,[4] Binbin Zhou,[5] G. Ellrod,[6] P. Minnis,[7] S. Trier,[3] S. Griffin,[8] Seong. S. Yum,[9] B. Gharabaghi,[10] W. Feltz,[8] M. Temimi,[11] Zhaoxia Pu,[12] L. N. Storer,[4] P. Kneringer,[13] M. J. Weston,[14] Hui-ya Chuang,[15] L. Thobois,[16] A. P. Dimri,[17] S. J. Dietz,[18] Gutemberg B. França,[19] M. V. Almeida,[19] and F. L. Albquerque Neto[19]

Abstract—This review paper summarizes current knowledge available for aviation operations related to meteorology and provides suggestions for necessary improvements in the measurement and prediction of weather-related parameters, new physical methods for numerical weather predictions (NWP), and next-generation integrated systems. Severe weather can disrupt aviation operations on the ground or in-flight. The most important parameters related to aviation meteorology are wind and turbulence, fog visibility, aerosol/ash loading, ceiling, rain and snow amount and rates, icing, ice microphysical parameters, convection and precipitation intensity, microbursts, hail, and lightning. Measurements of these parameters are functions of sensor response times and measurement thresholds in extreme weather conditions. In addition to these, airport environments can also play an important role leading to intensification of extreme weather conditions or high impact weather events, e.g., anthropogenic ice fog. To observe meteorological parameters, new remote sensing platforms, namely wind LIDAR, sodars, radars, and geostationary satellites, and in situ instruments at the surface and in the atmosphere, as well as aircraft and Unmanned Aerial Vehicles mounted sensors, are becoming more common. At smaller time and space scales (e.g., < 1 km), meteorological forecasts from NWP models need to be continuously improved for accurate physical parameterizations. Aviation weather forecasts also need to be developed to provide detailed information that represents both deterministic and statistical approaches. In this review, we present available resources and issues for aviation meteorology and evaluate them for required improvements related to measurements, nowcasting, forecasting, and climate change, and emphasize future challenges.

Key words: Fog and precipitation visibility, aviation meteorology, ice microphysics, wind shear and gust, nowcasting and forecasting.

1. Introduction

Numerous studies have shown that weather severely impacts both civilian and defense aviation operations (e.g., Cook et al. 2009, Rudra et al. 2015;

[1] MRD, ECCC, Toronto, ON, Canada. E-mail: Ismail.gultepe@canada.ca

[2] Faculty of Engineering and Applied Science, Ontario Technical University, Oshawa, ON, Canada.

[3] National Center for Atmospheric Research, Research Applications Laboratory, P.O. Box 3000, Boulder, CO 80307, USA. E-mail: sharman@ucar.edu; trier@ucar.edu

[4] Department of Meteorology, University of Reading, Earley Gate, Reading RG6 6BB, UK. E-mail: p.d.williams@reading.ac.uk; luke.storer@pgr.reading.ac.uk

[5] IMSG and EMC/NCEP/NOAA, College Park, MD, USA. E-mail: Binbin.Zhou@noaa.gov

[6] EWxC, LLC, Granby, CT, USA. E-mail: gary.ellrod@gmail.com

[7] Science Systems and Applications, Inc., One Enterprise Parkway, Suite 200, Hampton, VA 23666, USA. E-mail: p.minnis@nasa.gov

[8] Space Science and Engineering Center, Cooperative Institute for Meteorological Satellite Studies (CIMSS), University of Wisconsin-Madison, 1225 W. Dayton, Madison, WI 53706, USA. E-mail: sarah.griffin@ssec.wisc.edu; wayne.feltz@ssec.wisc.edu

[9] Department of Atmospheric Sciences, Yonsei Univeristy, Seoul 03722, Korea. E-mail: seongsooyum@gmail.com

[10] Water Resources Engineering, School of Engineering, University of Guelph, 50 Stone Road East, Guelph, ON N1G 2W1, Canada. E-mail: bgharaba@uoguelph.ca

[11] Water and Environment Engineering Program, Masdar Institute, Khalifa University of Science and Technology, Abu Dhabi, UAE. E-mail: marouane.temimi@ku.ac.ae

[12] Department of Atmospheric Sciences, University of Utah, Salt Lake City, UT, USA. E-mail: zhaoxia.pu@utah.edu

[13] University Innsbruck, Innsbruck, Tirol, Austria. E-mail: philipp.kneringer@uibk.ac.at

[14] Department of Civil Infrastructure and Environmental Engineering, Khalifa University, Abu Dhabi, UAE. E-mail: mjweston@masdar.ac.ae

[15] NCEP Environmental Modeling Center, 5830 University Research Court, College Park, MD 20740, USA. E-mail: hui-ya.chuang@noaa.gov

[16] Leosphere Inc., 3 Rue Jean Rostand, 91400 Orsay, France. E-mail: lthobois@leosphere.com

[17] School of Environmental Sciences, Jawaharlal Nehru University, New Delhi, India. E-mail: apdimri@hotmail.com

Gultepe et al. 2014a, b, 2017a). The impact of atmospheric processes on aviation has been recognized since the 1900s. For example, Dines (1917) stated that "thus it appears that the demand of the airman on the meteorologist is that he shall be able to forecast wind and fog, and to less extent clouds, on the route, the airman is proposing to follow." Presently, his comments on aviation-related parameters such as wind speed (U_h) and visibility (Vis) are still valid (Figs. 1 and 2).

Weather conditions that cause or contribute to the aviation accidents include wind, visibility/ceiling, high density altitude, turbulence, carburetor icing, updrafts/downdrafts, precipitation, icing, thunderstorms, wind shear, thermal lift, temperature (T) extremes, and lightning (NTSB 2010). Figure 1a shows a bar plot of the statistics of weather-related conditions from 1994 to 2003 that affect near-surface aviation operations; they are mostly wind and visibility (NTSB 2010). In Fig. 1a, Vis, ceiling height (h_c), and precipitation related conditions occurred 485 times, wind and turbulence 1381 times, and icing and engine icing 150 times. This work suggested that from 2003 through 2007, there were 8657 aviation related accidents and weather was a factor in 1740 of these accidents. Figure 1b shows these parameters in percentiles also for 1994–2003 period; wind and visibility are still the most critical parameters. For small, non-commercial aircraft (Part 91 class) the primary cause of weather-related accidents from 2000 to 2011 was adverse winds, followed by low ceilings (h_c) (Fig. 1c). At cruising levels of commercial jet aircraft (Part 121 class), this picture is different, with over 70% of weather-related accidents from 2000 to 2011 being related to turbulence (Fig. 1d) (e.g., Sharman and Lane 2016).

Delays and damage to aircraft during landing, taking off and high-level cruise due to hazardous weather can often occur. Gultepe et al. (2017a, b) stated that a three-days freezing fog event in the UK resulted in about 50 M US$ in financial losses for businesses. Kessinger et al. (2006a, b) also stated that aviation hazards for oceanic flights could impact both safety, economic efficiency, and productivity that lead to a total cost of about 62.7 M US$ per year. Cook et al. (2009) quantified the impact of weather factors on flight delays and found that contributing factors are highly airport dependent. Because of its importance to aviation, the National Oceanic and Atmospheric Administration (NOAA) National Weather Service (NWS) operationally provides forecasts of cloud-top height, turbulence, lightning detection, precipitation, icing, low clouds and fog, and volcanic ash. These forecasts are available to the public through their website (https://www.aviationweather.gov/).

Strong wind regions within jet streams called jet streaks (JS) can also affect aircraft operations (Uccellini and Johnson 1979; Uccellini 1980). They studied transverse circulations in the left exit and right entrance regions of jet streaks (which are the divergence regions) using Numerical Weather Prediction (NWP) models. Rose et al. (2004) detailed the interactions between upper and lower level jets, and the development of severe convective system. They stated that "in the (jet) exit region, the geometry of this adjustment, combined with warm, moist, lower tropospheric air to the right and ahead of the jet streak and cool, dry air at the jet streak level, produced the differential advections that convectively destabilized the atmosphere". Results of their work suggest that development of severe convective storms can be influenced by mass and momentum fluxes along the propagation of an upper tropospheric jet streak. The development of a low-level jet (LLJ) usually was not considered for convection development in regard to synoptic-scale processes (Uccellini 1980). He stated that a systematic upper level flow pattern led to lee-side cyclogenesis or a leeside trough, and that produced the strong pressure gradient forces needed for the development of LLJs. These studies suggested that improved knowledge of the relationships between JS and LLJ are important for understanding aviation weather hazards. Clear-air turbulence (CAT) and eddies related to jet streams can also be extremely important for mountainous and high-level flights near jet streams (e.g., Ellrod and Knapp 1992; Brown 1973; Gultepe et al. 1995; Koch and Caracena 2002).

[18] University of Innsbruck, Innsbruck, Austria. E-mail: Sebastian.Dietz@uibk.ac.at

[19] Department of Meteorology, Federal University of Rio de Janeiro, As. Athos da Silveria Ramos, 274 Cidade Universitária, 22941-616 Rio de Janeiro, Brazil. E-mail: gutemberg@lma.ufrj.br; valdonel@lma.ufrj.br; albuquerque@lma.ufrj.br

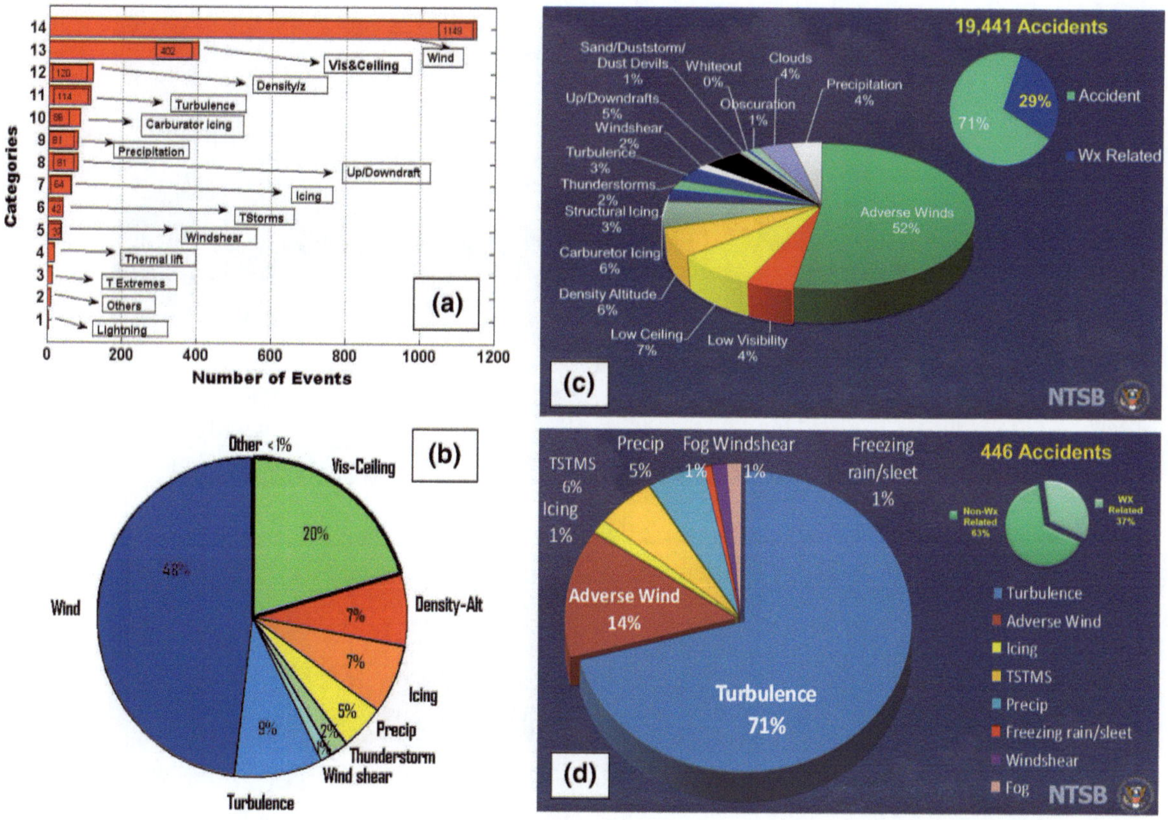

Figure 1

a, **b** Statistics for aircraft related accidents related to meteorological parameters from 1994 to 2000: **a** actual numbers of accidents and **b** probabilistic distributions in pie chart (data from NTSB 2010; 1996). **c**, **d** Part 91-Weather as cause/factor during all accidents for the period of 2000–2011. The NTSB (National Transportation Safety Board) based statistics which resulted in 19,441 accidents and 29% of these accidents were related to weather conditions (**c**) (Eick 2014) and Part 121-air carrier weather related cause/factors for 2000–2011 (**d**)

Reduced Vis, commonly caused by fog and precipitation (Tardif and Rasmussen 2007; Gultepe et al. 2009), is also caused by dust and ash (Hodges and Pu 2016) (Fig. 2). The transport of fine-grained dust by strong winds can occur over a broad range of time and spatial scales (Hadley et al. 2004) and impact not only Vis but have also been associated with aircraft engine failure.

Another critical hazard for aviation is aircraft icing (Isaac and Schemenauer 1979; Guttman and Jeck 1987; Politovich 1989; Tafferner et al. 2003; Vivekanandan et al. 2001). Prediction of icing and deicing conditions (at the airports) is important for flight safety (Rasmussen et al. 2001; Black and Mote 2015). The accumulation of ice on aircraft before takeoff has long been recognized as one of the most significant safety hazards affecting the aviation industry. As little as 0.1 mm of clear ice over 2 min on a wing surface can increase drag and reduce airplane lift by 25–30% and reduce critical angle of attack by 8° (https://aircrafticing.grc.nasa.gov/1_1_3_2.html; TC 2004).

Recently it has been realized that prevention of aircraft engine power losses due to high ice water content (HIWC) at higher elevations is essential for safe aviation operations (Leroy et al. 2016, 2017). The HIWC conditions, typically associated with deep convection, can cause engine power loss and air data events (affecting altitude and airspeed measurements) on commercial aircraft and need to be better characterized (Mason et al. 2006).

Improvements in aviation nowcasts and forecasts require better measurements and predictions of atmospheric parameters such as T, RH, wind and

Figure 2
Various weather conditions representing visibilities related to warm fog, moisture, rain, freezing fog, ice fog, snow, blowing snow, and ashes (from top left to the right boxes)

turbulence, icing, and Vis. These parameters are also related to climatic change over various scales. Therefore, for future research, the impact of climate change on meteorological parameters needs to be evaluated. For example, state-of-the-art research done by Goodman and Griswold (2017) suggest that temperature and pressure are important for aviation applications because they can impact aircraft drag and lift. Their work suggested suggests that aviation related impacts due to climate change need to be evaluated.

State-of-the-art global climate models were used in the Coupled Model Inter-comparison Project Phase 5 (CMIP5) simulating climate extreme indices described by the Expert Team on Climate Change Detection and Indices (ETCCDI) Sillmann 2013a, b). In their work, the climatic indices based on daily T and precipitation amounts and rates were calculated using multi-model simulations. They found that the duration of cold and warm spells changed significantly since the 1950s. These works also suggest that climate change issues related to aviation operations are need to be evaluated.

Global and regional weather prediction systems are needed to characterize the atmospheric state in details required to diagnose aviation related hazards. Accurate predictions of the most dangerous processes related to Vis, wind and turbulence, icing, and convection are essential to reduce in-flight injuries, structural damage, ground preparations for flights, and flight delays (e.g., Meckalski et al. 2007, Sharman et al. 2012; Sharman and Lane 2016; Karstens et al. 2018). Therefore, forecasting and nowcasting of aviation related meteorological parameters require a better understanding of the underlying meteorological processes. In this respect, knowledge of dynamics of upper-level and low-level jets, gusts, eddies, large-amplitude gravity waves, wind shear (including low level wind shear (LLWS) as well as physics and thermodynamics of cloud, fog, and precipitation can be very important for aviation mission planning and reducing financial losses (Uccellini and Johnson 1979; Zhou et al. 2004; NWPSD 2004; FAA 1988 Thobois et al. 2018).

The goal of this review is to provide a summary of meteorological parameters, processes, and their prediction issues that are critical for aviation operations and that create challenges for forecasters. The critical parameters, processes, and analysis techniques of importance for aviation meteorology over various scales can be listed as

- Weather physics
  1. Clouds and fog
  2. Ice and droplet microphysics
  3. Precipitation type and rate
  4. Freezing precipitation
  5. Icing, and ground deicing
  6. Frost
  7. Visibility and ceiling
  8. Blowing snow
  9. Convective activities and high ice water content
  10. Lightning
  11. Ash and dust
  12. Physical parameterizations
- Weather Dynamics
  1. Turbulence and energy dissipation
  2. Wind gusts
  3. Waves
  4. Jet streak intensity
  5. Low Level Wind Shear (LLWS)
  6. Microburst and tornados
  7. Temperature and RH (relative humidity)
- Weather Forecasting (Multi-scale)
  1. NWP time and space scale issues
  2. Data integration
  3. Artificial intelligence
  4. Deterministic versus probabilistic methods for nowcasting
  5. Integrated systems
  6. Climate change modeling for aviation applications
  7. Contrails and climate change
  8. Reporting systems for aviation meteorological applications

Considering the above items, NWP and climate change models play an important role on their own or integrated with observations to improve nowcasting and long-term forecasting of high impact weather events. Weather-related events such as fog, precipitation, clear-air and in-cloud turbulence, wind shear, gust, or icing may be related to changing climate conditions; if this is the case, for the next 50 years and beyond, aircraft flying conditions need to be considered to improve future aviation operations in response to extreme weather conditions.

In this overview, observations will first be summarized in the next sections and followed by discussions of NWP issues. Then, integrated methods will be introduced for aviation operation applications. In the final section, future issues and possible challenges related to observations, NWP, and climate change will be provided.

2. *Observations*

Observations are needed to discern and monitor meteorological parameters required for safe and efficient aviation operations. These are clearly needed in the terminal area of major airports and at cruise levels along flight routes, but also over wide areas to support a range of operational activities at smaller airfields and other remote locations. Helicopter air ambulance operation in remote areas, for example, represents a particular aviation weather need that creates significant forecasting challenges. Observations are important since some parameters are poorly diagnosed by NWP models at the scales needed by the aviation community and they are also used for model initializations. Clearly, observations are critical for providing monitoring and nowcasting related to aviation meteorological parameters for operations, and that cannot be predicted accurately with NWPs alone for lead times less than a few hours.

2.1. *Ground Based In Situ Observations*

The Aviation Automated Weather Observation System (AV-AWOS) originally suggested by Mandel (1975) provides meteorological parameters critical for aviation operations. Their proposed AWOS included additional parameters such as Vis and sky conditions. Later, Wade (2003) used the Automated Surface Observing Systems (ASOS) with a the light-emitting diode weather identifier (LEDWI) sensor to discriminate drizzle from other particles because precipitation rate from drizzle is usually less than 0.25 mm h^{-1}. AWOS and ASOS were designed for unattended locations and use multi-sensor based algorithms (Ramsay 1999). Presently, these types of systems are continuously being improved by adding ultrasonic 3D wind systems, fog and precipitation sensors, as well as a portable ceilometer and a microwave radiometer (Ware et al. 2013; Gultepe et al. 2018). In addition to individual ground based in situ sensors, compact meteorological platforms were also developed for visibility, icing, fog, and precipitation type studies and applications such as ones developed by Landolt et al. (2010) and Gultepe et al. (2018). Sims et al. (2000) used an integrated icing diagnostic algorithm (IIDA) that extensively utilized ground based in situ systems with other observational and prediction systems. Similarly, supersites having various high-resolution ground based in situ sensors (see next subsection) were also designed and used for operations.

2.2. *Meteorological Supersites*

Meteorological supersites with various sensors can be designed for specific goals related to atmospheric-hydrologic-oceanographic applications. Presently, these supersites with extensive ground based in situ and remote sensing platforms have been used by researchers (Rasmussen et al. 2012, Gultepe et al. 2015, 2018; Ralph et al. 2013). They were designed for aviation and forecasting validations/operations and included mainly observations representing Vis, U_h, turbulence, fog (FG), precipitation amount (PA) and intensity (PR), h_c, and cloud base height (h_b), and atmospheric profiling of meteorological parameters. Although supersites usually are located at a single point, representing a small area around it, Gultepe et al. (2018) used their supersites with satellite sites (e.g., smaller weather observing station) located within a 1 km perimeter radius. These observations provided statistical advantages to evaluate the scale issues related to measurements (Gultepe et al. 2018). The fact that most accidents occurring around airports emphasizes that high resolution observations and areal representation of observations as done in above work are needed to improve aviation operations/NWP simulations.

2.3. Aviation Weather Reports

Detailed airport observations can be used for making flight decisions by pilots and airport authorities, and for verifying forecasts. These observations usually are reported at 30-min or 60-min intervals as Meteorological Aviation Routine Weather Reports (METARs). At some airports, observations are also performed by using weather sensors autonomously (Gultepe et al. 2017a, b). Weather forecasts for aviation operations are provided 2 h in advance based on trend-type forecasts (TRENDs) and over 2 h as terminal aerodrome forecasts (TAFs) (Jacobs and Maat 2005). Both TAFs and TRENDs contain information on horizontal wind (Uh), Vis, cloud amount (C_a), h_b, and PR in the vicinity of the airport (Lynn 1997). The TAF can be issued as often as 12 times a day, including eight short TAFs with lead times from 1 to 10 h, and four long TAFs with lead times from 8 to 26 h (Jacobs and Maat 2005). In fact, most aircraft landing and takeoff decisions are made based on Vis and h_b, as well as wind conditions at the airports (Thobois et al. 2018).

PIREPS (voice-transmitted pilot reports) are commonly used to report icing and turbulence conditions and they are easily available but the PIREP reporting system was not proposed to be used for research or forecasting applications (Schultz and Politovich 1992; Kelsch and Wharton 1996; Schwartz 1996). These studies stated that inspection of PIREPS on turbulence and icing conditions reveal serious shortcomings that limit their usefulness. Lately, Bernstein et al. (2005) also stated that icing algorithms developed by many others overestimated icing amounts compared to PIREPS. These works suggested that the PIREP reporting systems need to be improved. Schultz and Politovich (1992) suggested that the FAA should consider creating a standard PIREPS system similar to the standard airways report made by surface weather reporting stations.

The TAFs and TRENDs are a good source of meteorological short time predictions and provide important support for forecasters. The TAF and TREND guidance and the AUTOTAF encoding software have been developed and implemented in close cooperation with the German company Meteo Service Weather Research (Knüppfer 1997; Jacobs and Maat 2005). Automated TAFs (Lynn 1997; Kilpinen 1994) use both direct NWP model outputs and model observation soundings (MOS; Glahn and Lowry 1972; Hart et al. 2004). The MOS-based methods usually use local data as in Jacobs and Maat (2005). In their work, an approach using observations from upstream locations and local data were used. Recently, using PIREPS, blended algorithms for aviation operations have been suggested for improving short term forecasts (Bailey et al. 2014). A combined physical–statistical approach is often applied to low clouds and Vis predictions because of NWP short-term unreliable prediction issues. Lately, Herzegh et al. (2015) developed an expert system for Vis and cloud base (h_b) forecasts based on blended numerical model and observational data. In their work, current and historical METARs, GOES observations, NWP output from the Rapid Update Cycle (RUC), National Center for Environment Prediction (NCEP), and global forecasting system (GFS) runs, MOS forecasts, as well as observational based rule methods for short term predictions were used. These works indicated importance of airport and pilot reports to be used for aviation operations.

2.4. Aircraft In Situ Observations for Icing

Research aircraft observations have been used for meteorological research extensively and here we only provided icing related research that generated significant improvements for icing measurements and predictions (Politovich 1989; Cooper et al. 1984; Sand et al. 1984; Isaac and Schemenauer 1979; Isaac et al. 2005; Reehorst et al. 2005). For example, icing research using aircraft platforms has been done by NASA Glenn, Canadian National Research Council (NRC) and Environment Canada (EC), NCAR, and University of Wyoming (Politovich 1989, 1996; Isaac et al. 2005; Serke et al. 2008; Reehorst et al. 2005). Bernstein et al. (2005) developed a Current Icing Product (CIP) that utilized satellite, radar, surface, lightning and PIREPS observations together with 3D model hourly diagnostics to retrieve icing and Supercooled Large Drop (SLD) potentials. Figure 3a shows the conceptual diagram of their CIP algorithm that was adapted by FAA to be used operationally. Figure 3b shows an example of the splintering

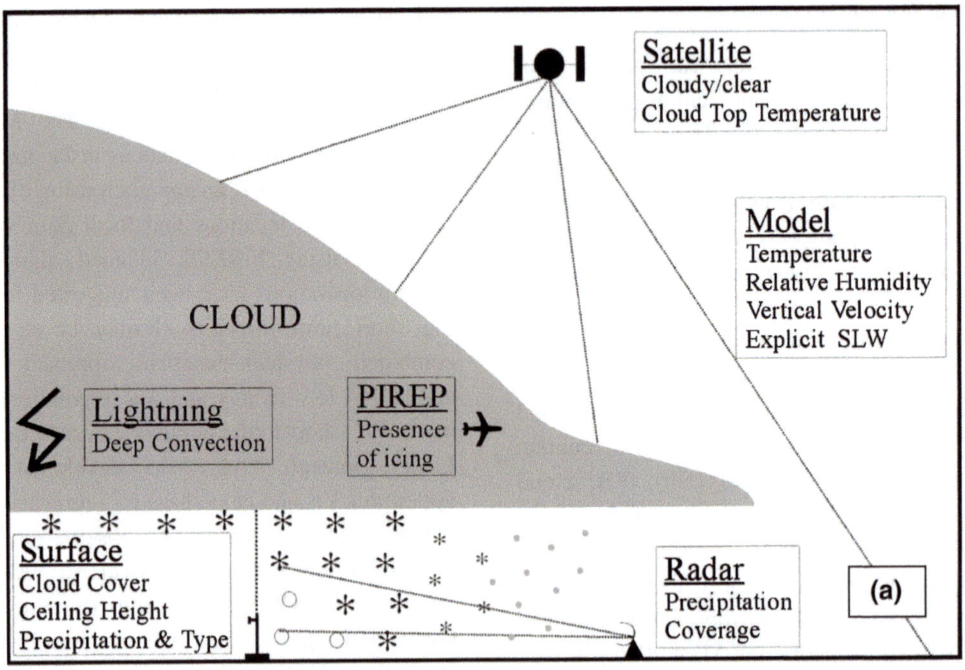

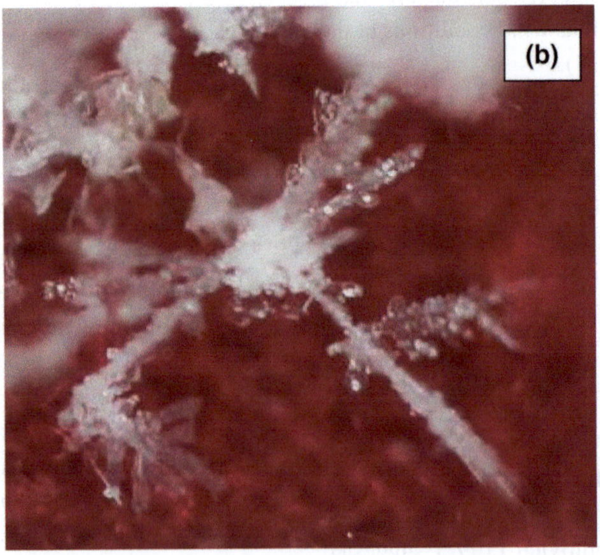

Figure 3
CIP (Current Icing Potential) project conceptual diagram (**a**). Precipitation types: snow (asterisks), rain (large open circles), and freezing drizzle (small gray circles). (Adapted from Bernstein et al. 2005) © American Meteorological Society. Used with permission. **b** shows a picture of splintering mechanism, which generated many small droplets and then quickly froze occurred during a wet snow case at $T = -3$ °C on March 15 2013 during SAAWSO project (Gultepe et al. 2017a, b) took place over St. Johns' area, NL, Canada

mechanism which generated many small droplets, then these quickly froze at T = − 3 °C on March 15 2013 during Satellite Applications for Arctic Weather and SAR (Search And Rescue) Operations (SAAWSO) project took place over St. Johns' area, NL, Canada (Gultepe et al. 2018).

Based on aircraft in situ observations of cloud microphysical parameters, super cooled large droplets (SLD) impact on aviation certification issues was emphasized by Politovich (1989), Isaac et al. (2005), and Bernstein et al. (2005). Politovich (1989) noted that substantial loss in rate of climb capability occurs in less than 10 min when droplet number concentration, $N_d < 0.1$–$1\ cm^{-3}$ and mean volume diameter (MVD) at about 30–400 μm. Various icing types occurring at the surface of aircraft (Politovich 1996; Newton 1947, 1978) can significantly impact aircraft lift and drag forces that can lead to aircraft accidents. Lately, high ice water content is also found to be an important factor for aircraft engine icing (Leroy et al. 2016; Haggerty et al. 2018) and that needs to be further researched for operational applications.

2.5. Satellite Observations

Observations from passive radiometric imagers (e.g. the Advanced Baseline Imager, ABI) mounted on operational satellites, such as the Geostationary Operational Environmental Satellites (GOES), are used extensively for evaluation of meteorological events e.g., fog and convection (Pavolonis 2010a, b; Heidinger 2010; Minnis et al. 2011; Bedka et al. 2010), as well as for nowcasting and data assimilation algorithms used in the NWP modeling systems (Miller et al. 2014; Illingworth et al. 2015; Meckalski et al. 2007). Both imager and sounder spectral radiances and relevant parameters derived from radiances have been used extensively for these purposes. Examples of these products and potential applications are discussed below and in following sections.

In addition to passive radiometric imagers on satellites, active remote sensing platforms are also starting to be used on the satellites such as Cloudsat (Sun-Mack et al. 2017; Deng et al. 2010) and CALIPSO (Smith 2014). The EarthCARE satellite mission is planned for 3D weather predictions by blending data from satellite based active sensors together with simulations from a 3D NWP model (Miller et al. 2014; Illingworth et al. 2015; Barker et al. 2011). The use of active sensors on these satellite platforms can provide insight information on cloud vertical structure, cloud top properties, and both microphysics and precipitation that can be blended with NWP simulations and ground based remote sensing platform data to be used for aviation applications.

A variety of cloud properties are retrieved in near-real time using the Satellite Clouds and Radiation Property retrieval System (SatCORPS; https://satcorps.larc.nasa.gov) applied to global geostationary and polar orbiting satellite imager data (Minnis et al. 2008; Minnis et al. 2016) using adaptations of the algorithms of Minnis et al. (2011). The parameters retrieved include, among others, the scene identification of the pixel as either cloudy or clear, cloud top phase, cloud top and height temperatures and altitudes, cloud optical depth, and cloud particle effective radius. Recent additions include multi-layered cloud detection and properties, aerosol concentration over ocean, and convective cloud overshooting top identification (e.g., Bedka et al. 2016). These products are currently used for a number of aviation applications such as aircraft icing (Bernstein et al. 2006; Smith et al. 2012), engine icing (Yost et al. 2017), and cloud ceiling, and aviation and severe weather NWP model data assimilation (Benjamin et al. 2016; Jones et al. 2016). Satellite observations retain significant potential for additional use by aviation systems. Improvements and enhancements in the satellite based retrieved products can expand that potential. For example, determining the separation between multilayer clouds (Sun-Mack et al. 2017) will likely improve chances of flying in clear air in multilayered conditions. Knowing the thickness of an icing layer and a thick ice cloud location above the supercooled droplets segment (e.g., Smith 2014) can further provide more choices for pilots faced with a hazard and few options.

Another source of cloud information is the algorithm developed by Sieglaff et al. (2011), which classifies each GOES pixel into groups of clear, liquid water, supercooled liquid water, mixed phase, opaque ice, non-opaque ice, and multilayered ice cloud (ice cloud is the highest cloud layer). These groups are described in the work of Pavolonis and Heidinger (2004) and Pavolonis et al. (2005). Details of the algorithm can be found in Pavolonis (2010a, b) and Heidinger (2010). Similarly, Liu et al. (2009) also defined cloud types based on China's FY-2C multichannel images and a neural network method. The satellite cloud analyses utilize clear-sky background correction, satellite zenith angle, and sensor

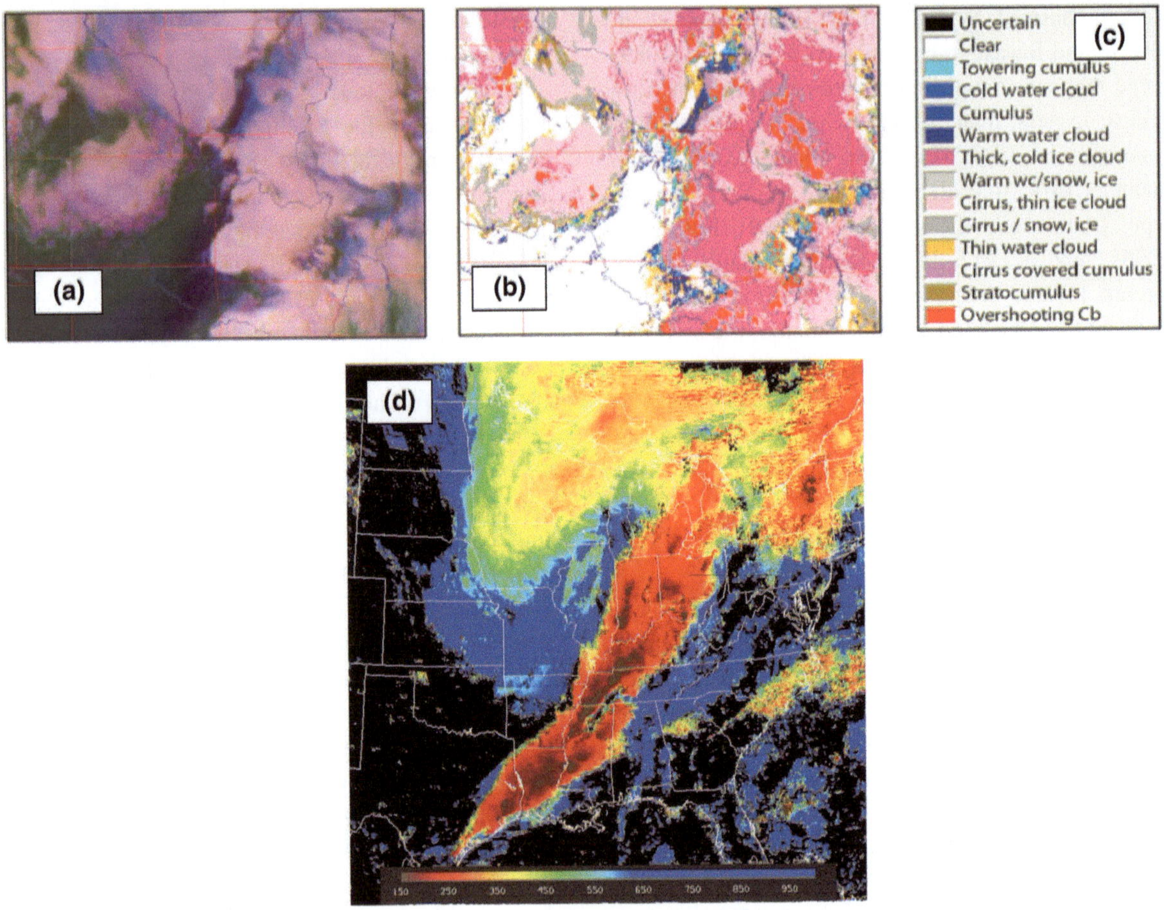

Figure 4
Example of texture-based clustering of GOES imagery on Dec. 5 2003: Three-band red, green, blue (**RGB**) image (**a**) and color-coded clustering classifier (**b**). Color bar shows cloud types (**c**). An example of the GOES-12 imager-derived cloud top pressure product at 1800 U T C 15 Nov 2005 is shown in **d** (Adapted from Meckalski et al. 2007). © American Meteorological Society. Used with permission

spectral response functions. These were then used for the cloud mask and type algorithms to be portable to many sensors. The cloud-type algorithm relies on an upfront cloud mask algorithm to determine which pixels contain cloud (Heidinger 2010). The satellite observations can usually lead to detection of cloud phase, fog, icing, and turbulence from spectral radiance measurements (Minnis et al. 2011; Pavolonis 2010a, b, Pavolonis and Heidinger 2004; Smith et al. 2012; Gultepe et al. 2007a, b; Ellrod and Pryor 2019). Integrated systems based on in situ observations, NWP model output, as well as GOES satellite products can improve monitoring fog conditions that include fog area and its intensity (defined by optical depth).

A number of efforts have focused on detecting and analyzing overshooting tops (indicating convective activity) using GOES products from Visible (VIS) and Infrared (IR) imagery (Nair et al. 1998; Meckalski et al. 2007; Bedka et al. 2010). In addition to garnering information directly from the VIS reflectance and IR brightness temperatures, and their differences with other channels were used to detect overshooting tops. Texture analyses have allowed the discrimination of convective cloud features within a shield of stratus-like clouds that often cover a convective complex. Figure 4a–c show the texture-based clustering analysis results based on Meckalski et al. (2007). In this figure, large-scale convective structures are visible, along with smaller cumulus and stratiform clouds. Figure 4d shows a cloud-top pressure product developed through the combination of height estimates from the CO_2 ratio technique for the upper clouds and from the older "IR window" technique for lower clouds (Schreiner et al. 1993; Donovan et al. 2008). The IR window technique uses the 11 μm thermal IR temperature to estimate T at cloud top, and combines it with an atmospheric sounding to get T to a corresponding pressure and height. The isolated overshooting cumulus tops were also detected as in red. Improvements on this technique have been reported by Bedka and Khlopenkov (2016).

The reasons for recent advances in aviation safety due to meteorological satellites are their ability to provide critical information of weather conditions, the enhanced speeds of data transmission and computer analysis, better calibrated high-resolution satellites, algorithm maturity, and more reliable validations using observations. Therefore, hazards, such as turbulence, in-flight icing, convective storms, and volcanic ash, can be diagnosed from satellite-based observations, and are used extensively in national aviation forecasting systems (Meckalski et al. 2007). The algorithms are always subject to improvement based on validation studies using in situ observations, passive (e.g., microwave) and active (radar and lidar) remote sensing retrievals, and integrated observations.

The aviation products in NOAA NWS operations include volcanic ash, cloud-top height, turbulence, lightning detection, precipitation, icing, and low clouds and fog (Schmit et al. 2017; Goodman et al. 2012). Recently, with the introduction of GOES-16 (16th Geostationary Operational and Environmental Satellite) products, lightning detection for aviation applications became possible. New sensors available on GOES-16 include total lightning detection and mapping of in-cloud and cloud-to-ground flashes (from the Geostationary Lightning Mapper (GLM; Goodman et al. 2012; Goodman et al. 2013). It is expected that the use of GLM data for severe weather research will improve aviation weather forecasts related to convection and storm physics.

Airborne ash from volcanic eruptions can also be a major threat to aviation safety (Casadevall 1994; Miller and Casadevall 1999 ; Hufford et al. 2000; Simpson et al. 2000). VIS satellite imagery can be used to detect ash clouds, but not to discriminate ash particles from natural water and ice clouds. Ash detection and discrimination is accomplished via multispectral infrared image analysis. The multispectral technique uses a strong 12-μm absorption signature (e.g., Prata 1989; Hufford et al. 2000; Meckalski et al. 2007), leading to a negative value of the 11–12 μm brightness temperature difference. As discussed Prata (1989), this technique is often termed as the reverse-absorption method. Volcanic eruptions usually release high concentrations of SO_2 having strong absorption bands at 7.3 and 8.6 μm that can aid the ash detection algorithms. SO_2 absorption channels are presently available on the high-resolution Moderate Resolution Imaging Spectroradiometer (MODIS) imager and the Atmospheric Infrared Sounder (AIRS) sounding instrument in polar orbit on NASA's Earth Observing System (EOS) satellites, and with the next-generation operational geostationary satellites (e.g., Himawari-8, GOES-16) (Goodman et al. 2012; Hutchison et al. 2008; Meckalski et al. 2007; Schmit et al. 2005).

The National Aeronautics and Space Administration (NASA) developed an Advanced Satellite Aviation-Weather Products (ASAP) system that transfers new satellite-observing products into operational use based on a collaborative effort with the FAA's existing Aviation Weather Research Program (AWRP) (Meckalski 2002; Meckalski et al. 2007). The ASAP ash detection algorithms are currently being developed at UW-CIMSS. The standard reverse absorption technique (Prata 1989) that uses channel differencing between 11 and 12 μm BBT (Black Body Temperature), is supplemented with VIS at 0.65 μm and near-IR at 3.75 μm channels for volcanic ash detection (Pavolonis et al. 2006). This new algorithm does not depend on a negative value of 11–12 μm brightness temperature difference because this difference is often absent in tropical eruptions and it is used globally. Lately, there is significant new research on automated detection of explosive volcanic eruptions that is based on a Cloud Growth Anomaly (CGA) technique obtained from geostationary satellite IR data, and it is suggested that this new method significantly reduced False Alarm rates (Pavolonis et al. 2018). The CGA technique, can be applied to any geostationary satellite (day and night) but is particularly valuable when applied to next generation satellites such as Himawari-8 and GOES-R. The CGA technique utilizes infrared measurements to identify cloud objects and compute cloud vertical growth rates from two successive images.

Hadley et al. (2004) studied the rare resuspension of volcanic ash and dust from the Katmai area volcanic eruption by strong winds on 20–21 September 2003 and found that the ash/dust cloud created severe disruptions to aviation operations while aircraft were attempting to avoid dust/ash clouds. Another volcanic ash event occurred from Mount Cleveland, Alaska when it erupted on 19 February 2009 and led to several severe disruptions of commercial aircraft traffic. The use of satellites for ash detection has recently become very common but issues with data analysis are still important, and more work needs to be done. Simpson et al. (2000) also suggested that detailed volcanic ash PIREPs are needed to improve real-time awareness for aviation routing.

Two major regions of dust originate from Asia (Gobi and Mongolian deserts) and Africa (Sahara Desert). Dust events usually occur when surface winds exceeding 5 m s^{-1} (Hadley et al. 2004) loft the dust particles into the atmosphere where they can travel across either the North Pacific or the tropical Atlantic Oceans, respectively (Gillette 1978). Dust can also affect radiative forcing and world climate (Myhre and Stordal 2001), as well as aviation safety (Simpson et al. 2003).

Satellite-based wind vectors can be used to infer wind shear, turbulence, and convective activity (Hubert and Whitney 1971; Meckalski et al. 2007; Perrie et al. 2008 ; Bedka et al. 2009). In the northern latitudes, polar winds are needed for better aviation flight planning because of lack of other observations and issues with NWPs through assimilation of observations (Key et al. 2003). Polar winds from satellite based retrievals (Turner and Warren 1989; Herman 1993) can be highly effective for aviation management because of the lack of airport availability after long flight times over Arctic regions. Santek (2010), using the National Centers for Environmental Prediction (NCEP) Global Forecast System (GFS), ran an experiment during August and September 2004, with and without the Terra satellite based MODIS (Moderate Resolution Imaging Spectroradiometer) Polar winds. From the five cases examined, it was determined that the addition of the polar winds modifies the mass balance in synoptic-scale waves near the polar jet streams that can affect flight planning.

2.6. *LIDAR Observations*

LIDAR-based observations are used to obtain Doppler wind speed (Vd) based on motion of air particles and backscattering ratio as well as depolarization rate. Excluding deep and heavy in-cloud water content conditions, they can provide 3D wind components from combination of LIDARS based on various scanning modes. At low elevation angles, LIDARs can also provide horizontal wind profiles to analyze turbulence and gust, and aerosol loading at lower levels.

LIDAR-based wind retrievals using the Doppler concept became important in the 1980s. Bilbro and Vaughan (1978) and Bilbro et al. (1984) used a coherent Doppler system to obtain 2-D wind components from an installation onboard the NASA Convair 990. For a recent overview of the use of LIDARs in atmospheric research see Reitebuch (2012). Recently, LIDARs have been used for aviation research and operations (Gultepe et al. 2017a, 2014b; Fuertes et al. 2014; Tucker et al. 2009; Sathe et al. 2011). Tucker et al. (2009) used a High Resolution Dial LIDAR (HRDL) at 2.022 μm to study mixing height, turbulence, shear, and aerosol profiles. The HRDL velocity and backscattered signal measurements collected during a ship-based field campaign were used in wind analysis. In their work, wind air velocity variance profiles were used for mixing height estimation. This was a significant development of LIDAR use in the aviation research. Their results are shown in Fig. 5 which shows time-height cross sections of σ_w^2 (vertical air velocity variance) and boundary layer height (h_o) (a), horizontal wind shear height (h_{spd}) (b), and composite velocity variance (σv^2) based on shear and mixing height (c). Their work also suggested that a motion-stabilized scanning coherent Doppler LIDAR can perform azimuth scans, elevation scans, and zenith stares to obtain the velocity field in three dimensions from a moving platform and that can improve aviation forecasts over the coastal regions.

Lately, Doppler LIDAR wind measurements have been used for aircraft operations at airports (Thobois

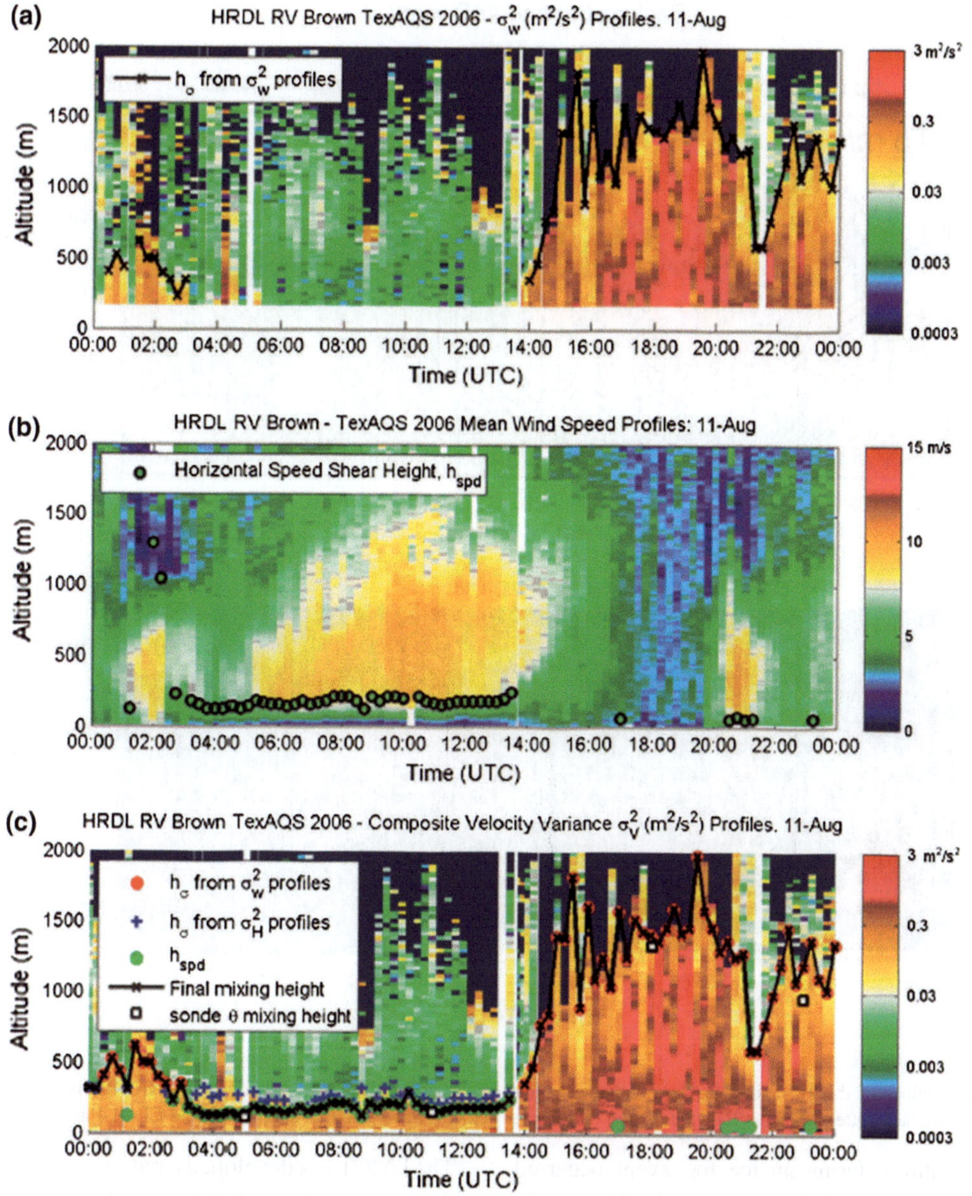

Figure 5

The time-height cross sections of HRDL (lidar) vertical air velocity variance σ_w^2 (**a**), horizontal mean wind speed shear height (h_{spb}) (**b**), and composite velocity variance (σv_2) (**c**) for a 24-h period on 11 Aug 2006 during TexAQS 2006 project (Adapted from Tucker et al. 2009)

et al. 2018; Chan and Shao 2007; Chan et al. 2006) to diagnose wind shear, gusts, microbursts, and aerosol loading, which can be serious issues for aviation operations (Sharman et al. 2012; Kessinger et al. 2006a b, Wong et al. 2013; Kim et al. 2015, 2016; Gultepe et al. 2018). But, presently Doppler lidars are used mostly in an experimental mode at airports. Among Doppler LIDARs commercially available are (1) Halo LIDAR (Gultepe et al. 2016), (2) Leosphere Cube LIDAR (Thobois et al. 2018), and (3) Lockheed Martin Coherent Technologies (LMCT) LIDAR (http://www.lockheedmartin.com). These LIDARs provide Doppler wind measurements in various scanning modes. The Halo LIDAR observations were used to obtain backscatter ratio and VAD (Velocity Azimuth Display) winds, and compared to CL51 ceilometer backscatter

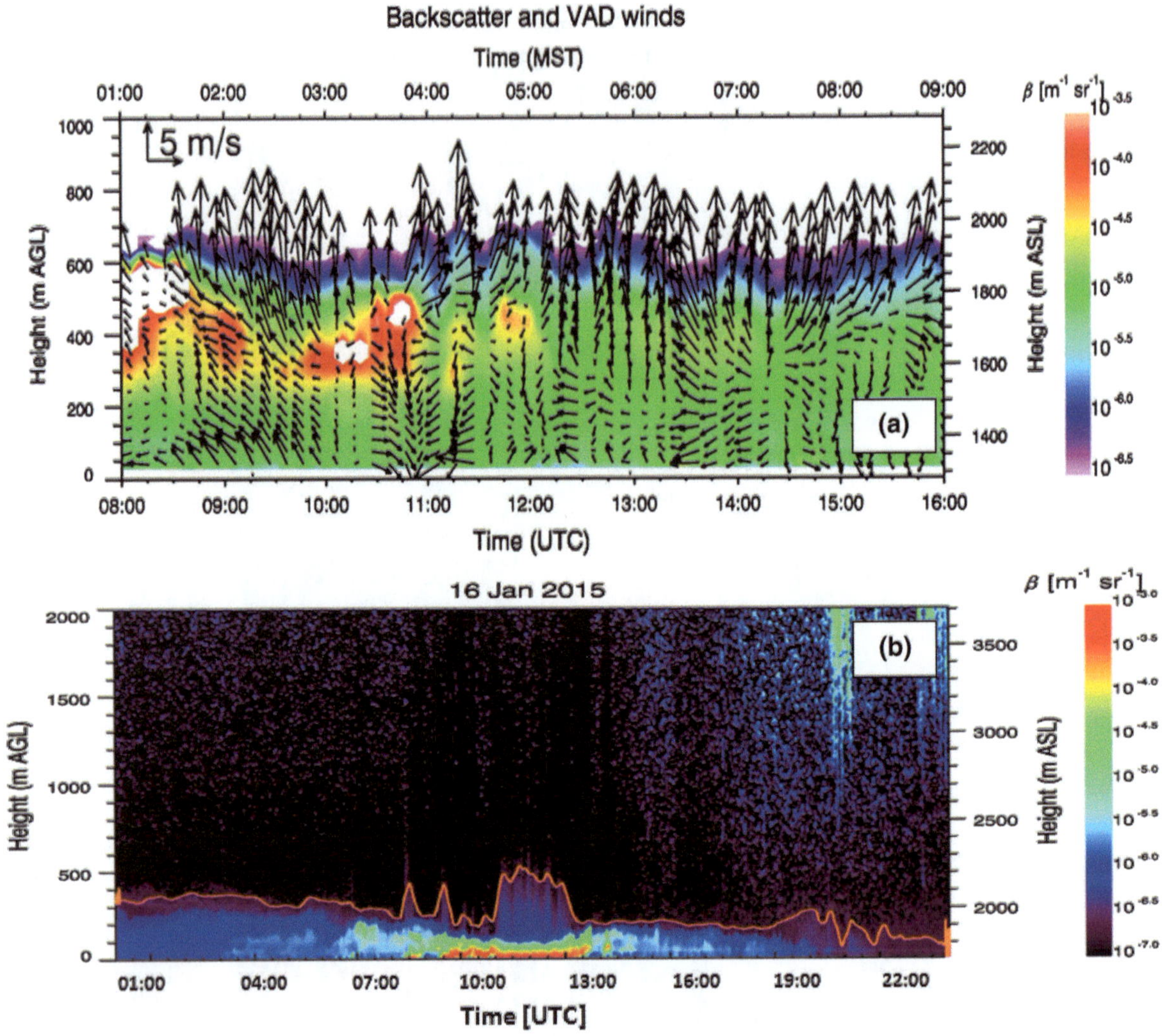

Figure 6

Example observations of the 16 January 2015, mountain ice fog event that occurred during the MATERNHORN project over the Utah Mountains; Halo backscatter ratio and 3D winds based on Doppler velocities and VAD technique (**a**), and CL-31 Backscatter ratio (β) (**b**)

ratio observations during an ice fog event occurred nearby Heber City, UT, are shown in Fig. 6a, b, respectively. The Cube Doppler LIDAR from Leosphere Inc is designed specifically for aviation applications and provides related aviation wind and turbulence parameters continuously (Thobois et al. 2018). Examples of turbulence estimation from LIDARs can be found in Frehlich and Sharman (2010), Hill et al. (2010), Chan (2016), and Vrancken (2016).

LMCT LIDAR at 2 μm (latest one at 1.6 μm) uses pulsed laser light to detect particles and varying weather conditions in the atmosphere (Bluestein et al. 2010; DeWekker et al. 2012 ; Bluestein et al. 2014). The LMCT has developed a pulsed coherent 2 limiting diode-pumped solid-state LIDAR receiver on an injection-seeded, Q-switched, 2 micron laser that meets Navy requirements for remote sensing, moderate range, high spatial resolution wind field measurements around air stations and aircraft carriers.

2.7. *Sodar Observations*

The SOnic Detection And Ranging (SODAR) is often used for atmospheric profiling. SODAR uses the Doppler concept to estimate 3D wind profiles from the propagation characteristics of high frequency sound waves to investigate the atmospheric boundary layer and has been used since 1980s. For example, SODAR data were used to investigate the boundary layer dynamics in detail by Motta et al. (2005), Sumner and Mason (2006), Gottschall and Peinke (2008), and Van den Berg (2008). In these works, atmospheric stability, wind shear, and turbulence intensity were analyzed for wind energy production. Gerz et al. (2009) and Chan (2014) used SODAR observations specifically for nowcasting applications to investigate aviation related meteorological events, such as fog, gust, and wind shear, at the Frankfurt International Airport and at the Hong Kong International Airport, respectively. Silva et al. (2016) developed a conceptual model to investigate wind profile changes using 234 runway cases that utilized a SODAR and surface meteorological station data from the Guarulhos International Airport (GIA) São Paulo, Brazil. França et al. (2018) also used SODAR data for training a neural network model for nowcasting the low level wind profiles at the lead times of 45 min for the GIA Airport.

2.8. *Radar Observations*

Doppler and dual-polarized radars (and conventional ones) such as X (8–12 GHz), C (4–8 GHz), and S (2–4 GHz) band radars have been used to detect storm convective activity and intensity, as well as icing conditions within clouds to identify particle microphysical characteristics (Merritt, 1969; Rasmussen et al. 1992; Ryzhkov et al. 2002; Schuur et al. 2012; Smith et al. 2016; Hubbert et al. 2018). Storm precipitation type and intensity are related to Vis, icing levels, and convective intensity, and these are all important for aviation operations. Smith et al. (2016) developed the Multi-Radar Multi-Sensor (MRMS) system that is operational at the National Centers for Environmental Prediction (NCEP). The MRMS system consists of the Warning Decision Support System–Integrated Information (WDSS-II; Lakshmanan and Smith 2009) suite. The severe weather and aviation products, as well as the quantitative precipitation estimation (QPE) products, are created by the National Mosaic and Multi-Sensor QPE (NMQ; Zhang et al. 2011) system.

Dual polarized Doppler radars such as the Weather Surveillance Radar-1988 (WSR-88D) radar and radars with dual polarization capabilities such as S-band radars (S-Pol) can improve rainfall estimates, discriminate ice and rain, identify hail cores and updraft regions, and as a result in general increase data quality (Hubbert 2017; Hubbert et al. 2018; Chandrasekar et al. 2013; Kumjian 2012, Kumjian 2013a,b). Hubbert et al. stated that the Weather Surveillance Radar-1988 Doppler polarimetric (WSR-88DP) is not able to scan along vertical planes (RHIs), and although this may be acceptable for operational purposes, it does limit research objectives. The S-Pol has increased range resolution using 1-μs transmit pulse length that corresponds to 150-m compared to WSR-88DP's 250-m resolution. The result is that S-Pol is able to obtain high-resolution, high-data-quality measurements related to storm structure and microphysics. In the end, improved storm physics and dynamical information can further be used for aviation operations that concern with wind and turbulence, gust, icing, high IWC, and low visibilities, as well as lightning and particle type and phase.

Dual polarized radars can effectively be used for melting layer and particle phase and type detection (Schuur et al. 2012; Ryzhkov et al. 2002, 2013; Van Den Broeke 2016; Hubbert et al. 2018). The sensitivity of this radar decreases with reducing particle size but retrieval of particle phase and 3D wind structure can improve forecasting and nowcasting algorithms related to gust and low Vis. Van Den Broeke et al. (2016) stated that quasi-vertical profiles of the polarimetric radar variables could improve short term forecasts in winter. Hubbert et al. (2018) summarized polarimetric variables used in S-Pol data interpretation. Figure 7 shows range-height plots of reflectivity (Z), and V_D, differential reflectivity (Z_{DR}), and differential phase (φ_{DP}) parameters from a convective storm. Using Z (indicates core of storms), Z_{DR} (reflectivity weighted particle mean axis ratio of particle distribution), V_D (particle motion), and φ_{DP} (particle microphysics), LDR is defined as linear depolarization ratio (related to particle shape), K_{DP} as specific differential phase (particle type and shape), ρ_{hv} as co-polar correlation coefficient (particle type and phase), and ρco-x as co- to cross-channel correlation coefficient (canted crystals and cloud electrification). A combination of these

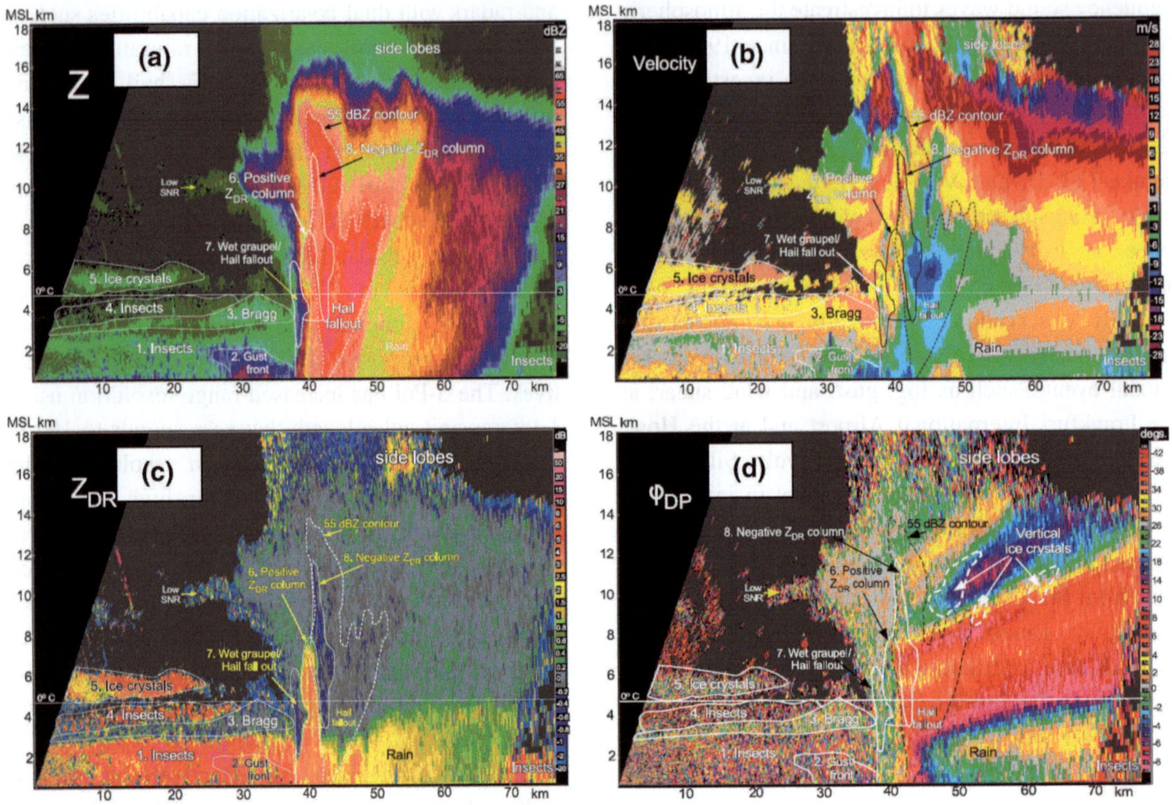

Figure 7

A reflectivity RHI of a large convective cell gathered at 0005:23 UTC 26 Jun 2015. Reflectivity field (Z) is shown in **a** where the dashed line marks the 55-dBZ contour; 60 dBZ is seen up to 13 km MSL, indicating the likely presence of large hail, Doppler velocities in **b**, the various regions based on S-Pol radar Z_{DR} marked with white contour lines in (**c**), and differential phase ϕ_{DP} is shown in **d**; starting offset is set at about 0 °C and over marked regions of vertical ice crystals, ϕ_{DP} is decreasing in range (d) Adapted from Hubbert et al. (2018), © American Meteorological Society (AMS). Used with permission

variables can be used for evaluating storm physics and dynamics (Hubbert et al. 1998, 2018; Ryzhkov et al. 2002, 2013; Van Den Broeke 2016) that can further be used for operation applications.

Doppler radar applications for fog and Vis work can be used to provide accurate nowcasting products for aviation weather. Hamazu et al. (2003) used a 35-GHz scanning Doppler radar (Ka band, 27–40 GHz) for fog observations and prediction. They used the three-dimensional structure of reflectivity and velocity field in a fog event for short term predictions. The characteristics and performance of an 8.6-mm-wavelength Doppler radar were described in observing sea fog.

The Doppler radar measurements were also used for eddy dissipation rate predictions. Aircraft based turbulence measurements with well-coordinated Doppler radar spectral width measurements were used to estimate energy dissipation rates within thunderstorms anvils (Meischner et al. 2001; Cohn 1995). The lower limit detectable dissipation rate with the C-band Doppler radar is found to be about 10^{-3} m^2 s^{-3}. Techniques for deriving Doppler radar estimates of energy dissipation rate were also described by Williams and Meymaris (2016), and comparisons to in situ aircraft measurements were provided in Dehghan et al. (2014). Doppler spectrum width as measured by operational weather radars was used for both turbulence and wind shear estimation and could also possibly be used for severe weather warnings (e.g., Hocking and Hocking 2018).

The K-band radars [Ku (12–18 GHz), K (18–27 GHz), Ka (27–40 GHz)] (Matrosov 2005;

Löffler-Mang et al. 1999), and W-band radars (40–300 GHz, Gossard et al. 1997; Mead et al. 1989) have been used for cloud microphysical structure but usually they are not used operationally for Aviation operations. When large water mass content exists within clouds, these radar beams can be attenuated significantly. Presently, Ku and Ka band radars are part of active sensors used on GPM satellite that focuses on precipitation from the space (Skofronick-Jackson et al. 2017). These cloud radars can provide specific information for cloud and fog properties, as well as precipitation information but they are operationally expensive and mostly they are used for experimental research.

2.9. Atmospheric Thermodynamic Profiling

2.9.1 Water Vapor and Temperature Profiling

Water vapor profiling is critical for weather forecasting because of the detail it can provide about cloud formation, nucleation processes, and storm development (Heymsfield et al. 2005). Ferrare et al. (1995) provided a summary of water vapor measurements that included satellite-based retrievals (Soden et al. 1994), microwave (Han et al. 1994), differential absorption LIDAR (DIAL) (Ismail and Browell 1994), and Raman LIDAR (Ansmann et al. 1997; Whiteman et al. 1992). The difference between the Raman LIDAR and DIAL LIDAR is that the former measures only vapor mixing ratio (qv) and the latter measures both qv and T. The work of Wulfmeyer (1998, 1999) used a DIAL system developed at the MaxPlanck Institute (MPI) in Hamburg for the measurement of absolute humidity profiles and this has been improved through the work of many others (Wulfmeyer et al. 1995; Wulfmeyer and Bosenberg 1996; Wulfmeyer 1998; Wulfmeyer and Bosenberg 1998). Eichinger et al. (1999) used a Raman LIDAR to study ABL vapor profiles. The DIAL systems have some inherent advantages over Raman LIDARs profiles. The DIAL systems are generally smaller, lighter, and use considerably less energy than Raman LIDAR systems. They can be used for aircraft based research because atmospheric backscatter increases with a range in a downward-looking system, and thus partially compensates the decrease in signal strength with range. Because of the strong atmospheric attenuation of both UV and near-UV light channels, DIAL systems in the near-IR are better suited for deep atmospheric water vapor sounding profiles. Recent overviews of the DIAL can be found in Wirth (2012), and Fix (2012).

The radar–Radio Acoustic Sounding System (RASS system) was developed at the University of Hamburg for measuring wind profiles in the lower troposphere and virtual temperature profiles in the boundary layer (Peters et al. 1988; Peters 1990). The turbulent variables measured with the DIAL system and the radar–RASS include vertical profiles of water vapor, vertical wind variance, and latent heat flux that uses the eddy correlation technique. Neely and Thayer (2011) also used a similar method to study high level moisture profiles over Greenland, which can be used to detect thermal and dynamical instabilities leading to particle formation and eventually cloud formation.

The Profiling microwave radiometers (PMWRs) have been also used for measuring q_v and T, as well as Liquid Water Content (LWC) which can be important for supercooled droplet detection and icing conditions (Solheim et al. 1998; Gultepe et al. 2014a, b). The Radiometrics Inc. MP-3000A profiling radiometer is used to retrieve the profiles of T, RH with respect to water (RH_w), and LWC over Whistler Mountain (Ware et al. 2013; Gultepe et al. 2014a, b). The MP-3000A observes 21 K-band (22–30 GHz) and 14 V-band (51–59 GHz) microwave channels at multiple elevation angles, one zenith infrared (9.6–11.5 μm) channel, and surface temperature, humidity and pressure sensors. The atmosphere is semi-transparent in the K-band and lower V-band channels during non-precipitating conditions, receiving emission from the atmosphere in addition to cosmic background radiation. The PMWR provides T, qv, RH_w, and LWC in the vertical (Bianco et al. 2005) but it needs to be validated using in situ observations; this may be a challenge for stratiform clouds because of a weak signal from cloud water content compared to convective clouds' water content. Integrated liquid water retrieval from microwave radiometer observations near the 22.2 GHz water vapor molecular resonance is well established (Westwater 1978; Turner et al. 2007). Integration of above remote sensing platforms can cover atmospheric thermodynamic conditions over the airports, and when they are combined with LIDAR and Radar observations, as well as with surface in situ observations, both clear air and cloud regions nearby airports can be analyzed for evaluating aviation hazards.

Atmospheric thermodynamic profiling for operational icing research is important to prevent aviation accidents. Serke et al. (2008, 2014) developed a compact small platform attachable to a radiosonde balloon that is similar to the radiosonde unit. With this system they were able to measure liquid water content profile within the clouds. This state-of-the-art icing platform promises to develop future balloon based systems to be used for icing and thermodynamic profiling.

In Arctic regions, profiles of water vapor can be important for aviation nowcasting applications, such as fog and low clouds, and storm development. The Arctic LIDAR Technology (ARCLITE) facility, having a Rayleigh/Mie/Raman LIDAR system, has been in operational use since 1993 at the Sondrestrom Upper Atmospheric Research Facility, Kangerlussuaq, Greenland (Neely and Thayer 2011). In their research, molecular and aerosol backscatter is measured at 532 nm to retrieve T profiles, and a Raman channel for molecular nitrogen (608 nm) to determine aerosol extinction values. A Raman water vapor channel at 661 nm was used to measure water vapor mixing ratio profiles into the upper troposphere. They suggested that use of Raman LIDAR technology for ABL water vapor soundings is available and is comparable with balloon based profiles. Because measurements of water vapor profiles in Arctic environments are rare, DIAL and Raman LIDAR based methods for water vapor profiling can improve fog, cloud, and Vis for short term predictions.

2.9.2 UAVs Thermodynamic Profiling

Unmanned aerial vehicles (UAVs) are becoming very popular for meteorological and environmental applications (e.g., Bottyán et al. 2013, 2016; Gultepe et al. 2018). UAVs can be accurately controlled by ground based pilots to move the UAVs in certain flight paths that represent constant altitude flights and profiling models. Measurements from the UAVs include various meteorological parameters similar to a radiosonde sounding. In addition to T, RH_w, pressure (P), altitude (z), Vis, U_h and direction (Gultepe et al. 2018; Bates et al. 2013). UAVs may be used for air quality monitoring of trace gases such as O_3, CO_2, CO, and PM2.5, and earth surface morphology, including snow coverage, water accumulation, and surface temperature. Measurements from UAVs have also been used to estimate temperature and velocity turbulence levels (e.g., Lawrence and Balsley 2013).

For aviation operations, UAVs can be used near airports to gather weather information that can help to improve the NWP's initial conditions and monitor weather conditions in real time. Although they have limitations on flying time (e.g., quadcopter UAVs), measuring 3D wind components, and flying into the clouds and fog, UAVs are being developed for extreme weather conditions (especially with fixed wing UAVs, Griffin and Velden 2018). In the future, UAVs can fly into the cloud systems (Wick et al. 2018; Wurman et al. 2012), and may provide continuous and accurate weather measurements. Uncertainty in UAV measurements can be significant for wind related parameters because of the impact of the rotors or body of the UAV on airflow. Therefore, additional work is needed to improve wind measurements from UAVs (Reineman et al. 2013; Gultepe et al. 2018). Various meteorological parameters from the UAVs have been used by Jonassen et al. (2012) for boundary layer (BL) research. These parameters include aerosol number concentrations (N_a), fog droplet number concentration (N_d), Vis, particle spectra, and air quality parameters (Boer et al. 2017; Gultepe et al. 2017a, b).

The UAVs can be used to improve NWP predictions for meteorological applications that include wind speed, Vis, RH and T. Jonassen et al. (2012) and Reineman et al. (2016) stated the importance of UAVs observations to improve NWP simulations, and this area of research is needed for further development of aviation products.

2.9.3 Aircraft Based In Situ Systems

Commercial aircraft flights can provide additional observations of meteorological parameters related to aviation operations. Automated meteorological reports from on board sensors mounted on major commercial air carriers are generally referred to internationally as Aircraft Meteorological Data Relay (AMDAR) reports. In the USA, they are called Aircraft Communication Addressing and Reporting System (ACARS) reports (Schwartz and Benjamin 1995), and the meteorological data is referred to as Meteorological Data Collection and Reporting System (MDCRS) (Moninger et al. 2003). The MDCRS reports are essential for NWP model initializations, especially for rapidly updated models such as the NOAA's WRF-RAP model (Benjamin et al. 2016). Also, the commercially available Tropospheric Airborne Meteorological Data Reporting system (TAMDAR), utilizing commercial flight data collection systems, provides a data base for aviation parameters such as wind, turbulence, temperature, etc. (Moosakhanian et al. 2006). The TAMDAR data (Moninger et al. 2010; Benjamin et al. 2006a, b; Fournier 2006) used with the RUC model over 3 year period revealed that TAMDAR data significantly improved RUC forecasts. The meteorological parameters obtained from AMDAR and TAMDAR are similar to radiosonde measurements. TAMDAR also provides EDR (defined as energy dissipation rate to the 1/3 power) estimates and icing detection amount. Moninger et al. (2010) over two 10-day periods provided results on meteorological parameters such as T, RH_w, and wind representing winter and summer conditions. Their assessment showed that TAMDAR's impacts on 3-h RUC forecasts of temperature, relative humidity, and wind are found to be positive and, for temperature and relative humidity, the improvements were substantial.

Estimates of turbulence are also available from selected AMDAR flights as either derived equivalent gust velocity (DEVG), or EDR. Both of these turbulence estimation algorithms are summarized in Sharman (2016); for more detailed discussions see Sharman et al. (2014), Cornman (2016), Cornman et al. (1995), and Kim et al. (2017). Stickland (1998) reevaluated DEVG and EDR estimations for aviation applications. Even though both estimators are independent of aircraft type, he recommended the use of EDR for aviation operations. This recommendation has been followed up by requiring EDR to be the international standard for aircraft turbulence intensity reporting (ICAO 2005). EDR can also be estimated from second-order structure functions of the AMDAR winds (Frehlich and Sharman 2010), and possibly through Mode-S and Automatic Dependent Surveillance-Broadcast (ADS-B) messages (Kopeć et al. 2016). So far though, these techniques have not been implemented operationally.

Another experimental commercial aircraft onboard system is the Backscatter Cloud Probe (BCP) that provides information about the concentration and size of ice crystals in clouds. The probe has been tested as part of the European Research Infrastructure program, In-service Aircraft for a Global Observing System (IAGOS). Beswick et al. (2014) demonstrated that the BCP measurements could be used to diagnose HIWC conditions, providing pilots with an onboard warning of an HIWC hazard. The sensor output could also be broadcast as part of other systems such as TAMDAR. Additional research is needed for use of the BCP operationally.

A number of experiments for evaluating high ice water content regions (HIWC), including the EU High Altitude Ice Crystals (http://www.haic.eu) and North American HIWC Projects (Strapp et al. 2016), were developed and executed using remote sensing and airborne in situ probes. Leroy et al. (2016, 2017). Leroy et al. studies characterized the ice particle size distributions in HIWC regions in the atmosphere using in situ measurements while Protat et al. (2014) used airborne C-band polarimetric radar data to profile the cloud IWC. Yost et al. (2017) developed a prototype method for detecting HIWC conditions using geostationary satellite data coupled with in situ total water content (TWC) observations obtained during aircraft-related icing projects. They developed three satellite-derived parameters that were used for determining high HIWC probability conditions. These include (1) the overshooting convective updraft or textured anvil cloud, (2) tropopause-relative infrared brightness T, and (3) daytime-only cloud optical depth. Their results are consistent with aircraft flight reports obtained near deep convective storms and cirrus anvils (Lawson et al. 1998; Mason et al. 2006; Bravin et al. 2015). Prediction of HIWC conditions needs to be improved by combining NWP model simulations and remote sensing results. Both require more in situ observations to establish reliable statistics (Gultepe and Heymsfield 2016; Haggerty et al. 2018).

Overall, observations from various observational platforms summarized above can contribute to improvements of aviation nowcasts, weather reports, and NWP predictions extensively, and they may eventually lead to improvements in aviation operations, but their limitations should also be considered.

3. NWP Models

NWP models are important for improving aviation related nowcasts (defined here as lead times < 3 h) and forecasts over medium and large lead times (> 3 h). Their short-term success in nowcasting is strongly related to the incorporation of observations (Kober et al. 2012). A NWP model's ability to accurately simulate atmospheric dynamical and physical processes depends critically on several initialization parameters and PBL characteristics (Jonassen et al. 2012). The spatial grid resolution and the parameterization schemes used to represent processes related to clouds, radiation, precipitation, and turbulence are most crucial ones (Pleim and Xiu 1995; Alapaty et al. 2001; Teixeira et al. 2008; Lin et al. 1983; Tomita 2008; Khairoutdinov and Kogan 2000; Ferrier et al. 1995; Harrington et al. 2013a, b; Pu et al. 2018).

The quality of the data used to initialize and force the model for predictions is also essential for the success of numerical model simulations. Many models are initialized from global atmospheric analyses or forecasts [e.g., from the Global Forecast System (GFS) or the European Centre for Medium-Range Weather Forecasts (ECMWF) with low resolutions, typically being 15–50 km in the horizontal and 3–6 h in time. Therefore, scale issues play an important role for predictions over short times scales (Gultepe and Starr 1995). These atmospheric data may not be accurate enough for high-resolution simulations of local features such as fog or turbulence, and these can be sensitive to small errors in the large-scale flow (Nance and Durran 1997; Bélair et al. 1998; Khairoutdinov and Randall 2006; Kücken et al. 2012; Selz and Craig 2015). Because of these errors, data assimilation techniques using detailed and accurate observations are necessary for generating more accurate initial conditions for NWP models for all scales. The quality of the analysis data for model initialization depends mainly on the data assimilation techniques and quality of observations, which compile the short-range forecasts (from a model to generate the first guess), observations, and their error statistics (Kalnay 2003; Kalnay et al. 1996). The quality and coverage of the observations used to create the analysis have also influence on the accuracy of the analysis (e.g., Langland et al. 1999). Observations can be particularly sparse over areas such as the world's oceans (Herzegh et al. 2002), Arctic, and Antarctic, (Herman 1993) and that can likely lead to large uncertainties in predictions.

3.1. Downscaling

Because of scale resolution issues, numerical model simulations can include large uncertainties in the predicted parameters. The winds can be to some extent reproduced by numerical downscaling of a state-of-the-art ECMWF operational analysis using the WRF model (Jonassen et al. 2012). By assimilating profile data obtained from Unmanned Airplane Systems (UAS) at Eyrarbakki in southwest Iceland into a NWP model, substantial improvements of wind, T, and RH predictions in the region were achieved. Using automatic nonlinear/linear methods that use NWP Model Output Statistics (MOS) data, Vislocky and Fritsch (1995) also developed a statistical model to predict meteorological sounding parameters that are important for aviation. Their method was time consuming but, with increasing computer resources, their method could improve integrated predictive systems over various scales.

3.2. Initial Conditions

Initialization of NWP forecasts requires assimilation of various observations for regional models as well as short-range NWP model outputs (Kalnay 2003, Kalnay et al. 1996; Warner 2011). For instance, inputs into the operational MRMS-Severe/Aviation system (Smith et al. 2016; Wen et al. 2017; Benjamin et al. 2010) include radar data from the U.S. WSR-88D network, GOES radiances and cloud properties

(Minnis et al. 2008), surface terrain elevation information, the National Lightning Detection Network (Orville 2008; Smith et al. 2016), and hourly surface and upper-air analyses from the Rapid Refresh model (Benjamin et al. 2010). Several individual and automated algorithms have been developed using the MRMS system to obtain an integrated forecasting and analysis system that provides real-time products applicable to severe weather and aviation nowcasting. In this system, automated algorithms based on data from multiple radars provided better in-formation with greater temporal resolution and spatial coverage than a single-radar.

3.3. Turbulence

Since operational NWP models are too coarse to even begin to resolve turbulence scales relevant for aircraft, post-processing algorithms applied to NWP model output are commonly used to infer regions of significant turbulence. These inferences are based on "diagnostics" of turbulence derived typically from spatial gradients of selected NWP model output variables. In the past, various post-processing turbulence diagnostics have been proposed, and some are used operationally. For example, the TI (Turbulence Index) diagnostic developed by Ellrod and Knapp (1992), Ellrod (1985) uses NWP model output velocity fields to derive a diagnostic based on the product of horizontal wind deformation and vertical wind shear. The TI has been used by the Aviation Weather Center (NOAA/AWC) (Behne 2008), the Air Force Weather Agency (Brooks and Oder 2004), the Meteorological Office in the United Kingdom (Turp 2008), and the Canadian Meteorological Centre (Turcotte and Verret 1999). The physical basis of TI was considered initially due to frontogenesis through the process of dynamical deformation (Mancuso and Endlich 1966, 1964; Ellrod and Knapp 1992; Ellrod et al. 2002: Ellrod et al. 2015). The hypothesis suggested that stronger horizontal thermal gradients caused by frontogenesis can lead to an increase in the vertical shear of the horizontal wind through the thermal wind relation, and indicates that a higher potential for clear air turbulence (CAT) occurs via the local reduction in Richardson number (Ri) and consequent production of Kelvin–Helmholtz instability (Dutton and Panofsky 1970). Commercial aircraft encounter severe or greater turbulence about 5000 times each year according to available pilot reports. These incidents result in tens of millions of dollars in injury claims per year (Sharman et al. 2006). In fact, these statistics were obtained only for the US airspace and are not a global representation. A significant limitation for the forecasting of all aviation turbulence types is to identify the source of gravity waves (McCann 2001; McCann et al. 2012; Knox et al. 2008). Observations from early field projects (Sorenson 1964; Colson and Panofsky 1965) indicated that two flow regimes associated with CAT are possible: strongly cyclonic and strongly anticyclonic flows. Because the TI neglects anticyclonic shear or curvature in its derivation, Ellrod and Knox (2010) suggested that methods should be taken into account for the anticyclonic shear (Knox 1997).

The Graphical Turbulence Guidance (GTG) product uses an ensemble mean of many different turbulence diagnostics, which seems to improve the statistical performance of the turbulence forecasts (Sharman et al. 2006; Kim et al. 2011; Sharman and Pearson 2017; Kim et al. 2017). Example turbulence diagnostics typically used for clear-air and mountain wave sources include vertical and horizontal wind shears, static stability, wind speed, horizontal deformation, frontogenesis, ageostrophic indicators, turbulent kinetic energy (TKE), and Richardson number (Ri) (e.g., Knox et al. 2016; Lee et al. 1984; Trier et al. 2012). These diagnostics are designed to capture grid-scale processes (10–100 km) that may lead to sub-grid scale turbulence.

The GTG product does provide short-term forecasts, but its nowcast usage is limited by a 1-h update cycle, latency in receiving the underlying NWP-model data and the time required to compute the turbulence diagnostics. Therefore, forecasts are usually not available until at least 2 h beyond the valid time. Pinto et al. (2015) and Sharman and Pearson (2017) also stated that because of the highly transient and small spatial scales of turbulence associated with convective storms, neither the NWP model nor the turbulence-forecasting post-processing algorithms are particularly skillful at forecasting turbulence associated with convection. To address these issues,

Pearson and Sharman (2017) developed a GTG nowcast system (GTGN) which merges turbulence observations with short-term GTG forecasts to produce more timely and accurate information for tactical turbulence avoidance. Their results suggested that turbulence nowcasts integrated with observations considerably outperforms the corresponding turbulence forecasts.

3.4. Low Level Wind Shear (LLWS)

Low level wind shear (LLWS) at airports can occur due to fronts, thunderstorms, inversions, and surface obstructions (FAA-P-8740-40 2008). Traditionally, surface in situ wind sensors are used to identify LLWS cases, but recently, remote sensing platforms and aircraft based measurements are also used for detection. The surface sensors are located at certain distances along the flight paths (Thobois et al. 2018; Oude Nijhuis et al. 2018), but do not cover higher levels; therefore, Doppler lidar (for cloud free conditions) and radars (cloudy air conditions) can be used for LLWS detection (Chan and Shao 2007). Low level wind shear prediction using NWP models depends on model time and space resolutions (especially within the PBL), model physics, large scale forcing conditions, and initial conditions. Using NWP model output, and both QuikSCAT space borne scatterometer (QSCAT) and buoy observations Kara et al. (2007, 2008) stated that wind speed errors near ocean boundaries can be up to 5 m s^{-1}. For high resolution NWP forecasts, the root mean square error (RMSE) in wind speed can be as high as 4 m s^{-1}. Banta et al. (2017) suggested that wind speed error in NWPs were larger than the required accuracy overhigh resolution areas and suggested the use of LIDAR observations to improve forecasts (McCarty et al. 2017). Gao et al. (2012) and Moninger et al. (2010) stated that errors in AMDAR measurements can be as high as 4–6 m s^{-1}, and suggested some corrections. Wind speed directional errors can be up to 35 degrees for AMDAR measurements and usually increase toward to the surface (Gao et al. 2012). Therefore, the accuracy of wind speed and direction measurements as well as their simulation from NWP models need to be improved since these parameters are directly used in LLWS calculations (Chun et al. 2017).

3.5. Physical Processes for Aviation Nowcasts

The production of aviation meteorological forecasts is based mainly on forecasters using numerical weather prediction (NWP) model data in combination with available observations (e.g., Jacobs and Maat 2005; Fahey et al. 2016; Bright et al. 2016). Their work suggests that NWP modeling has not yet reached a state where physical processes such as clouds and precipitation can be resolved at the spatial and temporal resolutions necessary for reliable aviation weather forecasts. Jacobs and Maat (2005) also emphasized that various physical processes associated with fog and low stratus clouds are not adequately described in NWP models due to the complexity of the underlying physical processes and the lack of sufficient vertical resolution in the atmospheric BL. Because of this, sudden weather changes on small time and spatial scales can only be evaluated and predicted if the forecaster has access to detailed observations concerning current weather changes. The results of their work strongly emphasized that quality of short-term forecasts, up to 6 h, depends mainly on the availability of local and upstream observations.

3.6. Icing

Most if not all, current icing algorithms include empirical relations (e.g., temperature dependency) as described in Thompson et al. (1997). In their work, inflight icing potential was predicted using algorithms developed by the National Center for Atmospheric Research (NCAR), the National Weather Service's National Aviation Weather Advisory Unit, and the Air Force Global Weather Center. The numerical model data from the Eta, MAPS, and MM5 models were used in their research. As part of the WISP94 field program (Thompson et al. 1997), detailed evaluations of icing algorithms were conducted. Most of the icing algorithms used in NWPs are only functions of T (Noh et al. 2013; Tan and Storelvmo 2016; Ødegaard 1997). Thompson et al. (1997) stated that icing algorithms all predict a flat or increasing frequency of icing at decreasing temperatures.

Statistical studies of pilot-reported icing (Rasmussen et al. 1992) suggested that the number of icing reports decreases with decreasing T. Note that saturation vapor pressure also diminishes with decreasing T. The decreases usually occur at lower temperatures (<−15 °C) because of increasing ice nucleation processes at cold T and less available vapor content. In the Thompson et al. (1997) work, all algorithms predicted increasing or stable icing at lower temperatures as expected because of nature of the parameterizations used in the models. These models simply diagnose icing conditions empirically where clouds can occur as a function of RH and T interval between 0 °C and − 20 °C. In fact, the large uncertainty in RH, as high as 10–15% (Gultepe et al. 2016), may lead to substantial differences in icing rates. Similar issues also exist for airport ground operations where de-icing calculations are strongly related to T and Vis (Thompson et al. 1997, Brown et al. 1997). Traditionally, the need for ground de-icing was assessed based on horizontal Vis, but research by Rasmussen et al. (2001, 2003) showed that the icing hazard was more dependent on the liquid equivalent of snowfall rate and particle shape. This discovery led to the development of NCAR's Weather Support to De–Icing Decision Making System (WSDDM, Rasmussen et al. 2001).

The above studies and Thompson et al. (2017) signify that icing research is presently immature and need to be improved for NWP model simulations for aviation operations.

3.7. Visibility Reduction Due to fog and Precipitation

Visibility reduction due to fog and precipitation is critical for aircraft operations. In fact, fog after wind represents one of the most hazardous weather events affecting aviation activities. Over last decade, progress has been made in the study of fog processes (e.g., MATERHORN-fog, Gultepe et al. 2016) and climatology (Hodges and Pu 2015; Albers 1977; Tardif and Rasmussen, 2007; Dorman et al. 2017). NWP models were also used for simulating various types of fog events (e.g., Bergot et al. 2007; Guedalia and Bergot 1994; Bott et al. 1990; Pu et al. 2016; Lin et al. 2017; Chachere and Pu 2018; Pithani et al. 2018) and indicated that numerical simulation of fog is sensitive to various physical parameterization schemes (Gultepe et al. 2006; Stoelinga and Warner 1999), initial model conditions (Jones 1965; Anderson 1996), and land surface processes (Guedalia and Bergot 1994). Specifically, numerical prediction of the fog presents one of the challenges in NWP due to uncertainties in model physics and Vis algorithms (Gultepe et al. 2009; Guedalia and Bergot 1994; Roquelaure and Bergot 2007; Lin et al. 2017). Also, significant errors in near-surface atmospheric variables prevent accurate prediction of fog and precipitation (especially snow) in NWPs (e.g., Chachere and Pu 2018; Pu et al. 2016; Pu 2017). These errors, therefore, need to be reduced when using NWP models for Vis predictions.

3.8. Satellite Data Assimilations

Satellite-based wind vectors can be used to infer wind shear, turbulence, and convective activity. Convective clouds identified by the CCM (convective cloud mask, Meckalski et al. 2007) were supported by the work of Velden et al. (1997, 1998, 2005) that developed the atmospheric motion vector (AMV) algorithm. In this way, GOES-based wind products can be utilized for wind-related event analysis and in data assimilation that can be used for NWP initial conditions and nowcasting applications. Short-term (0–1 h) convective storm nowcasting remains a problem for operational weather forecasting and poses a significant financial risk for the aviation industry (Sieglaff et al. 2011). The output from NWP models and meteorological observations, including these from radars, are extensively used for short-term convective forecasting but all these have shortcomings. Geostationary imagers data can help reduce some uncertainty, and that can be valuable for convective initiation predictions. The University of Wisconsin Convective Initiation (UWCI) nowcasting algorithm provides an objective, satellite-based decision support tool (Meckalski et al. 2007). Lately, winds are being obtained from satellites (e.g., Aeolus satellite, Baker et al. 2014) based active platforms such as Doppler LIDARs that can also be used for data assimilations techniques but currently they are used in research method and their data can be used cautiously for aviation operations. Folger and Weissmann (2014) investigated pressure height corrections

Table 1

FAA based flight classification that is based on cloud ceiling height and horizontal visibility (Adapted from Verlinden and Bright 2017)

Flight classification	Cloud ceiling (ft)	Horizontal visibility (miles)
IFR	< 1000	< 3
MVFR	≥ 1000 and ≤ 3000	≥ 3 and ≤ 5
VFR	> 3000	> 5

operational atmospheric motion vectors (AMVs) from the geostationary satellites Meteosat-9 and -10 with cloud-top heights retrieved from LIDAR observations by the polar orbiting Cloud–Aerosol Lidar and Infrared Pathfinder Satellite Observations (CALIPSO) satellite. They found out that 700 mb pressure errors were reduced up to 20%.

3.9. Cloud Base Height

Cloud base height (h_c) is an important parameter for aviation operations (Table 1), but without proper prediction of cloud microphysical parameters, h_c cannot be estimated accurately. Its estimation is related to cloud total water content (CTWC) and relative humidity (as well as dew point temperature depression (Guttman and Jeck 1987). The formation of cloud at low levels is also related to Planetary Boundary Layer (PBL) physical and dynamical conditions. Unfortunately, cloud TWC from NWPs can be subject to large uncertainties in the PBL because of issues related to model resolution at low levels and turbulence/physical processes interactions. For example, a warm bias of about 3.3 °C in sea surface temperature (SST) can lead to a q_v bias of 1.2 g kg^{-1} that was simulated by ECMWF (Sun et al. 2014). Therefore, improved NWP prediction of RH and T in the PBL can lead to more accurate TWC prediction and that can be used for improving forecasting of h_c levels.

3.10. Microphysical Schemes for Fog, Cloud, and Precipitation

NWPs need better physically-based algorithms to improve fog, cloud and precipitation processes. Detailed and accurate microphysical algorithms are required at various scales to obtain reliable precipitation amount and types, Vis, wind, and turbulence, and cloud radiative properties. Recently, new microphysical algorithms based on improving the single particle growth history were developed (Harrington et al. 2013a, b) rather than assumed size distributions for each particle phase (Gultepe et al. 2018; Ferrier 1994, Lin et al. 1983) improved predictions. Pu et al. (2018) found that the simulation of clouds within mesoscale convective systems is sensitive to the various microphysical schemes. They commented that ice hydrometeors play an important role in the accurate numerical prediction of clouds and precipitation. Lately, there has been a shift in the way ice-phase hydrometeors are represented in microphysics schemes, moving from predefined hydrometeor categories with prescribed physical characteristics (e.g., bulk density) and focusing on the prediction of the particle physical evolution instead (Harrington et al. 2013a, b; Morrison and Milbrandt 2015). This has led to the smoother evolution of ice crystals during growth and avoids the artificial process of ‘‘conversion’’ between ice categories and auto-conversion processes. These results suggest that further improvements of the microphysical algorithms can lead to the development of better weather warning and decision making systems.

The NCEP and UKMet Office are working together to create World Area Forecast (WAF) guidance for the aviation weather community worldwide. NCEP has a plan to upgrade its GFS to a Finite Volume—Version 3 model in 2019 (FV3: see https://www.gfdl.noaa.gov/fv3/fv3-documentation-and-references), which will employ an advanced cloud-allowed microphysical scheme, so that better cloud-related predictions of Vis, ceiling, and reflectivity can be provided to the global aviation weather community.

As stated on the various challenging topics described above, NWP predictions for nowcasting applications related to aviation operations may include large uncertainties; therefore, integrated methods should be considered for short term forecasts (e.g., lead time < 6 h), and these are described in next section.

4. Integrated Systems

4.1. Numerical Guidance Systems for Aviation

A numerical guidance system (NGS) for aviation applications that includes post-processing NWP model output, and local and upstream observations representing high-resolution topography can be used to provide information on changing weather conditions at airports (Hansen et al. 2008). The NGS, supported by detailed predictions of Vis, ceiling, wind, and precipitation observations as well as TAF and TREND type weather forecasts, can be used more efficiently for aviation weather nowcasts.

The NGS can be supplied with high-resolution remote sensing observations from platforms such as radar, LIDAR, and wind profilers. For example, Luce et al. (2010) used both LIDAR and radar to study CAT conditions below a cirrus cloud system. A statistical approach, known as the Graphical Turbulence Guidance (Sharman et al. 2006; Sharman and Pearson 2017), was developed to predict turbulence using a weighted regression of multiple turbulence diagnostics. This product and others such as convection, icing, ceiling and visibility, and surface wind gusts are available operationally on NOAA's Aviation Digital Data Service website (https://www.aviationweather.gov/).

4.2. Integrated Systems for Nowcasting and Test Bed Sites (Supersites)

Integrated weather nowcast systems rely on both in situ and remote sensing observations, as well as NWP model forecasts. Test bed sites (also called supersites) are an integral part of meteorological research and operations. Ralph et al. (2013) emphasized that *they (supersites) can foster new forecast innovations and their transition into operations.* The supersites present new opportunities for businesses and agencies to improve their products and services. Their work extensively studied the importance of test beds for research and operations. They also stated that: "Test beds tend to be "outsiders" relative to either the core mission of forecasting or the core mission of research. In spite of this, they enable more rapid improvements in forecast services and demonstrate tangible relevance of research centers to forecast services in the future". A supersite, planned by NWS as the "Operations Proving Ground (OPG)" in Kansas City, MO, will provide full integration testing of new tools and methods in an operational environment. These OPGs require a well-established research community focused on performing exploratory research and development. Ralph et al. (2013) also suggested that transformational research can provide breakthrough advances for forecast/operation services in the future. The objectives of test beds can accelerate the translation of research and development into operations, services, and decision making for short lead time periods. Overall, a test bed site can be used for operational systems, and provide better use of data in forecasts and applications to improve economic/public safety benefits (Ralph et al. 2013; Gultepe et al. 2018).

A new integrated system for aviation operations called The Short-Term Prediction Research and Transition (SPoRT) program using NASA, NOAA, and DoD satellite data and research capabilities is available to the operational weather community to improve short term weather forecasts on regional and local scales (Ralph et al. 2013; Stano et al. 2010; Ellrod and Gultepe 2007). The SPoRT focuses on weather and aviation related problems that include the timing and location of severe weather, changing weather conditions influenced by topography, visibility, land–ocean boundaries, and the monitoring weather in remote areas. The SPoRT involves forecasters in the entire process who help develop product training materials and help assess the utility of the products.

The NOAA NWS Advanced Weather Interactive Processing System (AWIPS, Argyle et al. 2017; Raytheon, 2016) program is a complex network of systems that ingest and integrate meteorological, hydrological, satellite, and radar data for display at Weather Forecast and River Forecast Centers. Weather forecasters then use the data to provide accurate weather, water, and climate predictions and highly reliable warnings and advisories (Kelly and Ghirardelli 1998). As stated in their work, the AWIPS results are extensively used in aviation operations and nowcasting, including time-sensitive, high-impact warnings to protect life and property (Ghirardelli and Glahn 2010). This work emphasized that the Localized Aviation MOS (Model Observed Sounding) Program (LAMP) consisting of analyzing observations, advective models, and statistical methods can improve the longer-range MOS forecasts based on the GFS model.

One of the earlier integrated systems developed was the CAN-Now project, and its primary objective was to provide a four-season forecasting/nowcasting system at the major Canadian airports with detailed nowcasts and forecasts (Isaac et al. 2014). The CAN-Now output allows airport related decision makers, including pilots, dispatchers, de-icing crews, ground operators, and air traffic controllers to make accurate decisions to improve safety and efficiency. The prototype nowcasts rely on existing routinely available weather information including NWP model output, site climatologies, remote sensing and lightning network observations, and in situ measurements of wind, precipitation, visibility, ceiling, and temperature. The integrated nowcast system called The Adaptive Blending of Observations and Models (ABOM) system was developed by Bailey et al. (2009) to be included in the CAN-Now system. The prototype system was used in a nowcasting mode for detecting weather hazards and providing forecasts out to about 3–6 h for most phenomena, and out to 36 h for some subsets of phenomena.

There are several other types of integrated systems that are being used in Europe. The integrated nowcasting through comprehensive analysis (INCA) (Haiden et al. 2011, Kann et al. 2015) and AROME-NWC (Auger et al. 2015; Seity et al. 2011) are common ones. These systems include downscale processes and updates NWP predictions using the latest observations from surface in situ observations and high-resolution (1 km) orography data. A verification of INCA simulations of T, RH, and wind analyses was performed against high-resolution network observations in Austria by Kann et al. (2015) and Haiden et al. (2011, 2014). The AROME was developed in France that uses a non-hydrostatic mesoscale model version simulated for the forecast time range of 0–30 h (Auger et al. 2011; Seity et al. 2011). This model is initialized by using a 3-D variational data assimilation scheme (3DVar), and that provides the initial fields relevant for an accurate nowcast; The AROME system was related to the ALADIN–France model (Fischer et al. 2005a, b). Integrated turbulence nowcasts have also been developed to be used for short-term NWP forecasts nudged by in situ and radar observations of turbulence (Pearson and Sharman 2017).

Another example of the use of integrated systems to detect and nowcast hazardous weather is provided by the low-level wind shear (LLWS) alert systems (Thobois et al. 2018). It is well-known that thunderstorms can be accompanied by intense updrafts and downdrafts. The strong downdrafts or downbursts below cloud base on or near the ground can force low flying aircraft downward. Then, strong divergent flow, producing low-level (horizontal) wind shear, can lead to unexpected rapid changes in aircraft airspeed and heading. The outflow winds associated with downbursts have horizontal extents > 4 km and can typically persist from 5 to 30 min. During the Joint Airport Weather Studies project (McCarthy et al. 1982), Doppler radar examination of numerous thunderstorm downdrafts and outflows indicated that wind shear particularly hazardous to aircraft occurred on much smaller temporal and spatial scales, and Fujita (1981, see also McCarthy and Serafin 1984, Wilson and Wakimoto, 2001) termed this smaller scale, with more hazardous downdrafts/outflows, as "microbursts". The damaging outflow winds associated with microbursts have horizontal extents of $\leq$ 4 km and can persist from 2 to 10 min, typically (Wilson et al. 1984; Wilson and Wakimota 1982).

The scale and suddenness of microbursts make them particularly hazardous to aircraft departing or approaching an airport. Thunderstorm outflow or microburst wind shear is known to have caused 21 aircraft accidents with 438 fatalities in the United States between 1975 and 1994 (Wolfson et al. 1994). In response to these accidents, the Federal Aviation Agency (FAA) developed and deployed three ground-based low altitude wind-shear detection systems: the Low Altitude Wind Shear Alert System (LLWAS) (e.g., Linden and Simpson 1985; Wilson and Gramzow 1991), Terminal Doppler Weather Radar (TDWR) (e.g., Wilson et al. 1984; Michelson et al. 1990), and Airport Surveillance Radar Weather Systems Processor (ASR-9 WSP) (Weber and Stone 1995; Cho 2015). The FAA-sponsored Integrated

Terminal Weather System (ITWS) uses TDWR reflectivity data with short-term NWP model nowcasts to predict microburst intensity and location (Wolfson et al. 1994). Since the deployment of these systems, along with enhanced pilot training, commercial aircraft LLWS accidents have dropped to nearly zero in the U.S. The dramatic decrease in these accidents testifies to the safety benefits provided by these detection systems.

Enhanced physical understanding of the microburst phenomenon has been provided by high-resolution simulations (e.g., Proctor 1988, 1989, Orf et al. 1996, Orf and Anderson 1999, Nicholls et al. 1993), and laboratory studies (e.g., Ferrero et al. 2014). Because of its importance to aviation, LLWS is routinely forecasted using the vector wind difference of wind between 2000 ft AGL (619.5 m) and the surface (e.g., NOAA NWS Instruction 10-813, 2016), although these forecasts are mainly driven by resolvable non-convective sources, e.g., frontal passages, low-level jets, lee side mountain effects, sea breeze fronts, etc. As NWP model resolution increases to become convection-resolving, routine microburst nowcasts using both observations and NWP forecasts can become routine.

4.3. *Artificial Intelligence*

As done for other scientific research involving large data sets, increasing data volume and resolution of observations, as well as model simulation output for weather forecasting, led to the application of artificial intelligence for aviation (AIA) research and guidance. Because current NWP models runs have increased time and space resolutions (McGovern et al. 2017; Weygandt et al. 2009), as well as new observing systems, including in situ platforms, weather radars, LIDARs, and GOES-16 also generate data at high time and space scale resolutions (Stano et al. 2010; Goodman et al. 2012). Forecasters and users do not usually have time to make rapid decisions if data are not somehow prepared for easy evaluation in advance (Karstens et al. 2015). These data sets are called "big data." Artificial intelligence (AI) methods (Pasini and Marzban 2008) use various techniques to process big data sets and then apply the results to weather forecasting issues, which are crucial for creating timely weather reports (McGovern et al. 2017). They stated that AI techniques based on physical understanding of the environment could improve prediction skill of high-impact weather situations. The AI approach Glahn et al. (2017) expands information available over MOS techniques (Glahn and Lowry 1972) for deriving probabilistic, categorical, and deterministic forecasts available from NWP models.

McGovern et al. (2017) provided an extensive summary of AI techniques used in meteorological applications. Haupt and Delle Monache (2014) also provided an overview of AI techniques applicable to artificial neural networks (ANNs), including decision tree algorithms, genetic algorithms (Allen et al. 2007), fuzzy logic, and principal component analysis (Elmore and Richman 2001). In this area, using a decision tree technique, Burrows et al. (2005) developed lightning detection algorithms for mid-latitudes. Williams (2014) used a random forest approach to diagnose convectively-induced turbulence. These works suggest that AI methods are starting to be used extensively in post-processing of NWP output and in situ observations.

The AI systems and statistical based neural network systems can help transfer knowledge related to aviation forecasts into the aviation operations when fast computer systems become easily available in the future. But, basic understandings of physical and dynamical processes still need to be developed for the atmosphere that includes cloud, fog, aerosols, as well as wind and turbulence.

5. *Future Challenges and Issues*

There are various challenging topics related to aviation meteorology that can significantly impact aviation operations, and these are summarized below.

5.1. *Visibility and Ceiling Issues*

Jacobs and Maat (2005) show that for lead times greater than 4 h, the TAF guidance provides more accurate cloud h_c and Vis forecasts than those derived manually by forecasters. For shorter lead times (< 3 h), their work suggested that the differences in the comparisons of aviation related parameters are found to be small. For aviation applications, the economic value of reliable weather forecasts is very high (Hansen et al. 2009; Gultepe et al. 2016). For this reason, large occurrence of high impact weather related to h_c and Vis at an airport can reduce airport capacity that can lead to enormous economic costs. The cloud ceiling (defined in meters for cloud cover $\geq$ 6/10 of the sky) predictions can be performed based on NWP predictions using the lowest level of cloud total condensed water content (TCWC). The fuzzy logic-based analog forecasting systems (Hansen 2007; Bankert et al. 2004) can also be used for this purpose. A TAF (Terminal Aerodrome Forecast) provides weather conditions and their most probable time of occurrence at the airports. The C_h and Vis (defined by Glickman 2000) are the two variables that together determine flight category [e.g., instrumented flight rules (IFR where $C_h < 1000$ ft or Vis < 3 miles) or visual flight rules (VFR) (see Table 1 for flight condition definitions)].

The meteorological and aviation scientific communities recently studied meteorological parameters important for aviation operations that are related to ceiling height and Vis forecasts (Rudack and Ghirardelli 2010). These forecasts are valuable for making economical decisions and societal context (e.g. the loss of life). Gultepe et al. (2007a, b) suggested that reductions in Vis can be caused by several hydrometeor types including fog, haze, mist, rain, snow, and blowing snow. Figure 2 shows the meteorological events that contribute to the integrated Vis estimations. Because of Vis complexity and difficulty in measurements especially in cold climates, its use in NWPs can still be difficult and includes large uncertainties (Gultepe et al. 2006, 2018).

5.2. *Contrails and Frost*

Contrail formation, development, and dissipation are important for aviation and climate sciences but studies on these are limited. Contrails form when water vapor condenses on IN (ice nuclei) during which vapor freezes on aerosols coming from the exhaust of aircraft engines at T less than typically − 40 °C (Schumann et al. 2012a, b; Appleman 1953). Although contrails at − 40 °C occur at high levels in mid-latitudes and tropical weather, they can also occur near the surface during Arctic winter (Gultepe et al. 2015). Heymsfield et al. (2010, 2011, 2005) also stated the importance of contrail microphysical conditions for better understanding of ice cloud nucleation processes. Contrails can indicate higher values of moisture at the cold temperatures that may cause frost and light snow precipitation at high northern latitudes (Gultepe et al. 2016, 2017a, b). Figure 8 shows heavy frost conditions occurred on an ice particle counter during an Arctic project (Gultepe et al. 2015, 2018). Frost formation on aircraft surfaces is also a condition for deicing at the airports which is a required by the FAA (2017).

Over the northern latitudes, contrails, including aircraft-produced ice particles (APIPs), can be visible within the ABL over the airports (Gultepe and Heymsfield 2016; Heymsfield et al. 2011; Schumann and Heymsfield 2017; Gultepe et al. 2014a, b, 2015; Woodley et al. 1991; Langmuir et al. 1948; Ludlam 1956). Contrails may modify the atmospheric environment in several ways. For example, Heymsfield et al. (2011) stated that holes in clouds are also due to inadvertent seeding of clouds with IN particles generated by aircraft and these are produced through spontaneous freezing of cloud droplets in air (Heymsfield and Sabin 1993; Heymsfield and Sabin 1989) and that is cooled as it flows around aircraft propeller tips or jet aircraft wings. Their work also suggested that polar clouds are particularly susceptible to the APIPs effects through modifications in radiative processes. This may also suggest that ice crystals sampled by aircraft probes over the Arctic environment may include APIPs and these need to be researched. Contrails are considered as prototype cirrus which impacts the energy budget of the atmosphere by reflecting incoming SW radiation and trapping outgoing IR radiation (Markowicz and Witek 2011, Schumann et al. 2012; Spangenberg et al. 2013; Minnis et al. 2013). How contrails from an aircraft can contribute to global climate change was studied in detail by Minnis et al. (1998, 1999), Schumann et al. (2012), Schumann and Heymsfield (2017). Whether or not contrails would contribute to global warming or cooling is still not clear, and probably cannot be ascertained without resort to high-

Figure 8
Frost formation on ice particle counter sensor which occurred during the FRAM ice fog project in Barrow, AL, on April 16 2008

resolution climate models (Schumann and Mayer 2017).

5.3. *Climate Change Impact on Aviation*

Future research on the combined impacts of climate change and climate variability on aviation operations can help airlines and other aircraft operators for long term planning. For example, Goodman and Griswold (2017) investigated future density-altitude (DA) trend evaluations. Their work stressed the importance of future investigations on the impacts of ENSO and Atlantic Oscillations (AO) on DA to be used in seasonal-scale planning of aviation operations. The weight restriction is an important factor in planning future flight operations, therefore, needs to be evaluated. For this reason, airlines may need to be rescheduled out of the hottest parts of the day. Other potential impacts include anticipated jet stream changes with consequent impacts to CAT (e.g., Williams and Joshi 2013; Williams 2017; Storer et al. 2017), and changes to convection (e.g., Del Genio et al. 2007) and other extreme events (Puempel and Williams 2016). Both Irvine et al. (2016) and Williams (2016) suggested that the effect of wind changes as a potential impact of climate change on aviation can be important for future aviation operations. This is of particular interest for trans-Atlantic flights, where the pattern of upper-level winds over the north Atlantic, in particular the location and strength of the jet stream, strongly influences both the optimal flight route and the resulting flight time. Some of these effects may be mitigated by "climate optimized routing" procedures (e.g., Matthes et al. 2012).

As an example, Fig. 9 shows the time series of cold spell T and warm spell T indexes obtained from CMIPS models (Sillmann 2013a, b) where increasing warm days and slightly decreasing cold days are seen from 1948 to 2005. Therefore, increasing convective activities are likely expected and that can be related to shear, turbulence, and heating processes and these can play an important role for planning procedures of the aviation applications.

5.4. *Operational Satellites*

Lately, geostationary satellites with more than the traditional five imaging channels are being used for weather analysis and operations, and these satellites

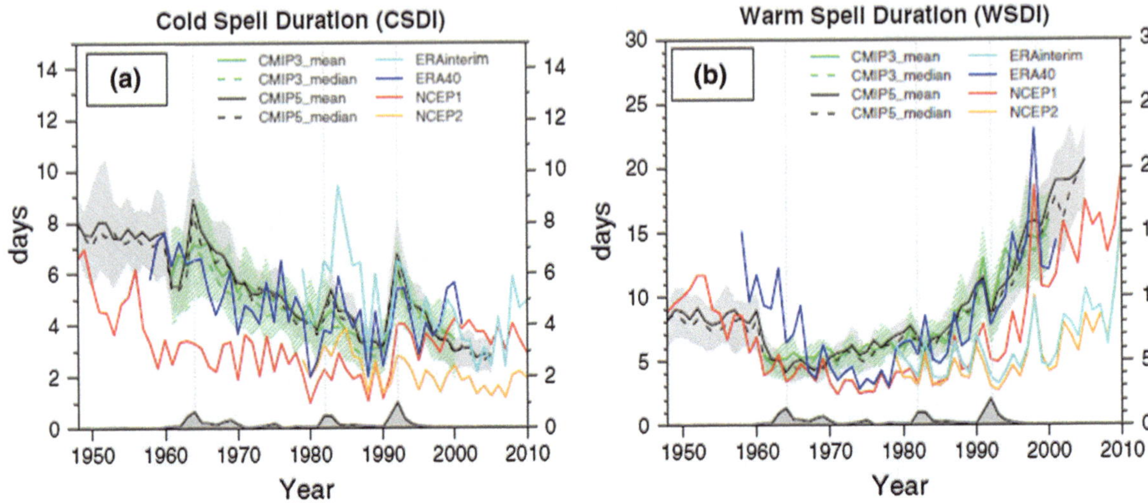

Figure 9
Time series of percentile indices from 1948 to 2005 of the ensemble mean (solid) and median (dashed) of 31 CMIP5 models (black) and 18 CMIP3 models (green). The shading indicates the interquartile ensemble spread (range between the 25th and 75th quantiles). Note that the percentile indices from the reanalysis ERA40 (blue) from 1958 to 2001 and NCEP1 (red) from 1948 to 2005 are calculated with a different base period (1961 to 1990) than those from ERA-Interim (cyan) and NCEP2 (orange) with a base period from 1979 to 2008. Displayed are global averages over all land for CSDI (**a**) and for WSDI (**b**). Grey shading along the horizontal x-axis indicates the evolution of globally averaged volcanic forcing according to Sato et al. (1993) (Adapted from Sillmann 2013a, b). © American Meteorological Society. Used with permission

carry additional channels in the IR windows. A review paper on geostationary satellites is given in this special issue by Ellrod and Pryor (2018); therefore, limited information is provided in this sub-section. The ABI is used on the Geostationary Operational Environmental Satellite-R Series (GOES-16) and that views Earth with 16 spectral bands (compared to five on previous GOES), including two visible channels, four near-infrared channels, and ten infrared channels. Himawari 8 is a Japanese weather satellite and carries an Advanced Himawari Imager (AHI) with also a 16 channel multispectral imager to capture visible and infrared images of the Asia–Pacific region (Berndt et al. 2018). The instrument was designed and built by Exelis Geospatial Systems (now Harris Space & Intelligence Systems). The FY-4 (FengYun-4) is the China Meteorological Administration (CMA) second-generation three-axis stabilized, geostationary meteorological satellite developed by CAST (China Academy of Space Technology) (Yang et al. 2017). Two variants of spacecraft of the FY-4, with one carrying optical sensors and the other carrying microwave sensors were developed. FV-4 carries an Advanced Geosynchronous Radiation Imager (AGRI) with 14 channels, Visible and Infrared Spin-Scan Radiometer (VISSR) 5 channels, Geostationary Interferometric Infrared Sounder (GIIRS), Lightning Mapping Imager (LMI, Goodman et al. 2013), and a Space Environment Package (SEP). All these satellites can improve global aviation products in more detail.

New generation imagers are also being deployed on polar-orbiting satellites, which provide aviation information over high-latitude regions, which are only partially observed by geostationary orbiters. These new imagers, such as the Visible Infrared Imaging Radiometer Suite (VIIRS) on the Suomi National Polar-orbiting Partnership (NPP) satellite and NOAA-20 series (the first one of four highly advanced polarorbiting environmental satellites), provide a large number of channels that are similar to or more expansive than their geostationary satellite counterparts (Menzel et al. 2018). These imagers will greatly enhance the satellite information useful for aviation over Polar Regions. Inclusion of spectral IR capability for geostationary satellites such as Chinese GIIRS (Menzel et al. 2018) can improve knowledge obtained for atmospheric stability, providing information on convective weather warnings and forecasts (Chiodi and Harrison 2010). Spectral IR capability has been used in polar orbiting satellites e.g., the Atmospheric Infrared Sounder (AIRS), Cross-Track Infrared Sounder (CrIS; Zang et al. 2016 ; Wang et al. 2012) and the Infrared Atmosphere Sounding Instrument (IASI) on Aqua, Suomi-NPP, NOAA-20, and Meteorological Operation (METOP) satellites (Mittaz and Harris 2011; Blumstein et al. 2007; Wang and Cao 2008) to obtain T, RH, and wind vertical structure in the region with no clouds and above the cloud tops. More extensive development of spectral IR techniques can revolutionize technology related to aviation weather forecasts.

To improve nowcasting techniques for detecting, tracking, and monitoring the early development of small convective clouds, convective initiation forecasts over the CONUS and possibly over marine environments were studied in detail using satellite observations (Meckalski et al. 2007; Walker et al. 2012). Historically, most of the convective weather predictions used weather radars (e.g., Mueller et al. 2003). However, radars usually do not see the early stages of convective cloud development prior to precipitation development. Since the 1990s, NCAR began to incorporate satellite information into convective storm nowcasting system (the Auto-Nowcaster). Satellite feature detection algorithms (e.g., Bankert 1994; Bankert et al. 2009; Roberts et al. 1999; Tag et al. 2000, and Roberts and Rutledge 2003) were usually used to classify cloud types, identify surface convergence boundaries, and monitor the cloud growth based on the changes in their IR cloud-top T. These developments were positive for nowcasting systems but additional data sets from GOES-16 channels can further provide information related to cloud types and related physics and dynamical processes, and need to be researched.

5.5. *Measurement Uncertainties*

5.5.1 *T and RH*

A better understanding of the uncertainties in observations can be important for decision making systems, and that can be used for obtaining better physical parameterizations for NWP simulations and

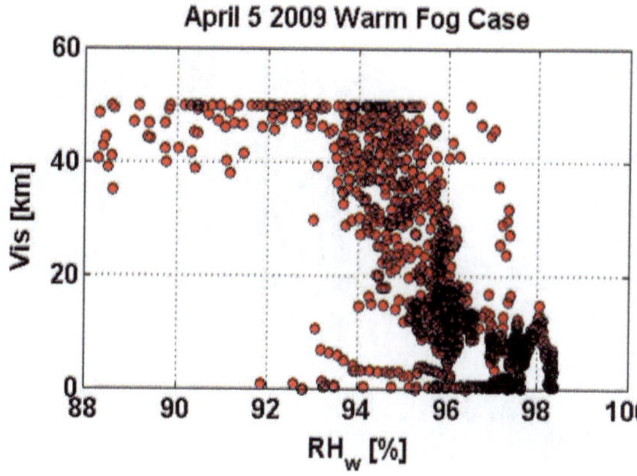

Figure 10
Vis versus RH_w from a severe fog case event at the FRAM site during in Snow-V10 project

improvements in data assimilation techniques (e.g. 4-D Var). Both T and RH uncertainties in NWP predictions can be very large and that affect the prediction of cloud formation, visibility, and convective intensity, and surface-air interactions through turbulent fluxes using NWPs Feingold et al. (1999). Figure 10 shows RH_w measurements obtained during the FRAM project (Gultepe et al. 2014a, b; Gultepe 2015). This plot shows that Vis is strongly related to RH_w; increasing RH_w results in decreasing Vis. A small change in RH_w at about 4–5% can lead to Vis changes from 50 km down to few meters. This means the possible errors in RH_w from NWP predictions can lead to significant issues for aviation forecasts, e.g. Vis and cloud types and that can affect operations significantly.

5.5.2 *Visibility and Ceiling Height*

The present challenges of Vis and h_c measurements are related to precipitation/fog hydrometeors types and scale issues (Gultepe and Isaac 2004, 2006). They suggested that discrimination of droplets from precipitation and its usage in Vis parameterizations are critical for NWP Vis predictions. Figure 11a shows a scatter plot of Vis measurements from the most commonly used sensors (Vaisala FD12p and Sentry sensors), and ice fog crystals (Fig. 11b) occurring over Yellowknife International Airport (CYK) during the FRAM ice fog project (2010–2011 winter, Trier 2018; Gultepe et al. 2016). Ice fog crystal sizes were found to

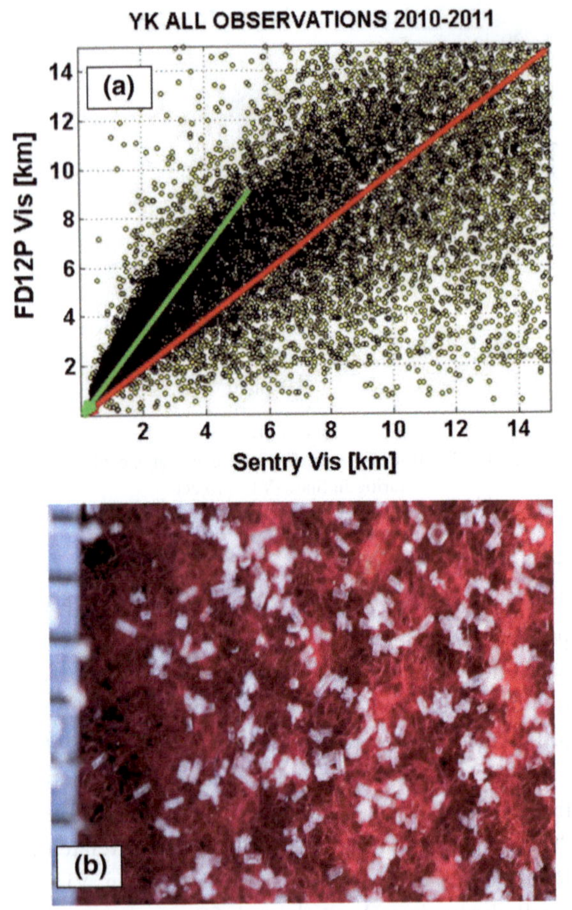

Figure 11
Vis from FD12P sensor against Sentry Vis (**a**) for all precipitation types and ice fog, and ice fog crystals (**b**) collected during FRAM-IF project took place over Yellowknife International Airport (2010–2011 Winter)

be usually < 300 μm. The concentration of the Vis observations (along green line) in Fig. 11a is seen above the 1:1 line (red line), indicating that FD12p instrument Vis values were usually larger than Sentry Vis by about 50%. For this reason, in automated observing systems, a double Vis sensor approach (Fig. 11a) should be used for model validations and comparisons with METAR observations. Especially, this needs to be improved for cold climate applications where warming trend can cause severe cold fog events (Fig. 12).

Based on fog occurrence, forecasting Vis over short time intervals (0–6 h) is challenging and most NWP models do not explicitly predict Vis (Chmielecki and Raftery 2011); therefore, Vis forecasts must first be derived from other meteorological parameters such as cloud water content (CWC), RH_w, and precipitation. Roquelaure and Bergot (2008, 2009) were the first to use Bayesian model averaging (BMA) analysis in Vis forecasting. Zhou et al. (2009) also described the use of a short-range ensemble forecast system to generate probabilistic visibility forecasts (Ryerson and Hacker 2014). The UK Met Office Global and Regional Ensemble Prediction System (MOGREPS) and NOAA NCEP Ensemble Prediction Model (Zhou et al. 2009; Zhou and Du 2010) have already been used for probabilistic aviation weather predictions (Gill 2012; Gill and Buchanan 2014; Chun et al. 2017; Kim et al. 2015).

Operationally, aviation-related interests in the US typically use two types of ceiling and Vis forecasts provided by the NWS Global Forecast System (GFS) model: (1) GFS MOS forecasts and (2) GFS Localized Aviation Model Output Statistics Program (LAMP) forecasts. Both the GFS MOS and GFSLAMP produce probabilistic forecasts for seven ranges of Vis and return to the user a categorical forecast corresponding to the most likely range of values. These existing methods do not provide a general framework for generating a full predictive probability density function (PDF) for Vis (Chmielecki and Raftery 2011). They stated that predictive PDFs are attractive for Vis forecasts because both thresholds of ceiling height and Vis correspond directly to the conditions governing flight rules. A predictive PDF allows the user to determine the probability of Vis falling below any threshold of interest rather than a single pre-specified threshold. Therefore, PDF based approaches for ensemble prediction of Vis and ceiling height need more research.

Cloud ceiling measurements are usually performed by ceilometers or are based on manual observations of sky conditions at airports. It is defined as the height of the lowest layer of clouds above the surface that is either broken or overcast, but not thin. The broken and overcast conditions are measured by the "octals", which are eight equal segments of the sky (NOAA 2012; Free and Sun 2013). The ceilometers measurements usually

Figure 12
Ice fog occurrence over the Arctic Ocean on Aug 24 2010 (Permission by G. Toth)

provide cloud conditions at three levels e.g. low clouds, middle clouds, and high level clouds, but they may not have the same meaning as man-made observations. Therefore, these measurements should be properly compared and integrated for aviation applications.

5.5.3 Turbulence and EDR

Measurements of 3D wind components and prediction of EDR and gust conditions are also critical to aviation operations (Sharman and Trier 2018; Gultepe et al. 2018; Kaplan et al. 2005a, b). Although gust values usually are presented based on the horizontal wind components, the vertical component also plays a significant role in gust and EDR estimation. These can be obtained using in situ ultrasonic sensors (Gultepe et al. 2018) or from AMDAR or TAMDAR EDR reports obtained from commercial flights, and also likely from remote sensing platforms such as LIDAR and radar, and inferences for satellite features (Benjamin et al. 2007). In all events, EDR is an estimated quantity that is not actually "measured", so it is particularly difficult to determine the uncertainty in the estimate. Pearson and Sharman (2017) studied EDR from onboard aircraft estimates versus nearby METARs wind speed and wind gust measurements (Fig. 13a, b, respectively). The fits for each represent large variability in data. The scatter around the mean curves was very large. This is not surprising since turbulence is a microscale phenomenon with large spatial and temporal variability. But this implies that uncertainties in the use of dynamical parameterizations for EDR versus gust can be large and NWPs need probabilistic approaches as well as better diagnostic approaches.

5.5.4 Low Level Wind Shear (LLWS) Measurements

As provided previously low level wind shear measurements can have large uncertainties based on measurement field designs and instrument type. For example, ultrasonic wind sensors (2D or 3D) can have issues when icing or precipitation occur (Gultepe et al. 2018). If they are heated for icing conditions, additional issues may arise due to heating the tips of sensing units. On the other hand, icing may also affect conventional wind measurements using regular anemometers. Use of Doppler LIDAR based techniques may work nicely to obtain good results for clear air and light cloud conditions but under the heavy precipitation and large optical thickness conditions they cannot penetrate in-cloud conditions and will fail for wind measurement retrievals (Thobois et al. 2018). Table 2 (ICAO 2005) provides criteria for low LLWS definitions with required accuracy for wind speed and direction measurements.

5.6. Convection Parameterization and Prediction

Convection parameterizations developed based on the data obtained from NWP models, observations, and integrated methods can play an important role for

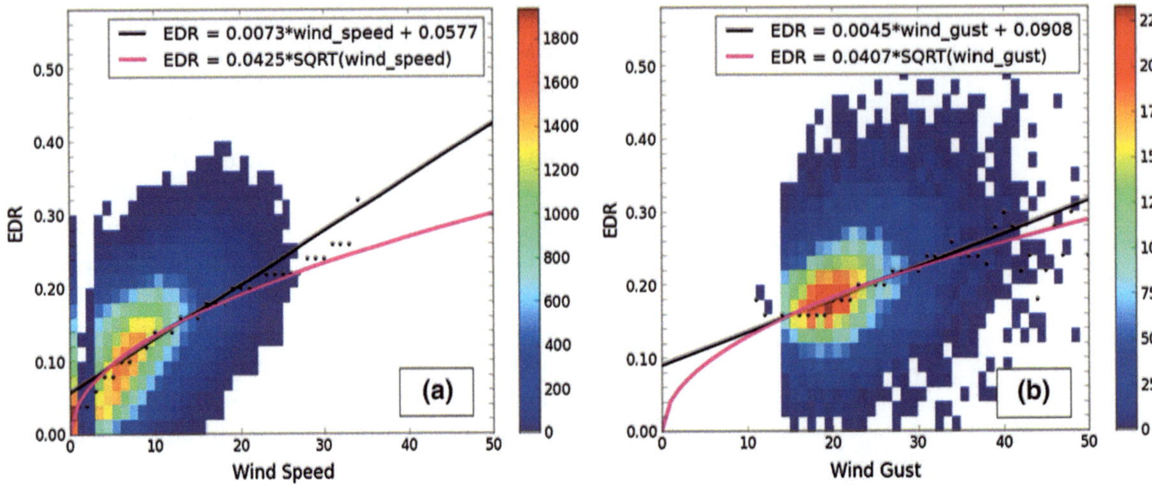

Figure 13

Plots of 117,069 wind speed (**a**) and 15,382 wind gust observations in Knots (**b**) matched with in situ Eddy dissipation rates (EDR) reports within 5 km, 5 min, and 1000 ft of each other. Median speed and gust values within each 1-kt bin are shown as black dots. Fit lines to the median values are shown in black (linear) and purple (square root) (Adapted from Pearson and Sharman 2017). © American Meteorological Society. Used with permission

aviation operations (Wang and Seeman 1997; Megenhardt et al. 2000; Grell and Devenyi 2002; Dupree et al. 2009; Yang et al. 2012). Wang and Seeman (1997) used four cumulus parameterization schemes (CPSs) representing the Anthes–Kuo, Betts–Miller, Grell, and Kain–Fritsch schemes based on The Pennsylvania State University–National Center for Atmospheric Research mesoscale model. The model's precipitation forecast skill is found to be better in rainfall amount compared to the areal coverage or the peak amount. They also stated that CPSs in warm-season cases with moist downdrafts were able to predict the surface features, such as pressure centers and gust front, more accurately but improving mesoscale quantitative precipitation forecasts overall remain a very challenging problem, even when a model is equipped with a sophisticated subgrid-scale convective scheme. This suggests that use of forecaster contributions for providing convective outlooks, mesoscale concepts, and severe weather warnings such as heavy precipitation, hail, and tornado watches can improve short term predictions for aviation operations. Usually, aviation forecasts need a forecast warning time of less than a 1 h time-period. The Storm Prediction Center (SPC) of NOAA NWS has also a human component for severe weather prediction (Karstens et al. 2018; Cohen et al. 2017). The human component needs to be considered because accumulated errors of NWP model physical components for predictions can be significant when simulation times are less than 1 h. In fact, synoptic scale environments associated with severe weather conditions (Doswell 1980) can complicate short term predictions but human based knowledge can help to improve short term predictions.

Convection-allowing numerical model (CAM) ensembles can provide extensive information related to storm intensity, location, and evolution but do not forecast accurately maximum hail size at the surface (McGovern et al. 2017). Based on the large variability related to meteorological observations and NWP scale Issues, deterministic parameterizations as well as probabilistic analysis of aviation meteorological predictions are needed to improve short term forecasts. For example, Fig. 14 shows the observed frequency of hail amount for sizes > 25 mm versus forecast probability using the CAM Analysis and Prediction of Storms (CAPS) ensemble predictions (McGovern et al. 2017). Verification results and a single forecast case in Fig. 14 are given for the machine-learning hail forecasts and other storm surrogate probability forecasts, including HAILCAST (1-D Hail Forecasting model, Jewell and Brimelow 2009), column total graupel, and updraft

Table 2

Synthesis of the ICAO guidelines for observing low level wind shears (ICAO 2005)

Features	Alerts	Warnings
Coverage	3 NM extension to runways is commonly used whatever the wind shear equipment (LLWAS, Lidars, TDWR/Radars)	Up to 9.55 km for a glide slope of 3Deg in order to monitor up to the altitude of 500 m
Update frequency	LLWAS: update every 30 s Radar/lidar: update between 1 to 6 min	Typically 5 min are used for radars and lidars
Resolution	LLWAS: 1 NM between 2 anemometers RADAR: 100–250 m Lidar: 100–200 m	Same as alerts
Scanning patterns	One horizontal scan (PPI) at 3° scans per approach are used	Several products can be used from 2 to 5 PPI scans
Methods for computing alerts	Wind shear alerts: Headwind/tailwind changes over 1 NM or along the runway superior to 15 Knots Microburst alerts: Headwind/tailwind changes over 1 NM or along the runway superior to 40 Knots LLWAS: NCAR Algorithm RADAR: Runway-oriented wind shears algorithms like the ones developed by MIT-Lincoln Lab for TDWR, HKO algorithm, products of commercial software like RAINBOW5 developed by Selex and IRIS developed by Vaisala	Additional products can be computed from radial radar or lidar data according to local needs like shear products, gust front detection, wind reconstruction

TDWR terminal Doppler weather radar, *LLWAS* low level wind shear, *PPI* plan position indicator, *HKO* Hong Kong Observatory

helicity (McGovern et al. 2017). The Random Forest (RF) analysis (Ahijevych et al. 2016) used for this experiment was trained using CAPS ensemble forecasts during May–June 2014, and evaluated based on CAPS ensemble forecasts. The performance diagram in Fig. 14a shows that for a given probability threshold, the machine-learning models tend to have fewer false alarms, a lower frequency bias, and higher overall accuracy than other methods. The attributes diagram in Fig. 14b indicates that the probabilities from the machine-learning models and updraft helicity are generally reliable, while other methods tend to produce probabilities that are overconfident. These results suggest that probabilistic approaches for convection prediction need to be developed in the future that are based on RF or other AI methods.

The demand for accurate nowcasts of convective precipitation that includes heavy precipitation and hail has led to development of the high-resolution data assimilation and rapid cycling numerical weather prediction system (e.g., Sun et al. 2014). In their work, they reviewed the recent progress on the use of NWP for nowcasting convective precipitation and provided future expected challenges and opportunities. They emphasized that NWP models generally produce larger quantitative precipitation forecasts (QPFs) than nowcasting systems beyond a few forecast hours. Therefore, they suggest blending radar echo extrapolation with a numerical model run to generate a seamless 0–6-h forecast. Nowcasting and Initialization for Modeling Using Regional Observation Data System (NIMROD; Golding 1998) was the first system that used blended radar echo extrapolation with a NWP output. For the first hour nowcast, the extrapolation of the observed precipitation field was given full weight, and it was gradually relaxed with increasing lead time to where the model eventually received full weight. This method was better than earlier radar echo tracking systems. The blending of the corrected model forecasts with extrapolation forecasts allows for a smooth transition from the extrapolation to model forecasts (Sun et al. 2014). Similar results have been shown over the years starting with Browning (1980), Doswell (1986), and Austin et al. (1987).

A recent paper by Sokol and Zacharov (2012) described a new blending method that assimilates the extrapolated radar reflectivity with a nudging technique. The decrease in skill by extrapolation related to the size and organization of the precipitation were

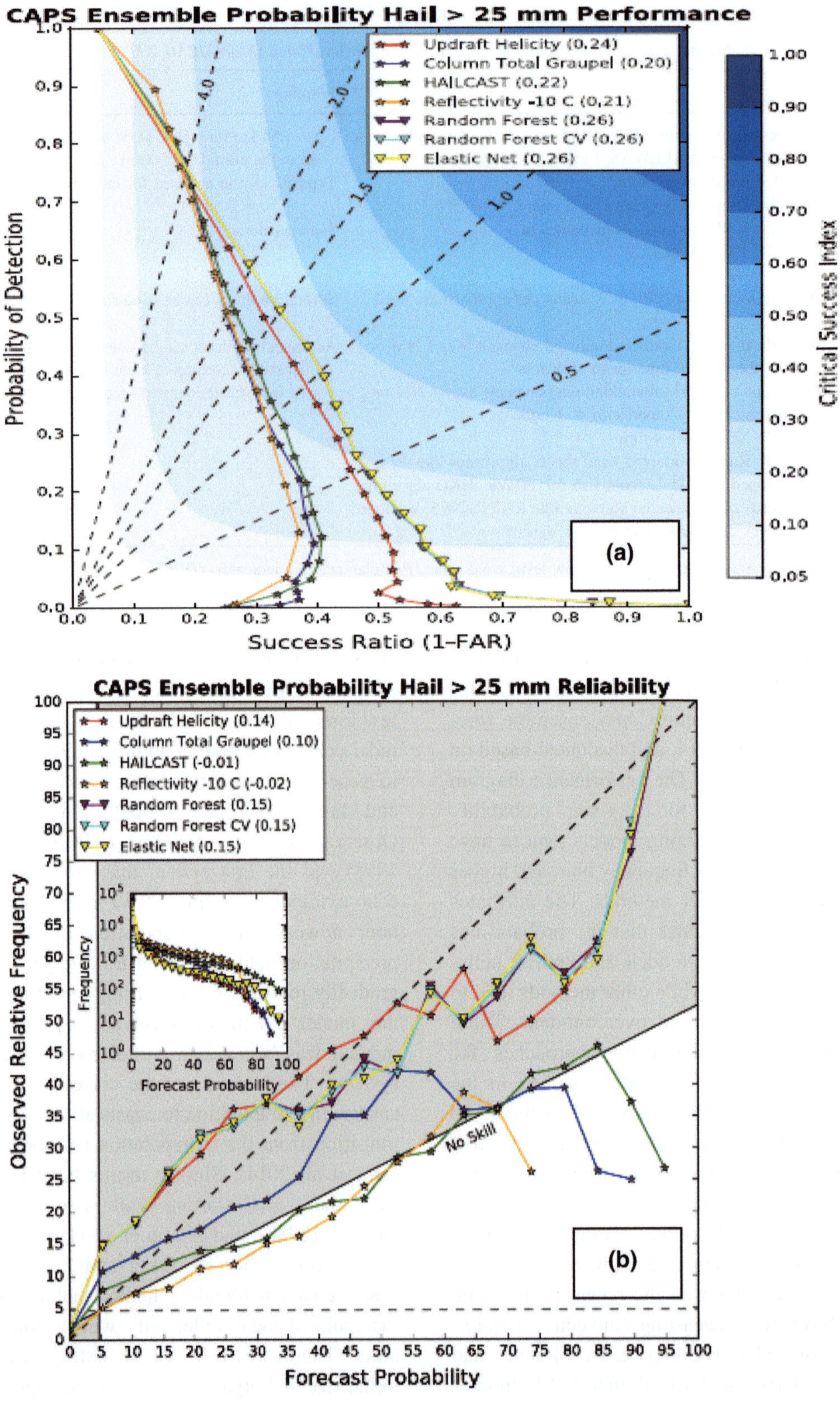
CAPS Ensemble Probability Hail > 25 mm Performance
Updraft Helicity (0.24)
Column Total Graupel (0.20)
HAILCAST (0.22)
Reflectivity -10 C (0.21)
Random Forest (0.26)
Random Forest CV (0.26)
Elastic Net (0.26)
Probability of Detection
Success Ratio (1-FAR)
Critical Success Index
(a)
CAPS Ensemble Probability Hail > 25 mm Reliability
Updraft Helicity (0.14)
Column Total Graupel (0.10)
HAILCAST (-0.01)
Reflectivity -10 C (-0.02)
Random Forest (0.15)
Random Forest CV (0.15)
Elastic Net (0.15)
Observed Relative Frequency
Forecast Probability
Frequency
No Skill
(b)

◀Figure 14
Performance diagram comparing different hail forecasting methods (**a**). Attributes diagram indicating the reliability of different forecasting methods (**b**) (Adapted from McGovern et al. 2017).

also emphasized by Wilson (1966) and Wilson et al. (1998). Based on a large set of predictor fields, and inserting the location of boundary layer convergence lines in NCAR's AutoNowcaster (ANC) (Mueller et al. 1993), Wilson et al. (2004) predicted storm initiation up to 1 h in advance that shows the importance of BL processes. The ANC system uses fuzzy logic to combine predictor fields that reflect the atmospheric environmental conditions and boundary layer forcing based on observations and numerical model runs.

To meet the needs of nowcasting, numerical models have to be run at resolutions less than a few kilometers. Wilson and Roberts (2006) stated that the 10-km Rapid Update Cycle (RUC10) 3-h forecasts issued every 3 h were correct at predicting areas of convective initiation only 13% of the time. Sun et al. (2014) stated that possible factors limiting the model's ability to predict precipitation initiation was likely insufficient model resolution in addition to many other factors. Models with high resolutions can enable the explicit representation of convective processes without cumulus parameterization schemes. These models are called as "convection-permitting" or "convection-allowing" NWP (e.g., Sun et al. 2014). Many other studies show that forecasts from the convection-permitting models produced more skillful guidance than those from a coarser-resolution model employing convective parameterization (e.g., Done et al. 2004; Kain et al. 2010; Weisman et al. 2008; Clark et al. 2009). Verification of convection forecasts is challenging, and innovative methods are needed to account for small errors in model derived location and timing (e.g., Pinto et al. 2015).

Overall, various algorithms and systems have been used in storm convection prediction. The FAA Aviation Weather Research Program (AWRP) has led to very successful development of forecasts of both convective systems and winter storms, using heuristic and numerical models for aviation applications (Wolfson et al. 2008; Wolfson and Clark 2006). The FAA effort consolidating the storm prediction systems led to the establishment of collaboration between MIT Lincoln Laboratory (MIT LL), the National Center for Atmospheric Research (NCAR) Research Applications Laboratory (RAL), the NOAA Earth Systems Research Laboratory (ESRL) Global Systems Division (GSD) and NASA, called the Consolidated Storm Prediction for Aviation (CoSPA; Wolfson et al. 2008). The CoSPA is funded under the FAA's Aviation Weather Research Program (AWRP) (Dupree et al. 2009). One of the goals of the Next Generation Air Transportation System (NextGen) project (Stobie et al. 2008) is to consolidate the redundant and sometimes conflicting forecast systems into a Single Authoritative Source (SAS) for aviation uses (Dupree et al. 2009). The current CoSPA prototype for 0–6 h forecasts is part of the NextGen Initial Operational Capability (IOC) in 2013.

5.7. *Observational Methods*

The accurate testing of nowcasting products is strongly related to observations and their analysis. Golding (1998) suggested that products should be assessed against both point observations and analyses. Nowcasting systems usually integrate observations with Numerical Weather Prediction (NWP) model products for short term predictions up to six hours ahead. In Golding (1998), precipitation, cloud, and visibility were the main conditions to be considered in the analysis. The precipitation rate in the analysis used a combination of the processed radar and satellite data, surface reports, and NWP simulation output. The precipitation type was also diagnosed in their work using NWP fields. Improving nowcasts is a strong function of the integrated systems that are used in the observations and model outputs (Bailey et al. 2009). Developments of these nowcasting systems in the near future will likely be improved by artificial intelligence (AI) methods and using new observational platforms such as S-Pol radars and LIDARs.

Generally, NWP prediction assessments against point observations are only reported for analyses because the representativeness errors of meteorological parameters are of similar magnitude to the forecast errors. In general, assessments of predictions against analyses performed over restricted geographical areas are considered good because of observational coverage. If not considered, then the verification statistics can be very uncertain. Because of possible natural variability in observations, use of supersites with satellite stations (Gultepe et al. 2018; Ralph et al. 2013) should to be further researched for data assimilation and model output validations. Verifications performed by Vislocky and Fritsch (1997) showed that the observations-based methods were not only far superior to persistence climatology at all lead times (5–20% improvement) but also outperformed the MOS-based technique at the 1- and 3-h lead time projections with skill increases averaging four percentage points. This suggests that for nowcasting applications, observations should be weighted more than MOS techniques.

In the future, although expanded computational techniques can be expected to reduce the uncertainty in the predicted parameters related to aviation meteorology, improvements in individual models and physical parameterizations, as well their careful verification are paramount (e.g., Yano et al. 2018 and references therein). Physical parameterizations related to aviation meteorology forecasting are not well-represented at NWP scales presently in use; therefore, new scale-dependent physical parameterizations will be needed to improve the NWPs predictions in deterministic or probabilistic approaches.

Forecasting accuracy of high impact weather parameters are strongly dependent on availability of integrated data sets (both observations and predictions) and should take advantage of probabilistic forecast methods (Benjamin et al. 2009, 2010; Zhou et al. 2009). Unfortunately, errors related to components of the NWP ensemble simulations do not usually cancel out each other but tend to accumulate. As shown in this review, both measurements and statistical approaches, as well as newly developed scale-dependent physical (e.g. Vis) and dynamical parameterizations (e.g. EDR, CAT) Brown (1973) are needed to improve the accuracy of the NWP model predictions that are related to core physical and dynamic processes.

Overall, the issues with predictions of gusts and EDR, Vis, precipitation amount and rate, fog, icing/deicing, and lightning as well as convective intensity require more advanced designs for the meteorological supersites and physically-based approaches for parameterizations of severe weather parameters under various climatic environments. In this respect, in situ observations at the surface and within the atmospheric layer in both cloud and cloud-free regions are required to improve the aviation nowcasting and also for improving the accuracy of the long-range forecasts.

In response for the needs of global aviation community, NCEP and the UK Meteorological Office are currently assigned by the International Civil Aviation Organization (ICAO) to create World Area Forecast (WAF) guidance for the aviation weather community worldwide. Both centers are now providing operational global icing and turbulence forecasts based on their global forecast systems. NCAR's fuzzy logic icing and GTG turbulence algorithms have already been implemented into NCEP's Unified Post Processor (UPP) to generate both icing and turbulence forecasts (Chuang et al. 2019). Results of this collaboration indicate that aviation meteorology issues are not only locally recognized but also globally, and need to be evaluated and integrated.

Acknowledgements

This review paper is funded by the various institutions representing co-authors given in the title, and received technical and funding support from ECCC and SAR offices in Canada that were related to fog and visibility issues. S. S. Yum is supported by the Research and Development Program for KMA Weather, Climate and Earth System Services (#2016-3100) of National Institute of Meteorological Sciences (NIMS). We also would like to thank for the reviewers for their comments to improve the manuscript, and specifically to one of the reviewers who made specific comments on satellite and radar based platforms to be used for aviation meteorology.

References

Ahijevych, D., Pinto, J. O., Williams, J. K., & Steiner, M. (2016). Probabilistic forecasts of mesoscale convective system initiation using the random forest data mining technique. *Weather Forecasting, 31,* 581–599. https://doi.org/10.1175/waf-d-15-0113.1.

Alapaty, K., Seaman, N. L., Niyogi, D. S., & Hanna, A. F. (2001). Assimilating surface data to improve the accuracy of atmospheric boundary layer simulations. *Journal of Applied Meteorology, 40,* 2068–2082.

Albers, H. W. (1977). ICAS items. *Bulletin American Meteorology Society, 58,* 342–343.

Allen, C. T., Haupt, S. E., & Young, G. S. (2007). Source characterization with a genetic algorithm–coupled dispersion–backward model incorporating SCIPUFF. *Journal of Applied Meteorology Climatology, 46,* 273–287.

Anderson, J. L. (1996). Selection of initial conditions for ensemble forecasts in a simple perfect model framework. *Journal of Atmospheric Science, 53,* 22–36.

Ansmann, A., Mattis, I., Wandinger, U., Wagner, F., Reichardt, J., & Deshler, T. (1997). Evolution of the pinatubo aerosol: Raman lidar observations of particle optical depth, effective radius, mass, and surface area over Central Europe at 53.4°N. *Journal of Atmospheric Science, 54,* 2630–2641.

Appleman, H. (1953). The formation of exhaust condensation trails by jet aircraft. *Bulletin American Meteorology Society, 34,* 14–20.

Argyle, E. M., Gourley, J. J., Flamig, Z. L., Hansen, T., & Manross, K. (2017). Toward a User-Centered Designof a Weather Forecasting Decision-Support Tool. *Bulletin of the American Meteorological Society, 98,* 373–382. https://doi.org/10.1175/BAMS-D-16-0031.1.

Auger, L., Dupont, O., Hagelin, S., Brousseau, P., & Brovelli, P. (2015). AROME–NWC: A new nowcasting tool based on an operational mesoscale forecasting system. *Quarterly Journal Royal Meteorological Society, 141,* 1603–1611.

Austin, G. L., Dionne, P., & Roch, M. (1987). On the interaction between radar and satellite image nowcasting systems and mesoscale numerical models. In *Proceedings mesoscale analysis and forecasting*, Vancouver, BC, Canada, European Space Agency, pp. 225–228.

Bailey, M. E., Isaac, G. A., Driedger, N., & Reid, J. (2009). Comparison of nowcasting methods in the context of high-impact weather events for the Canadian Airport Nowcasting Project. In *International symposium on nowcasting and very short range forecasting*, 30 August–4 September 2009, Whistler, British Columbia.

Bailey, M. E., Isaac, G. A., Gultepe, I., Heckman, I., & Reid, J. (2014). Adaptive blending of model and observations for automated short-range forecasting: Examples from the Vancouver 2010 Olympic and Paralympic Winter Games. *Journal of Pure and Applied Geophysics, 171,* 257–276. https://doi.org/10.1007/s00024-012-0553-x.

Baker, W. E., Atlas, R., Cardinali, C., Clement, A., Emmitt, G. D., Gentry, B. M., et al. (2014). Lidar-measured wind profiles: The missing link in the global observing system. *Bulletin American Meteorology Society, 95,* 543–564.

Bankert, R. L. (1994). Cloud classification of avhrr imagery in maritime regions using a probabilistic neural network. *Journal of Applied Meteorology, 33,* 909–918.

Bankert, R. L., Hadjimichael, M., Kuciauskas, A. P., Thompson, W. T., & Richardson, K. (2004). Remote cloud ceiling assessment using data-mining methods. *Journal of Applied Meteorology, 43,* 1929–1946.

Bankert, R. L., Mitrescu, C., Miller, S. D., & Wade, R. H. (2009). Comparison of GOES cloud classification algorithms employing explicit and implicit physics. *Journal of Applied Meteorology Climatology, 48,* 1411–1421.

Banta, R., Pichugina, Y., Brewer, A., James, E., Olson, J., Benjamin, S., et al. (2017). Evaluating and improving NWP forecast models for the future: How the needs of offshore wind energy can point the way. *Bulletin American Meteorology Society.* https://doi.org/10.1175/bams-d-16-0310.

Barker, H. W., Jerg, M. P., Wehr, T., Kato, S., Donovan, D. P., & Hogan, R. J. (2011). A 3D cloud construction algorithm for the EarthCARE satellite mission. *Quarterly Journal of the Royal Meteorological Society, 137,* 1042–1058. https://doi.org/10.1002/qj.824.

Bates, T. S., Quinn, P. K., Johnson, J. E., Corless, A., Brechtel, F. J., Stalin, S. E., et al. (2013). Measurements of atmospheric aerosol vertical distributions above Svalbard, Norway, using unmanned aerial systems (UAS). *Atmosphere Measure Technology, 6,* 2115–2120.

Bedka, K., Brunner, J., Dworak, R., Feltz, W., Otkin, J., & Greenwald, T. (2010). Objective satellite-based detection of overshooting tops using infrared window channel brightness temperature gradients. *Journal of Applied Meteorology Climatology, 49,* 181–202.

Bedka, K. M., & Khlopenkov, K. (2016). A probabilistic multispectral pattern recognition method for detection of overshooting cloud tops using passive satellite imager observations. *Journal of Applied Meteorology Climatology, 55,* 1983–2005. https://doi.org/10.1175/JAMC-D-15-0249.1.

Bedka, K. M., Velden, C. S., Petersen, R. A., Feltz, W. F., & Mecikalski, J. R. (2009). Comparisons of satellite-derived atmospheric motion vectors, rawinsondes, and NOAA wind profiler observations. *Journal of Applied Meteorology Climatology, 48,* 1542–1552.

Behne, D. (2008). NAM-WRF verification of subtropical jet turbulence. *Electronics Journal of Operational Meteorology,* Paper 2008-EJ3. http://www.nwas.org/ej/pdf/2008-EJ3.pdf.

Bélair, S., Lacarrère, P., Noilhan, J., Masson, V., & Stein, J. (1998). High-resolution simulation of surface and turbulent fluxes during HAPEX-MOBILHY. *Monthly Weather Review, 126,* 2234–2253.

Benjamin, S. G., et al. (2009). *Rapid Refresh/RUC project technical review.* NOAA/ESRL/GSD Internal Review. http://ruc.noaa.gov/pdf/RR-RUC-TR_11_3_2009.pdf.

Benjamin, S. G., Jamison, B. D., Moninger, W. R., Sahm, S. R., Schwartz, B. E., & Schlatter, T. W. (2010). Relative short-range forecast impact from aircraft, profiler, radiosonde, VAD, GPS-PW, METAR, and mesonet observations via the RUC hourly assimilation cycle. *Monthly Weather Review, 138,* 1319–1343.

Benjamin, S. G., Moninger, W. R., Smith, T. L., Jamison, B. D., & Schwartz, B. E. (2006a). TAMDAR aircraft impact experiments with the rapid update cycle. In Preprints, *10th Symposium on integrated observing and assimilation systems for atmosphere, oceans, and land surface*, Atlanta, GA, Amer. Meteor. Soc., 9.8. http://ams.confex.com/ams/pdfpapers/.

Benjamin, S. G., Moninger, W. R., Smith, T. L., Jamison, B. D., & Schwartz, B. E. (2006b) Impact of TAMDAR humidity, temperature, and wind observations in RUC parallel experiments. In

Preprints, 12th Conference on aviation, range, and aerospace meteorology, Atlanta, GA, American Meteorology Society, p. 4.5.

Benjamin, S. G., Moninger, W. R., Smith, T. L., Jamison, B. D., Szoke, E. J., & Schlatter, T. W. (2007). 2006 TAMDAR impact experiment results for RUC humidity, temperature, and wind forecasts. In *Preprints, 11th symposium on integrated observing and assimilation systems for the atmosphere, oceans, and land surface*, San Antonio, TX, *Amer. Meteor. Soc.*, 9.2. http://ams.confex.com/ams/pdfpapers/119903.pdf.

Benjamin, S., Weygandt, S., Brown, J., Hu, M., Alexander, C., Smirnova, T., et al. (2016). A North American hourly assimilation and model forecast cycle: The rapid refresh. *Monthly Weather Review, 144*(4), 1669–1694. https://doi.org/10.1175/mwrd-15-0242.1.

Bergot, T., Terradellas, E., Cuxart, J., Mira, A., Liechti, O., Mueller, M., et al. (2007). Intercomparison of single-column numerical models for the prediction of radiation fog. *Journal of Applied Meteorology Climatology, 46,* 504–521.

Berndt, E., Elmer, N., Schultz, L., & Molthan, A. (2018). A methodology to determine recipe adjustments for multispectral composites derived from next-generation advanced satellite imagers. *Journal of Atmosphere Oceanic Technology, 2,* 22. https://doi.org/10.1175/JTECH-D-17-0047.1.

Bernstein, B. C., McDonough, F., Politovich, M. K., Brown, B. G., Ratvasky, T. P., Miller, D. R., et al. (2005). Current icing potential: Algorithm description and comparison with aircraft observations. *Journal of Applied Meteorology, 44,* 969–986.

Bernstein, B. C., Wolff, C. A., & Minnis, P. (2006). Practical application of NASA-Langley advanced satellite products to in-flight icing nowcasts. In Proc. *44th AIAA Aerospace Sci. Mtg. & Exhibit,* Reno, NV, January 9–12, AIAA-2006-1220, p. 18.

Beswick, K., Baumgardner, D., Gallagher, M., Volz-Thomas, A., Nedelec, P., Wang, K.-Y., et al. (2014). The backscatter cloud probe – a compact low-profile autonomous optical spectrometer. *Atmos. Meas. Tech.*, *7,* 1443–1457. https://doi.org/10.5194/amt-7-1443-2014.

Bianco, L., Cimini, D., Marzano, F. S., & Ware, R. (2005). Combining microwave radiometer and wind profiler radar measurements for high-resolution atmospheric humidity profiling. *Journal of Atmospheric and Oceanic Technology, 22*(7), 949–965.

Bilbro, J., Fichtl, G., Fitzjarrald, D., Krause, M., & Lee, R. (1984). Airborne Doppler lidar wind field measurements. *Bulletin American Meteorology Society, 65,* 348–359.

Bilbro, J. W., & Vaughan, W. W. (1978). Wind field measurement in the nonprecipitous regions surrounding severe storms by an airborne pulsed Doppler lidar system. *Bulletin American Meteorology Society, 59,* 1095–1100.

Black, A. W., & Mote, T. L. (2015). Characteristics of winter-precipitation-related transportation fatalities in the United States. *Weather Climate Society, 7,* 133–145.

Bluestein, H. B., French, M. M., PopStefanija, I., Bluth, R. T., & Knorr, J. B. (2010). A mobile, phased-array Doppler radar for the study of severe convective storms: The MWR-05XP. *Bulletin American Meteorology Society, 91,* 579–600.

Bluestein, H. B., Houser, J. B., French, M. M., Snyder, J. C., Emmitt, G. D., PopStefanija, I., et al. (2014). Observations of the boundary layer near tornadoes and in supercells using a mobile, collocated, pulsed doppler lidar and radar. *Journal of Atmosphere Oceanic Technology, 31,* 302–325.

Blumstein, D., Tournier, B., Cayla, F. R., Phupin, R. F. T., Bull, C., & Ponce, G. (2007). In-flight performance of the infrared atmospheric sounding interferometer (IASI) on MetOp-A. Atmospheric and environmental remote sensing data processing and utilization III: Readiness for GEOSS. In M. D. Goldberg et al. (Eds.) *International Society for Optical Engineering (SPIE Proceedings* (vol. 6684, p. 66840).

Boer, G. D., Palo, S., Argrow, B., LoDolce, G., Mack, J., Gao, R.-S., et al. (2017). The pilatus unmanned aircraft system for lower atmospheric research. *Atmosphere Measure Technology Discussion, 8,* 11987–12023.

Bott, A., Sievers, U., & Zdunkowski, W. (1990). A radiation fog model with a detailed treatment of the interaction between radiative transfer and fog microphysics. *Journal of Atmospheric Science, 47,* 2153–2166.

Bottyán, Z., Tuba, Z., & Gyöngyösi, A. Z. (2016). Weather Forecasting System for the Unmanned AircraftSystems (UAS) Missions with the Special Regard to Visibility Prediction in Hungary. In L. Nádai & J. Padányi (Eds.), *Critical Infrastructure Protection Research* (Vol. 12). Cham: Springer. https://doi.org/10.1007/978-3-319-28091-2_2.

Bottyán, Z., Wantuch, F., Gyöngyösi, A. Z., Tuba, Z., Hadobács, K., Kardos, P., et al. (2013). Development of a complex meteorological support system for UAVs. *World Academy of Science, Engineering and Technology International Journal of Geology and Environment Engineering, 7*(4), 646–651.

Bravin, M., Strapp, J. W., & Mason, J. (2015). An investigation into location and convective lifecycle trends in an ice crystal icing engine database. *Tech. rep., SAE Technical Paper 2015-01-2130,* SAE International, Warrendale, Pennsylvania, USA. https://doi.org/10.4271/2015-01-2130.

Bright, D. R., Lack, S. A., & Sparks, J. A. (2016). A summary of turbulence forecasting techniques used by the national weather service. In R. Sharman & T. Lane (Eds.), *Aviation turbulence: Processes, detection, prediction* (pp. 213–226). Berlin: Springer.

Brooks, G. R., & Oder, A. (2004). Low level turbulence algorithm testing at-or-below 10,000 ft. In *Preprints, 11th conference on aviation, range, and aerospace meteorology*, Hyannis, MA, Amer. Meteor.Soc., P4.16. http://ams.confex.com/ams/pdfpapers/82122.pdf.

Brown, R. (1973). New indices to locate clear-air turbulence. *Meteorological Magazine, 102,* 347–361.

Brown, B. G., Thompson, G., Bruintjes, R. T., Bullock, R., & Kane, T. (1997). Intercomparison of in-flight icing algorithms. Part II: Statistical verification results. *Weather Forecasting, 12,* 890–914.

Browning, K. A. (1980). Local weather forecasting. *Proceeding Royal Society of London, A371,* 179–211.

Burrows, W. R., Price, C., & Wilson, L. J. (2005). Warm season lightning probability prediction for Canada and the Northern United States. *Weather Forecasting, 20,* 971–988.

Casadevall, T. J. (1994). The 1989–90 eruption of Redoubt volcano. Alaska: Impacts on aircraft operations. *Journal of Volcanology and Geothermal Research, 62,* 301–316.

Chachere, C., & Pu, Z. (2018). Numerical simulations of an inversion fog event in the salt lake valley during the MATERHORN-fog field campaign. *Pure and Applied Geophysics.* https://doi.org/10.1007/s00024-018-1770-8.

Chan, P. W. (2014). Performance and aviation applications of minisodars at Hong Kong International Airport. *Meteorological Applications, 21,* 62–73. https://doi.org/10.1002/met.1388.

Chan, P. W. (2016). LIDAR-based turbulence intensity for aviation applications. In R. D. Sharman & T. P. Lane (Eds.), *Aviation turbulence: Processes, detection, and prediction* (pp. 193–209). Berlin: Springer.

Chan, P. W., & Shao, A. M. (2007). Depiction of complex airflow near Hong Kong International Airport using a Doppler LIDAR with a two-dimensional wind retrieval technique. *Meteorologische Zeitschrift, 16,* 491–504.

Chan, P. W., Shun, C. M., & Wu, K. C. (2006). Operational LIDARLIDAR-based system for automatic wind shear alerting at the Hong Kong International Airport. In *12th conference on aviation, range, & aerospace meteorology*, American Meteorological Society, Atlanta, GA, USA, 29 January–2 February 2006.

Chuang, H.-Y., Mao, Y., & Zhou, B. (2019). R2O Transition of NCAR's Icingand Turbulence Algorithms into NCEP's Operations. *Pure and Applied Geophysics.* https://doi.org/10.1007/s00024-018-1975-x.

Chandrasekar, V., Keränen, R., Lim, S., & Moisseev, D. (2013). Recent advances in classification of observations from dual polarization weather radars. *Atmospheric Research, 119,* 97–111.

Chiodi, A. M., & Harrison, D. E. (2010). Characterizing warm-ENSO variability in the equatorial Pacific: An OLR perspective. *Journal of Climate, 23,* 2428–2439. https://doi.org/10.1175/2009JCLI3030.1.

Chmielecki, R. M., & Raftery, A. E. (2011). Probabilistic visibility forecasting using bayesian model averaging. *Monthly Weather Review, 139,* 1626–1636.

Cho, J. Y. N. (2015). Enhanced signal processing algorithms for the ASR-9 weather systems processor. *Journal of Atmospheric and Oceanic Technology, 32,* 1847–1859.

Chun, H., Kim, J., Lee, D., Kim, S., Strahan, M., Pettegrew, B., et al. (2017). Research collaborations for better predictions of aviation weather hazards. *Bulletin American Meteorology Society, 98,* ES103–ES107. https://doi.org/10.1175/BAMS-D-17-0010.1.

Clark, A. J., Gallus, W. A., Jr., Xue, M., & Kong, F. (2009). A comparison of precipitation forecast skill between small convection-permitting and large convection-parameterizing ensembles. *Weather and Forecasting, 24,* 1121–1140.

Cohen, A. E., Cavallo, S. M., Coniglio, M. C., & Brooks, H. E. (2017). Evaluation of multiple planetary boundary layer parameterization schemes in Southeastern U.S. Cold Season severe weather environments. *Weather and Forecasting, 32,* 1857–1884.

Cohn, S. A. (1995). Radar measurements of turbulent eddy dissipation rate in the troposphere: A comparison of techniques. *Journal of Atmosphere Oceanic Technology, 12,* 85–95.

Colson, D., & Panofsky, H. A. (1965). An index of clear-air turbulence. *Quarterly Journal of Royal Meteorology Society, 91,* 507–513.

Cook, L., Wood, B., Klein, A., Lee, R., & Memarzadeh, B. (2009). Analyzing the share of individual weather factors affecting NAS performance using the weather impacted traffic index. In *AIAA 2009-7017. 9th AIAA aviation technology, integration, and operations conference (ATIO)*, Hilton Head, SC, September 2009. https://doi.org/10.2514/6.2009-7017.

Cornman, L. B. (2016). Airborne in situ measurements of turbulence. In R. Sharman & T. Lane (Eds.), *Aviation turbulence: Processes, detection, prediction* (pp. 97–120). Berlin: Springer.

Cornman, L. B., Morse, C. S., & Cunning, J. (1995). Real time estimation of atmospheric turbulence severity from in situ aircraft measurements. *Journal of Aircraft, 32,* 171–177.

Cooper, W. A., Sand, W. R., Politovich, M. K., & Veal, D. L. (1984). Effects of icing on performance of a research aircraft. *Journal of Aircraft, 21,* 708–715.

Dehghan, A., Hocking, W. K., & Srinivasan, R. (2014). Comparisons between multiple in situ aircraft measurements and radar in the troposphere. *Journal of Atmospheric and Solar-Terrestrial, 118,* 64–77.

Del Genio, A. D., Yao, M.-S., & Jonas, J. (2007). Will moist convection be stronger in a warmer environment? *Geophysical Research Letters, 34,* L16703. https://doi.org/10.1029/2007gl030525.

Deng, M., Mace, G. G., Wang, Z., & Okamoto, H. (2010). Tropical composition, cloud and climate coupling experiment validation for cirrus cloud profiling retrieval using CloudSat radar and CALIPSO lidar. *Journal of Geophysics Research, 115,* D00J15. https://doi.org/10.1029/2009jd013104.

DeWekker, S. F. J., Godwin, K. S., Emmitt, G. D., & Greco, S. (2012). Airborne Doppler lidar measurements of valley flows in complex coastal terrain. *Journal of Applied Meteorology Climatology, 51,* 1558–1574.

Dines, W. H. (1917). Meteorology and aviation. *Monthly Weather Review, 45,* 401. https://doi.org/10.1175/1520-0493(1917)45%3c401b:MAA%3e2.0.CO;2.

Done, J., Davis, C. A., & Weisman, M. L. (2004). The next generation of NWP: Explicit forecasts of convection using the Weather Research and Forecasting (WRF) model. *Atmosphere Science Letter, 5,* 110–117.

Donovan, M. F., Williams, E. R., Kessinger, C., Blackburn, G., Herzegh, P. H., Bankert, R. L., et al. (2008). The identification and verification of hazardous convective cells over oceans using visible and infrared satellite observations. *Journal of Applied Meteorology Climatology, 47,* 164–184.

Dorman, C. E., Mejia, J. F., Korocin, D., & McEvoy, D. J. (2017). Worldwide marine fog occurrence and climatology. In R. Sugan (Ed.), *A chapter in the book of marine fog: Challenges and advancements in observations, modeling, and forecasting*. Berlin: Springer. https://doi.org/10.1007/978-3-319-45229-6_2.

Doswell, C. A. (1986). Short-range forecasting. In P. Ray (Ed.), *Mesoscale meteorology and forecasting* (pp. 689–719). Genesco: American Meteor Society.

Doswell, C. A. (1980). Synoptic-scale environments associated with high plains severe thunderstorms. *Bulletin of American Meterological Society, 61,* 1388–1400.

Dupree, W., Morse, D., Chan, M., Tao, X., Iskenderian, H., Reiche, C., & Wolfson, M., et al. (2009). The 2008 CoSPA Forecast demonstration (collaborative storm prediction for aviation). *89th AMS annual meeting ARAM special symposium on weather—air traffic phoenix*, AZ/11-15 January 2009. P1.1, p. 19.

Dutton, J. A., & Panofsky, H. A. (1970). Clear air turbulence: A mystery may be unfolding. *Science, 167,* 937–944.

Eichinger, W. E., Cooper, D. I., Forman, P. R., Griegos, J., Osborn, M. A., Richter, D., et al. (1999). The development of a scanning raman water vapor lidar for boundary layer and tropospheric observations. *Journal of Atmosphere Oceanic Technology, 16,* 1753–1766.

Eick, D. (2014) Turbulence related accidents and incidents. Presentation at NCAR Turbulence Impact Mitigation Workshop 2, 3–4 Sep 2014. https://ral.ucar.edu/sites/default/files/public/events/2014/turbulence-impact-mitigation-workshop-2/docs/eick-turbulencerelatedaccidents.pdf. Accessed 3 Jan 2019

Ellrod, G. P. (1985). *Detection of high level turbulence using satellite imagery and upper air data*. NOAA Tech. Memo. NESDIS 10, p. 30.

Ellrod, G. P., Connell, B. H., & Hillger, D. W. (2003). Improved detection of airborne volcanic ash using multispectral infrared satellite data. *Journal of Geophysical Research, 108,* 4356. https://doi.org/10.1029/2002jd002802.

Ellrod, G. P., & Gultepe, I. (2007). Inferring low cloud base heights at night for aviation using satellite infrared and surface temperature data. *Journal of Pure and Applied Geophysics, 164,* 1193–1205.

Ellrod, G. P., & Knapp, D. I. (1992). An objective clear-air turbulence forecasting technique: Verification and operational use. *Weather Forecasting, 7,* 150–165.

Ellrod, G. P., & Pryor, K. (2019). Applications of Geostationary Satellite Data to Aviation. *Pure Appl Geophys*. https://doi.org/10.1007/s00024-018-1821-1.

Ellrod, G. P., & Knox, J. A. (2010). Improvements to an operational clear-air turbulence diagnostic index by addition of a divergence trend term. *Weather Forecasting, 25,* 789–798.

Ellrod, G. P., Lester, P. F., & Ehernberger, J. (2002). Clear air turbulence. In J. R. Holton, et al. (Eds.), *Encyclopedia of the atmospheric sciences* (pp. 393–403). Oxford: Academic.

Ellrod, G.P., Knox, J.A., Lester, P.F., Ehernberger, L.J., 2015. Aviation: Clear Air Turbulence. In: Gerald R. North (editor-in-chief), J. Pyle and F. Zhang (Eds.), *Encyclopedia of Atmospheric Sciences*, 2nd edn, Vol 1, (pp. 177–186) Elsevier, Inc. https://doi.org/10.1016/B978-0-12-382225-3.00104-3.

Elmore, K. L., & Richman, M. B. (2001). Euclidean distance as a similarity metric for principal component analysis. *Monthly Weather Review, 129,* 540–549.

Endlich, R. M. (1964). The mesoscale structure of some regions of clear-air turbulence. *Journal of Applied Meteorology, 3,* 261–276.

FAA. (1988). *Advisory circular on pilot wind shear guide*. AFS-200. AC No: 00-54, p. 56.

FAA. (2017). *Continued operational safety (COS) report. Special category light-sport aircraft*, July 2004–Sep 2017, p. 50.

FAA-P-8740-40. (2008). *Wind shear*. HQ 101130, p. 8.

Fahey, T., Wilson, E., O'Loughlin, R., Thomas, M., & Klipfel, S. (2016). A history of weather reporting from aircraft and turbulence forecasting for commercial aviation. In R. Sharman & T. Lane (Eds.), *Aviation turbulence: Processes, detection, prediction* (pp. 31–58). Berlin: Springer.

Feingold, G., Cotton, W. R., Kreidenweis, S. M., & Davis, J. T. (1999). The impact of giant cloud condensation nuclei on drizzle formation in stratocumulus: Implications for cloud radiative properties. *Journal of Atmospheric Science, 56,* 4100–4117.

Ferrare, R. A., Melfi, S. H., Whiteman, D. N., Evans, K. D., Schmidlin, F. J., & Starr, D. O. (1995). A comparison of water vapor measurements made by Raman Lidar and radiosondes. *Journal of Atmosphere Oceanic Technology, 12,* 1177–1195.

Ferrero, E., Mortarini, L., Manfrin, M., Solari, M., & Forza, R. (2014). Physical simulation of atmospheric microbursts. *Journal of Geophysics Research Atmosphere, 119,* 6292–6305. https://doi.org/10.1002/2013jd021243.

Ferrier, B. S. (1994). A double-moment multiple-phase four-class bulk ice scheme. Part I: Description. *Journal of Atmosphere Science, 51,* 249–280.

Ferrier, B. S., Tao, W. K., & Simpson, J. (1995). A double-moment multiple-phase four-class bulk ice scheme. Part II: Simulations of convective storms in different large-scale environments and comparisons with other bulk parameterizations. *Journal of Atmosphere Science, 52,* 1001–1033.

Fischer, C., Montmerle, T., Berre, L., Auger, L., & Stefanescu, S. (2005a). Anover view of the variational assimilation in the ALADIN/France numerical weather-prediction system wave-driven circulation of the mesosphere. *Quarterly Journal Royal Meteorological Society, 131,* 3477–3492.

Fischer, A. S., Terray, P., Guilyardi, E., Gualdi, S., & Delecluse, P. (2005b). Two independent triggers for the Indian Ocean dipole/zonal mode in a coupled GCM. *Journal of Climate, 18,* 3428–3449.

Fix, A. (2012). Tunable light sources for lidar applications. In U. Schumann (Ed.), *Atmospheric physics—Background, methods, trends, prediction* (pp. 509–527). Berlin: Springer.

Folger, K., & Weissmann, M. (2014). Height correction of atmospheric motion vectors using satellite lidar observations from CALIPSO. *Journal of Applied Meteorology Climatology, 53,* 1809–1819.

Fournier, G. (2006). Development of the Canadian aircraft meteorological data relay (AMDAR) program and plans for the future. In *10th symposium on integrated observing and assimilation systems for the atmosphere, oceans, and land surface*, Atlanta, GA, Amer. Meteor. Soc. Annual Meeting. https://ams.confex.com/ams/pdfpapers. Accessed 3 Jan 2019

França, G. B., Almeida, M. V., Bonnet, S. M., & Albuquerque Neto, F. L. (2018). Nowcasting model of low wind profile based on neural network using SODAR data at Guarulhos Airport, Brazil. *International Journal of Remote Sensing, 39*(8), 2506–2517. https://doi.org/10.1080/01431161.2018.1425562.

Free, M., & Sun, B. (2013). Time-Varying Biases in US total cloud cover data. *Journal of Atmosphere Ocean Technology, 30,* 2838–2849.

Frehlich, R. G., & Sharman, R. (2010). Climatology of velocity and temperature turbulence statistics determined from rawinsonde and ACARS/AMDAR data. *Journal of Applied Meteorology Climatology, 49,* 1149–1169. https://doi.org/10.1175/2010jamc2196.1.

Fuertes, F. C., Iungo, G. V., & Porté-Agel, F. (2014). 3D turbulence measurements using three synchronous wind lidars: Validation against sonic anemometry. *Journal of Atmosphere Oceanic Technology, 31,* 1549–1556.

Gao, F., Zhang, X., Jacobs, N. A., Huang, X.-Y., Zhang, X., & Childs, P. P. (2012). Estimation of TAMDAR observational error and assimilation experiments. *Wea. Forecasting, 27,* 856–877.

Gerz, T., Holzäpfel, F., Gerling, W., Scharnweber, A., Frech, M., Wiegele, A., et al. (2009). The wake vortex prediction and monitoring system WSVBS—Part II: Performance and ATC integration at Frankfurt airport. *Air Traffic Control Quarterly, 17*(4), 323–346. https://doi.org/10.2514/atcq.17.4.323.

Ghirardelli, J. E., & Glahn, B. (2010). The meteorological development laboratory's aviation weather prediction system. *Weather Forecasting, 25,* 1027–1051.

Gill, P. G. (2012). Objective verification of World Area Forecast Centre clear air turbulence forecasts. *Meteorological Applications, 21,* 3–11. https://doi.org/10.1002/met.1288.

Gill, P. G., & Buchanan, P. (2014). An ensemble based turbulence forecasting system. *Meteorological Applications, 21,* 12–19. https://doi.org/10.1002/met.1373.

Gillette, D. (1978). A wind tunnel simulation of the erosion of soil: Effects of soil texture, sandblasting, wind speed, and soil consolidation on the dust production. *Atmospheric Environment, 12,* 1735–1743.

Glahn, H. R., & Lowry, D. A. (1972). The use of model output statistics (MOS) in objective weather forecasting. *Journal of Applied Meteorology, 11,* 1203–1211.

Glahn, B., Schnapp, A. D., Ghirardelli, J. E., & Im, J. (2017). A LAMP–HRRR MELD for improved aviation guidance. *Weather Forecasting, 32,* 391–405.

Glickman, T. S. (Ed.). (2000). *Glossary of meteorology* (2nd ed., p. 855). Geneseo: American Meteorological Society.

Golding, B. W. (1998). Nimrod: A system for generating automated very short range forecasts. *Meteorological Applications, 5,* 1–16.

Goodman, S. J., Blakeslee, R. J., Koshak, W. J., Mach, D., Bailey, J., Buechler, D., et al. (2013). The GOES-R geostationary lightning mapper (GLM). *Atmospheric Research, 125–126,* 34–49.

Goodman, C., & Griswold, J. (2017). Climate impacts on density altitude and aviation operations. *Journal of Applied Meteorology Climatology, 57,* 517–523. https://doi.org/10.1175/jamc-d-17-0126.1.

Goodman, S. J., Gurka, J., DeMaria, M., Schmit, T. J., Mostek, A., Jedlovec, G., et al. (2012). The GOES-R proving ground: Accelerating user readiness for the next-generation geostationary environmental satellite system. *Bulletin American Meteorology Society, 93,* 1029–1040.

Gossard, E. E., Snider, J. B., Clothiaux, E. E., Martner, B., Gibson, J. S., Kropfli, R. A., et al. (1997). The potential of 8-mm radars for remotely sensing cloud drop size distributions. *Journal of Atmosphere Oceanic Technology, 14,* 76–87.

Gottschall, J., & Peinke, J. (2008). How to improve the estimation of power curves for wind turbines. *Environ. Res. Lett., 3,* 7. https://doi.org/10.1088/1748-9326/3/1/015005.

Grell, G. A., & Devenyi, D. (2002). A generalized approach to parameterizing convection combining ensemble and data assimilation techniques. *Geophysics Research Letters, 29*(14), 1693. https://doi.org/10.1029/2002gl015311.

Griffin, S. M., & Velden, C. S. (2018). Hazard avoidance products for convectively-induced turbulence in support of high-altitude global hawk aircraft missions. *Pure Applied Geophysics.* https://doi.org/10.1007/s00024-018-1772-6.

Guedalia, D., & Bergot, T. (1994). Numerical forecasting of radiation fog. Part II: A comparison of model simulation with several observed fog events. *Monthly Weather Review, 122,* 1231–1246.

Gultepe, I. (2015). Mountain weather: Observations and modeling. *Advances in Geophysics, 56,* 229–312.

Gultepe, I., Agelin-Chaab, M., Komar, J., Elfstrom, G., Boudala, F., & Zhou, B. (2019). A meteorological supersite for aviation and cold weather applications. *Pure Applied Geophysics.* https://doi.org/10.1007/s00024-018-1880-3 (**this issue**).

Gultepe, I., Fernando, H. J. S., Pardyjak, E. R., Hoch, S. W., Silver, Z., Creegan, E., et al. (2016). An overview of the MATERHORN fog project: Observations and predictability. *Pure and Applied Geophysics, 173,* 9. https://doi.org/10.1007/s00024-016-1374-0.

Gultepe, I., & Heymsfield, A. J. (2016). Ice fog, ice clouds, and remote sensing; Introduction. *Pure and Applied Geophysics, 173,* 9. https://doi.org/10.1007/s00024-016-1380-2.

Gultepe, I., Heymsfield, A. J., Field, P. R., & Axisa, D. (2017a). Ice-phase precipitation. *Meteorological Monographs, 58,* 6.1–6.36. https://doi.org/10.1175/AMSMONOGRAPHS-D-16-0013.1.

Gultepe, I., Heymsfield, A. J., Gallagher, M., Ickes, L., & Baumgardner, D. (2017b). Ice fog: The current state of knowledge and future challenges. *Meteorological Monographs, 58,* 4.1–4.24. https://doi.org/10.1175/AMSMONOGRAPHS-D-17-0002.1.

Gultepe, I., & Isaac, G. A. (2004). An analysis of cloud droplet number concentration (Nd) for climate studies: Emphasis on constant Nd. *Quarterly Journal of Royal Meterological Society, 130*(602), 2377–2390.

Gultepe, I., Isaac, G. A., Joe, P., Kucera, P., Thériault, J., & Fisico, T. (2014a). Roundhouse (RND) mountain top research site: Measurements and uncertainties for winter alpine weather conditions. *Journal of Pure and Applied Geophysics.* https://doi.org/10.1007/s00024-012-0582-5.

Gultepe, I., Kuhn, T., Pavolonis, M., Calvert, C., Gurka, J., Isaac, G. A., et al. (2014b). Ice fog in Arctic during FRAM-IF project: Aviation and nowcasting applications. *Bulletin of American Meterological Society, 95,* 211–226.

Gultepe, I., Müller, M. D., & Boybeyi, Z. (2006). A new warm fog parameterization scheme for numerical weather prediction models. *Journal of Applied Meteorology, 45,* 1469–1480.

Gultepe, I., Pawgoski, M., & Reid, J. (2007a). Using surface data to validate a satellite based fog detection scheme. *Journal of Weather and Forecasting, 22,* 444–456.

Gultepe, I., Tardif, R., Michaelides, S. C., Cermak, J., Bott, A., Bendix, J., et al. (2007b). Fog research: A review of past achievements and future perspectives. *Journal of Pure and Applied Geophysics, 164,* 1121–1159.

Gultepe, I., Pearson, G., Milbrandt, J. A., Hansen, B., Platnick, S., Taylor, P., et al. (2009). The fog remote sensing and modeling (FRAM) field project. *Bulletin of American Meteorological Society, 90,* 341–359.

Gultepe, I., & Starr, D. O. C. (1995). Dynamical structure and turbulence in cirrus clouds: Aircraft observations during FIRE. *Journal of Atmospheric Science, 52,* 4659–4682.

Gultepe, I., Starr, D. O. C., Heymsfield, A. J., Uttal, T., Ackerman, T. P., & Westphal, D. L. (1995). Dynamical characteristics of cirrus clouds from aircraft and radar observations in micro and meso-gamma scales. *Journal of Atmospheric Science, 52,* 4060–4078.

Gultepe, I., Zhou, B., Milbrandt, J., Bott, A., Li, Y., Heymsfield, A. J., et al. (2015). A review on ice fog measurements and modeling. *Atmospheric Research, 151,* 2–19.

Guttman, N. B., & Jeck, R. K. (1987). Aircraft icing environment in low ceiling conditions near Washington, D.C. *Weather Forecasting, 2,* 114–126.

Hadley, D., Hufford, G. L., & Simpson, J. J. (2004). Resuspension of relic volcanic ash and dust from katmai: Still an aviation hazard. *Weather Forecasting, 19,* 829–840.

Haggerty, J., Defer, E., De Laat, A., Bedka, K., Moisselin, J., Potts, R., et al. (2019). Detecting Clouds Associated with Jet Engine Ice Crystal Icing. *Bulletin of the American Meteorological Society, 100,* 31–40. https://doi.org/10.1175/BAMS-D-17-0252.1.

Haiden, T., Kann, A., & Pistotnik, G. (2014). Nowcasting with INCA during SNOW-V10. *Pure and Applied Geophysics, 171–1,* 171–172. https://doi.org/10.1007/s00024-012-0547-8.

Haiden, T., Kann, A., Wittmann, C., Pistotnik, G., Bica, B., & Gruber, C. (2011). The integrated nowcasting through comprehensive analysis (INCA) system and its validation over the Eastern Alpine region. *Weather Forecasting, 26,* 166–183.

Hamazu, K., Hashiguchi, H., Wakayama, T., Matsuda, T., Doviak, R. J., & Fukao, S. (2003). A 35-GHz scanning doppler radar for fog observations. *Journal of Atmosphere Oceanic Technology, 20,* 972–986.

Han, Y., Snider, J. B., Westwater, E. R., Melfi, S. H., & Ferrare, R. A. (1994). Observations of water vapor by ground-based microwave radiometers and Raman lidar. *Journal of Geophysics Research, 99*(D9), 18695–18702.

Hansen, B. (2007). A fuzzy logic-based analog forecasting system for ceiling and visibility. *Weather Forecasting, 22,* 1319–1330.

Hansen, B., Gultepe, I., & Ling, A. (2009). Update on WIND-3: An analog forecasting system for ceiling and visibility. In *Joint session of the sixth conference on artificial intelligence applications to environmental science and the 13th conference on aviation, range and aerospace meteorology; 88th annual meeting of the american meteorological society,* New Orleans, LA, 20–24 January 2008. Oral presentation.

Harrington, J. Y., Sulia, K., & Morrison, H. (2013a). A method for adaptive habit prediction in bulk microphysical models. Part I: Theoretical development. *Journal of Atmospheric Science, 70,* 349–364.

Harrington, J. Y., Sulia, K., & Morrison, H. (2013b). A method for adaptive habit prediction in bulk microphysical models. Part II: Parcel model corroboration. *Journal of Atmospheric Science, 70,* 365–376. https://doi.org/10.1175/jas-d-12-0152.1.

Hart, K. A., Steenburgh, W. J., Onton, D. J., & Siffert, A. J. (2004). An evaluation of mesoscale-model-based model output statistics (MOS) during the 2002 olympic and paralympic winter games. *Weather Forecasting, 19,* 200–218.

Haupt, S. E., & Delle Monache, L. (2014). *Understanding ensemble prediction: How probabilistic wind power prediction can help in optimizing operations.* WindTech International. http://www.windtech-international.com/editorial-features/understanding-ensemble-prediction.

Heidinger, A. K. (2010). *ABI cloud mask algorithm theoretical basis document.* NOAA/NESDIS Center for Satellite Applications and Research, p. 67.

Herman, L. (1993). High frequency satellite cloud motion at high latitudes. In *Preprints, eighth symp. on meteorological observations and instrumentation,* Anaheim, CA, Amer. Meteor. Soc., pp. 465–468.

Herzegh, P., Wiener, G., Bateman, R., Cowie, J., & Black, J. (2015). Data fusion enables better recognition of ceiling and visibility hazards in aviation. *Bulletin American Meteorology Society, 96,* 526–532.

Herzegh, P. H., Williams, E. R., Lindholm, T. A., Mosher, F. R., Kessinger, C., Sharman, R., Hawkins, J. D., & Johnson, D. B. (2002). Development of automated aviation weather products for oceanic/remote regions: Scientific and practical challenges, research strategies and first steps. In *Preprints, 10th aviation, range and aerospace meteorology conference,* AMS, Portland, OR, 13–16 May 2002.

Heymsfield, A., Baumgardner, D., DeMott, P., Forster, P., Gierens, K., & Kärcher, B. (2010). Contrail microphysics. *Bulletin American Meteorology Society, 91,* 465–472.

Heymsfield, A. J., & Sabin, R. M. (1989). Cirrus crystal nucleation by homogeneous freezing of solution drops. *Journal of Atmospheric Science, 46,* 2252–2264.

Heymsfield, A. J., & Sabin, R. M. (1993). Homogeneous ice nucleation and supercooled liquid water in orographic wave clouds. *Journal of Atmospheric Science, 50,* 2335–2353.

Heymsfield, A. J., Schmitt, C., Bansemer, A., Twohy, C., Poellot, M., Fridland, A., et al. (2005). Homogeneous ice nucleation in subtropical and tropical convection and its influence on cirrus anvil microphysics. *Journal of Atmospheric Science, 62,* 41–64.

Heymsfield, A. J., Thompson, G., Morrison, H., Bansemer, A., Rasmussen, R. M., Minnis, P., et al. (2011). Formation and spread of aircraft-induced holes in clouds. *Science, AAAS, 333,* 77–81.

Hill, M., Calhoun, R., Fernando, H., Wieser, A., Dörnbrack, A., Weissmann, M., et al. (2010). Coplanar Doppler lidar retrieval of rotors from T-REX. *Journal of Atmospheric Science, 67*(3), 713–729.

Hocking, A., & Hocking, W. K. (2018). Tornado identification and forewarning with very high frequency windprofiler radars. *Atmosphere Science Letters, 19,* e795. https://doi.org/10.1002/asl.795.

Hodges, D., & Pu, Z. (2016). The climatology, frequency, and distribution of cold season fog events in northern Utah. *Pure and Applied Geophysics.* https://doi.org/10.1007/s00024-015-1187-6.

Hodges, D., & Pu, Z. (2018). Characteristics and variations of low level jets in the contrastingprecipitation extremes of 2006 and 2007 over the Southern Great Plains. *Theoretical and Applied Climatology.* https://doi.org/10.1007/s00704-018-2492-7.

Hubbert, J. C. (2017). Differential reflectivity calibration and antenna temperature. *Journal of Atmosphere Oceanic Technology, 34,* 1885–1906.

Hubbert, J., Bringi, V. N., Carey, L. D., & Bolen, S. (1998). CSU-CHILL Polarimetric Radar Measurementsfrom a Severe Hail Storm in Eastern Colorado. *Journal of Applied Meteorology, 37,* 749–775. https://doi.org/10.1175/1520-0450(1998)037<0749:CCPRMF>2.0.CO;2.

Hubbert, J. C., Wilson, J. W., Weckwerth, T. M., Ellis, S. M., Dixon, M., & Loew, E. (2018). S-Pol's polarimetric data reveal detailed storm features (and insect behavior). *Bulletin American Meteorology Society, 99,* 2045–2060.

Hubert, L. F., & Whitney, L. F. (1971). Wind estimation from wind estimation from geostationary –satellite pictures. *Monthly Weather Review, 99,* 665–672.

Hufford, G. L., Salinas, L. S., Simpson, J. J., Barske, E. G., & Pieri, D. (2000). Operational implications of airborne volcanic ash. *Bulletin American Meteorology Society, 81,* 745–755.

Hutchison, K. D., Iisager, B. D., Kopp, T. J., & Jackson, J. M. (2008). Distinguishing aerosols from clouds in global, multispectral satellite data with automated cloud classification algorithms. *Journal of Atmosphere Oceanic Technology, 25,* 501–518.

ICAO. (2005). *Manual On Low-Level Wind Shear And Turbulence First Edition*—2005. Doc 9817, AN/449. Publisher: International Civil Aviation Organization (ICAO). DOC-09817-001-01-E-P, p. 213.

Illingworth, A. J., Barker, H. W., Beljaars, A., Ceccaldi, M., Chepfer, H., Clerbaux, N., et al. (2015). The EarthCARE Satellite: The next step forward in global measurements of clouds, aerosols, precipitation, and radiation. *Bulletin American Meteorology Society, 96,* 1311–1332.

Irvine, E. A., Shine, Keith P., & Stringer, Marc A. (2016). What are the implications of climate change for trans-Atlantic aircraft routing and flight time? *Transportation Research Part D, 47,* 44–53.

Isaac, G.A., Ayers, J. K., Bailey, M., Bissonnette, L., Bernstein, B. C., Cober, S. G., & Driedger, N., et al. (2005). *43rd AIAA aerospace sciences meeting and exhibit 10–13 January 2005,* Reno, Nevada, AIAA 2005-252, p. 18.

Isaac, G. A., Bailey, M., Boudala, F., Burrows, W., Cober, S. G., Crawford, R. T., et al. (2014). The Canadian airport nowcasting system (CAN-Now). *QJRM Meterological Applications, 21,* 30–49.

Isaac, G. A., & Schemenauer, R. S. (1979). Large particles in supercooled regions of northern Canadian cumulus clouds. *Journal of Applied Meteorology, 18,* 1056–1065.

Ismail, S., & Browell, E. V. (1994). Recent Lidar technology developments and their influence on measurements of tropospheric water vapor. *Journal of Atmosphere Oceanic Technology, 11,* 76–84.

Jacobs, A. J., & Maat, N. (2005). Numerical guidance methods for decision support in aviation meteorological forecasting. *Weather Forecasting, 20,* 82–100.

Jewell, R., & Brimelow, J. (2009). Evaluation of Alberta Hail growth model using severe hail proximity soundings from the United States. *Weather Forecasting, 24,* 1592–1609. https://doi.org/10.1175/2009WAF2222230.1.

Jonassen, M. O., Ólafsson, H., Ágústsson, H., Rögnvaldsson, Ó., & Reuder, J. (2012). Improving high-resolution numerical weather simulations by assimilating data from an unmanned aerial system. *Monthly Weather Review, 140,* 3734–3756.

Jones, R. H. (1965). Optimal estimation of initial conditions for numerical prediction. *Journal of Atmospheric Science, 22,* 658–663.

Jones, T. A., Knopfmeier, K., Wheatley, D., Creager, G., Minnis, P., & Palikonda, R. (2016). Storm-scale data assimilation and ensemble forecasting with the NSSL Experimental Warn-on Forecast. Part 2: Combined radar and satellite data experiments. *Weather Forecasting, 31,* 297–327. https://doi.org/10.1175/waf-d-15-0107.1.

Kain, J. S., et al. (2010). Assessing advances in the assimilation of radar data within a collaborative forecasting-research environment. *Weather Forecasting, 25,* 1510–1521.

Kalnay, E. (2003). *Atmospheric modeling, data assimilation and predictability* (p. 341). Cambridge: Cambridge University Press.

Kalnay, E., et al. (1996). The NCEP/NCAR 40-year reanalysis project. *Bulletin of the American Meteorological Society, 77,* 437–471.

Kann, A., Schellander-Gorgas, T., & Wittmann, C. (2015). Enhanced short-range forecasting of sub-inversion cloudiness in complex terrain. *Atmospheric Science Letters, 16,* 1–9.

Kaplan, M. L., Huffman, A. W., Lux, K. M., Cetola, J. D., Charney, J. J., Riordan, A. J., et al. (2005a). Characterizing the severe turbulence environments associated with commercial aviation accidents. Part 1: A 44-case study synoptic observational analyses. *Meteorology and Atmospheric Physics, 88,* 129–152.

Kaplan, M. L., Huffman, A. W., Lux, K. M., Cetola, J. D., Charney, J. J., Riordan, A. J., et al. (2005b). Characterizing the severe turbulence environments associated with commercial aviation accidents. Part 2: Hydrostatic mesoscale numerical simulations of super gradient wind flow and streamwise ageostrophic frontogenesis. *Meteorology and Atmospheric Physics, 88,* 153–173.

Kara, A. B., Wallcraft, A. J., Barron, C. N., Hurlburt, H. E., & Bourassa, M. A. (2008). Accuracy of 10 m winds from satellites and NWP products near land-sea boundaries. *Journal of Geophysics Research, 113,* C10020. https://doi.org/10.1029/2007jc004516.

Kara, A. B., Wallcraft, A. J., & Hurlburt, H. E. (2007). A correction for land contamination of atmospheric variables near land-sea boundaries. *Journal of Physical Oceanography, 37,* 803–818.

Karstens, C. D., Correia, J., Jr., LaDue, D. S., Wolfe, J., Meyer, T. C., Harrison, D. R., et al. (2018). Development of a human-machine mix for forecasting severe convective events. *Weather and Forecasting, 33,* 715–737.

Karstens, C. D., et al. (2015). Evaluation of a probabilistic forecasting methodology for severe convective weather in the 2014 hazardous weather testbed. *Weather Forecasting, 30,* 1551–1570.

Kelly, D. S., & Ghirardelli, J. E. (1998). A general overview of methodology and applications of the Local AWIPS MOS Program (LAMP), a short-range forecast guidance product. In *Preprints, 16th conf. on weather analysis and forecasting,* Phoenix, AZ, Amer. Meteor. Soc., pp. 437–439.

Kelsch, M., & Wharton, L. (1996). Comparing PIREPs with NAWAU turbulence and icing forecasts: issues and results. *Weather Forecasting, 11,* 385–390.

Kessinger, C., et al. (2006a). The FAA AWRP Oceanic weather program development team. In *Preprints-CD, 12th aviation, range and aerospace meteorology conference, AMS,* Atlanta, GA, 30 Jan–2 Feb 2006.

Kessinger, C., Herzegh, P., Blackburn, G., Sharman, R., Wiener, G., Hendrickson, B., & Levesque, K., et al. (2006b). The FAA AWRP oceanic weather program development team. In *Preprints, AMS 12th conf. on aviation range and aerospace.* Poster 3.9/P1.19.

Key, J. R., Santek, D., Velden, C. S., Bormann, N., Riishojgaard, J.-N. L. P., Zhu, Y., et al. (2003). Cloud drift and water vapor winds in the polar regions from MODIS. *IEEE Transactions on Geoscience and Remote Sensing, 41,* 482–492.

Khairoutdinov, M., & Kogan, Y. (2000). A new cloud physics parameterization in a large-eddy simulation model of marine stratocumulus. *Monthly Weather Review, 128,* 229–243.

Khairoutdinov, M., & Randall, D. (2006). High-resolution simulation of shallow-to-deep convection transition over land. *Journal of Atmospheric Science, 63,* 3421–3436.

Kilpinen, J. (1994). Computer-aided weather forecasting system set to enter operation in Scandinavia. *ICAO Journal, 49*(8), 17–18.

Kim, J. H., Chan, W. N., Banavar, S., & Sharman, R. D. (2015). Combined winds and turbulence prediction system for automated air-traffic management applications. *Journal of Applied Meteorology Climatology, 54,* 766–784. https://doi.org/10.1175/jamc-d-14-0216.1.

Kim, J. H., Chan, W. N., Sridhar, B., Sharman, R. D., Williams, P. D., & Strahan, M. (2016). Impact of the North Atlantic Oscillation on transatlantic flight routes and clear-air turbulence. *Journal of Applied Meteorology Climatology, 55,* 763–771. https://doi.org/10.1175/jamc-d-15-0261.1.

Kim, S.-H., Chun, H.-Y., & Chan, P. W. (2017). Comparison of turbulence indicators obtained from in situ flight data. *Journal of Applied Meteorology Climatology, 56,* 1609–1623. https://doi.org/10.1175/jamc-d-16-0291.1.

Kim, J.-H., Chun, H.-Y., Sharman, R. D., & Keller, T. L. (2011). Evaluations of upper-level turbulence diagnostics performance using the Graphical Turbulence Guidance (GTG) system and pilot reports (PIREPs) over East Asia. *Journal of Applied Meteorology Climatology, 50,* 1936–1951. **(Corrigendum, 50, 2193)**.

Knox, J. A. (1997). Possible mechanisms of clear-air turbulence in strongly anticyclonic flow. *Monthly Weather Review, 125,* 1251–1259.

Knox, J. A., McCann, D. W., & Williams, P. D. (2008). Application of the lighthill-ford theory of spontaneous imbalance to clear-air turbulence forecasting. *Journal of Atmospheric Science, 65,* 3292–3304. https://doi.org/10.1175/2008JAS2477.1.

Knox, J. A., et al. (2016). Automated turbulence forecasting strategies. In R. Sharman & T. Lane (Eds.), *Aviation turbulence: Processes, detection, prediction* (pp. 243–260). Berlin: Springer.

Knüppfer, K. (1997). Automation of aviation forecasts. The projects AUTOTAF and AUTOGAFOR. In *Preprints, seventh conf. on aviation, range, and aerospace meteorology,* Long Beach, CA, Amer. Meteor. Soc., pp. 444–449.

Kober, K., Craig, G. C., Keil, C., & Dörnbrack, A. (2012). Blending a probabilistic nowcasting method with a high-resolution numerical weather prediction ensemble for convective precipitation forecasts. *Quarterly Journal Royal Meteorological Society, 138,* 755–768.

Koch, S. E., & Caracena, F. (2002) Predicting clear-air turbulence from diagnosis of unbalanced flow. In *Preprints, 10th conf. on aviation, range, and aerospace meteorology,* Portland, OR, Amer. Meteor. Soc., 10.4. http://ams.confex.com/ams/pdfpapers/39012.pdf.

Kopeć, J. M., Kwiatkowski, K., de Haan, S., & Malinowski, S. P. (2016). Retrieving atmospheric turbulence information from regular commercial aircraft using Mode-S and ADS-B. *Atmosphere Measure Technology, 9,* 2253–2265. https://doi.org/10.5194/amt-9-2253-2016.

Kücken, M., Hauffe, D., & Österle, H. (2012). A high-resolution simulation of the year 2003 for Germany using the regional model COSMO. *Journal of Applied Meteorology Climatology, 51,* 1889–1903.

Kumjian, M. R. (2012). Freezing of raindrops in deep convective updrafts: A microphysical and polari-metric model. *Journal of Atmospheric Science, 69,* 3471–3490.

Kumjian, M. R. (2013a). Principles and applications of dual-polarization weather radar. Part I: Description of the polarimetric radar variables. *Journal of Operational Meteorology, 1,* 226–242.

Kumjian, M. R. (2013b). Principles and applications of dual-polarization weather radar. Part II: Warm- and cold-season applications. *Journal of Operational Meteorology, 1,* 243–264.

Lakshmanan, V., & Smith, T. (2009). Data mining storm attributes from spatial grids. *Journal of Atmosphere Oceanic Technology, 26,* 2353–2365.

Landolt, S., Politovich, M., Rasmussen, R., & Gaydos, A. (2010). A comparison of an automated freezing drizzle algorithm to human observations. AMS annual meeting, Atlanta, Georgia, on 16–21 Jan 2010. In *14th conference on aviation, range, and aerospace meteorology,* p. 4.

Langland, R. H., et al. (1999). The North Pacific Experiment (NORPEX-98): Targeted observations for improved North American weather forecasts. *Bulletin American Meteorology Society, 80,* 1363–1384.

Langmuir, I., Schaefer, V.J., Vonnegut, B., Maynard, K., Smith-Johannsen, R., Blanchard, D., & Falconer, R. E. (1948). *Final reports, project cirrus,* RL140. General Electric Res. Lab. Req 21190 with the Depart. Of the Army Project: 3-99-07-022, p. 119.

Lawrence, D. A., & Balsley, B. B. (2013). High-resolution atmospheric sensing of multiple atmospheric variables using the DataHawk small airborne measurement system. *Journal of Atmosphere Oceanic Technology, 30,* 2352–2366. https://doi.org/10.1175/jtech-d-12-00089.1.

Lawson, R. P., Angus, L. J., & Heymsfield, A. J. (1998). Cloud 5 particle measurements in thunderstorm anvils and possible weather threat to aviation. *Journal of Aircraft, 35,* 113–121.

Lee, D. R., Stull, R. S., & Irvine, W. S. (1984). *Clear air turbulence forecasting techniques. Air Weather Service Tech. Note* AFGWC/TN-79/001 (REV), Air Force Global Weather Central, Offutt AFB, NE, p. 16.

Leroy, D., Fontaine, E., Schwarzenboeck, A., & Strapp, J. W. (2016). Ice crystal sizes in high ice water content clouds. Part I: On the computation of median mass diameter from in situ measurements. *Journal of Atmosphere Oceanic Technology, 33,* 2461–2476.

Leroy, D., Fontaine, E., Schwarzenboeck, A., Strapp, J. W., Korolev, A., McFarquhar, G., et al. (2017). Ice crystal sizes in high ice water content clouds. Part II: statistics of mass diameter percentiles in tropical convection observed during the HAIC/HIWC project. *Journal of Atmosphere Oceanic Technology, 34,* 117–136.

Lewis, W. (1947). *A flight investigation of the meteorological conditions conducive to the formation of ice on airplanes.* NACA TN 1393.

Lin, Y. L., Farley, R. D., & Orville, H. D. (1983). Bulk parameterization of the snow field in a cloud model. *Journal of Climatology Applied Meteorology, 22,* 1065–1092.

Lin, C.-Y., Zhang, Z., Pu, Z., & Wang, F. (2017). Numerical simulations of an advection fog event over the Shanghai Pudong Airport with the WRF model. *Journal of Meteorological Research, 31,* 874–889.

Linden, P. F., & Simpson, J. E. (1985). Microbursts: A hazard to aviation. *Nature, 317*(17), 601–602.

Liu, Y., Xia, J., Shi, C. X., & Hong, Y. (2009). An improved cloud classification algorithm for China's FY-2C multi-channel images using artificial neural network. *Sensor, 9,* 5558–5579.

Löffler-Mang, M., Kunz, M., & Schmid, W. (1999). On the performance of a low-cost K-band doppler radar for quantitative rain measurements. *Journal of Atmosphere Oceanic Technology, 16,* 379–387.

Luce, H., Nakamura, T., Yamamoto, M. K., Yamamoto, M., & Fukao, S. (2010). MU Radar and Lidar Observations of Clear-Air Turbulence underneath Cirrus. *Monthly Weather Review, 138,* 438–452.

Ludlam, F. H. (1956). Fall-streak holes. *Weather, 11,* 89–90.

Lynn, K. J. (1997) *International survey of TAF automation systems. Forecasting systems,* Met Office Internal Rep., p. 20.

Mancuso, R. L., & Endlich, R. M. (1966). Clear air turbulence frequency as a function of wind shear and deformation. *Monthly Weather Review, 94,* 581–585.

Mandel, E. (1975). An early look at the development of an unmanned automated surface aviation weather observation system. *Bulletin American Meteorology Society, 56,* 979–982.

Markowicz, K. M., & Witek, M. L. (2011). Simulations of contrail optical properties and radiative forcing for various crystal shapes. *Journal of Applied Meteorology Climatology, 50,* 1740–1755.

Mason, J. G., Strapp, J. W., & Chow, P. (2006). The ice particle threat to engines in flight. In: *44th AIAA aerospace sciences meeting,* Reno, Nevada, abstract number AIAA 2006-206, 9–12 January.

Matrosov, S. Y. (2005). Attenuation-based estimates of rainfall rates aloft with vertically pointing K_a-band radars. *Journal of Atmosphere Oceanic Technology, 22,* 43–54.

Matthes, S., Schumann, U., Grewe, V., Frömming, C., Dahlmann, K., Koch, A., et al. (2012). Climate optimized air transport. In U. Schumann (Ed.), *Atmospheric physics—Background, methods, trends, prediction* (pp. 727–746). Berlin: Springer.

McCann, D. W. (2001). Gravity waves, unbalanced flow, and clear air turbulence. *National Weather Digest, 25,* 3–14.

McCann, D. W., Know, J. A., & Williams, P. D. (2012). An improvement in clear-air turbulence forecasting based on spontaneous imbalance theory: the ULTURB algorithm. *Meteorological Applications, 19,* 71–78.

McCarthy, J., & Serafin, R. J. (1984). The microburst hazard to aircraft. *Weatherwise, 37,* 120–127.

McCarthy, J., Wilson, J. W., & Fujita, T. T. (1982). The joint airport weather studies (JAWS) project. *Bulletin American Meteorology Society, 63,* 15–22.

McCarty, J., Banta, R. M., Olson, J. B., Carley, J. R., Marquis, M. C., Brewer, W. A., et al. (2017). Assessment of NWP forecast models in simulating offshore winds through the lower boundary layer by measurements from a ship-based scanning Doppler lidar. *Monthly Weather Review, 22,* 22. https://doi.org/10.1175/mwr-d-16-0442.1.

McGovern, A., Elmore, K. L., Gagne, D. J., Haupt, S. E., Karstens, C. D., Lagerquist, R., et al. (2017). Using artificial intelligence to improve real-time decision-making for high-impact weather. *Bulletin American Meteorology Society, 98,* 2073–2090.

Mead, J. B., Mcintosh, R. E., Vandemark, D., & Swift, C. T. (1989). Remote sensing of clouds and fog with a 1.4-mm radar. *Journal of Atmosphere Oceanic Technology, 6,* 1090–1097.

Meckalski, J. R., et al. (2002). *NASA advanced satellite aviation-weather products (ASAP) study report.* NASA Tech. Rep., p. 65.

Meckalski, J. R., Feltz, W. F., Murray, J. J., Johnson, D. B., Bedka, K. M., Bedka, S. T., et al. (2007). Aviation applications for satellite-based observations of cloud properties, convection initiation, in-flight icing, turbulence, and volcanic ash. *Bulletin American Meteorology Society, 88,* 1589–1607.

Megenhardt, D., Mueller, C. K., Rehak, N., & Cunning, G. (2000). Evaluation of the national convective weather forecast product. In *Preprints. 9th conf on aviation, range, and aerospace meteorology*, AMS, Orlando, FL, pp. 171–176.

Meischner, P., Baumann, R., Höller, H., & Jank, T. (2001). Eddy dissipation rates in thunderstorms estimated by Doppler Radar in relation to aircraft in situ measurements. *Journal of Atmosphere Oceanic Technology, 18,* 1609–1627.

Menzel, W. P., Schmit, T. J., Zhang, P., & Li, J. (2018). Satellite-based atmospheric infrared sounder development and applications. *Bulletin American Meteorology Society, 99,* 583–603. https://doi.org/10.1175/BAMS-D-16-0293.1.

Merritt, L. P. (1969). Comparison of airborne and ground based weather radars. *Journal of Applied Meteorology, 8,* 963–974.

Michelson, M., Shrader, W. W., & Wieler, J. G. (1990). Terminal doppler weather radar. *Microwave Journal, 33,* 139–148.

Miller, T. P., & Casadevall, T. J. (1999). Volcanic ash hazards to aviation. In H. Sigurdsson, et al. (Eds.), *Encyclopedia of volcanoes* (pp. 925–930). Oxford: Academic.

Miller, Steven D., et al. (2014). Estimating three-dimensional cloud structure via statistically blended satellite observations. *Journal of Applied Meteorology Climatology, 53,* 437–455.

Minnis, P., Bedka, S. T., Duda, D. P., Bedka, K. M., Chee, T. L., Ayers, J. K., et al. (2013). Linear contrails and contrail cirrus properties determined from satellite data. *Geophysical Research Letters, 40,* 3220–3226. https://doi.org/10.1002/grl.50569.

Minnis, P., Bedka, K., Trepte, Q., Yost, C. R., Bedka, S. T., Scarino, B., & Khlopenkov, K., et al. (2016). *A consistent long-term cloud and clear-sky radiation property dataset from the Advanced Very High Resolution Radiometer (AVHRR). Climate Algorithm* Theoretical Basis Document (C-ATBD), CDRP-ATBD-0826 Rev 1 AVHRR Cloud Properties—NASA, NOAA CDR Program, 19 September, p. 159.

Minnis, P., Nguyen, L., Palikonda, R., Heck, P. W., Spangenberg, D. A., Doelling, D. R., & Ayers, J. K., et al. (2008) Near-real time cloud retrievals from operational and research meteorological satellites. In *Proc. SPIE 7108, Remote Sens. Clouds Atmos.* XIII, Cardiff, Wales, UK, 15–18 September, p. 8. https://doi.org/10.1117/12.800344

Minnis, P., Schumann, U., Doelling, D. R., Gierens, K. M., & Fahey, D. W. (1999). Global distribution of contrail radiative forcing. *Geophysical Research Letters, 26,* 1853–1856.

Minnis, P., Sun-Mack, S., Young, D. F., Heck, P. W., Garber, D. P., Chen, Y., et al. (2011). CERES Edition-2 cloud property retrievals using TRMM VIRS and Terra and Aqua MODIS data, Part I: Algorithms. *IEEE Transactions on Geoscience and Remote Sensing, 49*(11), 4374–4400. https://doi.org/10.1109/tgrs.2011.2144601.

Minnis, P., Young, D. F., Nguyen, L., Garber, D. P., Smith, W. L., Jr., & Palikonda, R. (1998). Transformation of contrails into cirrus during SUCCESS. *Geophysical Research Letters, 25,* 1157–1160.

Mittaz, J., & Harris, A. (2011). A Physical method for the calibration of the AVHRR/3 thermal IR channels. Part II: An in-orbit comparison of the AVHRR longwave thermal IR channels on board MetOp-A with IASI. *Journal of Atmosphere Oceanic Technology, 28,* 1072–1087. https://doi.org/10.1175/2011JTECHA1517.1.

Moninger, W. R., Benjamin, S. G., Jamison, B. D., Schlatter, T. W., Smith, T. L., & Szoke, E. J. (2010). Evaluation of regional aircraft observations using TAMDAR. *Weather Forecasting, 25,* 627–645.

Moninger, W. R., Mamrosh, R. D., & Pauley, P. M. (2003). Automated meteorological reports from commercial aircraft. *Bulletin American Meteorology Society, 84,* 203–216.

Moosakhanian, A., Schmidt, S., Dash, E. R., Daniels, T.vS., & Stough, P. (2006). FAA–NASA collaboration on automated aircraft weather observations—culminating in TAMDAR. In *12th conference on aviation range and aerospace meteorology,* Atlanta, GA, Amer. Meteor. Soc. Annual Meeting. https://ams.confex.com/ams/pdfpapers/104168.pdf.

Morrison, H., & Milbrandt, J. (2015). Parameterization of cloud microphysics based on the prediction of bulk ice particle properties. Part I: Scheme description and idealized tests. *Journal of Atmospheric Science, 72,* 287–311.

Motta, M., Barthelmie, R. J., & Vølund, P. (2005). The influence of non-logarithmic wind speed profiles on potential power output at danish offshore sites. *Wind Energy, 8,* 219–236. https://doi.org/10.1002/(issn)1099-1824.

Mueller, C., Saxen, T., Roberts, R., Wilson, J., Betancourt, T., Dettling, S., et al. (2003). NCAR auto-nowcast system. *Weather Forecasting, 18,* 545–561.

Mueller, C. K., Wilson, J. W., & Crook, N. A. (1993). The utility of sounding and mesonet data to nowcast thunderstorm initiation. *Weather Forecasting, 8,* 132–146.

Myhre, G., & Stordal, F. (2001). Global sensitivity experiments of the radiative forcing due to mineral aerosols. *Journal of Geophysics Research, 106,* 18193–18204.

Nair, U. S., Weger, R. C., Kuo, K. S., & Welch, R. M. (1998). Clustering, randomness, and regularity in-cloud fields. The nature of regular cumulus cloud fields. *Journal of Geophysics Research, 103,* 11363–11380.

Nance, L. B., & Durran, D. R. (1997). A modeling study of non-stationary trapped mountain lee waves. Part I: Mean-flow variability. *Journal of Atmospheric Science, 54,* 2275–2291.

Neely, R. R., & Thayer, J. P. (2011). Raman lidar profiling of tropospheric water vapor over Kangerlussuaq, Greenland. *Journal of Atmosphere Oceanic Technology, 28,* 1141–1148.

Newton, D. W. (1978). An integrated approach to the problem of aircraft icing. *Journal of Aircraft, 15,* 374–381.

Nicholls, M., Pielke, R., & Meroney, R. (1993). Large eddy simulation of microburst winds flowing around a building. *Journal of Wind Engineering and Industrial Aerodynamics, 46–47,* 229–237.

NOAA, 2012: *NWSChat Live User Manual,* p. 12.

NOAA NWS. (2016). *Terminal aerodrome forecasts.* Instruction 10-813. http://www.nws.noaa.gov/directives/.

Noh, Y., Seaman, C. J., Vonder Haar, T. H., & Liu, G. (2013). In situ aircraft measurements of the vertical distribution of liquid and ice water content in midlatitude mixed-phase clouds. *Journal of Applied Meteorology Climatology, 52,* 269–279.

NTSB. (1996). *Aircraft accident report.* Vol. 1. National Transportation Safety Board NTSB/AAR–96/01–PB96–910401, 322 pp. [Available from NTSB, 490 L'Enfant Plaza, S.W., Washington, DC 20594.].

NTSB (2010). *NASDAC Review of National Transportation Safety Board (NTSB) weather-related accidents (2003–2007).* http://www.asias.faagov/.

NWPSD (2004). *Operations and services, aviation weather services, NWSPD,* 10-8, 2004, NWS Instruction 10-813,Terminal Aerodrome Forecasts, Feb. 1, 2004, p. 51.

Ødegaard, V. (1997). Ice phase parameterization in a numerical weather prediction model. *Weather Forecasting, 12,* 127–139.

Orf, L. G., & Anderson, J. R. (1999). A numerical study of traveling microbursts. *Monthly Weather Review, 127,* 1244–1257.

Orf, L. G., Anderson, J. R., & Straka, J. M. (1996). A three-dimensional numerical analysis of colliding microburst outflow dynamics. *Journal of Atmospheric Science, 53,* 2490–2511.

Orville, R. E. (2008). Development of the national lightning detection network. *Bulletin American Meteorology Society, 89,* 180–190.

Oude Nijhuis, A., Thobois, L., Barbaresco, F., De Haan, S., Dolfi-Bouteyre, A., Kovalev, D., et al. (2018). Wind hazard and turbulence monitoring at airports with lidar, radar and Mode-S downlinks: The UFO Project. *Bulletin American Meteorology Society.* https://doi.org/10.1175/BAMS-D-15-00295.1.

Pasini, A., & Marzban, C. (2008). *Artificial intelligence methods in the environmental sciences* (p. 424). Berlin: Springer.

Pavolonis, M. J. (2010a). Advances in extracting cloud composition information from space borne infrared radiances: A robust alternative to brightness temperatures. Part I: Theory. *Journal of Applied Meteorology Climatology, 49,* 1992–2012.

Pavolonis, M. J. (2010b). *ABI cloud type/phase algorithm theoretical basis document.* NOAA/NESDIS/Center for Satellite Applications and Research (STAR), p. 60.

Pavolonis, M. J., Feltz, W. F., Heidinger, A. K., & Gallina, G. M. (2006). A daytime complement to the reverse absorption technique for improved automated detection of volcanic ash. *Atmosphere Oceanic Technology, 23,* 1422–1444.

Pavolonis, M. J., & Heidinger, A. K. (2004). Daytime cloud overlap detection from AVHRR and VIIRS. *Journal of Applied Meteorology, 43,* 762–778.

Pavolonis, M. J., Heidinger, A. K., & Uttal, T. (2005). Daytime global cloud typing from AVHRR and VIIRS: Algorithm description, validation, and comparisons. *Journal of Atmosphere Oceanic Technology, 44,* 804–826.

Pavolonis, M. J., Sieglaff, J., & Cintineo, J. (2018). Automated detection of explosive volcanic eruptions using satellite-derived cloud vertical growth rates. *Earth and Space Science, 5,* 903–928. https://doi.org/10.1029/2018EA000410.

Pearson, J. M., & Sharman, R. D. (2017). Prediction of energy dissipation rates for aviation turbulence. Part II: Nowcasting convective and nonconvective turbulence. *Journal of Applied Meteorology Climatology, 56,* 339–351. https://doi.org/10.1175/JAMC-D-16-0312.1.

Perrie, W., Zhang, W., Bourassa, M., Shen, H., & Vachon, P. W. (2008). Impact of satellite winds on marine wind simulations. *Weather Forecasting, 23,* 290–303.

Peters, G. (1990). Temperature and wind profiles from radar wind profilers equipped with acoustic sources. *Meteorological Rundsch, 42,* 152–154.

Peters, G., Hasselmann, D., & Pang, S. (1988). Radio acoustic sounding using an FM CW radar. *Radio Science, 23,* 640–646.

Pinto, J. O., Grim, J. A., & Steiner, M. (2015). Assessment of the high-resolution rapid refresh model's ability to predict mesoscale convective systems using object-based evaluation. *Weather Forecasting, 30,* 892–913.

Pithani, P., Ghude, S., Naidu, C. V., Kulkarni, R. G., Steeneveld, G., Sharma, A., et al. (2018). WRF model Prediction of a dense fog event occurred during WInter Fog EXperiment (WIFEX). *Pure and Applied Geophysics.* https://doi.org/10.1007/s00024-018-2053-0.

Pleim, J. E., & Xiu, A. (1995). Development and testing of a surface flux and planetary boundary layer model for application in mesoscale models. *Journal of Applied Meteorology, 34,* 16–31.

Politovich, M. K. (1989). Aircraft icing caused by large super-cooled droplets. *Journal of Applied Meteorology, 28,* 856–868.

Politovich, M. K. (1996). Response of a research aircraft to icing and evaluation of severity indices. *Journal of Aircraft, 33,* 291–297.

Prata, A. J. (1989). Observations of volcanic ash clouds in the 10-12 micrometer window using AVHRR/2 data. *International Journal of Remote Sensing, 10,* 751–761.

Proctor, F. H. (1988). Numerical simulations of an isolated microburst. Part I: Dynamics and structure. *Journal of Atmospheric Science, 45,* 3137–3160.

Proctor, F. H. (1989). Numerical simulations of an isolated microburst. Part II: Sensitivity experiments. *Journal of Atmospheric Science, 46,* 2143–2165.

Protat, A., Rauniyar, S., Kumar, V. V., & Strapp, J. W. (2014). Optimizing the probability of flying in high ice water content conditions in the tropics using a regional-scale climatology of convective cell properties. *Journal of Applied Meteorology Climatology, 53,* 2438–2456.

Pu, Z. (2017). Surface data assimilation and near-surface weather prediction over complex terrain. Book Chapter. In S. K. Park & L. Xu (Eds.), *Data assimilation for atmospheric, oceanic and hydrologic applications* (Vol. III, pp. 219–240). Berlin: Springer. https://doi.org/10.1007/978-3-319-43415-5_10

Pu, Z., Chachere, C., Hoch, S., Pardyjak, E., & Gultepe, I. (2016). Numerical prediction of cold season fog events over complex terrain: The performance of the WRF model during MATERHORN-fog and early evaluation. *Pure and Applied Geophysics, 22,* 22. https://doi.org/10.1007/s00024-016-1375-z.

Pu, Z., Lin, C., Dong, X., & Krueger, S. (2018). Sensitivity of numerical simulations of a mesoscale convective system to ice hydrometeors in bulk microphysical parameterization. *Pure and Applied Geophysics*. https://doi.org/10.1007/s00024-018-1787-z.

Puempel, H., & Williams, P. D. (2016). *The impacts of climate change on aviation: Scientific challenges and adaptation pathways*. ICAO Environmental Report, pp. 205–207. https://www.icao.int/environmental-protection/Pages/env2016.aspx. Accessed 3 Jan 2019

Ralph, F. M., Intrieri, J., Andra, D., Atlas, R., Boukabara, S., Bright, D., et al. (2013). The emergence of weather-related test beds linking research and forecasting operations. *Bulletin American Meteorology Society, 94,* 1187–1211.

Ramsay, A. C. (1999). A multi-sensor freezing drizzle algorithm for the automated surface observing system. In *Preprints, 15th Int. conf. on interactive information and processing systems for meteorology, oceanography, and hydrology,* Dallas, TX, Amer. Meteor. Soc., pp. 193–196.

Rasmussen, R., Baker, B., Kochendorfer, J., Meyers, T., Landolt, S., Fischer, A. P., et al. (2012). How well are we measuring snow: The NOAA/FAA/NCAR winter precipitation test bed. *Bulletin American Meteorology Society, 93,* 811–829.

Rasmussen, R., Dixon, M., Hage, F., Cole, J., Wade, C., Tuttle, J., et al. (2001). Weather support to deicing decision making (WSDDM): A winter weather nowcasting system. *Bulletin American Meteorology Society, 82,* 579–596.

Rasmussen, R., Dixon, M., Vasiloff, S., Hage, F., Knight, S., Vivekanandan, J., et al. (2003). Snow nowcasting using a real-time correlation of radar reflectivity with snow gauge accumulation. *Journal of Applied Meteorology, 42,* 20–36. https://doi.org/10.1175/1520-0450(2003)042%3c0020:snuart%3e2.0.co;2.

Rasmussen, R., Politovich, M., Marwitz, J., Sand, W., McGinley, J., Smart, J., et al. (1992). Winter icing and storms project (WISP). *Bulletin American Meteorology Society, 73,* 951–974.

Raytheon. (2016). *AWIPS CAVE-D2D user's manual: AWIPS II operational build 13.4.1. Raytheon Doc.* AWP.MAN.UM.A2-OB13.4.1, p. 609. www.unidata.ucar.edu/software/awips2/doc/CAVE_D2D_Manual.pdf.

Reehorst, A. L., Brinker, D. J., Ratvasky, T. P., Ryerson, C. C., & Koenig, G. G. (2005). *The NASA icing remote sensing system. NASA/TM*—2005-213591. Paper # 80776, p. 11.

Reineman, B. D., Lenain, L., & Melville, W. K. (2016). The use of ship-launched fixed-wing UAVs for measuring the marine atmospheric boundary layer and ocean surface processes. *Journal of Atmosphere Oceanic Technology, 33,* 2029–2052.

Reineman, B. D., Lenain, L., Statom, N. M., & Melville, W. K. (2013). Development and testing of instrumentation for UAV-based flux measurements within terrestrial and marine atmospheric boundary layers. *Journal of Atmosphere Oceanic Technology, 30,* 1295–1319.

Reitebuch, O. (2012). Wind lidar for atmospheric research. Atmospheric physics. In U. Schumann (Ed.), *Background, methods, trends, prediction* (pp. 487–507). Berlin: Springer.

Roberts, R. D., & Rutledge, S. (2003). Nowcasting storm initiation and growth using GOES-8 and WSR-88D data. *Weather Forecasting, 18,* 562–584.

Roberts, R. D., Saxen, T., Mueller, C., Wilson, J., Crook, A., Sun, J., & Henry, S. (1999). Operational application and use of NCAR's thunderstorm nowcasting system. In *Preprints, Int. conf. on interactive information and processing systems*, Dallas, TX, Amer. Meteor. Soc., pp. 158–161.

Roquelaure, S., & Bergot, T. (2007). Seasonal sensitivity on COBEL-ISBA local forecast system for fog and low clouds. *Pure and Applied Geophysics, 164,* 1283–1301.

Roquelaure, S., & Bergot, T. (2008). A local ensemble prediction system for fog and low clouds: Construction, bayesian model averaging calibration, and validation. *Journal of Applied Meteorology Climatology, 47,* 3072–3088.

Roquelaure, S., & Bergot, T. (2009). Contributions from a local ensemble prediction system (LEPS) for improving fog and low cloud forecasts at airports. *Weather Forecasting, 24,* 39–52.

Rose, S. F., Hobbs, P. V., Locatelli, J. D., & Stoelinga, M. T. (2004). A 10-yr climatology relating the locations of reported tornadoes to the quadrants of upper-level jet streaks. *Weather Forecasting, 19,* 301–309.

Rudack, D. E., & Ghirardelli, J. E. (2010). A comparative verification of localized aviation model output statistics program (LAMP) and numerical weather prediction (NWP) model forecasts of ceiling height and visibility. *Weather Forecasting, 25,* 1161–1178.

Rudra, R., Dickinson, W. T., Ahmed, S. I., Patel, P., Zhou, J., & Gharabaghi, B. (2015). Changes in rainfall extremes in Ontario. *International Journal of Environment Research, 9*(4), 1117–1372.

Ryerson, W. R., & Hacker, J. P. (2014). The potential for mesoscale visibility predictions with a multimodel ensemble. *Weather Forecasting, 29,* 543–562.

Ryzhkov, A. V., Kumjian, M. R., Ganson, S. M., & Zhang, P. (2013). Polarimetric radar characteristics of melting hail. Part II: Practical implications. *Journal of Applied Meteorology Climatology, 52,* 2871–2886.

Ryzhkov, A. V., Zrnic, D. S., Hubbert, J. C., Bringi, V. N., Vivekanandan, J., & Brandes, E. A. (2002). Polarimetric radar observations and interpretation of co-cross-polar correlation coefficients. *Journal of Atmosphere Oceanic Technology, 19,* 340–354.

Sand, W. R., Cooper, W. A., Politovich, M. K., & Veal, D. L. (1984). Icing Conditions Encountered by a Research Aircraft. *Journal of Climate and Applied Meteorology., 23,* 1427–1440. https://doi.org/10.1175/0733-3021-23.10.1427.

Santel, D. (2010). The impact of satellite-derived polar winds on lower-latitude forecasts. *Monthly Weather Review, 138,* 123–139.

Sathe, A., Mann, J., Gottschall, J., & Courtney, M. S. (2011). Can wind lidars measure turbulence? *Journal of Atmosphere Oceanic Technology, 28,* 853–868.

Schmit, T. J., Griffith, P., Gunshor, M. M., Daniels, J. M., Goodman, S. J., & Lebair, W. J. (2017). A closer look at the ABI on the GOES-R series. *Bulletin American Meteorology Society, 98,* 681–698.

Schmit, T. J., Gunshor, M. M., Menzel, W. P., Li, J., Bachmeier, S., & Gurka, J. J. (2005). Introducing the next-generation advanced baseline imager on GOES-R. *Bulletin American Meteorology Society, 86,* 1079–1096.

Schreiner, A. J., Unger, D. A., Menzel, W. P., Ellrod, G. P., Strabala, K. I., & Pellett, J. L. (1993). A comparison of ground and satellite observations of cloud cover. *Bulletin American Meteorology Society, 74,* 1851–1861.

Schultz, P., & Politovich, M. K. (1992). Toward the improvement of aircraft-icing forecasts for the continental United States. *Weather Forecasting, 7,* 491–500.

Schumann, U., Graf, K., Mannstein, H., & Mayer, B. (2012a). Contrails: Visible aviation induced climate impact. In U. Schumann (Ed.), *Atmospheric physics—Background, methods, trends, prediction* (pp. 239–257). Berlin: Springer.

Schumann, U., Mayer, B., Graf, K., & Mannstein, H. (2012b). A parametric radiative forcing model for contrail cirrus. *Journal of Applied Meteorology Climatology, 51,* 1391–1406.

Schumann, U., & Heymsfield, A. J. (2017). On the life cycle of individual contrails and contrail cirrus. *Meteorological Monographs, 58,* 3.1–3.24.

Schumann, U., & Mayer, B. (2017). Sensitivity of surface temperature to radiative forcing by contrail cirrus in a radiative-mixing model. *Atmospheric Chemistry Physics, 17,* 13833–13848.

Schuur, T. J., Park, H., Ryzhkov, A. V., & Reeves, H. D. (2012). Classification of precipitation types during transitional winter weather using the RUC model and polarimetric radar retrievals. *Journal of Applied Meteorology Climatology, 51,* 763–779.

Schwartz, B. (1996). The quantitative use of PIREPs in developing aviation weather guidance products. *Weather Forecasting, 11,* 372–384. https://doi.org/10.1175/1520-434(1996)011,0372:tquopi.2.0.co;2.

Schwartz, B. E., & Benjamin, S. G. (1995). A comparison of temperature and wind measurements from ACARS equipped aircraft and rawinsondes. *Weather Forecasting, 10,* 528–544.

Seity, Y., Brousseau, P., Malardel, S., Hello, G., Benard, P., Bouttier, F., et al. (2011). The AROME-France convective-scale operational model. *Monthly Weather Review, 139,* 976–991.

Selz, T., & Craig, G. C. (2015). Upscale Error growth in a high-resolution simulation of a summertime weather event over Europe. *Monthly Weather Review, 143,* 813–827.

Serke, D., Hall, E., Bogna, J., Jordan, A., Abdo, S., Baker, K., et al. (2014). Supercooled liquid water content profiling case studies with a new vibrating wire sonde compared to a ground-based microwave radiometer. *Atmospheric Research, 149,* 77–87.

Serke, D., Politovich, M., Reehorst, A., & Gaydos, A. (2008). The use of X-band radar to support the detection of in-flight icing hazards by the NASA Icing Remote Sensing System During AIRS-II. In *Proc. SPIE 7088, #18*, 2008.

Sharman, R. (2016). Nature of aviation turbulence. In R. Sharman & T. Lane (Eds.), *Aviation turbulence: Processes, detection, prediction* (pp. 3–30). Berlin: Springer.

Sharman, R., Cornman, L. B., Meymaris, G., Pearson, J., & Farrar, T. (2014). Description and derived climatologies of automated in situ eddy dissipation rate reports of atmospheric turbulence. *Journal of Applied Meteorology Climatology, 53,* 1416–1432. https://doi.org/10.1175/jamc-d-13-0329.1.

Sharman, R., & Lane, T. (2016). *Aviation turbulence: Processes, detection, prediction* (p. 523). Berlin: Springer.

Sharman, R., & Pearson, J. M. (2017). Prediction of energy dissipation rates for aviation turbulence. Part I: Forecasting non-convective turbulence. *Journal of Applied Meteorology Climatology, 56,* 317–337.

Sharman, R., Tebaldi, C., Wiener, G., & Wolff, J. (2006). An integrated approach to mid- and upper-level turbulence forecasting. *Weather Forecasting, 21,* 268–287.

Sharman, R. D., & Trier, S. B. (2018). Influences of gravity waves on convectively induced turbulence (CIT): A review. *Pure and Applied Geophysics*. https://doi.org/10.1007/s00024-018-1849-2.

Sharman, R., Trier, S. B., Lane, T. P., & Doyle, J. D. (2012). Sources and dynamics of turbulence in the upper troposphere and lower stratosphere: A review. *Geophysical Research Letters, 39,* L12803. https://doi.org/10.1029/2012gl051996.

Sieglaff, J. M., Cronce, L. M., Feltz, W. F., Bedka, K. M., Pavolonis, M. J., & Heidinger, A. K. (2011). Nowcasting convective storm initiation using satellite-based box-averaged cloud-top cooling and cloud-type trends. *Journal of Applied Meteorology Climatology, 50,* 110–126.

Sillmann, J., Kharin, V. V., Zhang, X., Zwiers, F. V., & Bronaugh, D. (2013a). Climate extremes indices in the CMIP5 multimodel ensemble. Part 1: Model evaluation in the present climate. *Journal of Geophysics Research Atmosphere, 118,* 1716–1733.

Sillmann, J., Kharin, V. V., Zwiers, F. W., Zhang, X., & Bronaugh, D. (2013b). Climate extreme indices in the CMIP5 multi-model ensemble. Part 2: Future projections. *Journal of Geophysics Researc, 22,* 22. https://doi.org/10.1002/jgrd.50188.

Silva, W. L., Albuquerque Neto, F. A., França, G. B., & Matschinske, M. (2016). Conceptual model for runway change procedure in guarulhos international airport based on SODAR data. *The Aeronautical Journal, 120*(1227), 725–734. https://doi.org/10.1017/aer.2016.33.

Simpson, J. J., Hufford, G. L., Pieri, D. C., & Berg, J. (2000). Failures in detecting volcanic ash from satellite-based technique. *Remote Sensing of Environment, 72,* 191–217.

Simpson, J. J., Hufford, G. L., Servranckx, R., Berg, J., & Pieri, D. (2003). Airborne Asian dust: Case study of long-range transport and implications for the detection of volcanic ash. *Weather Forecasting, 18,* 121–141.

Sims, D. L., Fidalgo, C. B., & Carty, T. C. (2000). Integrated Icing Diagnostic Algorithm assessment at regional airlines. In *9th conference on aviation, range, and aerospace meteorology*. AMS 2000 Annual Meeting. Preprints, #4.16.

Skofronick-Jackson, G., Petersen, W. A., Berg, W., Kidd, C., Stocker, E. F., Kirschbaum, D. B., et al. (2017). The global precipitation measurement (GPM) mission for science and society. *Bulletin American Meteorology Society, 98,* 1679–1695.

Smith, W. L. (2014). *4-D cloud properties from passive satellite data and applications to resolve the flight icing threat to aircraft.* PhD Dissertation, University of Wisconsin-Madison, July 22, p. 15.

Smith, T. M., Lakshmanan, V., Stumpf, G. J., Ortega, K. L., Hondl, K., Cooper, K., et al. (2016). Multi-radar multi-sensor (MRMS) severe weather and aviation products: initial operating capabilities. *Bulletin American Meteorology Society, 97,* 1617–1630.

Smith, W. L., Minnis, P., Fleeger, C., Spangenberg, D., Palikonda, R., & Nguyen, L. (2012). Determining the flight icing threat to aircraft with single-layer cloud parameters derived from operational satellite data. *Journal of Applied Meteorology Climatology, 51,* 1794–1810.

Soden, B. J., Ackerman, S. A., Starr, D. O. C., Melfi, S. H., & Ferrare, R. A. (1994). Comparison of upper tropospheric water vapor from GOES, Raman lidar, and cross-chain loran atmospheric sounding system measurements. *Journal of Geophysics Research, 99*(D10), 21005–21016.

Sokol, Z., & Zacharov, P. (2012). Nowcasting of precipitation by an NWP model using assimilation of extrapolated radar reflectivity. *Quarterly Journal of Royal Meteorological Society, 138,* 1072–1082.

Solheim, F., Godwin, J., & Ware, R. (1998). Passive ground-based remote sensing of atmospheric temperature, water vapor, and cloud liquid water profiles by a frequency synthesized microwave radiometer. *Meterologische Zeitschrift, 7,* 370–376.

Sorenson, J. E. (1964). *Synoptic patterns for clear air turbulence.* UAL Meteorology Circular 56, 64 pp. Dept. of Atmospheric Science, Colorado State University, Fort Collins, CO 80523.

Spangenberg, D. A., Minnis, P., Bedka, S. T., Palikonda, R., Duda, D. P., & Rose, F. G. (2013). Contrail radiative forcing over the Northern Hemisphere from 2006 Aqua MODIS data. *Geophysics Research Letters, 40,* 595–600. https://doi.org/10.1002/grl.50168.

Stano, G. T., Fuell, K. K., & Jedlovec, G. J. (2010). NASA SPoRT GOES-R Proving Ground activities. In *Preprints, Sixth annual symp. on future national operational environmental satellite systems: NPOESS and GOES-R,* Atlanta, GA, Amer. Meteor. Soc., 8.2. https://ams.confex.com/ams/90annual/techprogram/. Accessed 3 Jan 2019

Stickland, J. J. (1998). *An assessment of two algorithms for automatic measurement and reporting of turbulence from commercial public transport aircraft* (p. 42). Melbourne, Australia: Rep. to the ICAO METLINK Study Group, Bureau of Meteorology.

Stobie, J., Moosakhanian, A., Jackson, P., & Brown, W. N. (2008). Evolution of FAA's weather and radar processor (WARP) into the next generation air transportation system (NextGen). In *13th conference on aviation, range and aerospace meteorology,* 21–24 January 2008, New Orleans, Amer. Met. Soc.

Stoelinga, M. T., & Warner, T. T. (1999). Nonhydrostatic, mesobeta scale model simulations of cloud ceiling and visibility for an east coast winter precipitation event. *Journal of Applied Meteorology, 38,* 385–404.

Storer, L. N., Williams, P. D., & Joshi, M. M. (2017). Global response of clear-air turbulence to climate change. *Geophysical Research Letters, 44,* 9976–9984. https://doi.org/10.1002/2017GL074618.

Strapp, J. W., Korolev, A., Ratvasky, R., Potts, A., Protat, P., May, A., & Ackerman, A., et al. (2016). *The high ice water content study of deep convective clouds: Report on science and technical plan.* Final Report. DOT/FAA/TC-14/31, p. 92. http://www.tc.faa.gov/its/worldpac/techrpt/tc14-31.pdf. Accessed 3 Jan 2019

Sumner, J., & Mason, C. (2006). A turbulence-based model for resolving velocity and temperature profiles in the atmospheric surface layer. *Wind Energy, 30,* 317–340. https://doi.org/10.1260/030952406779295462.

Sun, J., Xue, M., Wilson, J. W., Zawadzki, I., Ballard, S. P., Onvlee-Hooimeyer, J., et al. (2014). Use of NWP for nowcasting convective precipitation: recent progress and challenges. *Bulletin American Meteorology Society, 95,* 409–426.

Sun-Mack, S., Minnis, P., Smith, W. L., Hong, G., & Chen, Y. (2017). Detection of single and multilayer clouds in an artificial neural network approach. In *Proc. SPIE 10424, remote sensing of clouds and the atmosphere* XXII, 1042408 (13 October 2017); https://doi.org/10.1117/12.2277397.

Tafferner, A., Hauf, T., Leifeld, C., Hafner, T., Leykauf, H., & Voigt, U. (2003). ADWICE: advanced diagnosis and warning system for aircraft icing environments. *Weather Forecasting, 18,* 184–203.

Tag, P. M., Bankert, R. L., & Brody, L. R. (2000). An AVHRR multiple cloud-type classification package. *Journal of Applied Meteorological, 39,* 125–134.

Tan, I., & Storelvmo, T. (2016). Sensitivity study on the influence of cloud microphysical parameters on mixed-phase cloud thermodynamic phase partitioning in CAM5. *Journal of Atmospheric Science, 73,* 709–728.

Tardif, R., & Rasmussen, R. M. (2007). Event-based climatology and typology of fog in the New York City region. *Journal of Applied Meteorology Climatology, 46,* 1141–1168.

TC. (2004). *Small and large aircraft; Aircraft critical surface contamination training for aircrew and ground crew.* TP 10643E, 12/2004. Ottawa, Ont., Canada, p. 138.

Teixeira, J., et al. (2008). Parameterization of the atmospheric boundary layer: A view from just above the inversion. *Bulletin American Meteorology Society, 89,* 453–458.

Thobois, L, Cariou, J. P., & Gultepe, I. (2018). Review of lidar based applications for aviation weather. *Pure and Applied Geophysics.* https://doi.org/10.1007/s00024-018-2058-8

Thompson, G., Bruintjes, R. T., Brown, B. G., & Hage, F. (1997). Intercomparison of in-flight icing algorithms. Part I: WISP94 real-time icing prediction and evaluation program. *Weather Forecasting, 12,* 878–889.

Thompson, G., Politovich, M. K., & Rasmussen, R. M. (2017). A numerical weather model's ability to predict characteristics of aircraft icing environments. *Weather Forecasting, 32,* 207–221.

Tomita, H. (2008). New microphysical schemes with five and six categories by diagnostic generation of cloud ice. Journal of the Meteorological Society of Japan. In: *Ser. II, Special Issue: The International Workshop on High-Resolution and Cloud Modeling*, 2006, 86A, pp. 121–142.

Trier, S. B., Sharman, R. D., & Lane, T. P. (2012). Influences of moist convection on a cold-season outbreak of clear-air turbulence (CAT). *Monthly Weather Review, 140,* 2477–2496.

Tucker, S. C., Senff, C. J., Weickmann, A. M., Brewer, W. A., Banta, R. M., Sandberg, S. P., et al. (2009). Doppler Lidar estimation of mixing height using turbulence, shear, and aerosol profiles. *Journal of Atmosphere Oceanic Technology, 26,* 673–688.

Turcotte, M.-F., & Verret, R. (1999). In-flight icing and turbulence forecasts for aviation. In *Proc. Sixth Workshop on Operational Meteorology*, Halifax, NS, Canada, Canadian Meteorological and Oceanographic Society, pp. 53–56.

Turner, D. D., Clough, S. A., Liljegren, J. C., Clothiaux, E. E., Cady-Pereira, K., & Gaustad, K. L. (2007). Retrieving liquid water path and precipitable water vapor from Atmospheric Radiation Measurement (ARM) microwave radiometers. *IEEE Transactions on Geoscience and Remote Sensing, 45,* 3680–3690. https://doi.org/10.1109/TGRS.2007.903703.

Turner, J., & Warren, D. E. (1989). Cloud track winds in the polar regions from sequences of AVHRR images. *International Journal of Remote Sensing, 10,* 695–703.

Turp, D., & Gill, P. (2008). Developments in numerical clear air turbulence forecasting at the U.K. Met Office. In *Preprints, 13th conf. on aviation, range and aerospace meteorology,* New Orleans, LA, Amer. Meteor. Soc., P3.10A. http://ams.confex.com/ams/pdfpapers/131782.pdf. Accessed 3 Jan 2019

Uccellini, L. W. (1980). On the role of upper tropospheric jet streaks and leeside cyclogenesis in the development of low-level jets in the great plains. *Monthly Weather Review, 108,* 1689–1696.

Uccellini, L. W., & Johnson, D. R. (1979). The coupling of upper and lower tropospheric jet streaks and implications for the development of severe convective storms. *Monthly Weather Review, 107,* 682–703.

Van den Berg, G. P. (2008). Wind turbine power and sound in relation to atmospheric stability. *Wind Energy, 11*(2), 151–169. https://doi.org/10.1002/we.240.

Van Den Broeke, M. S. (2016). Polarimetric variability of classic supercell storms as a function of environment. *Journal of Applied Meteorology Climatology, 55,* 1907–1925.

Van Den Broeke, M. S., Tobin, D. M., & Kumjian, M. R. (2016). Polarimetric radar observations of precipitation type and rate from the 2–3 March 2014 winter storm in Oklahoma and Arkansas. *Weather and Forecasting, 31,* 1179–1196.

Velden, C., Daniels, J., Stettner, D., Santek, D., Key, J., Dunion, J., et al. (2005). Recent innovations in deriving tropospheric winds from meteorological satellites. *Bulletin American Meteorology Society, 86,* 205–224.

Velden, C. S., Hayden, C. M., Nieman, S., Menzel, W. P., Wanzong, S., & Goerss, J. (1997). Upper-tropospheric winds derived from geostationary satellite water vapor observations. *Bulletin American Meteorology Society, 78,* 173–195.

Velden, C. S., Olander, T. L., & Wanzong, S. (1998). The impact of multispectral GOES-8 wind information on Atlantic tropical cyclone track forecasts in 1995. Part I: Dataset methodology, description, and case analysis. *Monthly Weather Review, 126,* 1202–1218.

Verlinden, K. L., & Bright, D. R. (2017). Using the second-generation GEFS reforecasts to predict ceiling, visibility, and aviation flight category. *Weather Forecasting, 32,* 1765–1780.

Vislocky, R. L., & Fritsch, J. M. (1995). Generalized additive models versus linear regression in generating probabilistic mos forecasts of aviation weather parameters. *Weather Forecasting, 10,* 669–680.

Vislocky, R. L., & Fritsch, J. M. (1997). An automated, observations-based system for short-term prediction of ceiling and visibility. *Weather Forecasting, 12,* 31–43.

Vivekanandan, J., Zhang, G., & Politovich, M. K. (2001). An assessment of droplet size and liquid water content derived from dual-wavelength radar measurements to the application of aircraft icing detection. *Journal of Atmosphere Oceanic Technology, 18,* 1787–1798.

Vrancken, P. S. (2016). Airborne remote detection of turbulence with forward-pointing LIDAR. In R. D. Sharman & T. P. Lane (Eds.), *Aviation turbulence: Processes, detection, and prediction* (pp. 443–464). Berlin: Springer.

Wade, C. G. (2003). A Multisensor Approach to Detecting Drizzle on ASOS. *Journal of Atmospheric and Oceanic Technology, 20,* 820–832. https://doi.org/10.1175/1520-0426(2003)020<0820:AMATDD>2.0.CO;2.

Walker, J. R., MacKenzie, W. M., Mecikalski, J. R., & Jewett, C. P. (2012). An enhanced geostationary satellite-based convective initiation algorithm for 0–2-h nowcasting with object tracking. *Journal of Applied Meteorology and Climatology, 51,* 1931–1949.

Wang, W., & Seaman, N. L. (1997). A Comparison Study of Convective Parameterization Schemes in aMesoscale Model. *Monthly Weather Review, 125,* 252–278. https://doi.org/10.1175/1520-0493(1997)125<0252:ACSOCP>2.0.CO;2.

Wang, L., & Cao, C. (2008). On-orbit calibration assessment of AVHRR longwave channels on MetOp-A using IASI. *IEEE Transactions on Geoscience and Remote Sensing, 46,* 4005–4013.

Wang, L., Han, Y., Tremblay, D., Weng, F., & Goldberg, M. (2012). Intercomparison of NPP/CrIS radiances with VIIRS, AIRS, and IASI: A post-launch calibration assessment. Earth Observing Missions and Sensors: Development, Implementation, and Characterization II. In H. Shimoda et al. Eds., *International Society for Optical Engineering* (*SPIE Proceedings*, Vol. 8528), 85280 J, https://doi.org/10.1117/12.978769.

Ware, R., Cimini, D., Campos, E., Giuliani, G., Albers, S., Nelson, M., et al. (2013). Thermodynamic and liquid profiling during the 2010 Winter Olympics. *Atmosphere Research., 132–133,* 278–290.

Warner, T. T. (2011). *Numerical Weather and Climate Prediction.* Cambridge University Press, Cambridge, p. 526.

Weber, M. E., & Stone, M. L. (1995). Low altitude wind shear detection using airport surveillance radars. *IEEE Aerospace and Electronic Systems Magazine, 10,* 3–9. https://doi.org/10.1109/62.387970.

Weisman, M. L., Davis, C., Wang, W., Manning, K. W., & Klemp, J. B. (2008). Experiences with 0–36-h explicit convective forecasts with the WRF-ARW model. *Weather Forecasting, 23,* 407–437.

Wen, Y., Kirstetter, P., Gourley, J. J., Hong, Y., Behrangi, A., & Flamig, Z. (2017). Evaluation of MRMS Snowfall Products over the Western United States. *Journal of Hydrometeorology, 18,* 1707–1713.

Westwater, E. (1978). The accuracy of water vapor and cloud liquiddeter minations by dual-frequency ground-based microwave radiometry. *Radio Sci., 13,* 677–685.

Weygandt, S., Smirnova, T., Benjamin, S., Brundage, K., Sahm, S., Alexander, C., & Schwartz, B. (2009). The High Resolution Rapid Refresh (HRRR): An hourly updated convection resolving model utilizing radar reflectivity assimilation from the RUC/RR. In *23rd Conf. on Weather Analysis and Forecasting/19th Conf. on Numerical Weather Prediction*, Omaha, NE, Amer. Meteor. Soc., 15A.6.

Whiteman, D. N., Melfi, S. H., & Ferrare, R. A. (1992). Raman lidar system for the measurement of water vapor and aerosols in the earth's atmosphere. *Applied Optics, 31,* 3068–3082.

Wick, G., Hock, T., Neiman, P., Vömel, H., Black, M., & Spackman, J. (2018). The NCAR/NOAA Global Hawk Dropsonde System. *Journal of Atmosphere Oceanic Technology, 22,* 222. https://doi.org/10.1175/jtech-d-17-0225.1.

Williams, J. K. (2014). Using random forests to diagnose aviation turbulence. *Machine Learning, 95,* 51–70. https://doi.org/10.1007/s10994-013-5346-7.

Williams, P. D. (2016). Transatlantic flight times and climate change. *Environmental Research Letters., 11*(2), 024008. https://doi.org/10.1088/1748-9326/11/2/024008.

Williams, P. D. (2017). Increased light, moderate, and severe clear-air turbulence in response to climate change. *Advances in Atmospheric Sciences, 34,* 576–586. https://doi.org/10.1007/s00376-017-6268-2.

Williams, P. D., & Joshi, M. M. (2013). Intensification of winter transatlantic aviation turbulence in response to climate change. Nature. *Climate Change., 3*(7), 644–648. https://doi.org/10.1038/nclimate1866

Williams, J. K., & Meymaris, G. (2016). Remote turbulence detection using ground-based Doppler radar, 2016: Aviation turbulence forecast verification. In R. Sharman & T. Lane (Eds.), *Aviation turbulence: Processes, detection, prediction* (pp. 149–177). Berlin: Springer.

Wilson, J. W. (1966). *Movement and predictability of radar echoes*. National Severe Storms Laboratory Tech. Memo. ERTM-NSSL-28, p. 30.

Wilson, J. W., Crook, N. A., Mueller, C. K., Sun, J., & Dixon, M. (1998). Nowcasting thunderstorms: A status report. *Bulletin American Meteorology Society, 79,* 2079–2099.

Wilson, J. W., Ebert, E., Saxen, T., Roberts, R., Mueller, C., Sleigh, M., et al. (2004). Sydney 2000 forecast demonstration project: Convective storm nowcasting. *Weather Forecasting, 19,* 131–150.

Wilson, F. W., & Gramzow, R. H. (1991). The redesigned low level wind shear alert system. In *Preprints, 4th Int. conf. on aviation weather systems*, Paris, France, pp. 370–375.

Wilson, J. W., & Roberts, R. D. (2006). Summary of convective storm initiation and evolution during IHOP: Observational and modeling perspective. *Monthly Weather Review, 134,* 23–47.

Wilson, J. W., Roberts, R. D., Kessinger, C., & McCarthy, J. (1984). Microburst structure and evaluation of Doppler radar for airport wind shear detection. *Journal of Climate and Applications Meteorology, 23,* 898–915.

Wilson, J. W., & Wakimoto, R. M. (1982). The discovery of the downburst: T. T. Fijita's contribution. *Bulletin American Meteorology Society, 82*(1), 49–62.

Wirth, M. (2012). Measuring water vapor with differential absorption lidar. In U. Schumann (Ed.), *Atmospheric physics—Background, methods, trends, prediction* (pp. 465–476). Berlin: Springer.

Wolfson, M. M., & Clark, D. A. (2006). Advanced aviation weather forecasts. *The Lincoln Laboratory Journal, 16*(1), 31–58.

Wolfson, M. M., Delanoy, R. L., Forman, B. E., Hallowell, R. G., Pawlak, M. L., & Smith, P. D. (1994). Automated microburst wind-shear prediction. *The Lincoln Laboratory Journal, 7*(2), 399–426.

Wolfson, M. M., Dupree, W. J., Rasmussen, R., Steiner, M., Benjamin, S., & Weygandt S. (2008). Consolidated storm prediction for aviation (CoSPA). In *AMS 13th conference on aviation, range, and aerospace meteorology*, New Orleans, LA, 2008.

Wong, M., Skamarock, W. C., Lauritzen, P. H., & Stull, R. B. (2013). A cell-integrated semi-lagrangian semi-implicit shallow-water model (CSLAM-SW) with conservative and consistent transport. *Monthly Weather Review, 141,* 2545–2560.

Woodley, W. L., Henderson, T. J., Vonnegut, B., Gordon, G., Breidenthal, R., & Holle, S. M. (1991). Aircraft-produced ice particles (APIPs) in supercooled clouds and the probable mechanism for their production. *Journal of Applied Meteorology, 30,* 1469–1489.

Wulfmeyer, V. (1998). Ground-based differential absorption lidar for water-vapor and temperature profiling: Development and specifications of a high-performance laser transmitter. *Applied Optics, 37,* 3804–3824.

Wulfmeyer, V. (1999). Investigation of turbulent processes in the lower troposphere with water vapor DIAL and radar–RASS. *Journal of Atmospheric Science, 56,* 1055–1076.

Wulfmeyer, V., & Bosenberg, J. (1996). Single-mode operation of an injection seeded alexandrite ring laser for application in water-vapor and temperature differential absorption lidar. *Optics Letters, 21,* 1150–1152.

Wulfmeyer, V., & Bosenberg, J. (1998). Ground-based differential absorption lidar for water-vapor profiling: Assessment of accuracy, resolution, and meteorological applications. *Applied Optics, 37,* 3825–3844.

Wulfmeyer, V., Lehmann, S., Senff, C., & Schmitz, S. (1995). Injection seeded alexandrite ring laser: Performance and application in a water-vapor differential absorption lidar. *Optics Letters, 20,* 638–640.

Wurman, J., Dowell, D., Richardson, Y., Markowski, P., Rasmussen, E., Burgess, D., et al. (2012). The second verification of the origins of rotation in tornadoes experiment: VORTEX2. *Bulletin American Meteorology Society, 93,* 1147–1170.

Yang, B., Qian, Y., Lin, G., Leung, R., & Zhang, Y. (2012). Some issues in uncertainty quantification and parameter tuning: a case study of convective parameterization scheme in the WRF regional climate model. *Atmosphere Chemistry Physics, 12,* 2409–2427. https://doi.org/10.5194/acp-12-2409-2012.

Yang, J., Zhang, Z., Wei, C., Lu, F., & Guo, Q. (2017). Introducing the new generation of chinese geostationary weather satellites, fengyun-4. *Bulletin American Meteorology Society, 98,* 1637–1658. https://doi.org/10.1175/BAMS-D-16-0065.1.

Yano, J., Ziemiański, M. Z., Cullen, M., Termonia, P., Onvlee, J., Bengtsson, L., et al. (2018). Scientific challenges of convective-scale numerical weather prediction. *Bulletin American Meteorology Society, 99,* 699–710. https://doi.org/10.1175/BAMS-D-17-0125.1.

Yost, C. R., Bedka, K. M., Minnis, P., Nguyen, L., Strapp, J. W., Palikonda, R., et al. (2017). A prototype method for diagnosing high ice water content probability using satellite imager data. *Journal of Atmosphere Measure Technology*. https://doi.org/10.5194/amt-2017-367.

Zhang, J., Howard, K., Langston, C., Kaney, B., Qi, Y., Tang, L., et al. (2016). Multi-Radar Multi-Sensor (MRMS) quantitative precipitation estimation: Initial operating capabilities. *Bulletin American Meteorology Society*, *97*, 621–637. https://doi.org/10.1175/bams-d-14-00174.1.

Zhang, J., Howard, K., Langston, C., Vasiloff, S., Kaney, B., Arthur, A., et al. (2011). National mosaic and multi-sensor QPE (NMQ) system: Description, results, and future plans. *Bulletin American Meteorology Society*, *92*, 1321–1338. https://doi.org/10.1175/2011bams-d-11-00047.1.

Zhou, B., & Du, J. (2010). Fog prediction from a multimodel mesoscale ensemble prediction system. *Weather Forecasting, 25,* 303–322.

Zhou, B., Du, J., McQueen, J., & Dimego, G. (2009). Ensemble forecast of ceiling, visibility, and fog with NCEP Short-Range Ensemble Forecast system (SREF). In *Preprints. Aviation, range, and aerospace meteorology special symp. on weather–air traffic management integration*, Phoenix, AZ, Amer. Meteor.Soc., 4.5. http://www.emc.ncep.noaa.gov/.mmb/SREF/AMS_2009CV_fog.pdf.

Zhou, B., Du, J., McQueen, J., Dimego, G., Manikin, G., Ferrier, B., & Toth, Z., et al. (2004). An introduction to NCEP SREF aviation project. In *11th conference on aviation range and aerospace*, Oct 4–8, Hyannis, MA, Amer. Meteor. Soc. Presentation, p. 10.

(Received June 6, 2018, revised January 7, 2019, accepted March 18, 2019, Published online May 6, 2019)

Pure Appl. Geophys. 176 (2019), 1923–1958

https://doi.org/10.1007/s00024-018-1849-2

Pure and Applied Geophysics

Influences of Gravity Waves on Convectively Induced Turbulence (CIT): A Review

ROBERT D. SHARMAN[1] and S. B. TRIER[1]

Abstract—Thunderstorms are known to produce turbulence. Such turbulence is commonly referred to as convectively induced turbulence or CIT, and can be hazardous to aviation. Although this turbulence can occur both within and outside the convection, out-of-cloud CIT is particularly hazardous, since it occurs in clear air and cannot be seen by eye or onboard radar. Furthermore, due to its small scale and its ties to the underlying convection, it is very difficult to forecast. Guidelines for out-of-cloud CIT avoidance are available, but they are oversimplified and can be misleading. In the search for more appropriate and physically based avoidance guidelines, considerable research has been conducted in recent years on the nature of the phenomenon, and in particular, its connection to gravity waves generated by the convection. This paper reviews the advances in our understanding of out-of-cloud CIT and its relation to convective gravity waves, and provides several detailed examples of observed cases to elucidate some of the underlying dynamics.

Key words: Convective gravity waves, convectively induced turbulence, Kelvin–Helmholtz instability, aviation hazards.

1. Introduction

Aircraft encounters with turbulence are an important safety and operational hazard for the aviation industry (e.g., Gultepe 2018 this issue). They are responsible for hundreds of injuries each year, occasionally cause aircraft damage, are the source of tens of millions of dollars of additional operational costs for commercial airlines, and thus indirectly increase the cost of air travel. Aircraft avoid turbulence, especially the severe events, whenever possible through the use of aviation weather forecasts and en route tactical avoidance procedures. Yet, turbulence owes its existence to a variety of atmospheric phenomena (e.g., jet streams, terrain-induced flows, and thunderstorms) and each of these requires its own unique forecasting or avoidance strategy tailored to the dynamics of the underlying source or sources.

Progress in improving avoidance methods for a number of these sources is hampered by an incomplete understanding of the governing atmospheric small-scale dynamics, especially with respect to convectively induced turbulence (CIT). CIT occurs within convective clouds, as well as in the clear air above or around the cloud, sometimes at substantial distances from the cloud boundaries. Although turbulence within the cloud can often be avoided through the use of visual cues or through available radar observations, out-of-cloud CIT is difficult to avoid, because it is invisible, and is, therefore, often the cause of unexpected turbulence encounters that may be significant. These out-of-cloud CIT encounters may occur as aircraft: (1) try to circumnavigate the high radar reflectivity regions of thunderstorms to minimize deviations from planned routes; (2) encounter convection that is either unexpected or appears benign from aircraft onboard radar; (3) encounter storm tops that rapidly rise into the aircraft's flight path; or (4) are inadvertently directed into turbulent regions by air traffic control (Hamilton and Proctor 2003).

Estimates of the relative frequency of CIT vary, depending on location, and correlation time and spatial windows used to compare a thunderstorm to a turbulence report. Pond (2015) estimates up to 7% of commercial flights over Europe are within 35 km of a thunderstorm. Meneguz et al. (2016) and Kim and Chun (2011) both estimate "moderate-or greater" CIT frequencies of about 10% over Europe and the northeastern Atlantic, and South Korea, respectively. Over the U.S., Kaplan et al. (2005) found that 86% of 44 "severe" turbulence cases examined were within 100 km of deep convection, while Cornman

[1] National Center for Atmospheric Research, P.O. Box 3000, Boulder, CO 80307-3000, USA. E-mail: sharman@ucar.edu

and Carmichael (1993) estimate CIT is responsible for as much as 60% of aircraft accidents. Wolff and Sharman (2008) estimate as much as 20% of turbulence reports issued by pilots are associated with nearby cloud–ground lightning. Figure 1 provides an alternative analysis of CIT frequencies over the U. S. based on 5 years of correlating 3D lightning stroke data with reports of turbulence from automated commercial aircraft reports (see Sect. 3) of at least "moderate" intensity. The relative frequency of CIT occurrence is $>40\%$ over large sections of the Great Plains and is $>60\%$ over parts of Texas, Florida, the Gulf of Mexico and the Atlantic. However, over the Northeast and Western half of the country, it is considerably smaller (10–20%). The lightning data in Fig. 1 were obtained from the Earth Networks lightning detection system (ENTLN) and include cloud-to-ground and in-cloud lightning pulses. The overall pattern of relative frequency of CIT occurrence is very similar to the pattern of deep convective events derived from the Atmospheric Infrared Sounder (AIRS) NASA Aqua satellite data (Hoffmann and Alexander 2010), from reanalysis data (Kang et al. 2017), and from high-resolution radiosonde data (Gong and Geller 2010; Geller and Gong 2010).

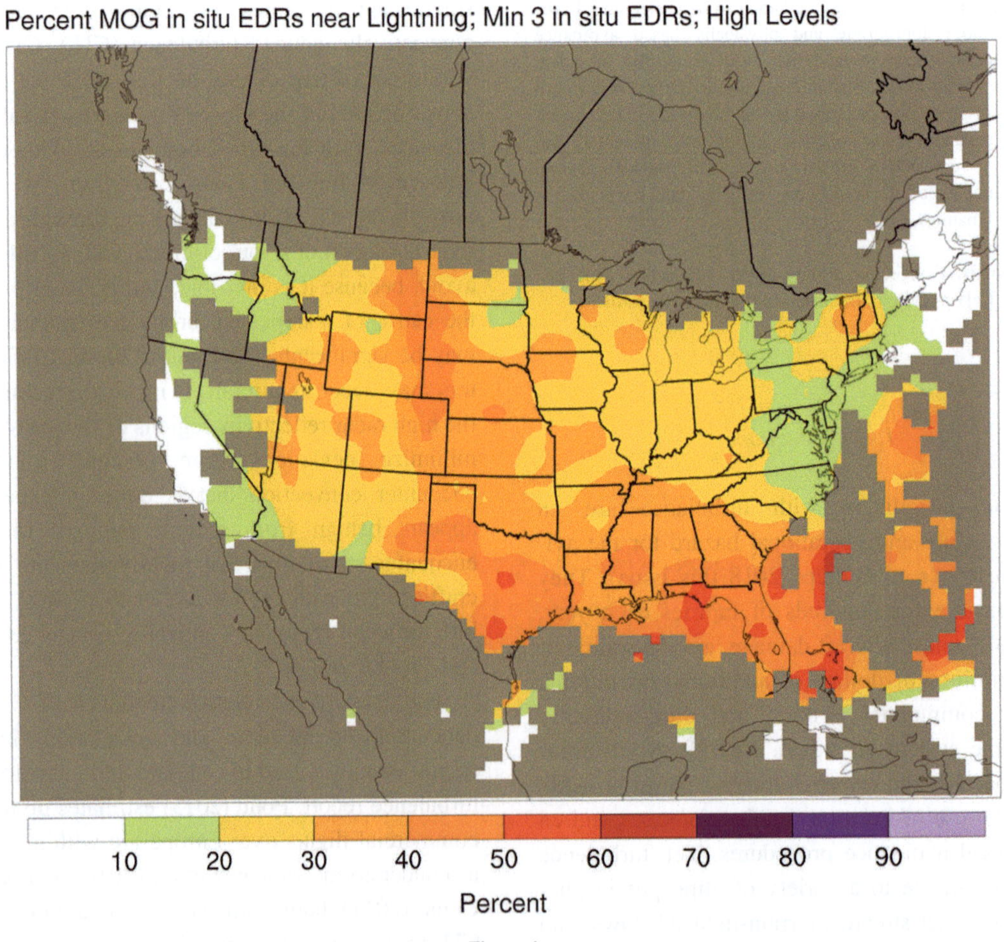

Figure 1
Frequency of occurrence turbulence in situ EDR reports in the vicinity of convection as indicated by ENTLN detected total (cloud-to-ground and in-cloud) lightning, as a percentage of total turbulence reports. Based on 5-years comparison of EDR reports $\geq 0.22\ m^{2/3}\ s^{-1}$ to total lightning reports. Reports are binned into $\sim$ 54-km squares and compared to lightning pulse data—that is at least one lightning pulse within $\pm$ 0.5 h and $\pm$ 50 km from the binned reports

Figure 2 shows a schematic of known out-of-cloud CIT generating regions in the upper troposphere and lower stratosphere (UTLS). The sketch in Fig. 2 provides an indication of where UTLS turbulence would most likely occur relative to the cloud boundaries, but exact locations may be transient and depend strongly on the particular storm type and its environment. This pattern is consistent with recent large eddy simulations (LES) of deep, moist convection (e.g., Lane and Sharman 2014; Verrelle et al. 2017). As shown in the diagram, one major source of out-of-cloud CIT is through convectively induced gravity waves propagating and breaking in the clear air away from the cloud (e.g., Lane et al. 2012). Enhanced shears above the cloud top (e.g., Lester 1993), cloud interfacial instabilities (e.g., Grabowski and Clark 1991; Lane et al. 2003), instabilities induced by sublimation of ice underneath anvils (Kudo 2013), and instabilities associated with cumulus turrets and overshooting tops (e.g., Bedka et al. 2010; Ahmad and Proctor 2011) can also lead to turbulence. Pseudo-obstacle effects may also generate turbulence above and in the downwind wake of the cloud (e.g., Burns et al. 1966; Lemon 1976; Pantley and Lester 1990; Bedard and Cunningham 1991; Bedard 1993; Fric and Roshko 1994). Some of these instabilities may extend well above (Prophet 1970) or below the visible anvil cloud. For more detailed discussions, see Lester (1993), Lane et al. (2012), and Sharman et al. (2012). Within the cloud, turbulence is usually most prevalent and most intense near deep convective updraft and downdraft regions, i.e., near strong vertical velocity gradients especially in the upper third of the cloud depth (Honomichl et al. 2013). Though usually lighter, turbulence also often occurs in the anvil of mature storms (e.g., Muhlbauer et al. 2014). Any of these regions may or may not coincide with the precipitation and thus strongest radar echo patterns as seen on onboard or ground-based radar. An example is shown in Fig. 3 indicating that, in least this case, the strongest turbulence is above, not within, the largest radar reflectivity regions.

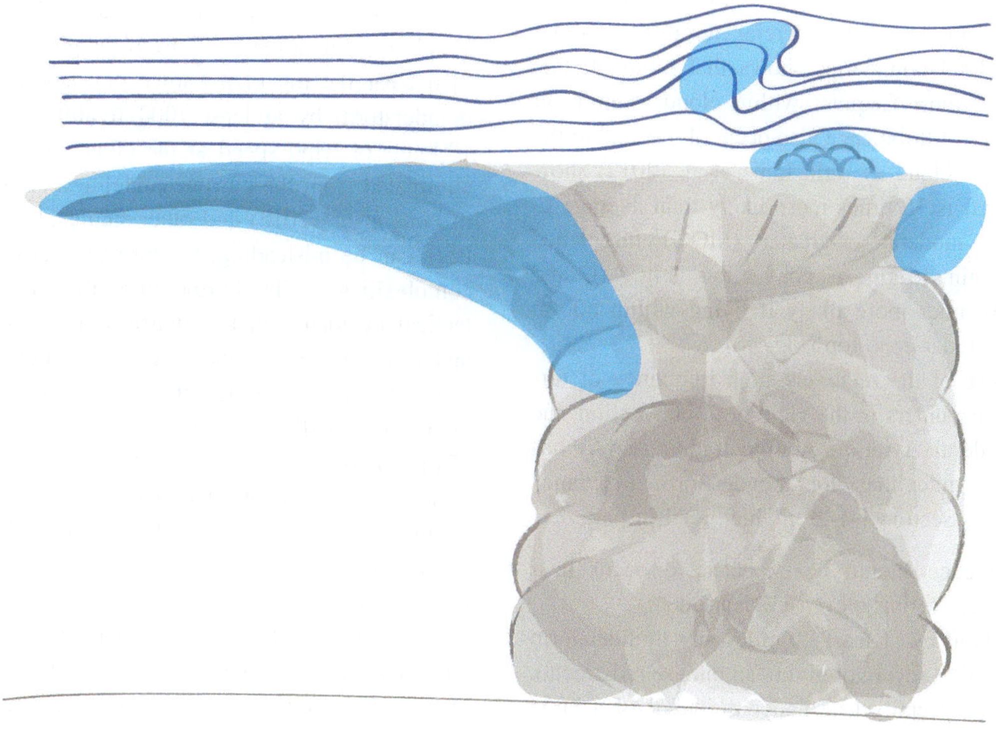

Figure 2
Schematic of common out-of-cloud CIT generation regions (blue areas) in a mature thunderstorm

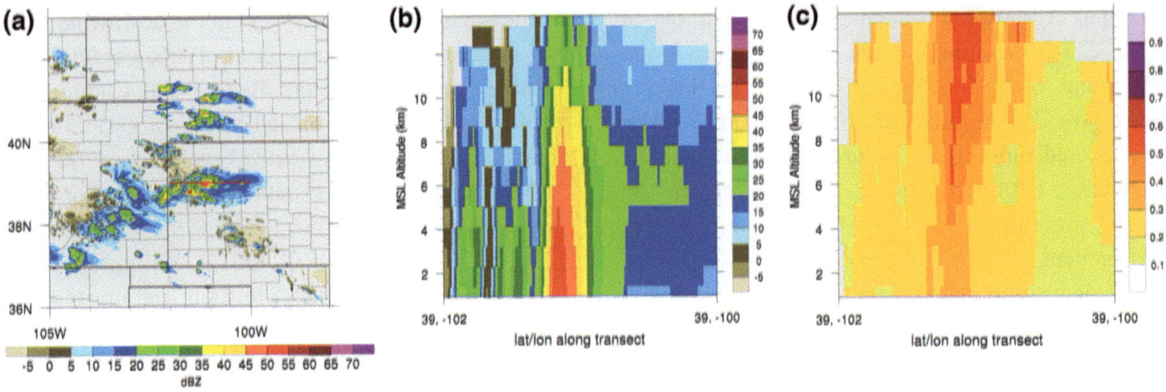

Figure 3
Comparison of vertical cross section of (**b**) radar reflectivity (dBZ) from a convective storm to (**c**) radar derived energy dissipation rate to the 1/3 power (EDR $m^{2/3}$ s^{-1}), using the algorithm described in Williams and Meymaris (2016) along the transect shown as a red line in (**a**). Date and time is Sept 28, 2007 at 2225 UTC

The frequency of out-of-cloud CIT is difficult to determine, since aircraft position relative to the cloud boundaries must be known to very good accuracy, but pilot reports (PIREPs), with a median position error of $\sim$ 40 km, have location uncertainties too large for such evaluations (Sharman et al. 2014). However, comparing more accurate automated turbulence reports from some commercial aircraft to radar derived cloud boundaries shows that although turbulence is most frequent within cloud, significant frequencies of turbulence also occur above and to the sides of the cloud as well. Lane et al. (2012) show that aircraft is 15 times more likely than average to encounter moderate-or-greater (MOG) turbulence when within 15 km of convection (Fig. 4a), and nearly 90 times more likely if flying within 800 m above the radar echo top (Fig. 4b).

Because of the relatively high frequency of turbulence encounters in the vicinity of convection, the U. S. Federal Aviation Administration (FAA) has developed guidelines for thunderstorm avoidance (FAA 2014 Section 7-1-29), which read in part,

- "Turbulence should be expected to occur near convective activity, even in clear air."
- "Do avoid by at least 20 miles any thunderstorm identified as severe or giving an intense radar echo. This is especially true under the anvil of a large cumulonimbus."
- "Don't attempt to fly under the anvil of a thunderstorm. There is a potential for severe and extreme clear air turbulence."
- "Do regard as extremely hazardous any thunderstorm with tops 35,000 ft or higher whether the top is visually sighted or determined by radar."

A strategy for turbulence avoidance above visible cloud given in earlier versions of the guidelines, viz., "Do clear the top of a known or suspected severe thunderstorm by at least 1000 ft altitude for each 10 knots of wind speed at the cloud top" has been eliminated from the current guidelines. As indicated in Lane et al. (2012) and other studies, these guidelines can be misleading, because they disregard the complexities of CIT. Furthermore, they may call for tactical avoidance that is unnecessary, wasting fuel and time. Although other visual indicators may be available [e.g., satellite imagery of overshooting cloud tops (Bedka et al. 2010; Monette and Sieglaff 2014), gravity-wave patterns embedded in the anvil, rapid anvil expansion, and banded structures along the outside edge of the anvil (Lenz et al. 2009)], these can provide only indirect inferences of the presence, location, and intensity of CIT. Improved forecasting and avoidance of turbulence near thunderstorms can only be realized through a better understanding of the conditions conducive to turbulence generation, and its extent and intensity. However, in recent decades, understanding of physical mechanisms influencing

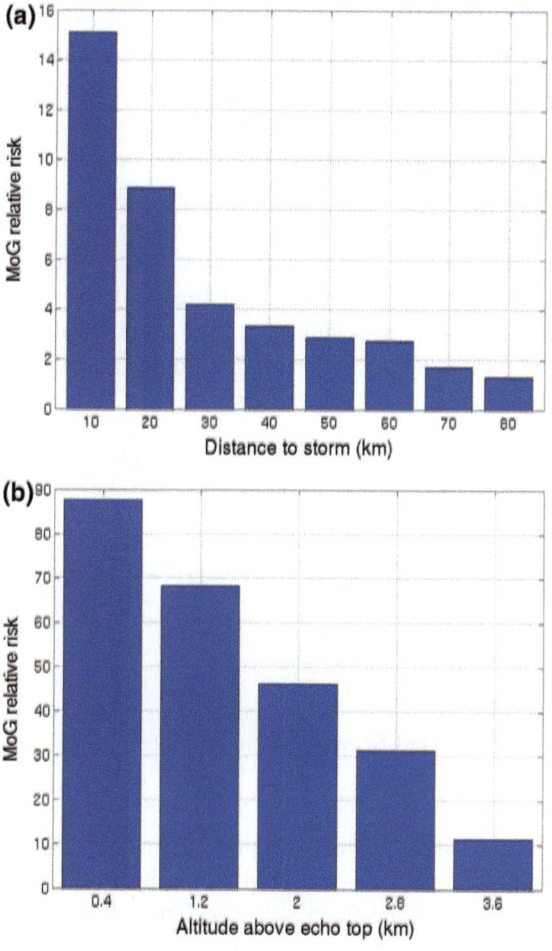

Figure 4
Relative risk of MOG turbulence derived from in situ EDR measurements as a function of: **a** distance to convection and (**b**) altitude above radar echo top in convective regions. The reports were between 7.6- and 12.8-km altitude (flight levels 25,000–42,000 ft) from May to October of 2004 and 2005. From Lane et al. (2012) **c** American Meteorological Society. Reprinted with permission

CIT has increased, mainly through the use of high-resolution numerical modeling of idealized scenarios and well-observed cases. Observations have also been improving, both in quantity and precision, so that it is now possible to correctly identify CIT cases, both within and out-of-cloud.

This article specifically reviews recent progress in understanding remote CIT, which we define as turbulence that is spatially and sometimes temporally removed from its parent deep convection. Since CIT may be remote we avoid the use of the terms "near-cloud turbulence" (NCT, Lane et al. 2012), "turbulence near thunderstorms" (TNT, Lester 1993) and "turbulence near thunderstorm tops" (TNTT, Pantley and Lester 1990). Thus, hereafter, remote CIT is referred to simply as CIT. Our examination of CIT in this review encompasses both observational and numerical modeling studies.

2. *Convective Gravity Waves*

In recent years, high-resolution numerical simulations have shown that in many cases, out-of-cloud CIT can be linked to gravity waves generated by convection. This is not surprising since gravity waves can be produced whenever air parcels are displaced vertically in a statically stable environment. References to early observational studies of convective gravity waves can be found in Gossard and Hooke (1975), Einaudi et al. (1978), Lu et al. (1984), Grachev et al. (1984), Alexander and Pfister (1995) and Hansen et al. (2002). The observations have been corroborated by theoretical studies (e.g., Townsend 1966, 1968; Stull 1976; Lin and Smith 1986; Chun and Baik 1994; Chun 1997; Baik and Chun 1996; Baik et al. 1999; for other early references see Lin 1994a, b; Romanova and Yakushkin 1995), laboratory experiments (e.g., Deardorff et al. 1969; Mclaren et al. 1973; Cerasoli 1978; Ansong and Sutherland 2010), and numerical simulations (Deardorff 1969; Fovell et al. 1992; Alexander et al. 1995; Gavrilov and Kshevetskii 2015). Additional references concerning results from simulations are given in Lane et al. (2012) and Lane (2016). Some of the techniques used to study internal gravity waves generated by penetrative convection in stellar interiors may also be relevant to the earth's atmosphere (see, e.g., Pinçon et al. 2016, and references therein). These and other studies have helped to better elucidate the dynamics of the waves and have identified three possible convective gravity-wave mechanisms:

1. thermal forcing and latent heat released within the cloud during condensation (e.g., Lin and Smith 1986; Bretherton 1988; Nicholls et al. 1991; Pandya et al. 1993; Pandya and Durran 1996; Pandya and Alexander 1999; Alexander and Barnet 2007; Chun and Baik 1998);

2. an obstacle effect (also termed "quasi-stationary forcing" by Fovell et al. 1992), whereby the cloud acts as a barrier to impinging airflow (e.g., Clark et al. 1986; Beres et al. 2002; Wang 2003);
3. a "mechanical oscillator" effect due to rapid cumulus growth impinging on a stable layer (e.g., the tropopause), overshooting its level of neutral buoyancy (termed penetrative convection) and initiating gravity waves (e.g., Pierce and Coroniti 1966; Stull 1976; Jones 1982; Fovell et al. 1992; Lane et al. 2001; Kumar 2007; Lane 2008).

For detailed explanations of these effects, see, e.g., Bradbury (1973), Clark et al. (1986), Fovell et al. (1992), Fritts and Alexander (2003), Kim et al. (2003), and Song et al. (2003). Fritts and Alexander (2003) indicate that the relative importance of these generating mechanisms in explaining observations depends on the particular environmental conditions. Numerical studies by Lane et al. (2001) and Song et al. (2003) also point to the importance of nonlinear advection effects in the forcing process.

The propagation characteristics of gravity waves are described in many excellent texts (e.g., Turner 1973; Gossard and Hooke 1975; Lighthill 1978; Gill 1982; Nappo 2002; Lin 2007) and review papers (Lin 1994a,b; Fritts and Alexander 2003). Unlike mountain lee waves, convective gravity waves are both mobile and transient, depending on the trajectory and life cycle of the generating cloud. The fundamental properties of buoyancy driven gravity waves can be demonstrated by a simple 2D linear, inviscid, adiabatic, nonhydrostatic, nonrotating, non-Boussinesq system, for which the following single governing equation for the wave induced density weighted vertical velocity $\tilde{w} = w' \exp(z/2H)$, where H is an atmospheric scale height, can be derived (e.g., Nappo 2002; Lin 2007):

$$(\partial_t + U\partial_x)^2 (\tilde{w}_{xx} + \tilde{w}_{zz}) + N^2 \tilde{w}_{xx} = \frac{g}{c_p T_0 U^2} Q. \tag{1}$$

Here, c_p is the specific heat at constant pressure, T_0 is the mean environmental temperature, Q is the heating rate (usually dominated by latent heat release), $U(z)$ is the mean flow velocity, and $N(z)$ is the Brunt–Väisälä or buoyancy frequency, $N^2 = (g/\theta_v)(\partial\theta_v/dz)$ where θ_v is the virtual potential temperature and g is gravity. Subscript notation is used for differentiation in (1), with x signifying the horizontal direction and z is the direction aligned with gravity, and t is time. If a monochromatic wave solution to (1) is assumed with $\tilde{w} = \hat{w}(z)e^{i(kx-\omega t)} = \hat{w}(z)e^{ik(x-ct)}$, where $k = 2\pi/\lambda$ is the horizontal wavenumber in the x direction (typically the west–east or zonal direction), with λ the corresponding wavelength, ω is a wave frequency, and c is the horizontal wave speed in the direction of propagation, then (1) becomes

$$\hat{w}_{zz} + (\ell^2 - k^2)\hat{w} = \frac{g}{c_p T_0 U^2} Q, \tag{2}$$

where

$$\ell^2 = \frac{N^2}{(U-c)^2} - \frac{U_{zz}}{(U-c)} - \frac{U_z}{H(U-c)} - \frac{1}{4H^2} \approx k^2\left[\frac{N^2}{\Omega^2} - \frac{U_{zz}}{k\Omega}\right], \tag{3}$$

is the so-called Scorer parameter, where

$$\Omega = \omega - kU, \tag{4}$$

is the intrinsic (or Doppler shifted) frequency, i.e., the frequency of the wave relative to the mean flow, whereas ω is the frequency in a frame of reference with respect to the ground. If $Q = 0$ and N and U are approximately uniform with height, then (2) and (3) admit homogeneous solutions with vertical dependence of the form:

$$\hat{w} = Ae^{+imz} + Be^{-imz}, \quad m = \sqrt{\ell^2 - k^2}, \quad \text{for } \ell^2 > k^2, \tag{5}$$

$$\hat{w} = Ce^{+\mu z} + De^{-\mu z}, \quad \mu = \sqrt{k^2 - \ell^2}, \quad \text{for } k^2 > \ell^2. \tag{6}$$

In (5), m is the vertical wavenumber and the + sign in the exponent (term with the coefficient A) corresponds to waves propagating energy upwards, and vice versa for the –sign (B term) (Eliassen and Palm 1961). For vertically propagating waves of the form (5), a dispersion relation that relates the frequency to the vertical and horizontal wavenumbers and the background buoyancy frequency N [ignoring the U_{zz} and non-Boussinesq terms for ℓ^2 in (3)] is

$$\Omega^2 = \frac{Nk^2}{k^2 + m^2} = N^2 \cos^2 \alpha, \tag{7}$$

where α is the angle between lines of constant phase and the vertical. As evident from (7), for vertically propagating waves, the intrinsic frequency Ω is always less than the buoyancy frequency N.

In (6), μ represents a growth rate (C term) or decay rate (D term) with height. In general, for propagation in unlimited domains, the first term is unphysical, and the wave amplitude decays with height. These are termed evanescent waves. In cases where ℓ^2 decreases with height, for example, a wave of a given k may be propagating at low levels, where $\ell^2 > k^2$ but may be evanescent at higher levels, where $\ell^2 < k^2$. In such cases, the waves are often described as "trapped" at levels where $\ell^2 - k^2 \approx 0$ (e.g., Lindzen and Tung 1976).

For the propagating solutions (5), the wave propagation or phase velocities C_p in the x and z directions are

$$\begin{aligned} C_{px} &= \frac{\omega}{k} = U \pm \frac{N}{\sqrt{k^2 + m^2}} \\ C_{pz} &= \frac{\omega}{m} = \frac{k}{m}\left(U \pm \frac{N}{\sqrt{k^2 + m^2}}\right). \end{aligned} \tag{8}$$

These relations may be helpful in the analysis of observations or output from simulations.

In reality, since the convection source is temporally and spatially dependent, a spectrum of gravity waves is generated, and since they are dispersive, different wavelengths travel at different velocities. In this case, for trapped waves, only a discrete set of wavelengths satisfy the criterion for trapping, and the waves are very well defined. More generally, the energy of groups of waves propagate according to the group velocity:

$$\begin{aligned} Cg_x &= \frac{\partial \omega}{\partial k} = U \pm \frac{Nm^2}{(k^2 + m^2)^{3/2}}, \\ Cg_z &= \frac{\partial \omega}{\partial m} = \mp \frac{Nkm}{(k^2 + m^2)^{3/2}}. \end{aligned} \tag{9}$$

For k, $m > 0$, the group or energy propagation is eastward and downward, so negative m corresponds to upward energy propagation. Note that the wave vector $\boldsymbol{k} = (k, m)$ is perpendicular to the group velocity ($\boldsymbol{k} \bullet \boldsymbol{C}_g = 0$), so that the group propagation is *along* the phase lines.

An interesting situation occurs at the altitudes where the wave speed c equals the background wind speed U, the so-called critical level z_c, where ℓ^2 is undefined (Eq. 3). As waves approach a critical level from the generation region the vertical wavelength $2\pi/m$ and vertical group velocity Cg_z in (9) become small, and the wave amplitudes can become large (Bretherton 1966). In this case, nonlinear processes become important, and may lead to wave steepening, overturning, and "breaking" into turbulence patches (e.g., Dörnbrack et al. 1995). Applying this concept to convective gravity waves propagating upward, waves with a downshear horizontal component of propagation may amplify and break when they approach a critical level z_c above the cloud; but upshear propagating waves do not encounter a critical level, and such waves are often found to decay with altitude (i.e., are evanescent). These two different situations are illustrated schematically in Fig. 5. Convective gravity-wave encounters with critical levels have been shown in a number of studies to lead to CIT (e.g., Teitelbaum et al. 1999; Lane et al. 2003; Moustaoui et al. 2004), so identification of critical levels is important for understanding many instances of CIT. For further details on critical levels and transient effects, see, e.g., Nappo (2002).

Wave encounters with critical levels are just one example of how gravity waves may perturb the environment such that the local gradient Richardson number $R_i = N^2/U_z^2$ becomes small enough to induce Kelvin–Helmholtz instabilities (e.g., Hodges 1967; Pavelin et al. 2001). The R_i criterion for instability depends on the environment, although theoretically $R_i > 1$ guarantees stability (e.g., Abarbanel et al. 1984; Miles 1986). In practice, the use of Ri as an instability diagnostic is hampered by its strong dependence on vertical differencing (e.g., Reiter and Lester 1968), and in simulation experiments, identification of regions of small R_i associated with gravity waves also requires that the gravity waves are properly resolved

Environmental shear, rotation, and 3D effects considerably complicate wave propagation characteristics. The introduction of three dimensions allows wave energy to spread horizontally, producing

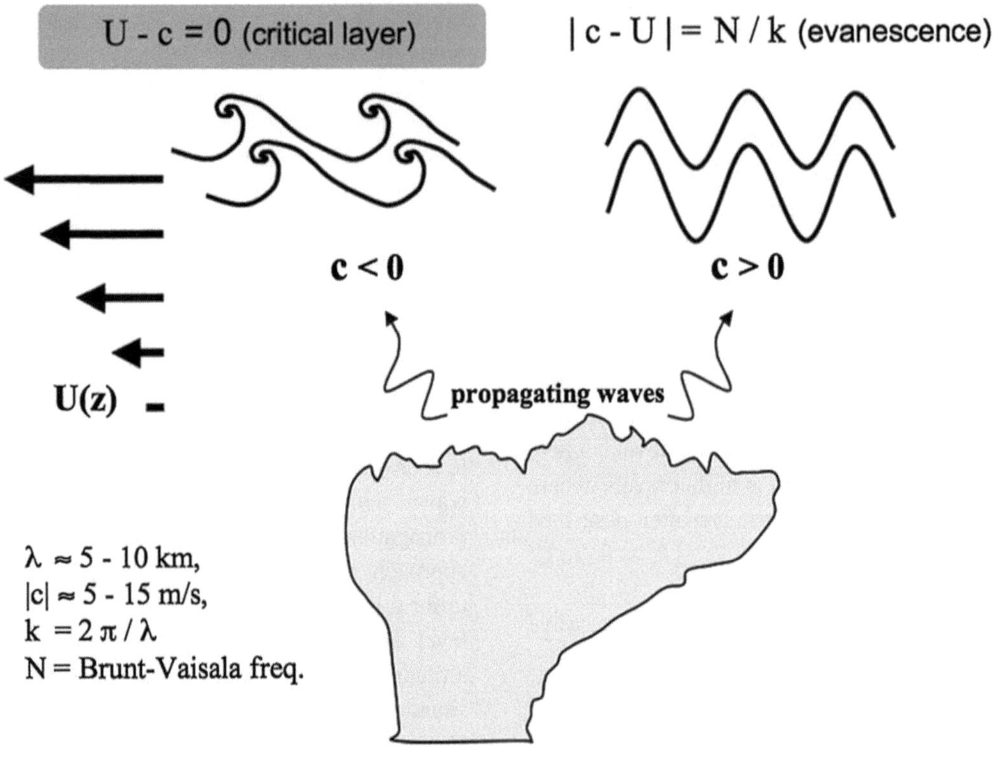

Figure 5
Schematic diagram of gravity-wave propagation and breakdown above deep convection in an environment with negative above-cloud vertical wind shear in a reference frame moving with the cloud-top wind. If the change in wind speed is sufficiently large, waves with negative phase speed c will encounter a critical layer where $U(z_c) = c$, and break down into turbulence, while waves with positive phase speed will become evanescent. Typical phase speeds and horizontal wavelengths at shown at bottom left. From Lane and Sharman (2008). **c** American Meteorological Society. Reprinted with permission

"ripples" resembling surface waves that are generated when a rock is dropped into a pond, (e.g., Piani et al. 2000; Lane et al. 2001; Lane 2016). However, in the vertical, surfaces of constant phase for a given $\omega < N$ will still take the form of a cone of half angle $\alpha = \cos^{-1}(\Omega/N)$ centered on the vertical axis with the apex at the gravity-wave source (Eq. 7). With no mean flow, the wave amplitudes $\propto J_1(kr)$, where r is radial distance and J_1 is the Bessel function of the first kind (Ansong and Sutherland 2010), but with substantial wind impinging on the cloud or overshooting top the generated gravity waves may produce parabolic-shaped wave patterns, which resemble 3D lee waves and ship waves (e.g. Sharman and Wurtele 1983; Lin 1986; Lin and Li 1988; Sharman and Wurtele 2004; Wang et al. 2010). Discussions of shear and rotation effects can be found, e.g., in Stobie et al. (1983) and Lin (2007). For example, studies involving topographically generated mountain waves point to important effects of wind turning with height (Broad 1995; Shutts 1998; Guarino et al. 2017), and (Rossby or Jones') singularities introduced by rotation at altitudes where $z = \pm f/kU_z$, where f is the Coriolis frequency (Jones 1967; Wurtele et al. 1996). Also an important nonlinear effect occurs when gravity waves break to form turbulence: the resulting disruption in stability can cause "secondary waves" to be generated, which can in turn propagate away from the mixing layer and possibly break at a different location than the primary waves. Examples of secondary waves generated by breaking of primary convective gravity waves can be found in Holton and Alexander (1999), Zhou et al. (2002), Snively and Pasko (2003), Chun and Kim (2008), and Lane and Sharman (2006).

In addition to these shorter wavelength convectively induced gravity waves, large amplitude, large wavelength (10–100 s km) waves or pulses, with compensating subsidence (e.g., Bretherton and Smolarkiewicz 1989), can also be generated by mesoscale regions of thunderstorms. Ruppert and Bosart (2014) contain a thorough review of the previous (mostly observational) work on this topic. Many case studies (typically based on examination of surface barograph parameters and analyses of radar and radiosondes) have shown that these mesoscale gravity waves can initiate new or modify existing deep convection. The dependence of gravity-wave characteristics on forcing scale was first demonstrated by Queney (1948) for the mountain wave problem, who showed that for mountains of scale $\sim$ 10 km or greater, the hydrostatic assumption is approximately satisfied [i.e., $k \rightarrow 0$ in Eqs. (2)–(9)]. For convectively generated mesoscale gravity waves, the resultant hydrostatic disturbance can be characterized as wave packets (e.g., Eom 1975), or solitary waves (e.g., Lin and Goff 1988) and may be long-lived, lasting for several hours or more (e.g., Trier and Sharman 2016). Since the wavelength and/or the gravity-wave lifetime are large, they may be influenced by the earth's rotation and Eqs. (2)–(9) must be modified to include rotational effects and are consequently termed "inertia–gravity waves" (see, e.g., Nappo 2002; Lin 2007).

Perhaps the first connections of short convective gravity waves to turbulence were proposed by Haman (1962) and Townsend (1966, 1968). Early observations of turbulence above cloud, which were probably related to convectively generated gravity waves, have been reported by Burns et al. (1966), Prophet (1970), Burnham (1970), and Pantley and Lester (1990). A 2D simulation case study of CIT by Keller et al. (1983), where the cloud was replaced by an obstacle, showed good agreement with a CIT encounter (Parks et al. 1985) over Hannibal, MO. A 3D cloud simulation study by Droegemeier et al. (1997) identified a connection between convectively generated gravity waves and turbulence. 2D and 3D numerical studies of a CIT encounter over Dickinson, ND by Lane et al. (2003) and Lane and Sharman (2006, 2008) demonstrated that the convective gravity waves were most likely due to the mechanical oscillator effect, and the turbulence was induced when the convective gravity waves encountered a cloud-relative critical level. Kim and Chun (2012), and others have confirmed this connection in another case. Since then, detection of convectively induced waves and turbulence in field programs, inferences from satellite imagery or radar, and from reports of turbulence through pilot experience and have become commonplace. An example of an extreme turbulence encounter above deep convection due to breaking convective gravity waves is provided in Fig. 6 from Kim and Chun (2012). The CIT occurred at a cruising altitude of 35 kft ($\sim$ 10.7-km MSL), and was marked by a change in vertical acceleration of 2.26 g during a 3-min interval. Further case study examples will be provided in Sect. 4. Emphasis there is not on the convective gravity waves themselves, but rather on how they may lead to turbulence.

3. Turbulence Metric and Observations

Research to better understand the nature and causes of aviation-scale turbulence in general, and CIT in particular, has been hampered in part by the lack of routinely available reliable observational data. Until recently, verbal pilot reports (PIREPs) have typically been the only routine source of information about the location and severity of turbulence. These reports are, however, incomplete (since reporting is voluntary), and highly subjective (Schwartz 1996). In particular, the turbulence intensity in PIREPs is noted only as "smooth", "light", "moderate", "severe" (or sometimes "heavy"), and "extreme", but these characterizations are aircraft dependent, and are influenced by the pilot's experience (what one pilot views as "moderate" might be perceived as "light" or "severe" by another), see Sharman (2016) for definitions of these intensities.

The intensity of atmospheric turbulent eddies within the range of eddy sizes that affect commercial aircraft ($\sim$ 100 m to a few km) is usually small enough to be perceived by occupants as "smooth", and Sharman et al. (2014) show that about 85–90% of PIREPs are in the smooth category. However, if the turbulent eddies within this size range are energetic enough, some level of aircraft bumpiness will be perceived by aircraft occupants. "Moderate" turbulence,

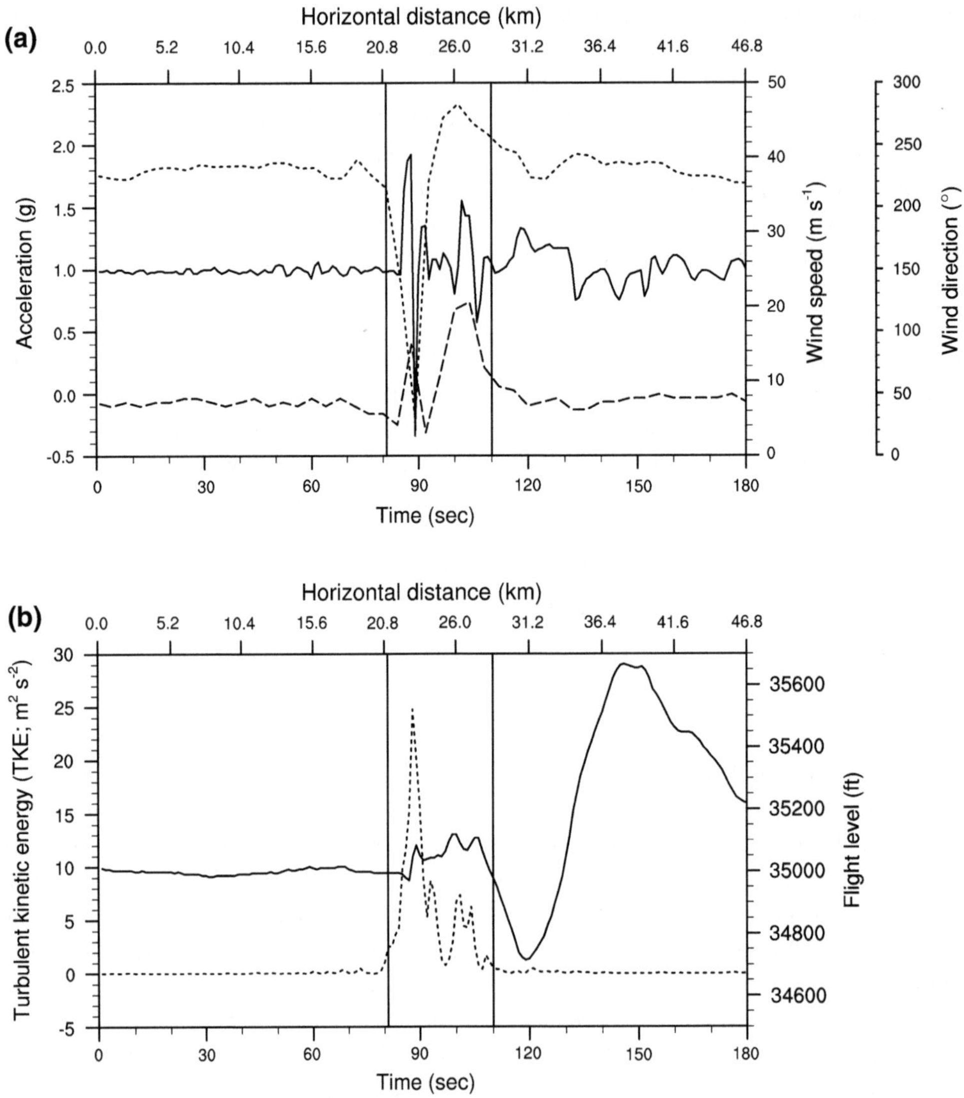

Figure 6

a Vertical acceleration (solid line), wind speed (dashed line), and wind direction (dotted line) and **b** TKE (dotted line) and flight level (solid line) as a function of time from 1033:10 to 1036:10 UTC 2 Sep 2007, derived from the Digital Flight Data Recorder. The area with the vertical lines from 89 to 109 s signify the duration of the turbulence event. From Kim and Chun (2012). **c** American Meteorological Society. Reprinted with permission

defined as a vertical acceleration magnitude of at least 0.5 g, may be enough for the pilot to seek changes in altitude or airspeed, but the aircraft remains in control at all times. "Severe" turbulence is defined as at least 1 g vertical acceleration, which may cause large, abrupt changes in the flight of the aircraft, and may even cause momentary loss of control.

For CIT investigations, probably more important than the subjective turbulence intensity reported in the PIREP is the associated position and time inaccuracy. Recent investigations into the accuracy of PIREPs have indicated an average position error of about 50 km (Sharman et al. 2014), which can easily be comparable to the size of a large individual thunderstorm. Thus, it becomes difficult if not impossible to determine if a report is in or out of cloud, although PIREPs may still be useful for identification of turbulence cases. To address the

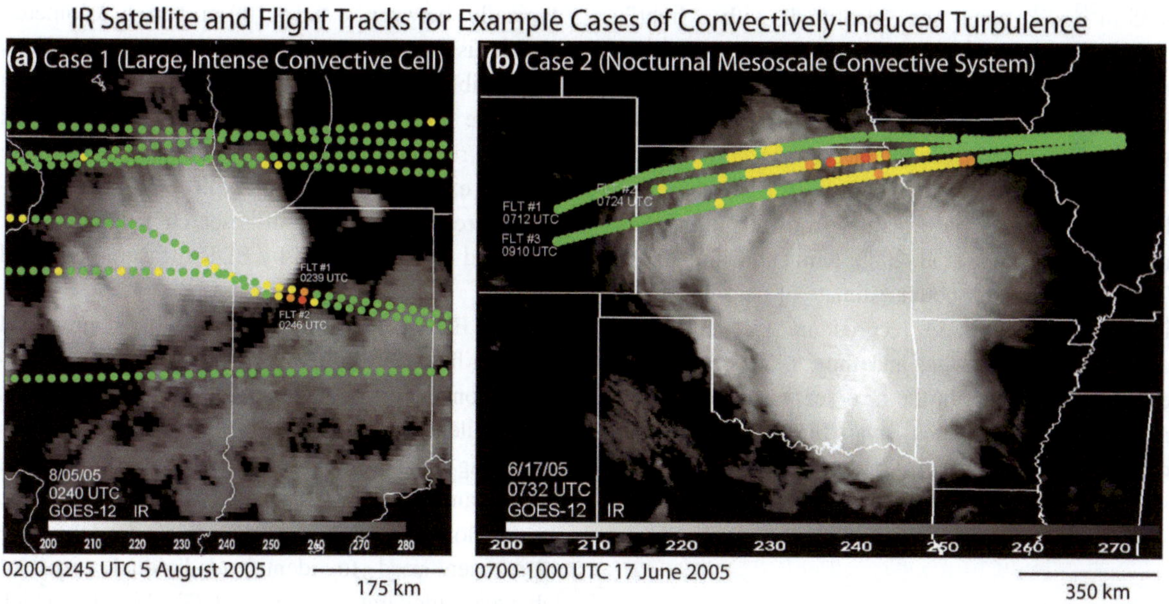

Figure 7
Infrared satellite images of two cases with in situ commercial aviation measurements of EDR during turbulence encounters. The flight tracks and turbulence intensities (green = smooth, yellow = light, orange = moderate, and red = severe) are overlaid on the images. Adapted from Lane et al. (2012)

accuracy problem associated with PIREPs, automated in situ reports of turbulence intensity are now being routinely generated onboard some aircraft, and this information is downlinked to the ground (Sharman et al. 2014). The strategy is to provide estimates of an atmospheric turbulence metric, the cube root of the energy dissipation ε. This quantity is commonly referred to as "EDR", and is the ICAO (2013) standard for turbulence reporting. The EDR estimates (in units of $m^{2/3}\ s^{-1}$) are provided by relating spectral levels based on analyses of time series of vertical acceleration, vertical velocity or true airspeed to EDR (MacCready 1964; Cornman et al. 1995; Sharman et al. 2014). EDR is directly related to the aircraft root-mean-square (RMS) normal acceleration σ_g (e.g., McCready 1964):

$$\sigma_g = V_T^{1/3}\left[\int |A(f)|^2 S_w(f)\mathrm{d}f\right]^{1/2} \varepsilon^{1/3}, \qquad (10)$$

where $A(f)$ is the aircraft specific response at frequency f and true airspeed V_T, and S_w is a specified atmospheric spectral model (e.g., Kolomogorov or von Karman) for the vertical wind w with a unit ε. Implicit in all onboard computational algorithms for EDR is the existence of an inertial subrange where $S_w \propto k^{-5/3}$ and k is the horizontal wavenumber. This has been shown by many investigators to be a good approximation in the range of wavenumbers or wavelengths that affect aircraft (e.g., Sharman 2016).

In the vertical wind-based algorithm described in Sharman et al. (2014), a 10-s sliding time window with a 5-s overlap is used to compute S_w and EDR, providing 12 EDR estimates each minute. The mean and peak of the 12 estimates are recorded and downlinked. The 1-min resolution corresponds to a horizontal scale of about 12 km at typical aircraft cruise speeds, providing a much better position of events for comparing to cloud observations than PIREPs. Figure 7 shows two examples of in situ EDR recordings outside of deep convection. Future in situ EDR implementations will identify the location of the peak EDR, providing about 1-km resolution. Useful guidelines for turbulence intensity as indicated by EDR are provided in Sharman et al. (2014). Based on comparisons to PIREPs, they find for a medium-weight category commercial aircraft (maximum takeoff 15,500–300,000 lbs) flying under typical cruise conditions that EDRs of ~ 0.15, 0.22,

0.35 $m^{2/3}$ s^{-1} could be identified with "light", "moderate", and "severe" PIREPs, respectively, although there is significant scatter in the data. These intensity thresholds are generally expected to be smaller for light aircraft and larger for heavy aircraft.

Some CIT case studies have used EDR estimates based on retrieved information from the flight data recorder (e.g., Lane et al. 2003; Kim and Chun 2012). This provides a still better estimate of peak EDR and its location, but these records are frequently not available to the public, and thus, most of the CIT examples presented in the next section use the automated EDR estimates available from selected commercial aircraft. As more aircraft implement some version of the EDR estimation algorithm, additional cases of CIT will be identified.

4. *Example CIT Types*

4.1. Overview

EDR measurements from commercial aircraft (Sect. 3) have helped characterize CIT, but are by themselves insufficient to determine its causes. However, Lane et al. (2012) describe how current research numerical weather prediction (NWP) models configured with fine-scale nesting can be used to diagnose different possible mechanisms explaining the origins of CIT. The horizontal grid spacing in the highest resolution domains of these models is typically between a few 100 m and a kilometer, which is not sufficient to resolve the motions responsible for the aircraft turbulence, but is often adequate to resolve the larger scale mechanisms directly responsible for its onset. In this section, we present examples of possible CIT initiation mechanisms from these types of research simulations of observed cases identified by EDR measurements and PIREPs.

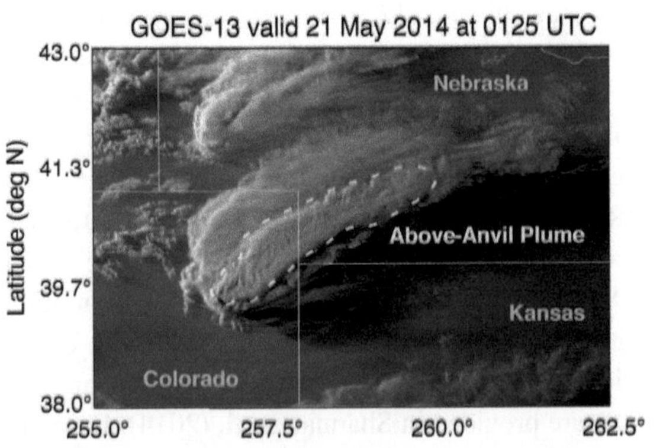

Figure 8
GOES 1-km visible satellite imagery for 0115 UTC 21 May 2014. State boundaries are indicated by the solid cyan lines and the thick dashed white line encircles an observed above-anvil cirrus plume. Adapted from Homeyer et al. (2017)

The EDR data from a subset of the commercial aviation fleet have been overlaid on displays of high-resolution satellite imagery to develop climatologies of turbulence near specific types of cloud patterns including banded cirrus in MCS anvils (Lenz et al. 2009) and relatively isolated overshooting deep convection (Bedka et al. 2010). Such displays have also been used to identify individual cases of observed moderate-or-greater CIT (Fig. 7), which have been simulated by high-resolution research models in an effort to better understand its different causes (e.g., Fovell et al. 2007; Trier and Sharman 2009; Lane et al. 2012; Trier et al. 2012).

These modeling studies have suggested that a variety of different mechanisms could be responsible for the onset of CIT in different situations. Though the overall mechanisms are different, a common aspect emphasized in this section is the potentially important role of gravity waves as either a direct or an indirect linkage between the deep convection and CIT onset. Supporting evidence for frequent gravity-wave activity in CIT environments has resulted from processing techniques used on imagery from new high-resolution satellite platforms (e.g., Wimmers et al. 2018).

4.2. Turbulence Occurring Above Deep Convection

Locations directly above either rapidly growing (e.g., Lane et al. 2003) or dissipating (e.g., Kim and Chun 2012) deep convection are common sites for CIT. Recall from Fig. 4 that the risk of moderate-or-greater turbulence is enhanced over the background risk by 50 times (25 times) for depths of approximately 2 km (3 km) above deep convection. In their high-resolution idealized simulation of a severe turbulence encounter above an isolated continental midlatitude thunderstorm, Lane et al. (2003) note that

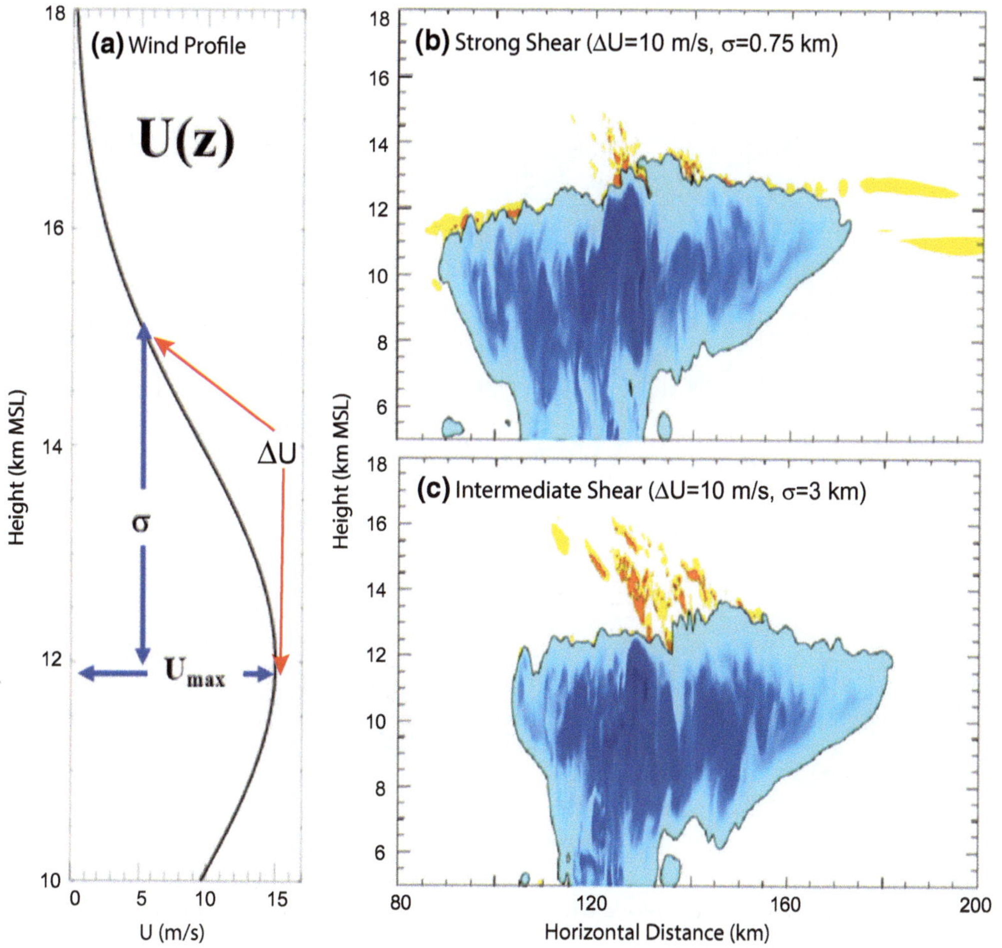

Figure 9

a Form of the idealized jet-like vertical profile of horizontal winds used in the high-resolution Lane and Sharman (2008) simulations, where $U_{\max}$ is the maximum horizontal wind near-cloud top and ΔU is the wind difference through a layer of depth σ. **b**, **c** vertical cross sections of model output from the Lane and Sharman (2008) simulations, where the blue colors are cloud condensate with the darker shades indicating larger cloud mixing ratios, and the shading above the cloud is resolved-scale turbulence kinetic energy (TKE) with gold, brown, and red colors indicating TKE values representative of light, moderate, and severe turbulence, respectively

the turbulence at different depths above the storm could result from a variety of different mechanisms. Turbulence can result from either strong vertical motions associated with vertical displacements of statically stable air induced by the storm below, small-scale shearing instabilities including Kelvin–Helmholtz instability (KHI) associated with strong deformation near the boundary of the cloud and surrounding atmosphere (e.g., Grabowski and Clark 1991), or the breaking of vertically propagating gravity waves initiated by the tropopause-penetrating convection (Fig. 5). Lane et al. (2003) note that turbulence events more than a kilometer above deep convection are likely dominated by the latter mechanism, which we focus on in the remainder of this subsection.

Though wave breaking above deep convection is often invisible, making it a significant hazard for aviation turbulence, such wave breaking is sometimes indicated by extensive cirrus plumes occurring above the deep convection (Levizanni and Setvak 1996; Wang 2003, 2007). The formation of these cirrus plumes (Fig. 8) is influenced by vertical displacements and resulting stratospheric–tropospheric exchange (STE) of water vapor during wave breaking. Homeyer et al. (2017) found that lower

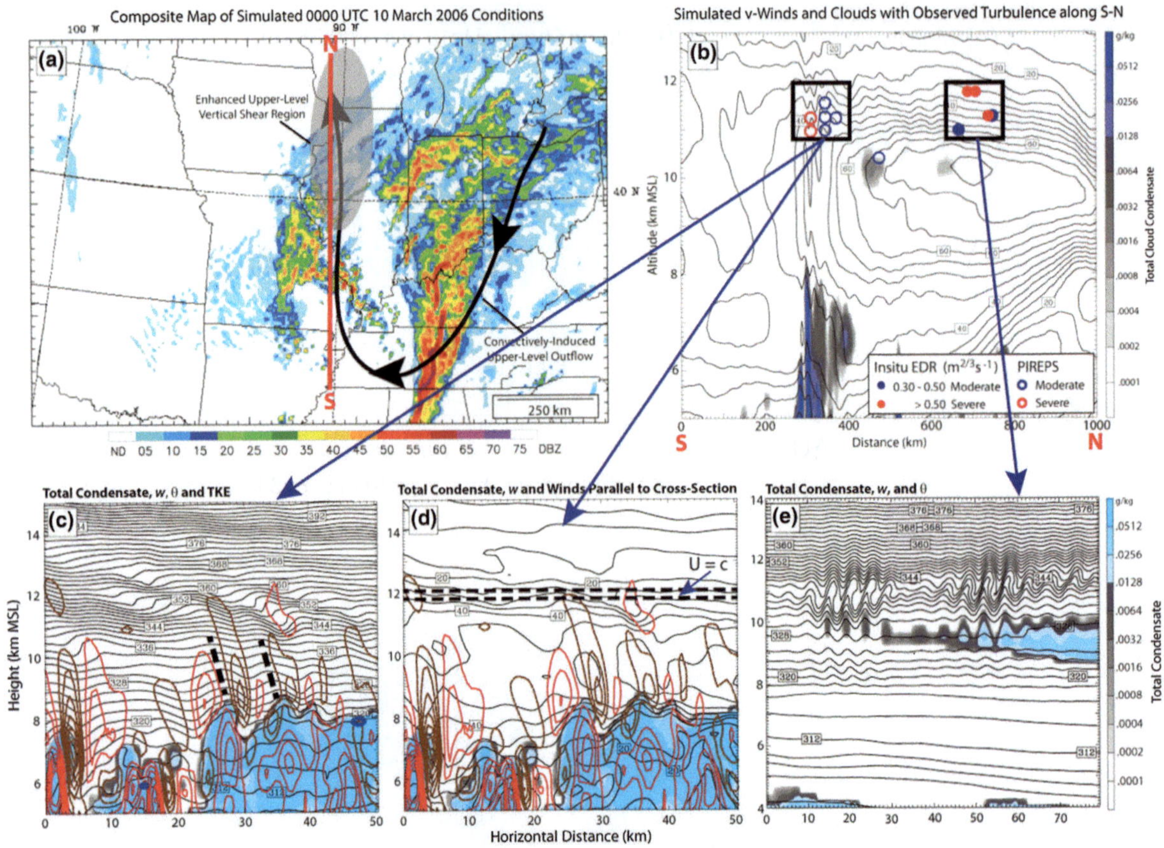

Figure 10

a Composite diagram of rainfall and convectively induced UTLS anticyclonic outflow (shown schematically by the bold curve with arrows) from a simulation described in Trier et al. (2012) using a model domain with 3.3-km horizontal grid spacing. **b** Vertical cross section of simulated meridional wind (5-m s^{-1} contour intervals) along transect SN in part (**a**) with time–space corrected observations of moderate-or-greater turbulence within 33.3 km of SN projected onto the cross section and described in the legend. **c**, **d** Vertical cross sections from a higher resolution simulation with 667-m horizontal grid spacing from the left inset region of part **b** of total cloud condensate (color shading; scale at right), vertical velocity (1-m s^{-1} contour interval; red lines ≥ 1 m s^{-1}; brown lines ≤ -1 m s^{-1}), where **c** has potential temperature (2-K contour intervals in black solid lines and **d** has winds parallel to the cross section (5-m s^{-1} contour intervals in black solid lines). The dashed vertically tilted lines in part **c** represent example phase lines of vertically propagating gravity waves, and the dashed horizontal lines in part **d** represent a range of estimates for the location of critical levels for vertically propagating waves of different frequencies. **e** Vertical cross section from the same higher resolution simulation as for parts **c** and **d**, but for the right inset region of part **b** of total cloud condensate (color shading; scale at right) and potential temperature (2-K contour intervals). From Trier (2016). Copyright Springer International Publishing

stratospheric cirrus plumes above deep convection are more likely to occur when there is strong flow relative to the horizontal motion of the deep convection, which may be indicative of strong vertical shear that would increase the likelihood of a critical level that leads to wave breaking.

In a numerical sensitivity study using 2D and 3D numerical models, Lane and Sharman (2008) found that wave breaking and turbulence is most intense and horizontally widespread in strong vertical shear conditions, but under such conditions is confined to locations relatively close to the top of the deep convection (Fig. 9b). However, Lane and Sharman also note that the depth through which wave breaking and turbulence can occur is maximized for intermediate values of vertical shear (Fig. 9c), which is germane to turbulence avoidance strategies. These two results from Lane and Sharman are consistent with the strength of the environmental vertical shear influencing the height of the critical level above the deep convection. Lane and Sharman also varied the strength of the static stability in their sensitivity experiments and found that small static stability generally facilitated wave breaking and development of turbulence. Homeyer et al. (2014) found that details in the vertical structure of static stability in different synoptic situations influence the depth of vertical mixing. Though their emphasis is on general aspects of STE, these results are also relevant to wave breaking and CIT.

Deep convection is much less common in mid-latitudes during the cold season (e.g., Kang et al. 2017), because conditional instability is often limited or nonexistent. However, despite its lesser frequency, several important environmental factors may contribute to wave breaking above deep convection being particularly hazardous to aviation during less frequent cold-season convective outbreaks. First, because of its association with strong synoptic disturbances in the winter, deep convection is often more widespread than in summer and, therefore, may be more difficult to avoid. Second, the height of the tropopause is lower during the cold season, particularly on the cyclonic side of jet streams, where stratospheric intrusions or tropopause folds are most common. Thus, commercial aviation cruising altitudes in the winter are more likely to be in the lower stratosphere (where vertical wave propagation is supported) than in summer. Third, the strong vertical shear above intense jets located at wintertime tropopause is more likely to lead to critical levels facilitating wave breaking above any deep convection that occurs.

Figure 10 summarizes the distribution of moderate-or-greater PIREPs and EDR reports and illustrates the mechanisms influencing CIT in a late-winter case on 9–10 March 2006 reported by Trier et al. (2012). This modeling study uses a large-domain strategy with multiple nests, which are typically employed in simulations of observed turbulence outbreaks, where it is necessary to both resolve the large-scale flow pattern conducive to turbulence and the fine-scale mechanisms directly responsible for its initiation. In these types of simulations, typical horizontal (vertical) spacings for the finest grid are less than 1 km (250 m), since accurate simulations of turbulence-producing phenomena, such as the breaking of vertically propagating internal gravity waves (e.g., Lane and Knievel 2005) and KHI (e.g., Trier and Sharman 2016), are sensitive to model resolution.

In this late-winter turbulence outbreak, there was widespread deep convection (Fig. 10a) and the turbulence occurred mostly above a strong upper tropospheric jet (Fig. 10b) that was oriented approximately south-to-north. In their simulations, Trier et al. (2012) found that the southerly jet was enhanced by $\sim 30\%$ due to the convectively induced anticyclonic UTLS circulation indicated schematically by the bold arrows in Fig. 10a.

A vertical cross section (Fig. 10b) along the transect SN in the $\Delta = 3.3$-km horizontal domain (Fig. 10a) indicates the most intense observed turbulence clustered in two regions including one from $x = 640$–760 km located directly above the jet maximum and another from $x = 300$–420 km located above the entrance region of the jet and relatively shallow moist convection. In the highest resolution horizontal domain with $\Delta = 667$ m, the moist convection in the jet entrance region (Fig. 10a, b) excites vertically propagating internal gravity waves when it impinges on the region of larger static stability beginning near 8-km MSL (Fig. 10c). These waves break while propagating downshear when they reach a critical level, where $U = c$ (Fig. 10d), similar to what is illustrated schematically in Fig. 5. Directly above the jet, overturning billows in the potential temperature field (Fig. 10e) implicate KHI as the turbulence mechanism. Though the KHI is horizontally remote from ongoing moist convection, note that it arises in response to strong vertical shear associated with convectively induced enhancement of the jet (Fig. 10a).

4.3. Turbulence Horizontally Displaced from Adjacent Deep Convection

In addition to occurring above cloud, CIT may also be displaced laterally from clouds associated with preexisting deep convection. In particular, the likelihood of moderate-or-greater turbulence is two times greater than the background occurrence within 75 km of deep convection (Fig. 4b). The ability of a current research model to simulate this phenomenon and its sensitivity to model resolution was analyzed by Barber et al. (2018). A specific example from

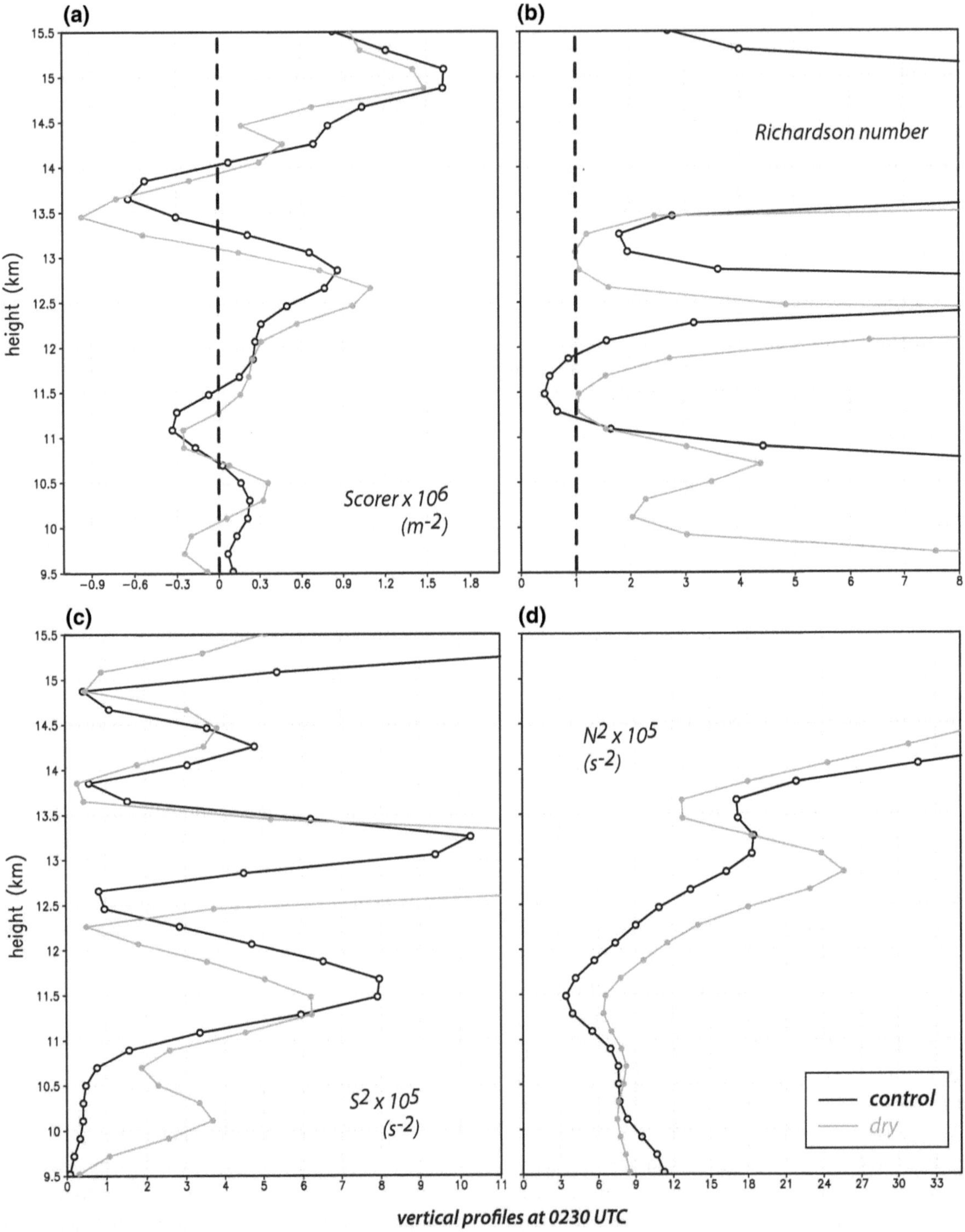

Figure 11
Vertical profiles from a location southeast of the storm from control and dry simulations of the observed case in Fig. 7a. (**a**) Scorer parameter calculated using horizontal winds parallel to the wave propagation vector and a 30-m s^{-1} horizontal phase speed; (**b**) Gradient Richardson number and its contribution from (**c**) the squared magnitude of the vertical shear S, and (**d**) the squared Brunt-Väisälä frequency. From Lane et al. (2012). Copyright American Meteorological Society. Reprinted with permission

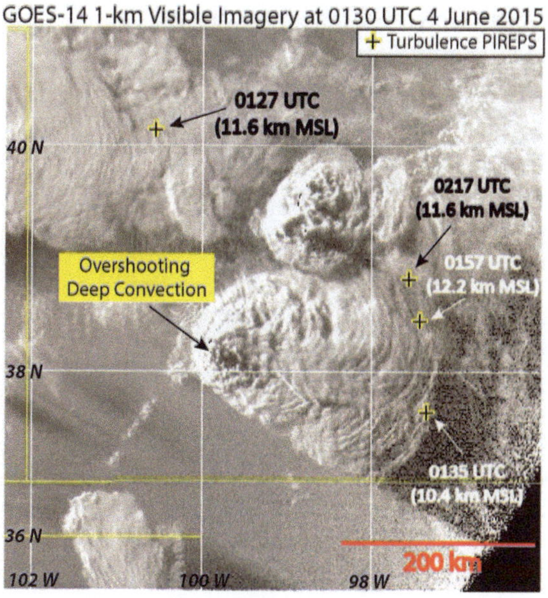

Figure 12
GOES 1-km visible satellite imagery from 0130 UTC 4 June 2015. The cross symbols indicate locations of turbulence PIREPs within 50 min of the time of the image

Fovell et al. (2007) and Lane et al. (2012) occurred on 5 August 2005. In this case, EDR measurements documented moderate and severe turbulences at 11.3–11.9-km MSL encountered by two commercial aircraft approximately 20–30-km southeast of a small but intense thunderstorm complex (Fig. 7a).

CIT horizontally adjacent to, but outside of, deep convection has been less frequently studied and is not as well understood as CIT occurring directly above deep convection. However, some candidate mechanisms that could account for such turbulence have been hypothesized based on results from numerical simulations. For instance, Fovell et al. (2007) and Lane et al. (2012) proposed that vertically trapped gravity waves triggered by deep convection could propagate outward from the storm and result in localized regions of subcritical Richardson number within larger scale environments, where the background R_i is small.

Fovell et al. (2007) and Lane et al. (2012) presented calculations of the Scorer parameter l^2 (Eq. 3) and, therefore, the susceptibility to wave trapping, for an idealized simulation based on the 5 August observed severe turbulence case at a downstream location a few tens of kilometers outside of the simulated cloud. These calculations are compared with those from a corresponding idealized simulation, where moist processes (e.g., deep convection) were excluded and both simulations have trapping layers centered in the upper troposphere near 11-km MSL and near the tropopause at around 13.5-km MSL (Fig. 11a). The full physics simulation that included moist processes also has a 0.5–1.0-km-deep layer of $R_\mathrm{i} < 1$ centered at 11.5-km MSL (Fig. 11b), which corresponds well with the altitude of the observed turbulence inferred from the EDR reports in Fig. 7a. The simulated Ri values at this altitude are smaller than the nearly subcritical values from the dry simulation of $R_\mathrm{i} \sim 1$ (Fig. 11b), and Fovell et al. (2007) and Lane et al. (2012) attribute these decreases to trapped internal gravity waves generated by upstream deep convection in the control (full physics) run. These R_i decreases in control – dry (Fig. 11b) near and just above the bottom trapping layer near 11.5 km (Fig. 11a) are related to the combination of increases in vertical shear magnitude (Fig. 11c) and reductions in static stability (Fig. 11d) compared to the background environment signified by the dry simulation.

In the above example, short-wavelength ($\lambda \sim 10$ km) trapped gravity waves are hypothesized to contribute to the onset of turbulence not through wave breaking but through shear instabilities resulting from local effects of the wave-induced perturbations on Ri. Figure 12 presents visual evidence of similarly trapped waves within an anvil extending downstream from a relatively isolated severe thunderstorm penetrating the tropopause in the central United States during 4 June 2015. In this case, there are turbulence PIREPs near wave fronts at the outer edge of the downstream anvil (Fig. 12). However, unlike in the 5 August 2005 case, only light intensities of turbulence were reported in the 4 June 2015 case.

Zovko-Rajak and Lane (2014) analyzed high-resolution idealized simulations to diagnose the relationship between a severe thunderstorm similar to the 4 June 2015 case (Fig. 12), and turbulence within its upper level outflow downstream of the overshooting deep convection. Like in the 4 June 2015 case, Zovko-Rajak and Lane (2014) also found short-wavelength internal gravity waves that were

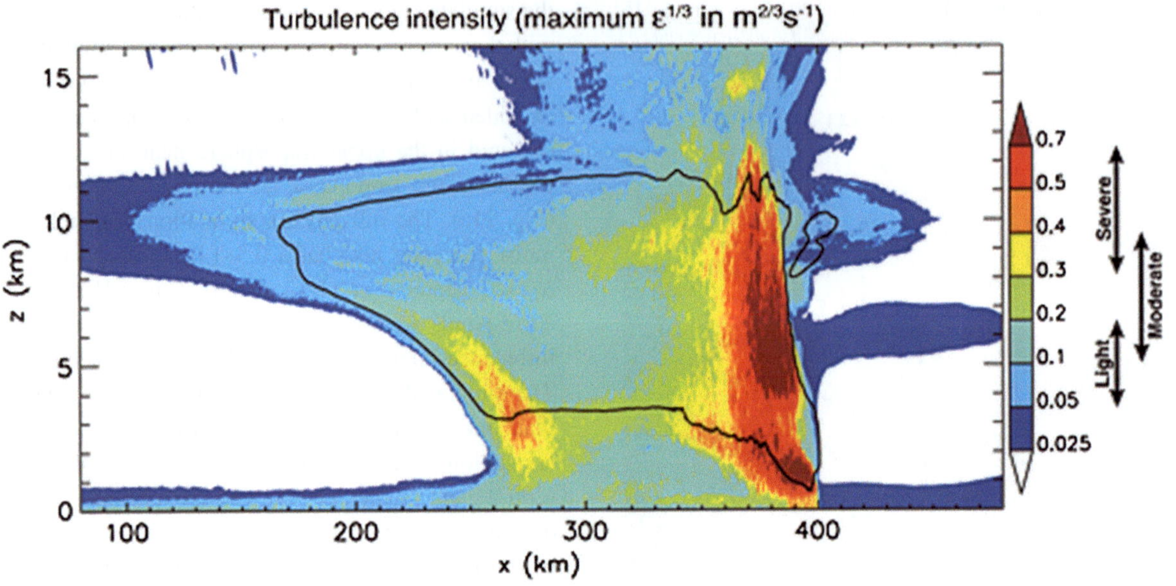

Figure 13
Spatial distribution of along-line maximum turbulence ($\varepsilon^{1/3}$) obtained during the final 30 min of the simulation reported in Lane and Sharman (2014). Also shown is the along-line mean cloud outline (black contours defined by the 0.1 g kg^{-1} cloud mixing ratio isopleth). The approximate turbulence categories are based on comparisons to pilot reports as described by Sharman et al. (2014). From Lane and Sharman (2014) Fig. 4. Copyright American Geophysical Union. Reprinted with permission

trapped near the jet level. However, unlike in the 4 June 2015 case (Fig. 12), their waves did not emanate from the region of deep convection. Instead, their gravity waves and CIT developed locally 50–100-km downstream from the parent deep convection, and were attributed to KHI located both above and below the jet layer in which the waves were trapped. Nevertheless, the gravity waves may have played an important role in their simulated CIT by providing a mechanism that allowed communication and phase locking between the regions of instability in vertical shear layers both above and below the outflow jet. Their analysis underscores the importance of both vertical shear and static stability modifications to the background environment in storm outflows, and the potentially complex interplay between waves and instabilities in such outflows on the development of CIT.

4.4. Turbulence Near and Within MCS Anvils

Mesoscale convective systems (MCSs), which are ensembles of thunderstorms producing a contiguous precipitation area of 100 km or more in at least one horizontal direction, constitute a significant hazard to commercial aviation. A high-resolution LES of a quasi-2D MCS (Lane and Sharman 2014) illustrates elevated levels of turbulence (EDR > 0.05) within deep convection near its leading edge ($x = 350$–400 km), wave breaking regions above deep convection (Sect. 4b), and within and near the cirrus anvil cloud that extends over 200 km from the deep convection region (Fig. 13).

As noted earlier, the heavily precipitating deep convection region in MCSs, where turbulence is typically most severe (Fig. 13), is easily avoided using on board and ground-based radar. The more extensive anvil regions, which themselves often contain significant turbulent vertical motions of order ~ 1 m s^{-1} (e.g., Petre and Verlinde 2004; Muhlbauer et al. 2014), can also typically be avoided through visual identification.

Research dropsonde and research aircraft measurements have documented widespread areas of either reduced or subcritical gradient *Ri* in both midlatitude continental MCS (e.g., Collander et al. 2006; Gultepe and Starr 1995) and tropical cyclone (e.g., Molinari et al. 2014; Duran and Molinari 2016) anvils. However, because of their size (which can sometimes exceed several hundred km) and the lesser

Figure 14
MODIS visible satellite image of transverse cirrus bands on the northern flank of a mesoscale convective system in the Midwestern United States at 1741 UTC 27 July 2008. Adapted from http://ge.ssec.wisc.edu/modis-today/images/terra/true_color/2008_07_27_209/t1.08209.USA3.143.250m.jpg

likelihood of severe turbulence than in deep convection, anvils are sometimes penetrated by commercial aviation. Thus, it is of interest to identify regions within anvils that are most susceptible to significant turbulence and to understand the variety of different mechanisms responsible for such turbulence.

The trailing anvil cloud and clear region beneath it (Fig. 13), respectively, comprise ascending and descending airstreams, which are driven by mesoscale buoyancy gradients within the MCS (Lafore and Moncrieff 1989; Weisman 1992). The region of enhanced EDR that extends downward beneath cloud base at the back edge of the anvil (Fig. 13) is associated with the descending mesoscale rear-inflow jet, which is a common feature in mature-to-dissipating MCSs (Augustine and Zipser 1987; Smull and Houze 1987). Collander et al. (2006) hypothesized that severe turbulence encounters within this region may be associated with KHI arising from strong vertical shear often found immediately beneath the rear-inflow jet. Furthermore, R_i can also be reduced by small static stability beneath the anvil cloud due to both subsidence and vertical gradients of diabatic cooling resulting from differences in the strength of sublimation or evaporation with depth beneath cloud base (e.g., Harris 1977; Knight et al. 2004).

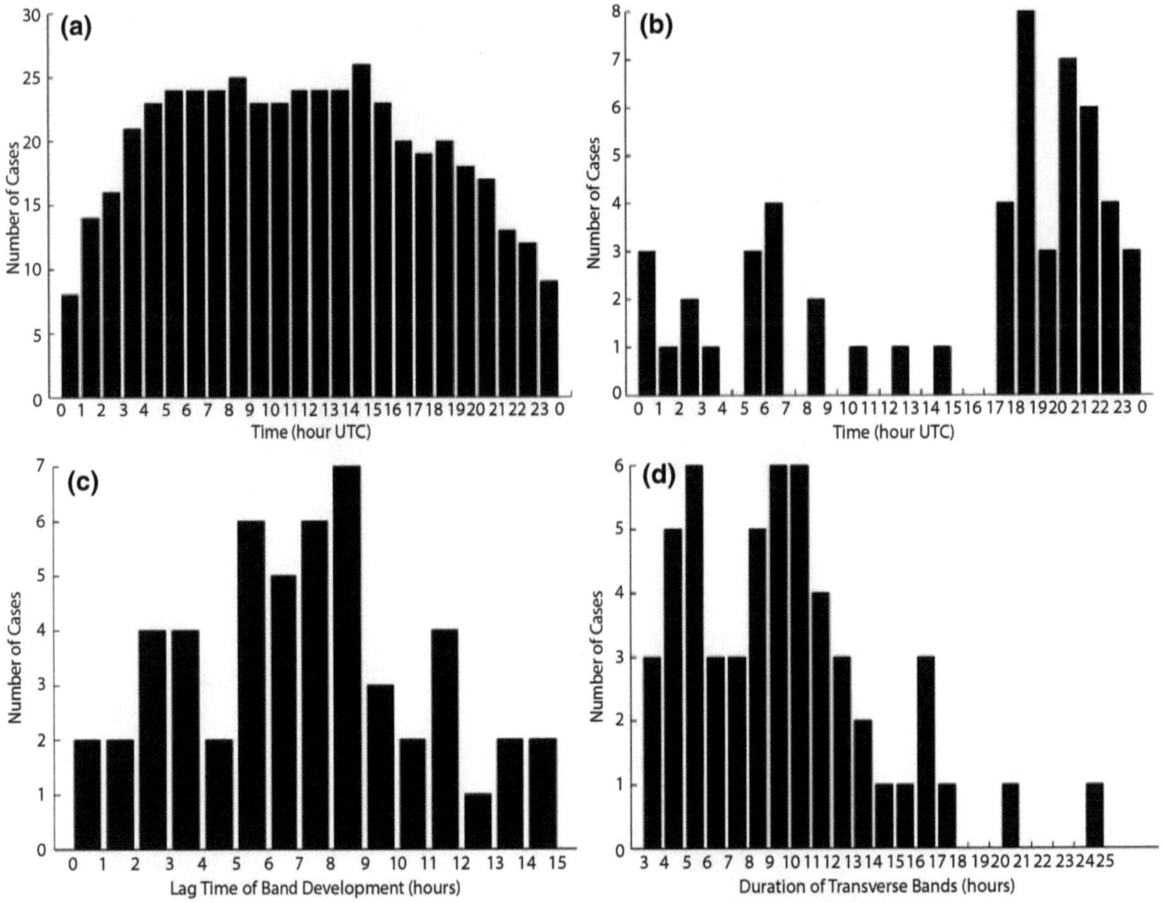

Figure 15
Histograms showing the **a** temporal distribution of transverse bands, **b** initiation time of associated deep convection, **c** time lag between initiation and transverse band development, and **d** transverse band duration. Adapted from Lenz et al. (2009)

Though they are most intense beneath the back edge of the anvil and in close proximity to the deep convection region, moderate-or-greater turbulence may occur with a shallow layer beneath cloud base along the entire horizontal extent of the anvil region in the Lane and Sharman (2014) simulation (Fig. 13). This is consistent with numerous studies of turbulence occurring underneath midlevel cirrus (e.g., Luce et al. 2010; Kudo 2013; Kudo et al. 2015), which has been attributed primarily to shallow regions of static instability developing from precipitation falling into approximately neutral subsaturated layers. In some cases (e.g., Luce et al. 2010), such turbulence may occur near or below mammatus clouds (Schultz et al. 2006) along the underside of the anvil.

At typical commercial aviation cruising altitudes (9–13-km MSL), EDR values consistent with light turbulence are found in the Lane and Sharman (2014) simulation extending nearly 100 km outside of the anvil outflow edge (Fig. 13). However, a shallow but horizontally extensive region of EDR values consistent with moderate turbulence is found near the top of the outer anvil (Fig. 13).

One common source of turbulence near the outer edge of the MCS anvil cloud shield is cirrus bands that extend radially outward from the storm center. Figure 14 presents an example from a high-resolution visible satellite image of this type of cloud

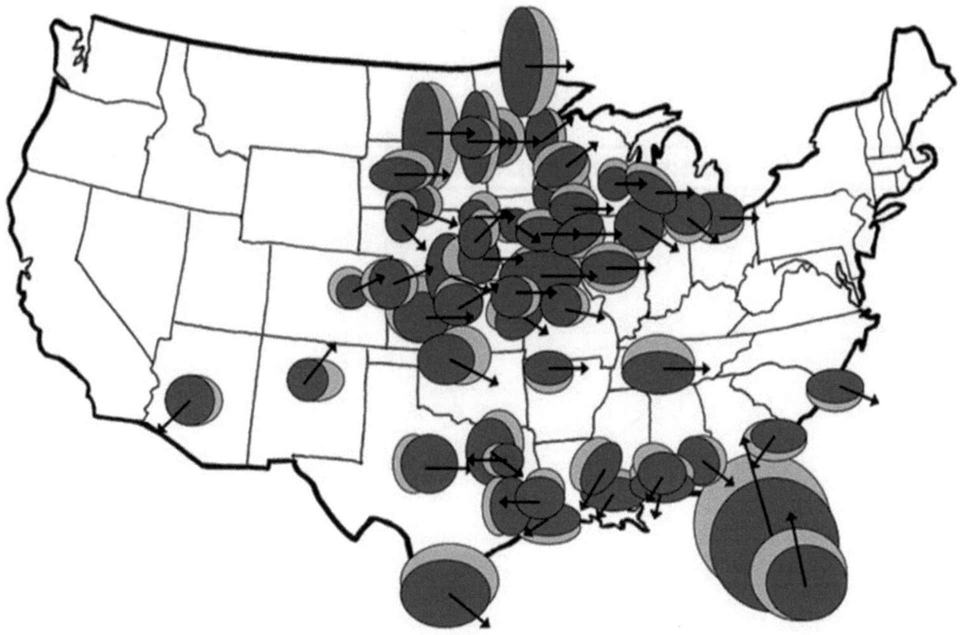

Figure 16
Geographic distribution of MCS anvils (dark gray ovals) for May–August 2006 with transverse band locations indicated by light gray shading. Orientation of arrows indicate the directions of MCS propagation. Adapted from Lenz et al. (2009)

organization, commonly referred to as "transverse bands" (Ellrod 1985; Lenz et al. 2009). Knox et al. (2010) note that transverse cirrus bands are also often present within tropical cyclone outflows and within upper level jet streams. The example from Knox et al. (2010) shows extensive banding with horizontal spacing between bands of less than 10 km and band lengths of about 100 km on the north side of deep convection within a small-to-medium sized MCS evident from the brighter white cloudiness (Fig. 14). These bands are nearly perpendicular (and hence transverse) to a strong 40–50-m s^{-1} precursor near-tropopause jet streak (not shown) that was likely further strengthened by anticyclonic outflow within the MCS.

Lenz et al. (2009) found that transverse band events lasted an average of 9 h in approximately ½ of a sample of 136 large MCSs over the central U.S. during the 2006 warm season. In their climatology, they found a similar 5–10-km spacing between bands to that in Fig. 14 and noted at least one observation of light (moderate) commercial aviation turbulence deduced from in situ EDR measurements within or near the bands for 93% (44%) of their cases. They also found transverse bands associated with MCSs to be most common overnight and during the morning (Fig. 15a) and were often long-lasting (Fig. 15d), sometimes persisting several hours beyond MCS dissipation. They further noted that the bands lagged convection initiation within the MCS by an average of 7 h (Fig. 15b, c) and were typically restricted to one side of the MCS cloud shield (Fig. 16).

These aspects suggest that such bands are closely linked to earlier deep convection within the MCS and its associated upper tropospheric outflow. Fritsch and Maddox (1981) discuss how a wind speed maximum typically occurs north and northeast of the origin of convection in MCSs, which is summarized schematically in Fig. 17 (Trier and Sharman 2009). In this diagram, upper tropospheric outflow trajectories from the parent convection within the MCS are indicated by the purple curves, which become anticyclonic due to Coriolis effects acting over the $t > 4$-h timescale of the outflow. Fritsch and Maddox (1981) point out that the strongest winds are often found where the outflow adds to the background flow, which is in the north and northeast portion of the MCS for the case of the climatological background westerly jet in the

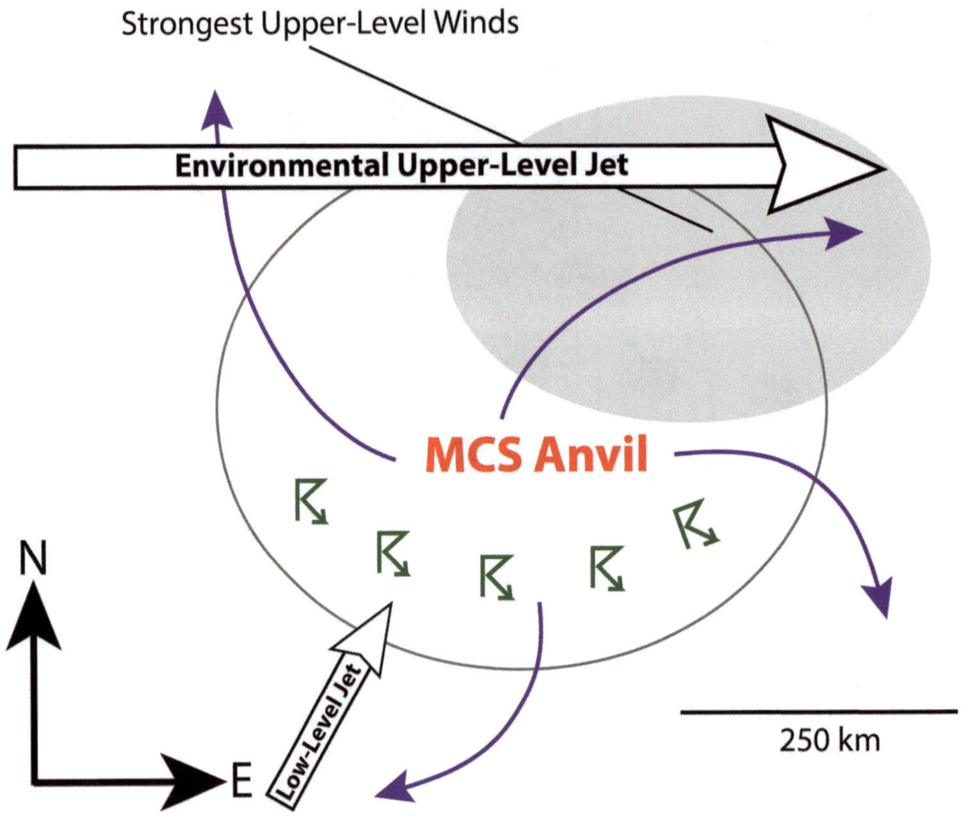

Figure 17
Schematic diagram illustrating spatial relationships among the environmental winds, the strongest deep convection (green thunderstorm symbols), MCS anvil cloud (thin oval), streamlines of MCS upper level outflow (thick violet curves with arrowheads) and the location of strongest upper level outflow winds (gray shading). Adapted from Trier and Sharman (2009)

midlatitude northern hemisphere (Fig. 17). While Lenz et al. (2009) found that transverse band formation was typically restricted to a single side of the MCS, there is not a clear bias in their cases towards the northern side (Fig. 16). However, the large variance and lack of a clear bias on the location of transverse bands in their observed northern hemispheric MCS cases could be associated with background flow differences related to synoptic variability among their different cases.

An example from 17 June 2005 of turbulence associated with transverse bands along the north side of a MCS is shown in Fig. 7b. Here, the observed bands in Fig. 7b appear less spectacular than in the Knox et al. (2010) example of Fig. 14, but this is largely an artifact of resolution differences (250 m versus 4-km pixel size) in the satellite images. Using a high-resolution horizontal nest with $\Delta = 600$ m, Trier et al. (2010) successfully simulated these bands occurring in the northern portion of the MCS upper tropospheric outflow (Fig. 18a, b). The simulated bands (Fig. 18b) originated from a region of moist static instability (Durran and Klemp 1982) and were oriented approximately parallel to the vertical wind shear through the depth of the outer anvil (not shown). This organization is similar to cumulus cloud streets in the boundary layer (e.g., LeMone 1973), which arise from thermal-shear instability (e.g., Asai 1970). Thermal-shear instability directly leading to upper tropospheric cirrus banding has also been reported in simulations of observed CIT associated with oceanic cyclones (Kim et al. 2014) and convectively enhanced synoptic jet streams (Trier and Sharman 2016). These model results concerning the connection between transverse bands and thermal-shear instability are consistent with earlier

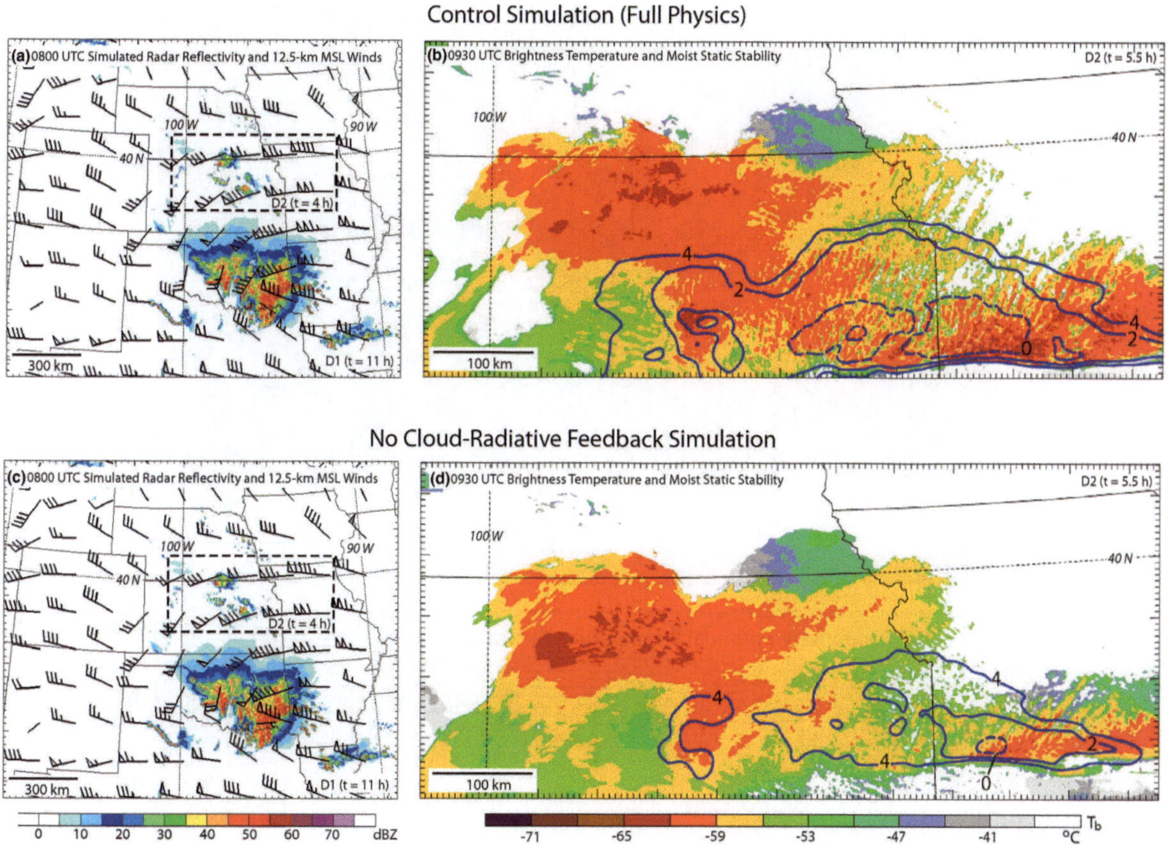

Figure 18

(Left) ARW–WRF model domains D1 (with 3-km horizontal grid spacing) and D2 (with 600-m horizontal grid spacing), 0800 UTC 17 June 2005 maximum radar reflectivity in a vertical column and 12.5-km horizontal winds for **a** control simulation and **c** simulation in which cloud-radiative feedbacks are disabled. (Right) Simulated brightness temperature and 12.1-km MSL smoothed moist static stability (2×10^{-5} s^{-1} with negative values dashed) at 0930 UTC 17 June 2005 in domain D2 for **b** control simulation and **d** simulation in which cloud-radiative feedbacks are disabled. Adapted from Trier et al. (2010)

speculation based on limited observations (Dixon et al. 2000).

Lenz et al. (2009) and Knox et al. (2010) indicate that in some cases, the cirrus bands occurring near the edges of MCS anvils are perpendicular to convectively generated gravity waves. Trier et al. (2010) found short-wavelength convectively generated gravity waves, which were trapped beneath the near-neutral layer in which the simulated transverse bands formed in their high-resolution simulation (Fig. 19). Here, local maxima in vertical velocity within the near-neutral layer are horizontally collocated with local minima in potential temperature within the gravity-wave layer below (Fig. 19). This suggests that convectively induced gravity waves could in some cases help release the moist static instability that evidently leads to transverse band formation. This hypothesis is consistent with observations of cirrus uncinus bands forming in near-neutral layers located above stable layers containing gravity-wave activity (Heymsfield 1975).

Trier et al. (2010) also found that both the moist static instability and the prominence of simulated cirrus banding were influenced by cloud-radiative feedbacks (cf. Fig. 18b, d) owing to differential radiative forcing related to longwave cooling at cloud top and warming at cloud base. However, this physical process does not explain the regionalization of the banding in the 17 June 2005 case and in the Lenz et al. (2009) climatology (Fig. 16), which is likely related to mesoscale asymmetries in the strength of the convectively induced upper

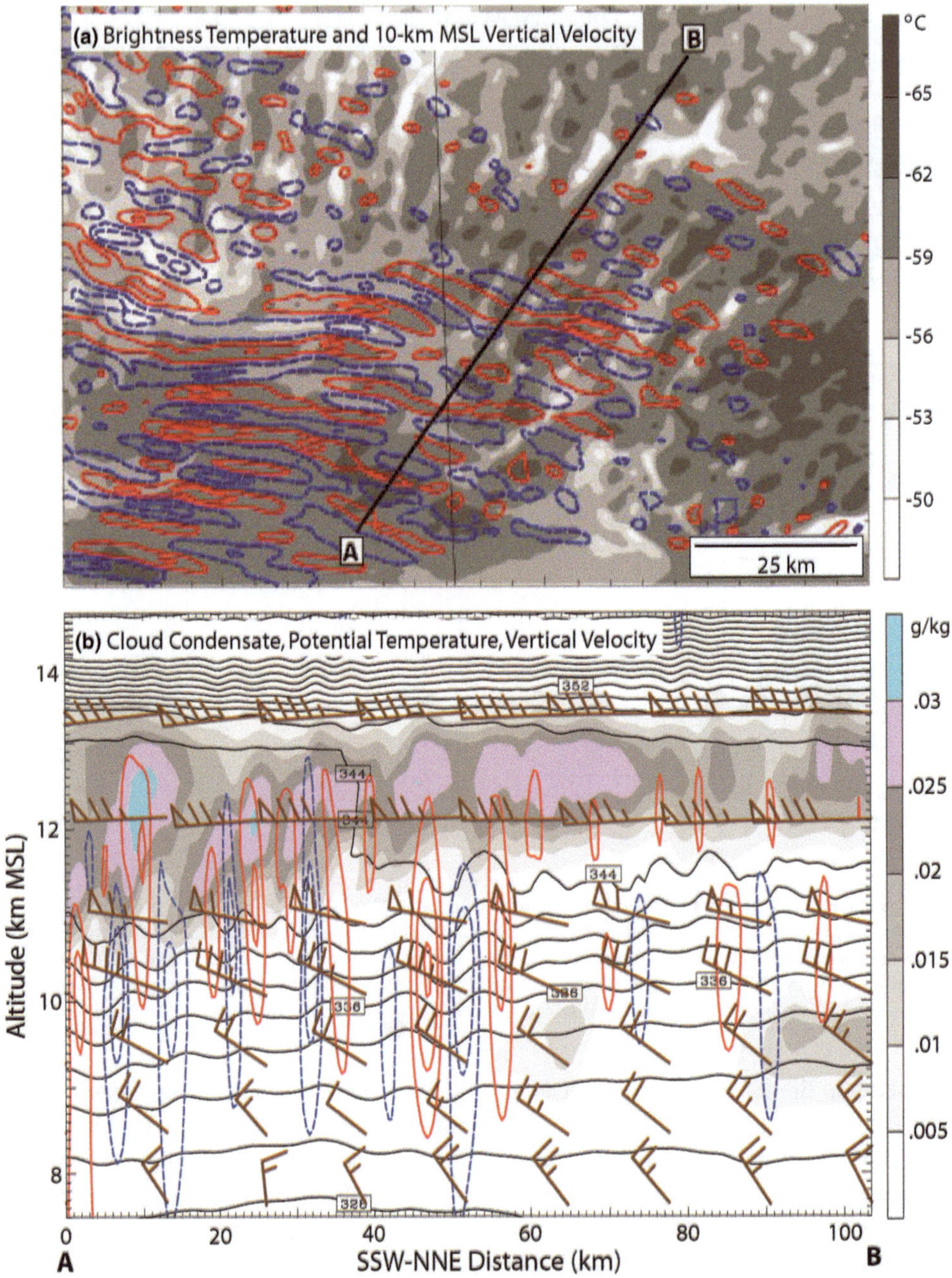

Figure 19
a Simulated brightness temperature and 10-km MSL vertical velocity contoured in 1.5-m s^{-1} (− 1.5 m s^{-1}) intervals with positive (negative) values in red (blue) starting at 0.75 (− 0.75) m s^{-1}. **b** Vertical cross section averaged for 6 km across transect AB in part **a** of total cloud condensate (color scale) horizontal winds, potential temperature contoured in 2-K intervals and vertical velocity contoured in 1-m s^{-1} intervals with positive (negative) values in red (dashed blue) starting at 0.5 (− 0.5) m s^{-1}. Adapted from Trier et al. (2010)

tropospheric outflow (Fig. 17) that affect the vertical shear and static stability. Similarly, recall that convectively induced outflow can lead to CIT initiated by KHI (e.g., Trier et al. 2012; Zovko-Rajak and Lane 2014) in cases where it is not statically unstable (i.e., $0 < R_i < 1$).

4.5. Remote Turbulence Associated with Convectively Enhanced Synoptic Jet Streams

Synoptic jet streams near the tropopause have long been recognized as locations predisposed to turbulence (e.g., Shapiro 1976; Kennedy and Shapiro 1975; Koch et al. 2005; Klostermeyer and Rüster 1980) for a variety of different reasons (Sharman et al. 2012). The downstream (exit) regions of such jets are often generation sites of mesoscale inertia–gravity waves (e.g., Uccellini and Koch 1987; Plougonven et al. 2003; Zhang 2004). Like smaller scale internal gravity waves, mesoscale inertia–gravity waves can be associated with turbulence (e.g., Ellrod et al. 2015; Lane et al. 2004; Koch et al. 2005). Plougonven and Zhang (2016) note that the substantial vertical tilt of these longer waves makes them particularly effective in modifying the environmental vertical shear and static stability in a manner conducive to turbulence. Inertia–gravity waves near jet streams have a variety of different possible generation mechanisms (see, e.g., Plougonven and Zhang 2016 and references therein). Among these different generation mechanisms, organized deep convection has been examined in recent observational (e.g., Hoffmann and Alexander 2010) and modeling (e.g., Wei and Zhang 2014) studies.

In the remainder of this subsection, we provide an example of how convectively generated mesoscale inertia–gravity waves can lead to turbulence at considerable distances from their source location in the exit region of a near-tropopause synoptic jet stream. The turbulence in transverse cirrus bands within MCS anvils discussed in the previous subsection provided an example of CIT occurring due to processes within the upper level outflow emanating from deep convection up to several 100-km away. In the forthcoming example from Trier and Sharman (2016), the turbulence was related to, but was even farther removed (both spatially and temporally) from deep convection that had occurred earlier in the anticyclonically sheared exit region of a synoptic jet.

In this case over the North Atlantic Ocean on 15 Nov 2011, episodes of turbulence occurred in the vicinity of separate mesoscale clusters of upper tropospheric cirrus bands. This is a common situation that aviation forecasters recognize in their turbulence forecasting procedures (Melissa Thomas, private communication). The mesoscale regions of cirrus bands of 15 Nov 2011 (Fig. 20a) were well simulated by Trier and Sharman (2016) and diagnosed to result directly from thermal-shear instability. A supporting simulation where cloud-radiative feedbacks were disabled (Fig. 20b) was used to infer mesoscale reductions in static stability prior to the onset of cirrus banding in their higher resolution full physics control run (cf. Fig. 20a, b). Both the regions of cirrus banding in the control run and the regions of lowered static stability in the supporting simulation had a mesoscale horizontal wavelength of 400–500 km (Fig. 20b) and had a structure consistent with inertia–gravity waves (Trier and Sharman 2016, their Fig. 23a).

The periodic mesoscale variations in upper tropospheric static stability in the control simulation are evident in a Hovmöller diagram (Fig. 21a). The three distinct episodes of cirrus banding A, B, and C formed about 4-h apart near −60° longitude (Fig. 20a) within eastward progressing regions of reduced static stability that originated farther west (Fig. 21a) over the eastern United States. The link to upstream deep convection is suggested by a comparison with a Hovmöller diagram from an otherwise identical dry simulation (Fig. 21b). Larger scale static stability variations occur in the dry simulation (Fig. 21b), including those associated with the passage of a synoptic trough, also evident in the full physics control simulation (Fig. 21a). However, the faster moving mesoscale stability minima (Fig. 21a), wherein the turbulent cirrus banding occurs, are absent in the simulation without upstream deep convection (Fig. 21b).

The wavelike mesoscale regions of reduced static stability in Fig. 20b were located rearward and approximately 90° out of phase with regions of negative potential vorticity. Using surface rain rate as a proxy for diabatic heating, these regions of negative potential vorticity are more directly linked to much earlier deep convection occurring at distances greater than 1000-km upstream over the eastern United States.

The linkage of upper tropospheric negative potential vorticity with earlier deep convection may be understood by recognizing that in the absence of

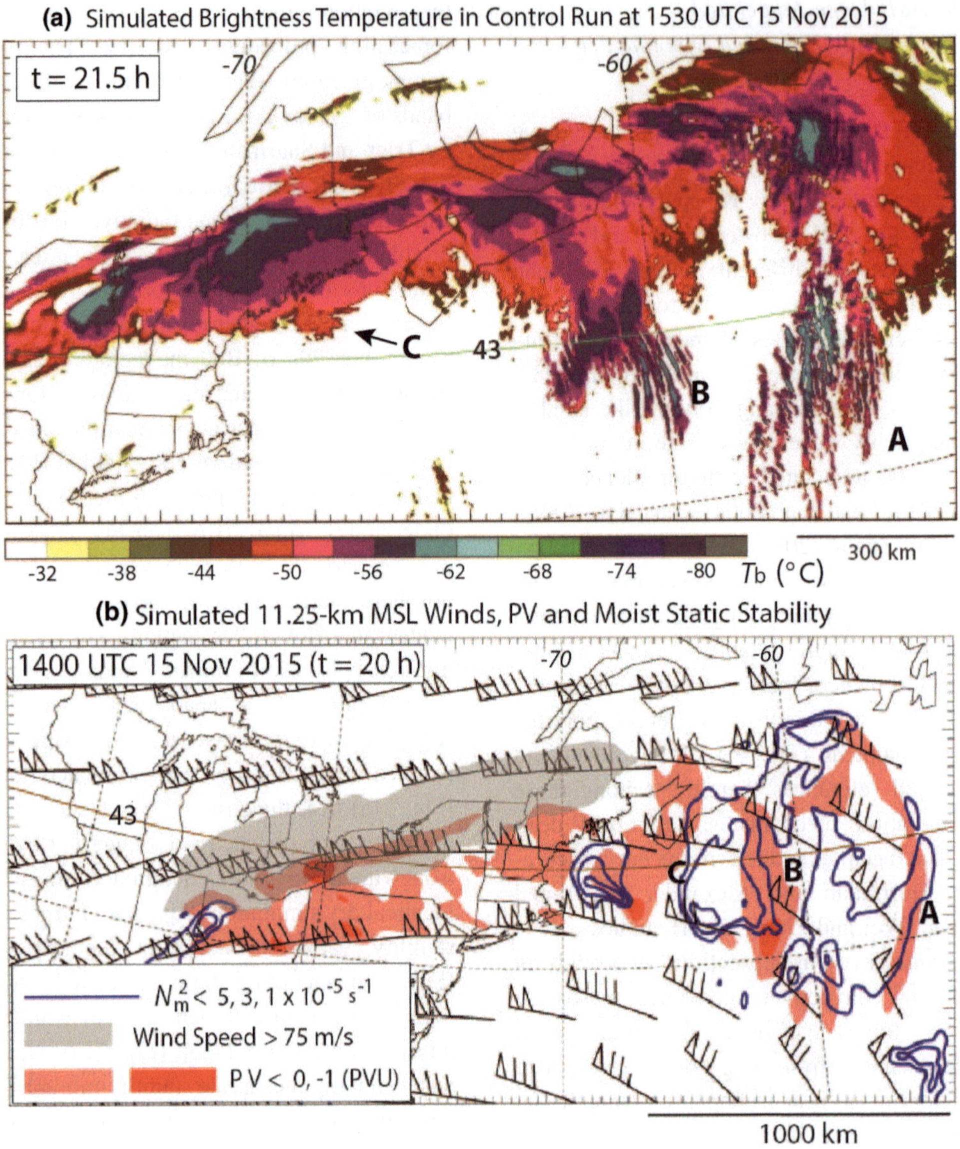

Figure 20

a Simulated infrared satellite brightness temperature within a control simulation with 1-km horizontal grid spacing at 1530 UTC 15 November 2011 and **b** 11.25-km MSL horizontal winds, moist static stability (N_m^2), areas of negative Ertel's potential vorticity (PV), and wind speed greater than 75 m s^{-1} (semitransparent gray shading) at 1400 UTC 15 November 2011. The annotations A, B, and C denote persistent negative PV features that were tracked using animations of model output having a frequency of 30 min. Corresponding regions of cirrus banding with turbulence that form in low moist static stability regions are indicated in part (**a**). Adapted from Trier and Sharman (2016)

friction, the rate of change of Ertel's potential vorticity, $P = (\zeta_a \cdot \nabla\theta)/\rho$, following an air parcel is

$$\frac{dP}{dt} = \frac{1}{\rho}(\zeta_a \cdot \nabla d\theta/dt).$$

The right side of the above equation indicates that diabatic heating ($d\theta/dt$) along the 3D absolute vorticity (ζ_a) vector is the physical process responsible for generating the negative potential vorticity. Hertenstein and Schubert (1991) demonstrated that vertical gradients of diabatic heating within large squall lines, such as those that occurred 12–18-h earlier over the eastern United States (Trier and

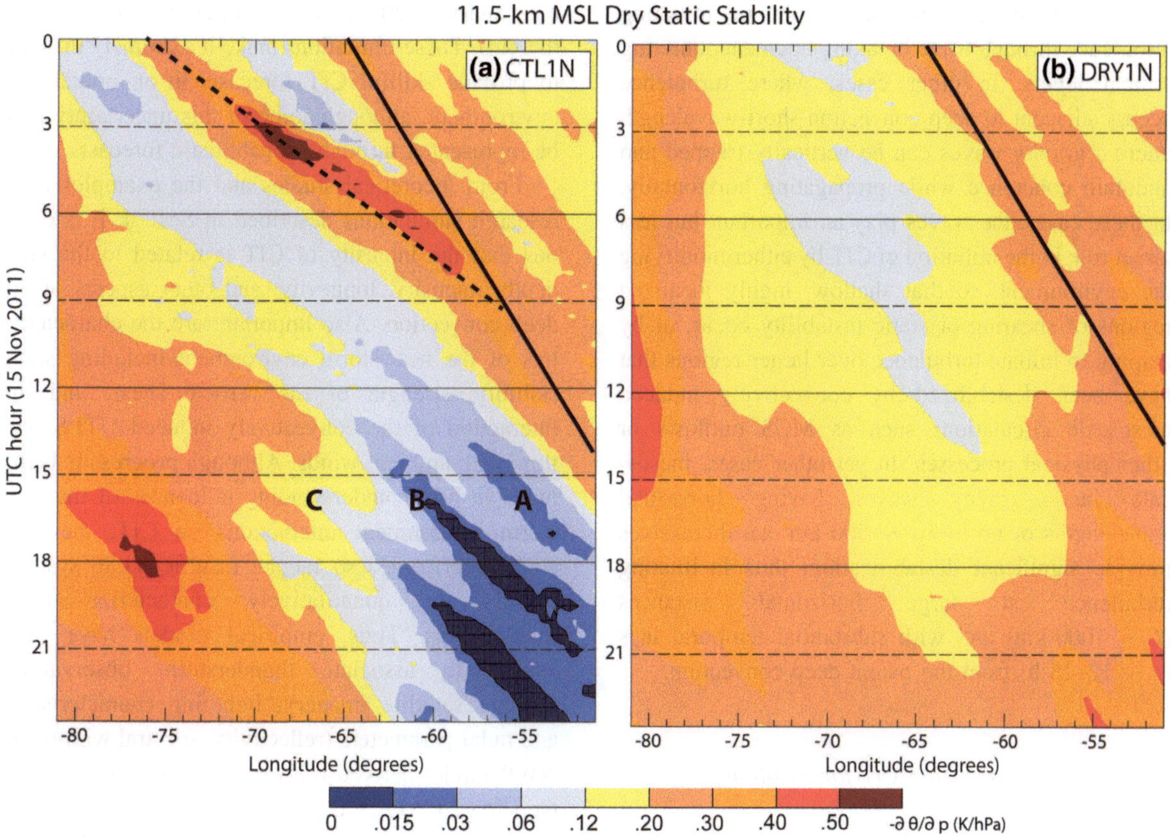

Figure 21
Time-longitude Hovmöller diagrams of 11.5-km MSL dry static stability [K $(\mathrm{hPa})^{-1}$] from **a** full physics control simulation, and **b** dry simulation reported by Trier and Sharman (2016). The sloping dashed line in **a** indicates the phase line of a leading high amplitude inertia–gravity wave from their higher resolution simulation, as shown in Fig. 20a. The annotations A, B, and C in part **a** indicate static stability minima in subsequent waves, which are associated with regions of cirrus banding shown in Fig. 20a. The sloping solid line in both parts **a**, **b** indicates the spatiotemporal location of a simulated upper level synoptic trough. Adapted from Trier and Sharman (2016)

Sharman 2016), can produce mesoscale regions of negative potential vorticity in the upper troposphere above where such heating is maximized. In the absence of subsequent diabatic sources, the negative potential vorticity is then advected downstream while undergoing substantial deformation in the exit region of the jet. Banded cirrus (Fig. 20a) with CIT develops in areas of reduced static stability that lags by ¼ wavelength the regions of negative potential vorticity (Fig. 20b).

4.6. Discussion

High-resolution numerical simulation research models have been used to investigate the onset of observed CIT in several cases influenced by diverse mechanisms. The turbulence in these cases is located either directly above or is horizontally displaced from tropopause-penetrating deep convection. In the latter, the CIT occurs at ranges from immediately adjacent to the parent deep convection to more than 1000-km away. A common aspect of each of these scenarios is that convectively generated gravity waves provide some link from the deep convection to subsequent CIT. However, both the horizontal scale of the gravity waves and the manner in which this linkage occurs varies substantially among different CIT types.

In the cases of turbulence occurring above deep convection, vertically propagating internal gravity waves of relatively short horizontal wavelengths

($\lambda_H \sim$ 5–10 km) play a direct role in initiating turbulence through wave breaking upon encountering critical levels. In other cases, where turbulence occurs adjacent to deep convection short-wavelength internal gravity waves can be vertically trapped and maintain coherence while propagating horizontally. In these cases, the waves play an important but less direct role in the initiation of CIT by either modifying the environment, so that shallow, highly localized regions of shearing or static instability occur, or by helping to initiate turbulence over larger regions that have been destabilized by convectively induced mesoscale circulations such as MCS outflows or other physical processes. In yet other cases, mesoscale inertia–gravity waves having horizontal wavelengths of up to $\lambda_H \sim 500$ km can themselves provide significant destabilization, thus facilitating turbulence at large horizontal distances ($L \sim 1000$ km) and with substantial temporal lags ($t \sim 12$–18 h) from the parent deep convection.

5. *Forecasting CIT for Aviation*

While Doppler weather radar can be used to detect in-cloud turbulence hazards (Williams and Meymaris 2016), it does not provide turbulence information in regions with low-signal-to-noise ratio (SNR) radar returns, contaminated data or poor radar coverage, and of course cannot be easily used as a forecasting tool. In addition, operational NWP models have limited utility in forecasting the location, strength, and type of convection due to latency issues, inadequate model resolution, and incomplete model physics (e.g., Lilly 1990; McNulty 1995; Bernardet et al. 2000; Weisman et al. 2008; Wakimoto and Murphey 2009). Furthermore, Lane and Knievel (2005) show a strong dependence of numerically simulated convective gravity-wave characteristics on model grid resolution. Finally, even if NWP models could properly forecast convection and convective gravity waves, explicit prediction of turbulence through direct numerical simulation (DNS) techniques or large eddy simulation (LES) techniques is computationally prohibitive, requiring semi-empirical (and less accurate) techniques to infer turbulence from the coarser resolution NWP model output (see, e.g., Sharman 2016; Knox et al. 2016 and references therein). These limitations make it extremely difficult to provide skillful CIT forecasts in an operational environment, although some of this uncertainty may be represented through probablitistic forecasts.

From theoretical studies and the examples from research simulations described in Sect. 4, it is obvious that the intensity of CIT is related to the size, depth, intensity, longevity, and other aspects of the deep convection. Also important are the characteristics of the near-storm environment including static stability, strength of the vertical shear, and the interaction of the convectively induced UTLS outflow with ambient winds. Although progress is being made in better understanding influences of thunderstorm–environment interactions on CIT, the case studies performed so far are insufficient to systematically or quantitatively characterize these relationships. Thus, empirical models have been sought to associate thunderstorm observations through satellite imagery, lightning characteristics, and radar parameters (reflectivity, spectral width) and NWP model derived relevant environmental parameters (e.g., CAPE, CIN, R_i), with observations of CIT.

One such empirical model, termed DCIT (Diagnosis of Convectively Induced Turbulence) is described in Williams (2014), where an artificial intelligence technique (viz., random forests) is used to fuse data from Doppler radar, geostationary satellites, total lightning data, and NWP model data to produce gridded deterministic and probabilistic CIT nowcasts at 6-km horizontal and 1000 ft vertical grid spacing updated every 15 min. The data fusion algorithm was trained using a retrospective dataset that includes in situ EDR reports and collocated predictor fields. The algorithm output was then evaluated on an independent test set using a common performance metric (receiver operating characteristic ROC curves, e.g., Gill 2016). Although the statistical evaluations show promising results for this technique, it is not yet available operationally.

Recent studies suggest that the lightning characteristics such as flash energy, length and extent densities may relate to turbulence intensity and location (Bruning and MacGorman 2013). Results based on comparisons of observational lightning to EDR data from Deierling and Williams (2016) suggest that the occurrence of moderate-or-greater turbulence in storms is related to storm total lightning flash rate, though the exact relationships depend on the storm type. The footprint of horizontal lightning flash extents is also observed to coincide with areas of moderate and greater turbulence.

Another approach is simply to infer CIT from CAT diagnostics (e.g., Sharman and Pearson 2017; Pearson and Sharman 2017) assuming that the NWP model captures large-scale perturbations due to convection. Similarly, cloud-top height (CTH) and other parameters can be used to produce a CIT likelihood (Kessinger et al. 2017). The CTH product maps the satellite cloud-top temperature (11 μm) to flight level using the global forecast system (GFS) model sounding at the cloud location. The convective diagnosis oceanic (CDO) product empirically determines the area of convective storms most hazardous for aviation by combining three satellite-based convection detection algorithms with global ground-based lightning strike data in a scaled and weighted data fusion scheme. Although not as sophisticated as the DCIT product, these products do provide global coverage, and also provide a means for uplinking the data to the cockpit on their electronic flight bag (EFB).

Empirical approaches such as these will only improve as higher resolution and cloud-resolving NWP models and other data sources (e.g., GOES-16 satellite) become available operationally. Convective gravity-wave parameterizations for use in larger scale NWP models are a continuing area of research (Beres 2004; Beres et al. 2004; Chun and Baik 1998, 2002; Kim et al. 2013; Stephan and Alexander 2015). Kim et al. (2017) have taken the next step of parameterizing CIT due to convective gravity-wave breaking. As our understanding of CIT increases through simulation case studies, these parameterizations will be become more sophisticated and likely more effective.

6. *Summary and Future Research Needs*

This review has attempted to demonstrate the utility of case studies based on actual CIT events, where high-resolution observations are used in conjunction with high-resolution numerical simulations. For the different mechanisms of CIT onset identified here, the high-resolution observations are provided either from digital flight recordings of an event or 1-min in situ EDR messages available from selected commercial aircraft, or to a lesser extant, PIREPs. Another approach would be to collect high quality atmospheric data from research aircraft involved in field campaigns, either through in situ measurements or remote sensing. Numerical simulation models must at a minimum be convection-resolving, which requires grid spacings of a few km or less (e.g., Bryan et al. 2003). This is usually sufficient to at least partly resolve convective generated gravity waves, and to provide an indication of conditions favorable to their breakdown into turbulence. However, much smaller grid spacings of about 100 m, would be required to start to resolve "turbulence" of relevance to aviation. Currently this can only be achieved using a grid nesting approach, but even then, the computational requirements can be severe.

The issue of CIT avoidance may be more pressing in the future, given that CIT frequencies could increase with increases in the frequency and intensity of thunderstorms implied by future climate projection models (e.g., Del Genio et al. 2007; Trapp et al. 2007; Diffenbaugh et al. 2013; Gensini et al. 2014; Romps et al. 2014; Prein et al. 2017). Climate change may affect clear-air turbulence as well (Williams 2017; Storer et al. 2017; Williams and Joshi 2013, 2016), further complicating the problem of turbulence forecasting.

Going forward, a multifaceted approach using dedicated research flights, expanded in situ EDR observations, and high-resolution modeling along with theoretical studies (to the extent possible) will continue to be required. Improvements in the representation of convective and turbulence processes within numerical simulation models will continue to be an active area of research. As computational resources increase, higher resolution grids can be accommodated over larger domains. High-resolution LES models that include the UTLS (Schalkwijk et al. 2015) have already been run operationally over limited horizontal domains.

Some fundamental research questions need to be addressed to promote enhanced understanding of CIT processes to ultimately be able to nowcast and possibly forecast CIT intensity, timing, and location. These questions include (from Sharman et al. 2016):

- What is the spectrum of convective gravity waves and how does the energy associated with the waves relate to the region of wavelengths important to aircraft?
- What is the relation of CIT to the convective cloud characteristics?
- What is the global climatology of CIT?
- How will climate change impact the intensity and frequency of CIT?
- What is the best way to accommodate uncertainty in CIT forecasts?
- What role do secondary gravity waves emitted from breaking convective gravity waves play in the turbulence process?
- What are the predictability limits of CIT using NWP models?
- What observations are needed to improve CIT?

Operationally, given the complexity of CIT, it does not appear that simple avoidance tactics as sought by the original FAA guidelines can be ever be developed. Knowledge of the thunderstorm structure and environmental parameters (e.g., the vertical distribution of potential temperature and wind throughout the UTLS) is required to provide a reasonable nowcast of CIT. Although such methods are beginning to be developed, a crucial aspect of tactical avoidance is to ensure timely transmission of these products to the cockpit. With onboard laptops or tablets the hazardous areas may be displayed for rapid evaluations, and hence facilitate better decisions for real-time avoidance and diversion strategies. Pilot training to promote a better understanding of CIT processes and the interpretation of uncertainty in forecasts is also required. This will undoubtedly reduce the number of turbulence encounters in the future. However, in the foreseeable future, improvements are likely to occur as only incremental steps towards fulfilling the operational goal of reducing the number of CIT encounters.

Acknowledgements

We thank Wiebke Deierling, Teddie Keller (each of NCAR), Gretchen Mullendore (University of North Dakota), and an anonymous reviewer for their comments of an earlier version of the manuscript, which helped improve the presentation. We also thank Julia Pearson (NCAR) for constructing Fig. 1 and Lara Ziady (NCAR) for Fig. 2. We acknowledge Earth Networks for providing ENTLN (lightning) data. This research is in response to requirements and funding by the Federal Aviation Administration (FAA). The views expressed are those of the authors and do not necessarily represent the official policy or position of the FAA.

References

Abarbanel, H. D. I., Holm, D. D., Marsden, J. E., & Ratiu, T. (1984). Richardson number criterion for the nonlinear stability of three-dimensional stratified flow. *Physical Review Letters, 52,* 2352–2355.

Ahmad, N. N., & Proctor, F. (2011). Large eddy simulations of severe convection induced turbulence. AIAA-2011-3201. *3rd AIAA Atmospheric Space Environments Conference*, Honolulu, HI.

Alexander, M. J., & Barnet, C. (2007). Using satellite observations to constrain parameterizations of gravity wave effects for global models. *Journal of the Atmospheric Sciences, 64,* 1652–1665.

Alexander, M. J., Holton, J. R., & Durran, D. R. (1995). The gravity wave response above deep convection in a squall line simulation. *Journal of the Atmospheric Sciences, 52,* 2212–2226.

Alexander, M. J., & Pfister, L. (1995). Gravity wave momentum flux in the lower stratosphere over convection. *Journal of Geophysical Research, 22*(15), 2029–2032.

Ansong, J. K., & Sutherland, B. R. (2010). Internal gravity waves generated by convective plumes. *Journal of Fluid Mechanics, 648,* 405–434.

Asai, T. (1970). Stability of a plane parallel flow with variable vertical shear and unstable stratification. *Journal of the Meteorological Society of Japan, 48,* 129–138.

Augustine, J. A., & Zipser, E. J. (1987). The use of wind profilers in a mesoscale experiment. *Bulletin of the American Meteorological Society, 68,* 4–17.

Baik, J.-J., & Chun, H.-Y. (1996). Effects of nonlinearity on the atmospheric flow response to low-level heating in a uniform flow. *Journal of the Atmospheric Sciences, 53,* 1856–1869.

Baik, J. J., Hwang, H.-S., & Chun, H.-Y. (1999). Transient, linear dynamics of a stably stratified shear flow with thermal forcing and a critical level. *Journal of the Atmospheric Sciences, 56,* 483–499.

Barber, K. A., Mullendore, G. L., & Alexander, M. J. (2018). Out-of-cloud convective turbulence: Estimation method and impacts of model resolution. *Journal of Applied Meteorology and Climatology, 57,* 121–136.

Bedard, Jr. A. J. (1993). Atmospheric turbulence aloft: A review of possible methods for detection, warning, and validation of predictive models. 31st *Aerospace Sciences Meeting & Exhibit, AIAA* 93-0847, Reno, NV, January 11–14.

Bedka, K. M., Brunner, J., Dworak, R., Feltz, W., Otkin, J., & Greenwald, T. (2010). Objective satellite-based overshooting top detection using infrared window channel brightness temperature gradients. *Journal of Applied Meteorology and Climatolology, 49*(2), 181–202. https://doi.org/10.1175/2009JAMC2286.1.

Beres, J. H. (2004). Gravity wave generation by a three-dimensional thermal forcing. *Journal of the Atmospheric Sciences, 61,* 1805–1815.

Beres, J. H., Alexander, M. J., & Holton, J. R. (2002). Effects of tropospheric wind shear on the spectrum of convectively generated gravity waves. *Journal of the Atmospheric Sciences, 59,* 1805–1824.

Beres, J. H., Alexander, M. J., & Holton, J. R. (2004). A method of specifying the gravity wave spectrum above convection based on latent heating properties and background wind. *Journal of the Atmospheric Sciences, 61,* 324–337.

Bernardet, L. R., Grasso, L. D., Nachamkin, J. E., Finley, C. A., & Cotton, W. R. (2000). Simulating convective events using a high-resolution mesoscale model. *Journal of Geophysical Research: Atmospheres, 105*(D11), 14963–14982. https://doi.org/10.1029/2000jd900100.

Bradbury, T. A. M. (1973). Glider flight in the lower stratosphere above cumulonimbus clouds. *Meteorological Magazine, 102,* 110–120.

Bretherton, F. P. (1966). The propagation of groups of internal gravity waves in shear flow. *Quarterly Journal of the Royal Meteorological Society, 92,* 466–480.

Bretherton, C. (1988). Group velocity and the linear response of stratified fluids to internal heat or mass sources. *Journal of the Atmospheric Sciences, 45,* 81–93.

Bretherton, C. S., & Smolarkiewicz, P. K. (1989). Gravity waves, compensating subsidence and detrainment around cumulus clouds. *Journal of the Atmospheric Sciences, 46*(6), 740–759.

Broad, A. S. (1995). Linear theory of momentum fluxes in 3-d flows with turning of the mean wind with height. *Quarterly Journal of the Royal Meteorological Society, 121,* 1891–1902.

Bruning, E. C., & MacGorman, D. R. (2013). Theory and observations of controls on lightning flash size spectra. *Journal of the Atmospheric Sciences, 70,* 4012–4029.

Bryan, G. H., Wyngaard, J. C., & Fritsch, J. M. (2003). Resolution requirements for the simulation of deep, moist convection. *Monthly Weather Review, 131,* 2394–2416.

Burnham, J. (1970). Atmospheric gusts—A review of the results of some recent research at the Royal Aircraft Establishment. *Monthly Weather Review, 98*(10), 723–734.

Burns, A., Harrold, T. W., Burnham, J., & Spavins, C. S. (1966). Turbulence in clear air near thunderstorms. *National Severe Storms Laboratory Technical Memorandum No. 30,* 20 pp.

Cerasoli, C. P. (1978). Experiments on buoyant-parcel motion and the generation of internal gravity waves. *Journal of Fluid Mechanics, 86*(2), 247–271.

Chun, H.-Y. (1997). Weakly non-linear response of a stably stratified shear flow to thermal forcing. *Tellus, 49A,* 528–543.

Chun, H.-Y., & Baik, J.-J. (1994). Weakly non-linear response of a stably stratified atmosphere to diabatic forcing in a uniform flow. *Journal of the Atmospheric Sciences, 41,* 3109–3121.

Chun, H.-Y., & Baik, J.-J. (1998). Momentum flux by thermally induced internal gravity waves and its approximation for large-scale models. *Journal of the Atmospheric Sciences, 55,* 3299–3310.

Chun, H.-Y., & Baik, J.-J. (2002). An updated parameterization of convectively forced gravity wave drag for use in large-scale models. *Journal of the Atmospheric Sciences, 59,* 1006–1017.

Chun, H.-Y., & Kim, Y.-H. (2008). Secondary waves generated by breaking of convective gravity waves in the mesosphere and their influence in the wave momentum flux. *Journal of Geophysical Research, 113,* D23107. https://doi.org/10.1029/2008JD009792.

Clark, T. L., Hauf, T., & Kuettner, J. P. (1986). Convectively forced internal gravity waves: results from two-dimensional numerical experiments. *Quarterly Journal of the Royal Meteorological Society, 112,* 899–925.

Collander, R., Jamison, B. D., Tollerud, E. I., Caracena, F., Lu, C., & Koch, S. E. (2006). Turbulence in MCS anvils: Observations and analyses from BAMEX. Preprints, 12th *Conference on Aviation, Range, and Aerospace Meteorology*, Atlanta, Georgia, American Meteorological Society, p. 7.1. Available online at http://ams.confex.com/ams/pdfpapers.104929.pdf.

Cornman, L. B., & Carmichael, B. (1993). Varied research efforts are under way to find means of avoiding air turbulence. *ICAO Journal, 48,* 10–15.

Cornman, L. B., Morse, C. S., & Cunning, G. (1995). Real-time estimation of atmospheric turbulence severity from in situ aircraft measurements. *Journal of Aircraft, 32,* 171–177. https://doi.org/10.2514/3.46697.

Bedard, A. J. Jr., & Cunningham, W. (1991). Potential aircraft hazards in the vicinity of convective clouds: A review from the perspective of a scale-model study. *Preprints,* 4*th Conference on Aviation Weather Systems* (pp. 66–70), Paris, FR, American Meteorological Society.

Deardorff, J. W. (1969). Numerical study of heat transport by internal gravity waves above a growing unstable layer. *Physics of Fluids, 12*(12), II-184. https://doi.org/10.1063/1.1692435.

Deardorff, J. W., Willis, G. E., & Lilly, D. K. (1969). Laboratory investigation of non-steady penetrative convection. *Journal of Fluid Mechanics, 35*(1), 7–31.

Deierling, W., & Williams, J. (2016). Relations between lightning and convective turbulence. In R. Sharman & T. Lane (Eds.), *Aviation turbulence: Processes, detection, prediction* (pp. 179–192). Switzerland: Springer.

Del Genio, A. D., Yao, M.-S., & Jonas, J. (2007). Will moist convection be stronger in a warmer climate? *Geophysical Research Letters, 34,* L16703. https://doi.org/10.1029/2007GL030525.

Diffenbaugh, N. S., Scherer, M., & Trapp, R. J. (2013). Robust increases in severe thunderstorm environments in response to greenhouse forcing. *Proceedings of the National academy of Sciences of the United States of America, 110*(41), 16361–16366. https://doi.org/10.1073/pnas.1307758110.

Dixon, R. S., Browning, K. A., & Shutts, G. J. (2000). The mystery of striated cloud heads in satellite imagery. *Atmospheric Science Letters, 1,* 1–13. https://doi.org/10.1006/asle.2000.0001.

Dörnbrack, A., Gerz, T., & Schumann, U. (1995). Turbulence breaking of overturning gravity waves below a critical level. *Applied Scientific Research, 54,* 163–176.

Droegemeier, K.K., Richardson, Y., Bassett, G. M., & Marroquin, A. (1997). Three-dimensional numerical simulations of turbulence generated in the near-environment of deep convective storms. *Preprints,* 7th *Conference on Aviation, Range, and Aerospace Meteorology*, 2–7 February, Long Beach, CA, American Meteorological Society, 169–174.

Duran, P., & Molinari, J. (2016). Upper-tropospheric low Richardson number in tropical cyclones: Sensitivity to tropical cyclone intensity and the diurnal cycle. *Journal of the Atmospheric Sciences, 73,* 545–554.

Durran, D. R., & Klemp, J. B. (1982). On the effects of moisture on the Brunt–Väisälä frequency. *Journal of the Atmospheric Sciences, 39,* 2152–2158.

Einaudi, F., Lalas, D. P., & Perona, G. E. (1978). The role of gravity waves in tropospheric processes. *Pure and Applied Geophysics, 117*(4), 627–663. https://doi.org/10.1007/BF00879972.

Eliassen, A., & Palm, E. (1961). On the transfer of energy in stationary mountain waves. *Geofysiske Publikasjoner, 22*(3), 1–23.

Ellrod, G. P. (1985). Detection of high level turbulence using satellite imagery and upper air data. *NOAA Technical Memorandum. NESDIS 10,* 30 pp.

Ellrod, G. P., Knox, J. A., Lester, P. F., & Ehernberger, L. J. (2015). Clear air turbulence. In G. R. North., J. Curry., J. Pyle., & F. Zhang (Eds.), *Encyclopedia of Atmospheric Sciences* (2nd edn, Vol. 1, pp. 177–186). Amsterdam: Academic Press, Elsevier Science Ltd.

Eom, J. K. (1975). Analysis of the internal gravity wave occurence 3 of 19 April 1970 in the Midwest. *Monthly Weather Review, 103,* 217–226.

Federal Aviation Administration (FAA) (2014). Aeronautical Information Manual (AIM). Available online at http://www.faa.gov/air_traffic/publications/media/AIM_Basic_4-03-14.pdf.

Fovell, R., Durran, D., & Holton, J. R. (1992). Numerical simulation of convectively generated stratospheric gravity waves. *Journal of the Atmospheric Sciences, 49,* 1427–1442.

Fovell, R. G., Sharman, R. D., & Trier, S. B. (2007). A case study of convectively-induced clear-air turbulence. Preprints, 12th *Conference on Mesoscale Processes,* American Meteorological Society. Waterville Valley, New Hampshire, Paper 13.4.

Fric, T. F., & Roshko, A. (1994). Vortical structure in the wake of a transverse jet. *Journal of Fluid Mechanics, 279,* 1–47.

Fritsch, J. M., & Maddox, R. A. (1981). Convectively driven mesoscale weather systems aloft. Part I: observations. *Journal of Applied Meteorology, 20,* 9–19.

Fritts, D. C., & Alexander, M. J. (2003). Gravity wave dynamics and effects in the middle atmosphere. *Reviews of Geophysics, 41*(1), 1003. https://doi.org/10.1029/2001RG000106.

Gavrilov, N. M., & Kshevetskii, S. P. (2015). Dynamical and thermal effects of nonsteady nonlinear acoustic-gravity waves propagating from tropospheric sources to the upper atmosphere. *Advances in Space Research, 56,* 1833–1843.

Geller, M. A., & Gong, J. (2010). Gravity wave kinetic, potential and vertical fluctuation energies as indicators of different frequency gravity waves. *Journal of Geophysical Research, 115,* D11111. https://doi.org/10.1029/2009JD012266.

Gensini, V. A., Ramseyer, C., & Mote, T. L. (2014). Future convective environments using NARCCAP. *International Journal of Climatology, 34,* 1699–1705.

Gill, A. E. (1982). *Atmosphere-ocean dynamics* (p. 662). San Diego: Academic Press.

Gill, P. G. (2016). Aviation turbulence forecast verification. In R. Sharman & T. Lane (Eds.), *Aviation turbulence: Processes, detection, prediction* (pp. 261–283). Berlin: Springer.

Gong, J., & Geller, M. A. (2010). Vertical fluctuation energy in United States high vertical resolution radiosonde data as an indicator of convective gravity wave sources. *Journal of Geophysical Research, 115,* D11110. https://doi.org/10.1029/2009JD012265.

Gossard, E. E., & Hooke, W. H. (1975). *Waves in the atmosphere.* New York: Elsevier.

Grabowski, W. W., & Clark, T. L. (1991). Cloud-environment interface instability: Rising thermal calculations in two spatial dimensions. *Journal of the Atmospheric Sciences, 48,* 527–546.

Grachev, A. I., Lobachevskiy, L. A., Matveyev, L. K., Mordukhovich, M. I., & Sergeyenko, O. S. (1984). Thick convective cloud cover as a source of internal waves in the troposphere and ionosphere. *Izvestiya, Atmospheric and Oceanic Physics, 20*(2), 127–129.

Guarino, M.-V., Teixeira, M. A. C., Keller, T. L., & Sharman, R. D. (2017). Mountain wave turbulence in the presence of directional wind shear over the Rocky Mountains. *Journal of the Atmospheric Sciences.* https://doi.org/10.1175/JAS-D-17-0128.1.

Gultepe, I., & Starr, D. O'C. (1995). Dynamical structure and turbulence in cirrus clouds: Aircraft observations during FIRE. *Journal of the Atmospheric Sciences, 52*(23), 4159–4182.

Gultepe, I., et al. (2018). A review of high impact weather on aviation meteorology applications. *Pure and Applied Geophysics,* this issue.

Haman, K. (1962). On some possible causes of clear air turbulence. *Acta Geophysica Polonica, 10,* 335–357.

Hamilton, D. W., & Proctor, F. H. (2003). An aircraft encounter with turbulence in the vicinity of a thunderstorm. *21st Applied Aerodynamics Conference,* 23–26 June 2003, Orlando, Florida, AIAA 2003-4075.

Hansen, A. R., Nastrom, G. D., Otkin, J. A., & Eaton, F. D. (2002). MST radar observations of gravity waves and turbulence near thunderstorms. *Journal of Applied Meteorology, 41,* 298–305.

Harris, F. I. (1977). The effects of evaporation at the base of ice precipitation layers: Theory and radar observations. *Journal of the Atmospheric Sciences, 34,* 651–672.

Hertenstein, R. F. A., & Schubert, W. H. (1991). Potential vorticity anomalies associated with squall lines. *Monthly Weather Review, 119,* 1663–1672.

Heymsfield, A. (1975). Cirrus uncinus generating cells and the evolution of cirriform clouds. Part II: The structure and circulations of the cirrus uncinus generating head. *Journal of the Atmospheric Sciences, 32,* 809–819.

Hodges, R. R., Jr. (1967). Generation of turbulence in the upper atmosphere by internal gravity waves. *Journal of Geophysical Research, 72,* 3455–3458.

Hoffmann, L., & Alexander, M. J. (2010). Occurrence frequency of convective gravity waves during the North American thunderstorm season. *Journal of Geophysical Research, 115,* D20111. https://doi.org/10.1029/2010JD014401.

Holton, J. R., & Alexander, M. J. (1999). Gravity waves in the mesosphere generated by tropospheric convection. *Tellus A-B, 51,* 45–58.

Homeyer, C. R., McAuliffe, J. D., & Bedka, K. M. (2017). On the development of above-anvil cirrus plumes in extratropical convection. *Journal of the Atmospheric Sciences, 74,* 1617–1633.

Homeyer, C. R., Pan, L. L., & Barth, M. C. (2014). Transport from convective overshooting of the extratropical tropopause and the role of large-scale lower stratospheric stability. *Journal of Geophysical Research, 119,* 2220–2240. https://doi.org/10.1002/2013JD020931.

Honomichl, S. B., Detwiler, A. G., & Smith, P. L. (2013). Observed hazards to aircraft in deep summertime convective clouds from 4–7 km. *Journal of Aircraft, 50*(3), 926–935. https://doi.org/10.2514/1.C032057.

International Civil Aviation Organization (ICAO) (2013). Meteorological service for international air navigation.—Annex 3 to the Convention on International Civil Aviation, Appendix 4 Technical Specifications Related to Aircraft Observations and Reports, Section 2.6 Turbulence, 18th Edition, 208 pp. Available online at http://store1.icao.int/index.php/publications/annexes/3-meteorological-service-for-international-air-navigation.html.

Jones, W. L. (1967). Propagation of internal gravity waves in fluids with shear flow and rotation. *Journal of Fluid Mechanics, 30,* 481–496.

Jones, T. B. (1982). Generation and propagation of acoustic gravity waves. *Nature, 299,* 488–489.

Kang, M.-J., Chun, H.-Y., & Kim, Y.-H. (2017). Momentum flux of convective gravity waves derived from an offline gravity wave parameterization. Part I: Spatiotemporal variations at source level. *Journal of the Atmospheric Sciences, 74*(10), 3167–3189. https://doi.org/10.1175/JAS-D-17-0053.1.

Kaplan, M. L., Huffman, A. W., Lux, K. M., Cetola, J. D., Charney, J. J., Riordan, A. J., et al. (2005). Characterizing the severe turbulence environments associated with commercial aviation accidents. Part 2: Hydrostatic mesoscale numerical simulations of supergradient flow and streamwise ageostrophic frontogenesis. *Meteorology and Atmospheric Physics, 88,* 153–173.

Keller, T. L., Ehernberger, L. J., & Wurtele, M. G. (1983). Numerical simulation of the atmosphere during a CAT encounter. Preprints, *Ninth Conference on Aerospace and Aeronautical Meteorology,* Omaha, Nebraska, American Meteorological Society, 316–319.

Kennedy, P. J., & Shapiro, M. A. (1975). The energy budget in a clear-air turbulence zone as observed by aircraft. *Monthly Weather Review, 103,* 650–654.

Kessinger, C., Megenhardt, D., Blackburn, G., Olivo, J., Lin, L., Hoang, V., et al. (2017). Displaying convective weather products on an electronic flight bag. *Journal of Air Traffic Control, 59*(3), 52–61.

Kim, Y.-H., Bushell, A. C., Jackson, D. R., & Chun, H.-Y. (2013). Impacts of introducing a convective gravity-wave parameterization upon the QBO in the Met Office Unified Model. *Geophysical Research Letters.* https://doi.org/10.1002/grl.50353.

Kim, J.-H., & Chun, H.-Y. (2011). Statistics and possible sources of aviation turbulence over South Korea. *Journal of Applied Meteorology and Climatology, 50,* 311–324.

Kim, J.-H., & Chun, H.-Y. (2012). A numerical simulation of convectively induced turbulence above deep convection. *Journal of Applied Meteorology and Climatology, 51,* 1180–1200.

Kim, J.-H., Chun, H.-Y., Sharman, R. D., & Trier, S. B. (2014). The role of vertical shear on aviation turbulence within cirrus bands of a simulated western Pacific cyclone. *Monthly Weather Review, 142,* 2794–2813.

Kim, S.-H., Chun, H. Y., Sharman, R. D., & Trier, S. B. (2017). Developing near-cloud aviation turbulence related to breaking of convective gravity waves. *Proceedings of the Autumn Meeting of Korean Meteorological Society (KMS),* 228–230.

Kim, Y.-J., Eckermann, S. D., & Chun, H.-Y. (2003). An overview of past, present and future of gravity-wave drag parameterization for numerical climate and weather prediction models. *Atmosphere Ocean, 41,* 65–98.

Klostermeyer, & Rüster, R. (1980). Radar observation and model computation of a jet stream-generated Kelvin-Helmholtz instability. *Journal of Geophysical Research, 85*(C5), 2841–2846.

Knight, C. A., Miller, L. J., & Hall, W. D. (2004). Deep convection and "first echoes' within anvil precipitation. *Monthly Weather Review, 132,* 1877–1890.

Knox, J. A., Bachmeier, A. S., Carter, W. M., Tarantino, J. E., Paulik, L. C., Wilson, E. N., et al. (2010). Transverse cirrus bands in weather systems: A grand tour of an enduring enigma. *Weather, 65,* 35–41.

Knox, J. A., Black, A. W., Rackley, J. A., Wilson, E. N., Grant, J. S., Phelps, S. P., et al. (2016). Automated turbulence forecasting strategies. In R. Sharman & T. Lane (Eds.), *Aviation turbulence: Processes, detection, and prediction* (pp. 243–260). Berlin: Springer.

Koch, S. E., Jamison, B. D., Lu, C., Smith, T. L., Tollerud, E. I., Girz, C., et al. (2005). Turbulence and gravity waves in an upper-level front. *Journal of the Atmospheric Sciences, 62,* 3885–3908.

Kudo, A. (2013). The generation of turbulence below midlevel cloud bases: The effect of cooling due to sublimation of snow. *Journal of Applied Meteorology and Climatology, 52,* 819–833.

Kudo, A., Luce, H., Hashiguchi, H., & Wilson, R. (2015). Convective instability underneath midlevel clouds: Comparisons between numerical simulations and VHF radar observations. *Journal of Applied Meteorology and Climatology, 54,* 2217–2227.

Kumar, K. K. (2007). VHF radar investigations on the role of mechanical oscillator effect in exciting convectively generated gravity waves. *Geophysical Research Letters, 34,* L01803. https://doi.org/10.1029/2006GL027404.

Lafore, J. P., & Moncrieff, M. W. (1989). A numerical investigation of the organization and interaction of convective and stratiform regions of tropical squall lines. *Journal of the Atmospheric Sciences, 46,* 521–544.

Lane, T. P. (2008). The vertical response to penetrative convection and the associated gravity-wave generation. *Atmospheric Science Letters, 9,* 103–110.

Lane, T. P. (2016). Processes underlying near-cloud turbulence. In R. Sharman & T. Lane (Eds.), *Aviation turbulence: Processes, detection, prediction* (pp. 317–334). Berlin: Springer.

Lane, T. P., Doyle, J. D., Plougonven, R., Shapiro, M. A., & Sharman, R. D. (2004). Observations and numerical simulations of inertia-gravity waves and shearing instabilities in the vicinity of a jet stream. *Journal of the Atmospheric Sciences, 61,* 2692–2706.

Lane, T. P., & Knievel, J. C. (2005). Some effects of model resolution on simulated gravity waves generated by deep, mesoscale convection. *Journal of the Atmospheric Sciences, 62,* 3408–3419.

Lane, T. P., Reeder, M. J., & Clark, T. L. (2001). Numerical modeling of gravity wave generation by deep tropical convection. *Journal of the Atmospheric Sciences, 58,* 1249–1274.

Lane, T. P., & Sharman, R. D. (2006). Gravity wave breaking, secondary wave generation, and mixing above deep convection in a three-dimensional cloud model. *Geophysical Research Letters, 33,* L23813. https://doi.org/10.1029/2006GL027988.

Lane, T. P., & Sharman, R. D. (2008). Some influences of background flow conditions on the generation of turbulence due to gravity wave breaking above deep convection. *Journal of Applied Meteorology, 47,* 2777–2796.

Lane, T. P., & Sharman, R. D. (2014). Intensity of thunderstorm-generated turbulence. *Geophysical Research Letters, 41,* 2221–2227.

Lane, T. P., Sharman, R. D., Clark, T. L., & Hsu, H.-M. (2003). An investigation of turbulence generation mechanisms above deep convection. *Journal of the Atmospheric Sciences, 60,* 1297–1321.

Lane, T. P., Sharman, R. D., Trier, S. B., Fovell, R. G., & Williams, J. K. (2012). Recent advances in the understanding of near-cloud turbulence. *Bulletin of the American Meteorological Society, 93*(4), 499–515.

Lemon, L. R. (1976). Wake vortex structure and aerodynamic origin in severe thunderstorms. *Journal of the Atmospheric Sciences, 33,* 678–685.

LeMone, M. A. (1973). The structure and dynamics of horizontal roll vortices in the planetary boundary layer. *Journal of the Atmospheric Sciences, 30,* 1077–1091.

Lenz, A., Bedka, K., Feltz, W., & Ackerman, S. (2009). Convectively-induced transverse band signatures in satellite imagery. *Weather and Forecasting, 24*(5), 1362–1373.

Lester, P. F. (1993). *Turbulence: A new perspective for pilots.* Englewood: Jeppesen Sanderson Inc.

Levizanni, V., & Setvak, M. (1996). Multispectral, high resolution satellite observations of plumes on top of convective storms. *Journal of the Atmospheric Sciences, 53,* 361–369.

Lighthill, J. (1978). *Waves in fluids* (p. 504). Cambridge: Cambridge University Press.

Lilly, D. K. (1990). Numerical prediction of thunderstorms—Has its time come? *Quarterly Journal of the Royal Meteorological Society, 116,* 779–798.

Lin, Y.-L. (1986). Calculation of airflow over an isolated heat source with application to the dynamics of V-shaped clouds. *Journal of the Atmospheric Sciences, 43,* 2736–2751.

Lin, Y.-L. (1994a). Airflow over mesoscale heat sources. Part I: Responses in a uniform flow. *Proceedings of the National Science Council, Republic of China. Part A, Applied Sciences, 18*(1), 1–32.

Lin, Y.-L. (1994b). Airflow over mesoscale heat sources. Part II: Responses in a shear flow. *Proceedings of the National Science Council, Republic of China. Part A, Applied Sciences, 18*(2), 119–150.

Lin, Y.-L. (2007). *Mesoscale dynamics* (pp. 630). Cambridge: Cambridge University Press.

Lin, Y.-L., & Goff, R. C. (1988). A study of a mesoscale solitary wave in the atmosphere originating near a region of deep convection. *Journal of the Atmospheric Sciences, 45,* 2401–2415.

Lin, Y.-L., & Li, S. (1988). Three-dimensional response of a shear flow to elevated heating. *Journal of the Atmospheric Sciences, 45,* 2987–3002.

Lin, Y.-L., & Smith, R. B. (1986). The transient dynamics of airflow near a local heat source. *Journal of the Atmospheric Sciences, 43,* 40–49.

Lindzen, R. S., & Tung, K. K. (1976). Banded convective activity and ducted gravity waves. *Monthly Weather Review, 104,* 1602–1617.

Lu, D., VanZandt, T. E., & Clark, W. L. (1984). VHF Doppler radar observations of buoyancy waves associated with thunderstorms. *Journal of the Atmospheric Sciences, 41*(2), 272–282.

Luce, H., Nakamura, T., Yamamoto, M. K., Yamamoto, M., & Fukao, S. (2010). MU radar and lidar observations of clear-air turbulence underneath cirrus. *Monthly Weather Review, 138,* 438–452.

MacCready, P. (1964). Standardization of gustiness values from aircraft. *Journal of Applied Meteorology, 3,* 439–449. https://doi.org/10.1175/1520-0450(1964)003,0439:SOGVFA.2.0.CO;2.

Mclaren, T., Pierce, A., Fohl, T., & Murphy, B. (1973). An investigation of internal gravity waves generated by a buoyantly rising fluid in a stratified medium. *Journal of Fluid Mechanics, 57*(2), 229–240. https://doi.org/10.1017/S0022112073001126.

McNulty, R. P. (1995). Severe and convective weather: A Central Region forecasting challenge. *Weather and Forecasting, 10,* 187–202.

Meneguz, E., Wells, H., & Turp, D. (2016). An automated system to quantify aircraft encounters with convectively induced turbulence over Europe and the Northeast Atlantic. *Journal of Applied Meteorology and Climatology, 55,* 1077–1089.

Miles, J. (1986). Richardson's criterion for the stability of stratified shear flow. *Physics of Fluids, 29*(10), 3470–3471.

Molinari, J., Duran, P., & Vallaro, D. (2014). Low Richardson number in the tropical cyclone outflow layer. *Journal of the Atmospheric Sciences, 71,* 3164–3179.

Monette, S. A., & Sieglaff, J. M. (2014). Probability of convectively induced turbulence associated with geostationary satellite–inferred cloud-top cooling. *Journal of Applied Meteorology and Climatology, 53*(2), 429–436.

Moustaoui, M., Joseph, B., & Teitelbaum, H. (2004). Mixing layer formation near the tropopause due to gravity wave-critical level interactions in a cloud-resolving model. *Journal of the Atmospheric Sciences, 61,* 3112–3124.

Muhlbauer, A., Kalesse, H., & Kollias, P. (2014). Vertical velocities and turbulence in midlatitude anvil cirrus: A comparison between in situ aircraft measurements and ground-based Doppler cloud radar retrievals. *Geophysical Research Letters, 41,* 7814–7821. https://doi.org/10.1002/2014GLO62279.

Nappo, C. J. (2002). *An introduction to atmospheric gravity waves* (p. 276). San Diego: Academic Press.

Nicholls, M. E., Pielke, R. A., & Cotton, W. R. (1991). Thermally forced gravity waves in an atmosphere at rest. *Journal of the Atmospheric Sciences, 48*(16), 1869–1884.

Pandya, R. E., & Alexander, M. J. (1999). Linear stratospheric gravity waves above convective thermal forcing. *Journal of the Atmospheric Sciences, 56,* 2434–2446.

Pandya, R. E., & Durran, D. R. (1996). The influence of convectively generated thermal forcing in the mesoscale circulation around squall lines. *Journal of the Atmospheric Sciences, 53,* 2924–2951.

Pandya, R., Durran, D., & Bretherton, C. (1993). Comments on "Thermally forced gravity waves in an atmosphere at rest". *Journal of the Atmospheric Sciences, 50*(24), 4097–4101.

Pantley, K. C., & Lester, P. F. (1990). Observations of severe turbulence near thunderstorm tops. *Journal of Applied Meteorology, 29,* 1171–1179.

Parks, E. K., Wingrove, R. C., Bach, R. E., & Mehta, R. S. (1985). Identification of vortex-induced clear air turbulence using airline flight records. *Journal of Aircraft, 22*(2), 124–129. https://doi.org/10.2514/3.45095.

Pavelin, E., Whiteway, J. A., & Vaughan, G. (2001). Observations of gravity wave generation and breaking in the lowermost stratosphere. *Journal of Geophysical Research, 106,* 5173–5179.

Pearson, J. M., & Sharman, R. D. (2017). Prediction of energy dissipation rates for aviation turbulence: Part I. Nowcasting convective and non-convective turbulence. *Journal of Applied Meteorology and Climatology, 56*(2), 339–351. https://doi.org/10.1175/JAMC-D-16-0312.1.

Petre, J. M., & Verlinde, J. (2004). Cloud radar observations of Kelvin–Helmholtz instability in a Florida anvil. *Monthly Weather Review, 132,* 2520–2523.

Piani, C., Durran, D., Alexander, M. J., & Holton, J. R. (2000). A numerical study of three-dimensional gravity waves triggered by deep tropical convection and their role in the dynamics of the QBO. *Journal of the Atmospheric Sciences, 57*(22), 3689–3702.

Pierce, A. D., & Coroniti, S. C. (1966). A mechanism for the generation of acoustic gravity waves during thunderstorm formation. *Nature, 18,* 1209–1210.

Pinçon, C., Belkacem, K., & Goupil, M. J. (2016). Generation of internal gravity waves by penetrative convection. *Astronomy and Astrophysics, A122,* 1-21, https://doi.org/10.1051/0004-6361/201527663.

Plougonven, R., Teitelbaum, H., & Zeitlen, V. (2003). Inertia-gravity waves generation by the tropospheric mid-latitude jet as given by the FASTEX radiosoundings. *Journal of Geophysical Research, 108,* 4686. https://doi.org/10.1029/2003JD003535.

Plougonven, R., & Zhang, F. (2016). Gravity waves generated by jets and fronts and their relevance for clear-air turbulence. In R. Sharman & T. Lane (Eds.), *Aviation turbulence: Processes, detection, and prediction* (pp. 385–406). Berlin: Springer.

Pond, S. R. (2015). Analysis of aircraft flights near convective weather over Europe. *Weather, 70*(10), 292–296.

Prein, A. F., Liu, C., Ikeda, K., Trier, S. B., Rasmussen, R. M., Holland, G. J., et al. (2017). Increased rainfall volume from future convective storms in the US. *Nature Climate Change, 7,* 880–884.

Prophet, D. T. (1970). Vertical extent of turbulence in clear air above the tops of thunderstorms. *Journal of Applied Meteorology, 9,* 320–321.

Queney, P. (1948). The problem of airflow over mountains. A summary of theoretical studies. *Bulletin of the American Meteorological Society, 29,* 16–26.

Reiter, E. R., & Lester, P. F. (1968). Richardson's Number in the free atmosphere. *Arch. Met. Geoph. Biokl. A, 17,* 1–7.

Romanova, N. N., & Yakushkin, I. G. (1995). Internal gravity waves in the lower atmosphere and sources of their generation (review). *Izvestiya, Atmospheric and Oceanic Physics, 31*(2), 151–172.

Romps, D. M., Seeley, J. T., Vollaro, D., & Molinari, J. (2014). Projected increase in lightning strikes in the United States due to global warming. *Science, 346,* 851–853.

Ruppert, J. H., Jr., & Bosart, L. F. (2014). A case study of the interaction of a mesoscale gravity wave with a mesoscale convective system. *Monthly Weather Review, 142,* 1403–1429.

Schalkwijk, J., Jonker, H. J. J., Siebesma, A. P., & Van Meijgaard, E. (2015). Weather forecasting using GPU-based large-eddy simulations. *Bulletin of the American Meteorological Society, 96,* 715–723. https://doi.org/10.1175/BAMS-D-14-00114.1.

Schultz, D. M., Kanak, K. M., Straka, J. M., Trapp, R. J., Gordon, B. A., Zrnic, D. R., et al. (2006). The mysteries of mammatus clouds: Observations and formation mechanisms. *Journal of the Atmospheric Sciences, 63,* 2409–2435.

Schwartz, B. (1996). The quantitative use of PIREPs in developing aviation weather guidance products. *Weather and Forecasting, 11,* 372–384. https://doi.org/10.1175/1520-434(1996)011,0372:TQUOPI.2.0.CO;2.

Shapiro, M. A. (1976). The role of turbulent heat flux in the generation of potential vorticity in the vicinity of upper-level jet stream systems. *Monthly Weather Review, 104*(7), 892–906

Sharman, R. (2016). Nature of aviation turbulence. In R. Sharman & T. Lane (Eds.), *Aviation turbulence: Processes, detection, and prediction* (pp. 3–30). Berlin: Springer.

Sharman, R. D., Cornman, L. B., Meymaris, G., Pearson, J., & Farrar, T. (2014). Description and derived climatologies of automated in situ eddy dissipation rate reports of atmospheric turbulence. *Journal of Applied Meteorology and Climatology, 53*(6), 1416–1432. https://doi.org/10.1175/JAMC-D-13-0329.1.

Sharman, R., Lane, T., & Schumann, U. (2016). Research needs. In R. Sharman & T. Lane (Eds.), *Aviation turbulence: Processes, detection, and prediction* (pp. 501–508). Berlin: Springer.

Sharman, R. D., & Pearson, J. M. (2017). Prediction of energy dissipation rates for aviation turbulence: Part I. Forecasting Non-convective turbulence. *Journal of Applied Meteorology and Climatology, 56*(2), 317–337. https://doi.org/10.1175/JAMC-D-16-0205.1.

Sharman, R. D., Trier, S. B., Lane, T. P., & Doyle, J. D. (2012). Sources and dynamics of turbulence in the upper troposphere and lower stratosphere: A review. *Geophysical Research Letters, 39,* L12803. https://doi.org/10.1029/2012GL051996.

Sharman, R. D., & Wurtele, M. G. (1983). Ship waves and lee waves. *Journal of the Atmospheric Sciences, 40,* 396–427.

Sharman, R. D., & Wurtele, M. G. (2004). Three-dimensional structure of forced gravity waves and lee waves. *Journal of the Atmospheric Sciences, 61,* 664–680.

Shutts, G. J. (1998). Stationary gravity-wave structure in flows with directional wind shear. *Quarterly Journal of the Royal Meteorological Society, 124,* 1421–1442.

Smull, B. F., & Houze, R. A., Jr. (1987). Rear inflow in squall lines with trailing stratiform precipitation. *Monthly Weather Review, 115,* 2869–2889.

Snively, J. B., & Pasko, V. P. (2003). Breaking of thunderstorm-generated gravity waves as a source of short-period ducted waves at mesopause altitudes. *Geophysical Research Letters, 30*(24), 2254. https://doi.org/10.1029/2003GL018436.

Song, I.-S., Chun, H.-Y., & Lane, T. P. (2003). Generation mechanisms of convectively forced internal gravity waves and their propagation into the stratosphere. *Journal of the Atmospheric Sciences, 60,* 1960–1980.

Stephan, C., & Alexander, M. J. (2015). Realistic simulations of atmospheric gravity waves over the continental U.S. using precipitation radar data. *Journal of Advances in Modeling Earth Systems, 7,* 823–835. https://doi.org/10.1002/2014MS000396.

Stobie, J. G., Einaudi, F., & Uccellini, L. W. (1983). A case study of gravity waves—convective storms interaction: 9 May 1979. *Journal of Atmospheric Science, 40*(12), 2804–2830.

Storer, L. N., Williams, P. D., & Joshi, M. M. (2017). Global response of clear-air turbulence to climate change. *Geophysical Research Letters.* https://doi.org/10.1002/2017GL074618.

Stull, R. B. (1976). Internal gravity waves generated by penetrative convection. *Journal of the Atmospheric Sciences, 33,* 1279–1286.

Teitelbaum, H., Moustaoui, M., Sadourny, R., & Lott, F. (1999). Critical levels and mixing layers induced by convectively generated gravity waves during CEPEX. *Quarterly Journal of the Royal Meteorological Society, 125,* 1715–1734.

Townsend, A. A. (1966). Internal waves produced by a convective layer. *Journal of Fluid Mechanics, 24*(2), 307–319.

Townsend, A. A. (1968). Excitation of internal waves in a stably-stratified atmosphere with considerable wind-shear. *Journal of Fluid Mechanics, 32*(1), 3145–3171.

Trapp, R. J., Diffenbaugh, N. S., Brooks, H. E., Baldwin, M. E., Robinson, E. D., & Pal, J. S. (2007). Changes in severe thunderstorm environment frequency during the 21st century caused by anthropogenically enhanced global radiative forcing. *Proceedings of the National academy of Sciences of the United States of America, 104*(50), 19719–19723.

Trier, S. B. (2016). Modeling studies of turbulence mechanisms associated with mesoscale convective systems. In Sharman R., Lane. T. (Eds.), *Aviation turbulence: Processes, detection, and prediction* (pp. 335-356). Berlin: Springer.

Trier, S. B., & Sharman, R. D. (2009). Convection-permitting simulations of the environment supporting widespread turbulence within the upper-level outflow of a mesoscale convective system. *Monthly Weather Review, 137,* 1972–1990.

Trier, S. B., & Sharman, R. D. (2016). Mechanisms influencing cirrus banding and aviation turbulence near a convectively enhanced upper-level jet stream. *Monthly Weather Review, 144,* 3003–3027.

Trier, S. B., Sharman, R. D., Fovell, R. G., & Frehlich, R. G. (2010). Numerical simulation of radial cloud bands within the upper-level outflow of an observed mesoscale convective system. *Journal of the Atmospheric Sciences, 67,* 2990–2999.

Trier, S. B., Sharman, R. D., & Lane, T. P. (2012). Influences of moist convection on a cold-season outbreak of clear-air turbulence (CAT). *Monthly Weather Review, 140,* 2477–2496.

Turner, J. S. (1973). *Buoyancy effects in fluids.* Cambridge: Cambridge University Press, 367 pp.

Uccellini, L. W., & Koch, S. E. (1987). The synoptic setting and possible energy sources for mesoscale wave disturbances. *Monthly Weather Review, 115,* 721–729.

Verrelle, A., Richard, D., & Lac, C. (2017). Evaluation and improvement of turbulence parameterization inside deep convective clouds at kilometer-scale resolution. *Monthly Weather Review, 145,* 3947–3967.

Wakimoto, R. M., & Murphey, H. V. (2009). Analysis of a dryline during IHOP: Implications for convection initiation. *Monthly Weather Review, 137,* 912–936.

Wang, P. K. (2003). Moisture plumes above thunderstorm anvils and their contributions to cross tropopause transport of water vapor in midlatitudes. *Journal of Geophysical Research, 108*(D6), 4194. https://doi.org/10.1029/2003JD002581.

Wang, P. K. (2007). The thermodynamic structure atop a penetrating convective thunderstorm. *Atmospheric Research, 83,* 254–262.

Wang, P. K., Su, S.-H., Setvak, M., Lin, H., & Rabin, R. M. (2010). Ship wave signature at the cloud top of deep convective storms. *Atmospheric Research, 97,* 294–302.

Wei, J., & Zhang, F. (2014). Mesoscale gravity waves in moist baroclinic jet-front systems. *Journal of the Atmospheric Sciences, 71,* 929–952.

Weisman, M. L. (1992). The role of convectively generated rear-inflow jets in the evolution of long-lived mesoscale convective systems. *Journal of the Atmospheric Sciences, 49,* 1826–1847.

Weisman, M. L., Davis, C., Wang, W., Manning, K. W., & Klemp, J. B. (2008). Experiences with 0-36-h explicit convective forecasts with the WRF-ARW model. *Weather and Forecasting, 23,* 407–437.

Williams, J. K. (2014). Using random forests to diagnose aviation turbulence. *Machine Learning, 95,* 51–70. https://doi.org/10.1007/s10994-013-5346-7.

Williams, P. D. (2017). Increased light, moderate, and severe clear-air turbulence in response to climate change. *Advances in Atmospheric Sciences, 34*(5), 576–586.

Williams, P. D., & Joshi, M. M. (2013). Intensification of winter transatlantic aviation turbulence in response to climate change. *Nature Climate Change, 3,* 644–648.

Williams, P. D., & Joshi, M. M. (2016). Clear-air turbulence in a changing climate. In Sharman R., Lane. T. (Eds.), *Aviation turbulence: Processes, detection, prediction* (pp. 465–480). Berlin: Springer.

Williams, J. K., & Meymaris, G. (2016). Remote turbulence detection using ground-based Doppler radar. Aviation turbulence forecast verification. In Sharman R., Lane. T. (Eds.), *Aviation turbulence: Processes, detection, and prediction* (pp. 335–356). Berlin: Springer.

Wimmers, A., Griffin, S., Gerth, J., Bachmeier, S., & Lindstrom, S. (2018). Observations of gravity waves with high-pass filtering in the new generation of geostationary imagers and their relation to turbulence. *Weather and Forecasting, 33,* 139–144.

Wolff, J. K., & Sharman, R. D. (2008). Climatology of upper-level turbulence over the continental Unites States. *Journal of Applied Meteorology and Climatology, 47*(8), 2198–2214. https://doi.org/10.1175/2008JAMC1799.1.

Wurtele, M. G., Datta, A., & Sharman, R. D. (1996). The propagation of gravity-inertia waves and lee waves under a critical level. *Journal of the Atmospheric Sciences, 53*(11), 1505–1523.

Zhang, F. (2004). Generation of mesoscale gravity waves in upper-tropospheric jet-front systems. *Journal of the Atmospheric Sciences, 61,* 440–457.

Zhou, X., Holton, J. R., & Mullendore, G. L. (2002). Forcing of secondary waves by breaking of gravity waves in the mesosphere. *Journal of Geophysical Research, 107*(D7), 4058. https://doi.org/10.1029/2001JD001204.

Zovko-Rajak, D., & Lane, T. P. (2014). The generation of near-cloud turbulence in idealized simulations. *Journal of the Atmospheric Sciences, 71,* 2430–2451.

(Received December 15, 2017, revised March 19, 2018, accepted March 20, 2018, Published online March 31, 2018)

Pure Appl. Geophys. 176 (2019), 1959–1976
© 2018 Crown
https://doi.org/10.1007/s00024-018-2058-8

Pure and Applied Geophysics

Review of Lidar-Based Applications for Aviation Weather

Ludovic Thobois,[1] Jean Pierre Cariou,[1] and Ismail Gultepe[2,3]

Abstract—Measurements collected by Leosphere Doppler lidars were reviewed to study meteorological processes such as wind shear, wind profiles, gust fronts, and wake vortices over airports. First, the basic concepts of lidar are discussed, then its use for wind environments with respect to high-impact weather events is presented. Issues related to previous definitions of wind-related algorithms and criteria are summarized to validate the use of Doppler lidar for clear-air environmental conditions. Based on International Civil Aviation Organization (ICAO) criteria that use a 500-m height threshold in the vertical for wind warning conditions, this work suggests that use of Doppler lidars can significantly improve the safety of flight environments along landing and takeoff corridors at airports by providing warnings to pilots and ground crew and optimizing air-traffic management. The wind measurements from the lidars are found to be accurate to 0.1 m s^{-1}, and use of Doppler lidars can increase the probability of detection of wind-related severe weather conditions by up to 50% beyond the 500 m of the atmospheric boundary layer (ABL).

1. Introduction

Doppler light detection and ranging (lidar) systems are becoming extensively used for measuring wind parameters, aerosol data, and cloud layers over airports and at high levels. They can cover large horizontal and vertical areas depending on weather conditions compared with ground-based in situ observations. For this reason, they may be more applicable for aviation operations. Using measurements of Doppler velocities (V_d) and backscatter fields (β), lidars can be used to obtain high-impact weather variables such as vertical air velocity, gust, low-level wind shear, clear-air turbulence (CAT), and wake vortices (Winckelmans et al. 2005; Frech and Holzäpfel 2008; Gerz et al. 2005). These parameters are very important for aviation operations (Gultepe et al. 2019; Luce et al. 2010).

Following the introduction of solid-state yttrium aluminum garnet (YAG) lasers in the 1980s, the first coherent Doppler lidars (CDLs) were developed. At the end of the 1990s, Coherent Technologies Inc. (CTI) developed the first commercialized lidar, based on emission wavelengths of 2 μm and 1.6 μm (Frehlich et al. 1997; Krishnamurthy et al. 2012). This lidar is presently commercialized by the company Lockheed Martin Coherent Technologies Inc. In the late 1990s, due to the rapid development of optical-fiber telecommunications technology, optoelectronic components including laser transmitters, optical amplifiers, and fast detectors operating at 1.5 μm started to be used because of the lower attenuation in optical fibers. These technologies, including CDL sensors with optical fiber amplifiers, have then been developed by various research laboratories including QinetiQ (UK) and Office National d'Études et de Recherches Aérospatiales (ONERA, France) (Cariou et al. 2006; Dolfi-Bouteyre et al. 2007). A CDL with a 1.5-μm optical fiber was commercialized by Leosphere Company (France), Halophotonics Company (UK), and Zephir (UK), and 1800 CDLs are being used worldwide for observations related to wind energy and meteorological applications. Presently, roughly 15 airports worldwide are equipped with Doppler lidars for measuring wind-related parameters (including low-level wind shear), aerosols, ash, or wake turbulence generated by aircraft.

A technological summary of lidars available for operations was performed by Steen et al. (2010), who provided information on the maturity, update rate, latency, coverage, range, accuracy, weather resilience, and cost of CDLs operating at wavelengths of

[1] Leosphere, 14-16 Rue Jean Rostand, 91400 Orsay, France.
[2] MRD, Environment and Climate Change Canada (ECCC), Toronto, ON, Canada. E-mail: ismail.gultepe@canada.ca
[3] Engineering and Applied Sciences, UOIT, Oshawa, ON, Canada.

1.5, 1.6, and 2 µm based on regular and solid-state lasers. It is known that the costs of CDLs based on optical-fiber lasers are significantly lower than those using solid-state lasers.

Using CDL technologies, dry wind shear and wake vortices have been studied extensively at airports recently; For example, the European Union Single European Sky Air Traffic Management (ATM) (SESAR) research project (https://ec.europa.eu/transport/modes/air/sesar) was developed to modernize existing weather observing systems at airports that use CDLs (Mutuel et al. 2014; Juge and Barbaresco 2014; Barbaresco 2015). Current wind observation systems at airports have remained almost the same for over 40 years (International Civil Aviation Organization (ICAO) Annex 3, 2007a, b), as stated by the World Meteorological Organization (WMO; Kulesa 2003). Both the WMO and ICAO have suggested that automated weather observing systems (AWOS) with detail surface observations of wind and visibility should be used at airports. These systems are dedicated to data acquisition, processing, dissemination, and display in real time of meteorological parameters that affect landing and takeoff operations. The surface parameters include surface wind, visibility, runway visual range, height of cloud base, air and dew-point temperatures, and atmospheric pressure, all of which affect flight conditions at the surface (Gultepe et al. 2009).

Early studies suggested that the majority of accidents occur during aircraft takeoff and landing, or near to airports. Many of these accidents are strongly related to winds, gust, wind shear, and turbulence conditions, and these usually represent major causes of aircraft accidents and delays at airports. A study performed in Europe (Eurocontrol 2013; Rooseleer and Treve 2015) showed that half of accidents according to the Aviation Safety Network (ANS) were the result of adverse weather conditions during the time period of 2010–2012. In terms of air-traffic control (ATC) delays and corresponding costs, Klein and Kavoussi (2009) and Kulesa (2003) stated that the impact of adverse weather on aviation operations accounted for 20 and 70% of delays in Europe and the USA, respectively. These adverse conditions include high-impact weather (HIW) conditions such as thunderstorms and convective weather, in-flight icing, and strong winds and turbulence, which were the reason for the majority of delays and extra costs associated with ATC.

The goal of this work is to provide information on the WindCube lidar technology used for aviation applications during several field projects developed by Leosphere. In this study, measurements and monitoring of wind-related parameters (e.g., gusts, eddies, wake vortices, and wind shear) are emphasized for aviation applications based on the three field projects described in Sect. 3. First, the WindCube lidar concept and corresponding equation are presented in Sect. 2. Then, project information, including sites and observations, are described in Sect. 3. The results and a discussion are then presented in Sects. 4 and 5, respectively, while future challenges related to application of lidar technology to aviation meteorology are given in Sect. 5. The conclusions of this work are finally presented in Sect. 6.

2. *Doppler Lidar Concept*

Lidar is an active remote-sensing tool used extensively for measuring atmospheric wind and turbulence, aerosol physical characteristics, and cloud profiles with optical thickness less than about 2–3 (Fuertes et al. 2014; Thobois et al. 2018). Lidar measurements have been extensively used for wind energy estimation, meteorological research or aviation weather, and aerosol research. A lidar system transmits a light signal that interacts with targets in the air, then the backscattered amount is received by the lidar receiver. The returned amount is corrected for attenuation/scattering conditions. Although they have some similarities with radars, their wavelength lies in the optical range from 0.350 to 10 µm. The power returned to a lidar is related to the particle surface area (D^2) rather than the power of D^6 for radar technology.

The choice of the emission wavelength (λ) depends on the application, being related to the atmospheric quantity to be measured, the availability of reliable and cost-effective laser sources, the optical atmospheric transmission, as well as eye safety issues. As shown in Fig. 1, the highest atmospheric transmission occurs in the visible range, where the

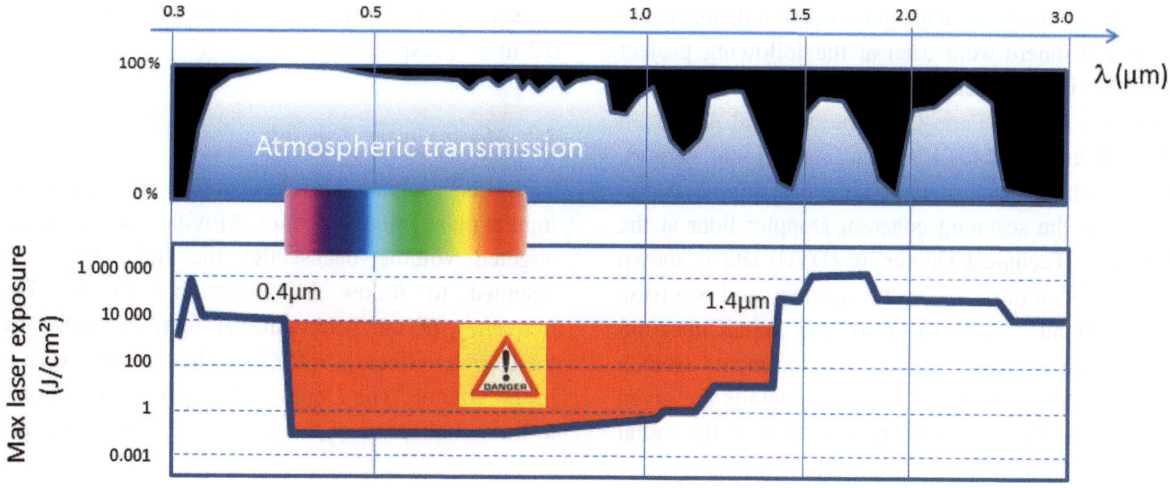

Figure 1
Atmospheric transmission and maximum laser exposure (adapted from Thobois et al. 2017)

maximum safe laser exposure is the lowest. To design an efficient lidar platform, λ should be selected to achieve the highest possible transmission while remaining eye-safe. In a lidar design, considering these aspects, the visible range should be avoided for automatic and operational applications. Generally, the ultraviolet (UV) spectral range below 0.4 µm and the near-infrared (near-IR) range (from 1.4 to 2.5 µm) are usually recommended for lidar designs.

The lidar equation for the WindCube lidar platform is first provided, then its use for obtaining meteorological parameters related to current field observations is discussed. The data availability is defined as the percentage of validated data for a given distance (ISO 28902-2 2017). In the data processing, a confidence index is derived from the signal-to-noise ratio definition and used to determine the data availability. The minimum detected power is limited by the photon noise. The power received by the lidar depends on the physical conditions and targeted weather physical parameters. The lidar equation is usually expressed as

$$\mathrm{Pr}(Z) = T \times P_{\mathrm{e}} \times \beta \times \frac{C\tau_{\mathrm{p}}}{2} \times \frac{\mathrm{Ar}}{Z^2} F(Z), \qquad (1)$$

where T is the global round-trip transmission, P_{e} is the emitted peak power, β is the atmospheric backscatter coefficient, C is the speed of light, τ_{p} is the pulse duration, Ar is the telescope effective area, Z is the distance from the lidar to the target, and $F(Z)$ is the instrumental overlap function, which depends on the focused distance and beam size. As a CDL emits light in the near-IR spectrum, the received power results from Mie scattering by aerosols or particles whose size is at least one-third of the wavelength (0.5 µm for 1.5 µm).

The backscattering and transmittance properties of a lidar laser are related to the atmospheric gases, aerosols, and hydrometeors. Both the received power and data availability decrease with the target distance (Z), while the data availability also decreases with weather conditions. As expressed by Eq. (1), the received power is reduced when the visibility (Vis) decreases because of atmospheric extinction (Gultepe et al. 2009). Rain, drizzle, and heavy fog conditions can absorb most of the light, resulting in reduced lidar returns (Gultepe et al. 2006; Chan 2008). The expected range of a WindCube400S lidar through the atmosphere, in the absence of precipitating or liquid clouds, is greater than 10 km over 80% of the time. Such performance is obtained for Vis > 10 km. In case of moderate or heavy rain, lidar-based measurements can be completely lost for ranges > 2 km.

3. *Field Projects and Measurements*

3.1. *Project Setup*

Lidar observations collected during three field projects were used to monitor and detect wind shear, gust, wake vortices, and updrafts. WindCube lidars from Leosphere were used at the following project sites: (1) the Denmark Tech University (DTU) site at Hovsore, DK, (2) Paris-CDG International Airport in Paris, France, and (3) Lanzhou International Airport, China. The project sites are shown in Fig. 2. As an example, the scanning coherent Doppler lidar at the Denmark Technical University (DTU) site is shown in Fig. 2a. In this project, the accuracy and precision of the wind measurements of the WindCube200S lidar were evaluated against measurements from a reference meteorological tower; the details are given in Sect. 4. The accuracy and precision of the radial velocity were estimated to be 0.1 m s^{-1} and 0.2 m s^{-1}, respectively.

3.2. Measurement Principle

In lidar technology, a laser source emits pulses of light at pulse frequency of ω_l. Moving particles in the targeted volume backscatter the light, which is assumed to follow Mie scattering theory. The frequency of the backscattered light is shifted by the Doppler effect ($f_d = 2V_r\lambda$) of the moving particles in the volume. The Doppler shift (f_d) is proportional to the radial speed of the particles (V_r), which is the

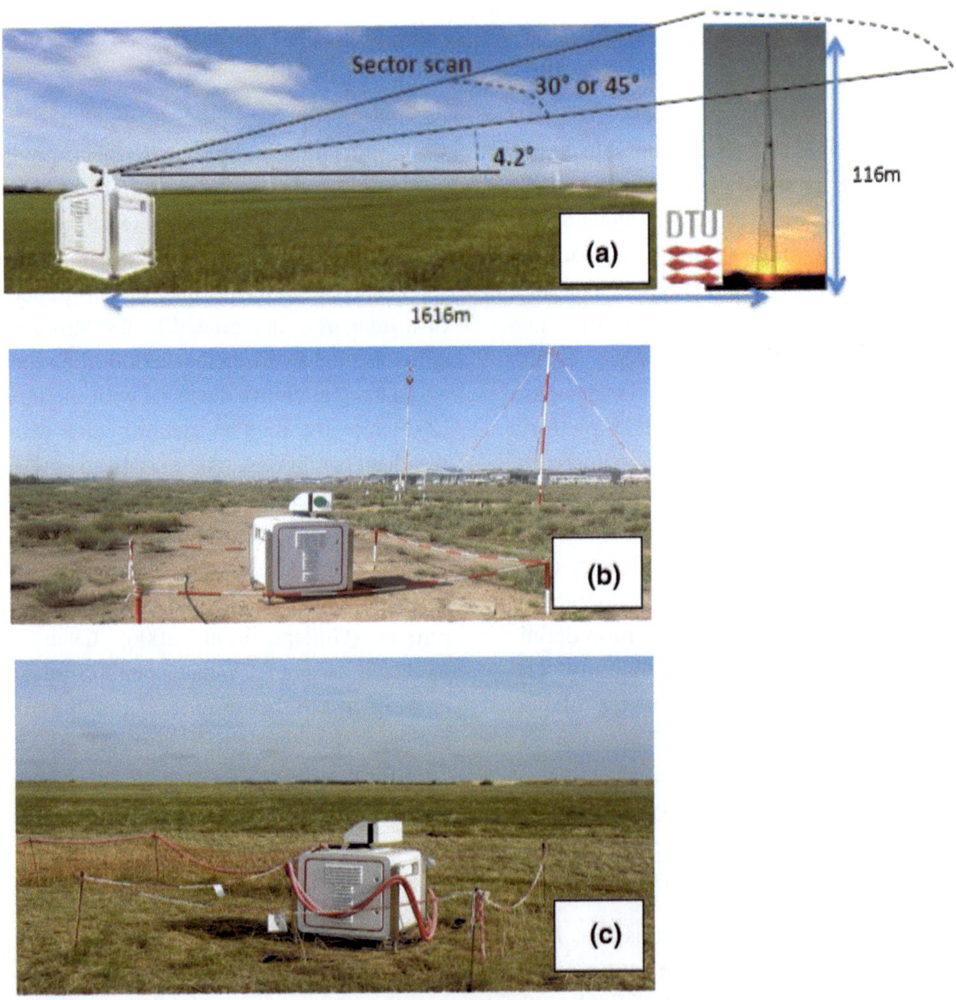

Figure 2
The WindCube200S lidar at the **a** Denmark Tech University (DTU) site in Hovsore, DK, **b** Lanzhou International Airport, and **c** Paris-Charles de Gaulle International Airport

projection of the wind vector along the line of sight of the laser. The backscattered signal is then mixed with a reference beam light in an interferometer, providing a heterodyne beat signal at the Doppler frequency f_d from which the radial wind speed can be accurately derived. The measurement principle of coherent Doppler lidar is shown in Fig. 3.

The measurement at a range x corresponds to the backscattered radiation received at time t after the laser pulse has travelled from the lidar to the particles located at range x, then back to the lidar. The position of the range gate is proportional to the time of flight according to $x = Ct/2$, where C is the speed of light. Assuming a Gaussian profile, the angular resolution Δz can be estimated from the convolution of the laser pulse profile $P(t)$ of duration τ_p with the time-equivalent range gate $W(t)$ of length τ_m (Fig. 4).

Δz can be estimated from τ_p (Fig. 4) and τ_m using the formula

$$\Delta z = \frac{C}{2}\sqrt{\tau_p^2 + \tau_m^2}. \tag{2}$$

The optimized trade-off between the range resolution and velocity resolution is achieved when $\tau_m \sim \tau_p$. The range resolution remains constant along the line of sight. For most CDLs, the pulse setting can be adjusted according to requirements.

CDLs measure the radial wind component, i.e., the projection of the wind vector along the line of sight (LOS). To retrieve the three components of the

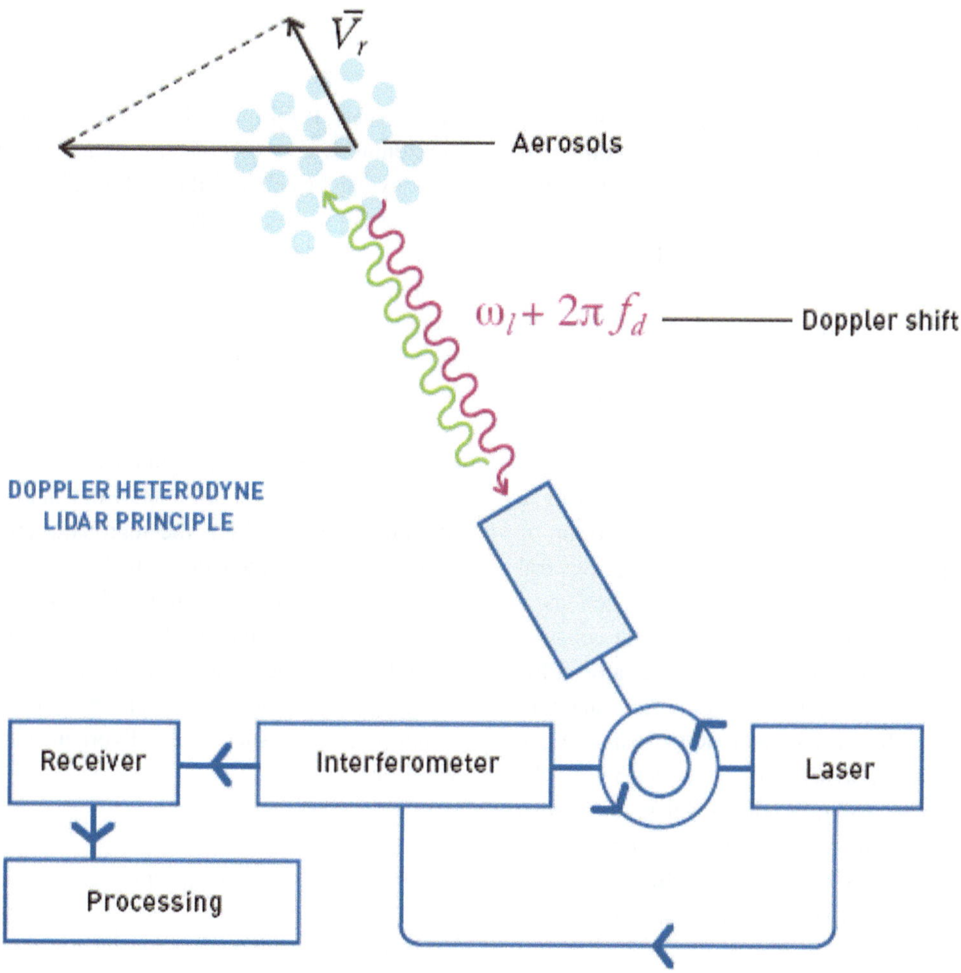

Figure 3
Measurement principle of a coherent Doppler lidar (Kongara et al. 2012; Krishnamurthy et al. 2012). Parameters defined in the text

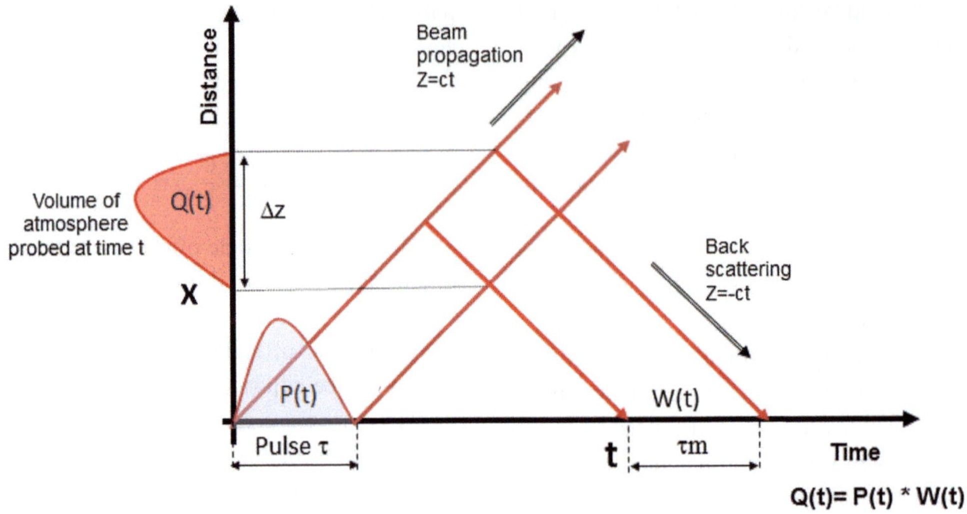

Figure 4
Time and distance diagram of a pulsed lidar. See text for symbol explanations

wind vector, viz. u, v, w, several laser beams aimed in several independent directions can be combined based on the assumption of the homogeneity of the horizontal wind (U_h). Different wind reconstruction techniques, known as Doppler beam swinging (DBS) and velocity azimuth display (VAD), can then be used to estimate the three-dimensional (3D) wind components (Gultepe et al. 1990).

In DBS mode, a CDL might use, for example, five beams, including one vertical LOS and four tilted LOSs pointing at 0°, 90°, 180°, and 270° with azimuthal angle of 15°. CDLs equipped with a scanning head can measure winds at up to 10 km or more, depending on the weather conditions. Several scanning scenarios are usually chosen, such as a single LOS or an azimuthal scan method at constant elevation, known as a plan position indicator (PPI) scan. A zenithal scan performed at constant azimuth is called a range height indicator (RHI) scan and can be used to obtain a vertical cross-section of the atmosphere. When using WindCube lidars, these scan techniques are commonly used to monitor the dynamical conditions of interest in 3D ranges and to provide information on wind and turbulence development.

When using the VAD technique, the continuity equation, assuming zero vertical air velocity at the ground and that the horizontal wind is linear at a given scan level, can be used to calculate the horizontal wind components, then the vertical air velocity can be calculated based on divergences using the lidar V_d values at various levels (Srivastava et al. 1986; Gultepe et al. 1990). Because of the use of PPI scanning and the uncertainty of up to 0.2 m s^{-1} in Doppler wind magnitudes, the error in the vertical air velocity obtained using the VAD technique can be as high as 0.5 m s^{-1}, thus it should be used with care when the atmosphere is dynamically stable and/or strong shear exists in the boundary layer. Under the assumption that the horizontal wind field at a given altitude is linear within the regions scanned by the lidar, the radial wind data can be fit using a sinusoidal curve as a function of azimuth for fixed elevations. Improvements of the VAD technique include the extended VAD (EVAD) (e.g., Srivastava et al. 1986; Matejka and Srivastava 1991), for extracting additional parameters such as the divergence and vertical wind fields.

4. Results

4.1. Accuracy of Wind Measurements

Use of lidar for wind measurements can be significantly hindered by severe weather conditions such as heavy fog, rain, snow, or convective conditions (Gultepe et al. 2007).

The coherent Doppler lidar characteristics and measurements have been evaluated over 10 years versus meteorological data obtained from masts. The meteorological parameters obtained from the lidar are sampled at 10-min intervals for averaged wind speed and direction. The lidar data were then evaluated at the reference test sites by independent institutions. Regression analysis was carried out between the calibrated cup anemometer and lidar data, and the results are shown in Fig. 5. The mean difference between the two platforms was estimated to be about 0.02 m s^{-1} by DNV-GL consulting company. The standard deviation (SD) wind speed was 0.13 m s^{-1}. Based on all the data, the typical accuracy was better than 0.1 m s^{-1} at all heights. Example scanning coherent Doppler lidar data obtained during the project designed by Denmark Technical University (DTU) are shown in Fig. 6. The campaign setup is shown in Fig. 2. These results suggest that the accuracy and precision of the radial velocity were 0.1 m s^{-1} and 0.2 m s^{-1}, respectively.

4.2. Causes of Low-Level Wind Shear

Wind shear is defined as a sudden and sustainable change of wind velocity and/or direction. Detection and forecast of this effect for airports are crucial for safe and efficient ATC. The general requirements for wind shear alerts are usually taken from ICAO Annex 3 (2005), but it does not provide detailed information or requirements for observation systems in terms of range, resolution, and wind shear. Setting these requirements at airport locations can be a difficult task as the wind shear changes continuously as a function of weather processes. Formation of low-level wind shear is strongly related to frontal systems, convection, microbursts, and thermal instabilities. The most dangerous wind shear occurs as a result of microbursts, because they exhibit the shortest temporal and spatial scales (< 15 min and < 10 km, respectively) with wind speed > 15 m s^{-1} (> 30 knots). Microbursts can significantly affect aircraft flight conditions during landing and taking off, and cannot be predicted accurately with numerical weather prediction (NWP) models because of their short-lived characteristics.

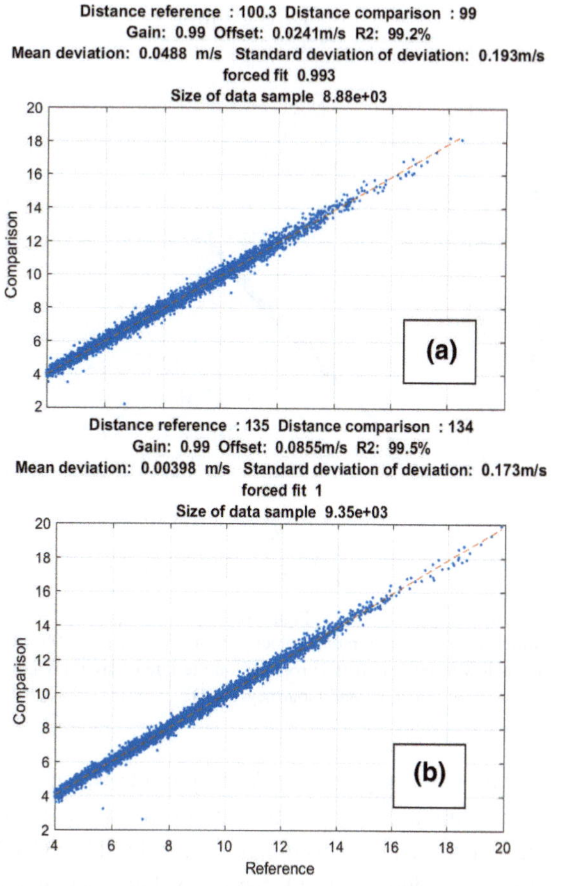

Figure 5
Regression analysis of measurements by a WindCube-v2 lidar short-range vertical profiler against calibrated cup anemometer measurements at altitude of 100 m (**a**) and 135 m (**b**)

4.3. ICAO Requirements for Airport Wind Applications

The ICAO recommendations and practices for low-level wind shear detection are detailed in Refs. (2005, 2007a, b). These recommendations and best practices were mainly written based on the use of low-level windshear alert systems (LLWAS) (using a network of anemometers) and terminal Doppler weather radar (TDWR) systems (using Doppler radars), but not for lidars. In the 1960s, low-level windshear alert systems (LLWAS) were developed using anemometers with separation of 20 m around airports. These systems mainly cover runways, but not takeoff paths or approaches. Doppler radar sensors have been developed for use in different bands (C, S, X) to detect wind shears, although they only capture cloud and precipitation conditions.

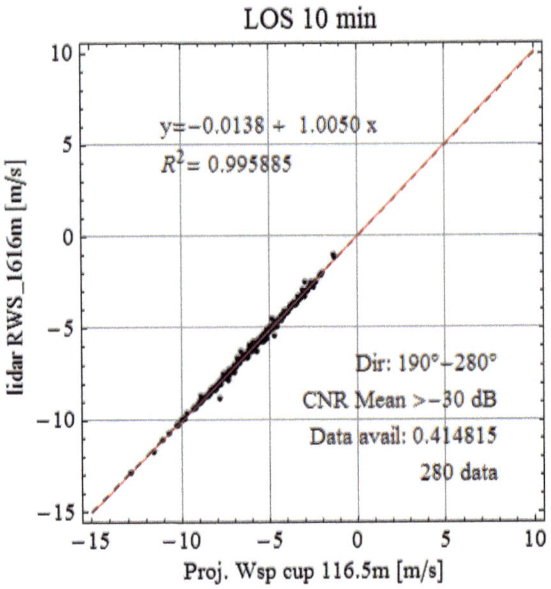

Figure 6
Comparison of 10-min radial velocity from a WindCube200S lidar versus tower measurements projected on the line of sight (Wagner and Courtney 2014)

The ICAO reports (2007a, b) provide two different requirements for aviation applications: (1) wind shear alerts and (2) warnings (Table 1). The ICAO (2007a, b) documents (Table 1) explain that, when wind shear is detected, wind shear alerts should be issued. However, the recommended coverage area for the wind shear (WS) alert is not provided. WS warnings are usually provided manually by meteorological officers based on various in situ observations and available forecasts. This deficiency can be covered by lidar observations. The current guidelines for wind shear observations based on ICAO-9817 (ICAO 2005) are presented in Table 2, suggesting that improvements in reporting of wind shear and gust conditions are required.

4.4. *Experimental Results from Lanzhou International Airport*

Lanzhou-Zhongchuan Airport (ICAO code ZLLL), Hong Kong, located in the North West region of China, was the site of WindCube lidar observations for detecting wind shear and gust detection. This site usually faces severe wind shear and gust conditions (Chan et al. 2006), with 15 cases of severe shear reported each year by pilots. The majority of wind events occur below altitude of 500 m when dry air conditions usually exist. These weather conditions do not allow efficient use of weather radars for WS detection. A project (Fig. 2) initiated by the Air Traffic Management Bureau (ATMB) of the Civil Aviation Administration of China (CAAC) was used to evaluate the capability of a Doppler lidar sensor to detect low-level wind shear conditions at ZLLL. The criteria for wind-related parameters were set at the beginning of the project

Table 1

Observed features for low-level wind shear (ICAO 2007a, b)

Feature	Alert	Warning
Coverage	Final approach and initial takeoff path	Approach + takeoff path + runways + circling approach
Update	frequency	At least 1 min suggested
None		
Altitude	Below 500 m	Below 500 m
Source	Automated, ground-based remote-sensing or detection equipment	Prepared by meteorological office based on ground-based, wind shear remote-sensing, or detection equipment, aircraft observations, other meteorological information (recommended)
Description	Wind shear involving headwind/tailwind changes of 15 knots or more which could adversely affect aircraft Wind shear (WS) alerts and warnings should include a specific reference to microbursts if observed (recommended) WS alerts shall (if practicable) relate to specific sections of runways and distances along approach or takeoff path (recommended) where existence of wind shear which could adversely affect aircraft is observed or expected	Observed or expected existence of wind shears which could adversely affect aircraft

Table 2

Synthesis of ICAO guidelines for observing low-level wind shear (ICAO 2005)

Feature	Alert	Warning
Coverage	3 nautical miles (NM) extension from runways is commonly used, whatever the wind shear equipment (LLWAS, lidars, TDWR/radars)	Up to 9.55 km for a glide slope of 3° to monitor up to altitude of 500 m
Update frequency	LLWAS: every 30 s Radar/lidar: between 1 and 6 min	Typically 5 min is used for radars and lidars
Resolution	LLWAS: 1 NM between two anemometers Radar: 100–250 m Lidar: 100–200 m	Same as alerts
Scanning patterns	One horizontal scan (PPI) at 3° per approach is used	Several products can be used, from two to five PPI scans
Methods for computing alerts	Wind shear alerts: headwind/tailwind changes over 1 NM or along the runway superior to 15 knots Microburst alerts: headwind/tailwind changes over 1 NM or along the runway superior to 40 knots LLWAS: National Center for Atmospheric Research (NCAR) algorithm Radar: runway-oriented wind shear algorithms such as those developed by MIT-Lincoln Lab for TDWR, Hong Kong Observatory (HKO) algorithm, commercial software such as Rainbow5 developed by Selex, or IRIS developed by Vaisala	Additional products can be computed from radial radar or lidar data according to local needs, such as shear products, gust front detection, or wind reconstruction

based on the local wind shear characteristics, because accurate wind shear observations were not available. Therefore, the lidar settings and configuration were first applied to increase the quality of the data applicable for proper operational wind shear detection.

A WindCube400S-AT coherent Doppler lidar provided by Leosphere was installed near a runway at Lanzhou ZLLL Airport for about 9 months in 2016. The goal was to maximize the performance of wind shear detection in both directions, viz. lateral and along the flight direction. Its proximity to the runway ensures that the Doppler velocity (V_r) is very close to the headwind and tailwind components. The configuration consisted of a series of full azimuthal scans at the glide slope angle, repeated every 3 min and spaced by full azimuthal PPI scans at various elevation angles and by an RHI vertical scan. The general principle behind this configuration is to provide a 3D wind plan with the following features:

- Wind shear alerts should be provided by PPI scans at 3° for landing and 6° for takeoff;
- Vertical wind profiles should be provided by PPI scans at higher elevations (> 45°);
- 3D monitoring of wind shear events should be provided by a combination of PPI and RHI scans.

The capability of the WindCube400S-AT lidar to provide relevant wind shear alerts was assessed using the probability of detection (POD, defined as the fraction of observed events that were correctly forecast) and the false alarm rate (FAR, the fraction of positive forecasts that were wrong) of wind shear events. Two methodologies were applied. First, the good detection ratio (GDR) of dry-air wind shear was defined as the ratio of valid wind shear alerts obtained using the Rainbow5 software developed by SELEX weather-radar company. The alerts were considered acceptable for changes of the headwind or tailwind speed along the profiles of above 15 knots. This parameter allows characterization of the performance of the detection algorithm.

Secondly, pilot reports provided by end users were used to determine the POD of the lidar measurements. The results suggested that the POD of the lidar data reached 83.3% (Table 3). Three events reported by pilots were not detected by the lidar, being cases with low-level turbulence and strong crosswind conditions. Note that, even though the POD when using lidar to detect wind shear was high, this does not mean that the FAR must be low. Indeed, the FAR was at least about 20% but could increase to more than 50% with increasing

Table 3

Probability of detection (POD) of wind shear events by the WindCube400S-AT lidar, estimated from 8 May to 24 November 2016 based on pilot reports

	Number of events	Probability of detection (%)
Total # of pilot reports	18	–
Events detected using the WindCube lidar	15	83.3
Events not detected using the WindCube lidar	3	16.7

precipitation, when the quality of lidar data quickly decreases. However, inclusion of lidar data into the analysis clearly made a good contribution to shear detection under clear environmental conditions.

4.5. Low-Level Dry Wind Shear Event

The WindCube400S CDL in PPI scan mode at an elevation angle of 3° was installed at ZLLL airport and tested during 31 May 2016. The wind direction was originally stable and from the south. The lidar provided a two-dimensional (2D) image of the radial wind speeds, as shown in Fig. 7, where positive radial wind speed (from yellow to red) is towards the lidar and negative radial wind speed (light to dark blues) is away from the lidar. The wind speed and direction on this day were highly variable; the wind direction changed from east below 200 m level to south above 200 m level. Eventually, the wind above 200 m turned to northwest during the morning. The northerly winds strengthened, resulting in cold air subsidence. These northerly and southerly winds at low altitudes led to a convergence zone that resulted in strong wind shear of 20 knots at ZLLL. An aircraft trying to land during this wind shear event was asked to modify its flight path due to strong tailwinds. Then, a pilot alert for wind shear was issued based on the lidar-detected wind shear conditions.

4.6. Gust Front Detection

The gust line follows the change in wind flow rate. This change can be used to identify the line of physical separation and discontinuity in the wind flow direction. The gust front detection algorithm indicated

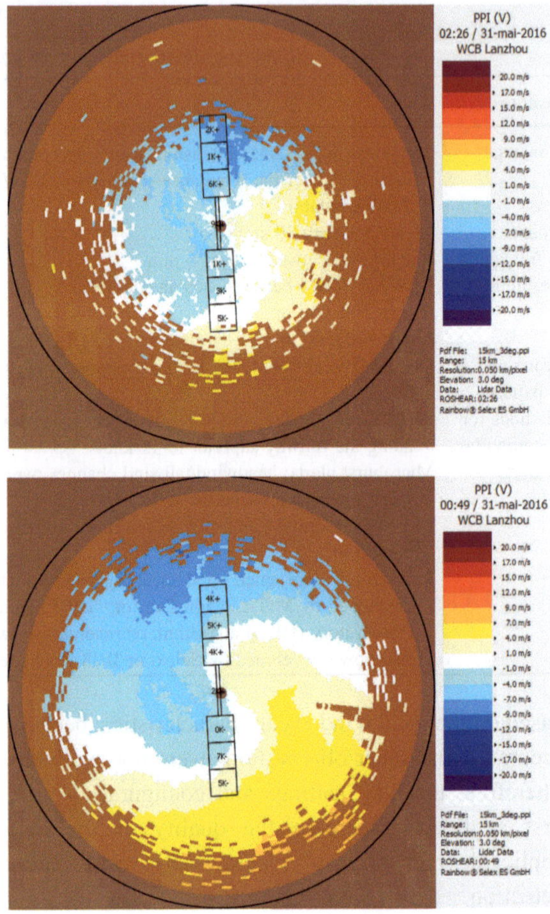

Figure 7
Temporal evolution of a dry wind shear event on 31 May 2016 as observed by the WindCube400S CDL installed at Lanzhou Airport in a PPI scan at elevation angle of 3°

the passage of a gust front based on lidar observations collected at ZLLL airport on 21 September 2016 by utilizing Rainbow5 software; the results are shown in Fig. 8 and described in Table 2. The results in Fig. 8 suggest that the location of the gust front and its time development were detectable.

4.7. VAD Technique and Vertical Air Velocity

The velocity azimuth display profile (VADP) algorithm is commonly used to monitor the vertical profile of 2D wind components with the vertical air velocity (w_a). Based on PPI scans, the mass continuity equation (Gultepe et al. 1990; Tian et al. 2015), assuming zero w_a at the surface, can be used to obtain the divergence/convergence field as a function of

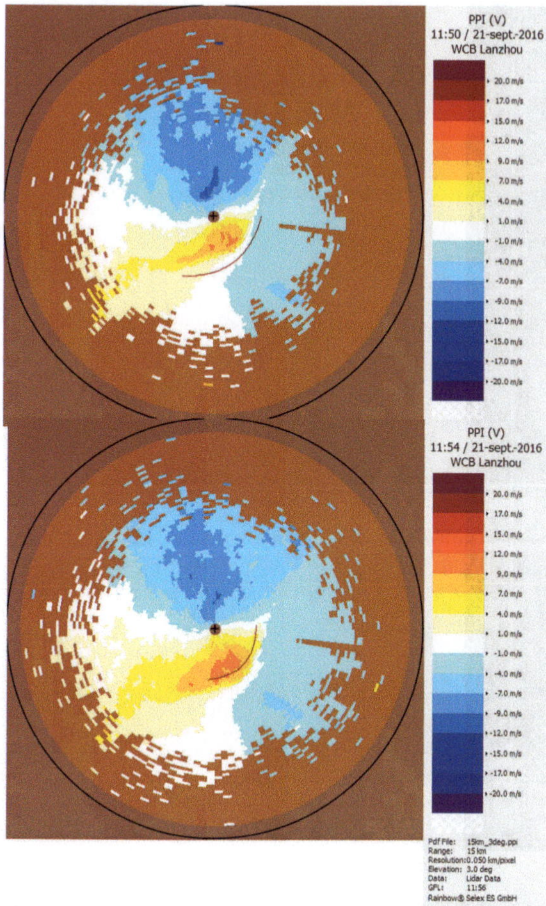

Figure 8
Example of gust front detection performed by Rainbow5 software using WindCube400S CDL data at Lanzhou-Zhongchuan Airport, Hong Kong during a wind shear event on 21 September 2016

height, which is then converted to w_a. Figure 9 shows the results of the VADP algorithm applied to a wind shear event on 21 September 2016, revealing strong wind shear at the surface and 2.1 km levels.

4.8. Wake Vortex Monitoring

Wake turbulence is generated by each flying aircraft and represents a risk for following aircraft. This is why minimum separations between aircrafts to avoid wake turbulence have been established by the ICAO, as described in the ICAO-Procedures for Air Navigation Services (PANS) wake turbulence scheme (ICAO-PANS-ATM 2007a, b). These regulations were established over 40 years ago to prevent the risk of a following aircraft encountering wake vortices generated by the aircraft ahead. These minimum separations to avoid wake turbulence are classified into three criteria based on aircraft weight and wingspan and represent one of the most important factors limiting airport capacity. In the early 2000s, lidar played an important role in detecting, localizing, and determining the intensity of wake vortices, which were studied during EU-supported projects (Speijker 2009; Sawyers 2010; Leweke 2005). Thanks to the in-depth understanding of wake turbulence behavior and decay, these studies revealed that the existing ICAO separation minima were overly conservative and should be optimized. The current ICAO regulations have been improved through a joint US–EU project called RECAT-WT (Re-Categorization of Wake Turbulence, 2014, available at http://www.recat-project.eu). In this project, the medium and heavy categories were subdivided into lower and upper groups. The European Aviation Safety Agency (EASA) accepted this new classification and implemented it at airports in Europe. Rather than the concept of a static separation distance, a dynamic distance separation concept called time-based separation (TBS) was also developed. The main concept behind TBS is to adjust the separation distance as a function of weather conditions, more specifically the headwind profile along the glide slope. However, for general dynamic distance separation concepts, wind, crosswind, and eddy dissipation rate (EDR) measurements or calculations are needed, because these parameters can affect the decay and lifetime of wake vortices (Singhal 2014). Therefore, Doppler lidar measurements can provide detailed information for such TBS applications.

Using the WindCube Doppler lidar at Paris-Charles de Gaulle International Airport, a wake vortex algorithm using wake vortex data from 2015 to 2017 was applied for risk monitoring after implementation of the new RECAT-EU scheme for optimizing the separation distance minima (Eurocontrol 2013). Measurements of wake vortices at the airport were performed using a specific configuration based on RHI scans to refresh the wake vortex data every 8 s with vertical (horizontal) resolution of 2 (5) m (Fig. 10). Because the altitude of aircraft is usually low (< 80 m), the vertical scan was fixed at 24° to allow interpretation and monitoring of wake

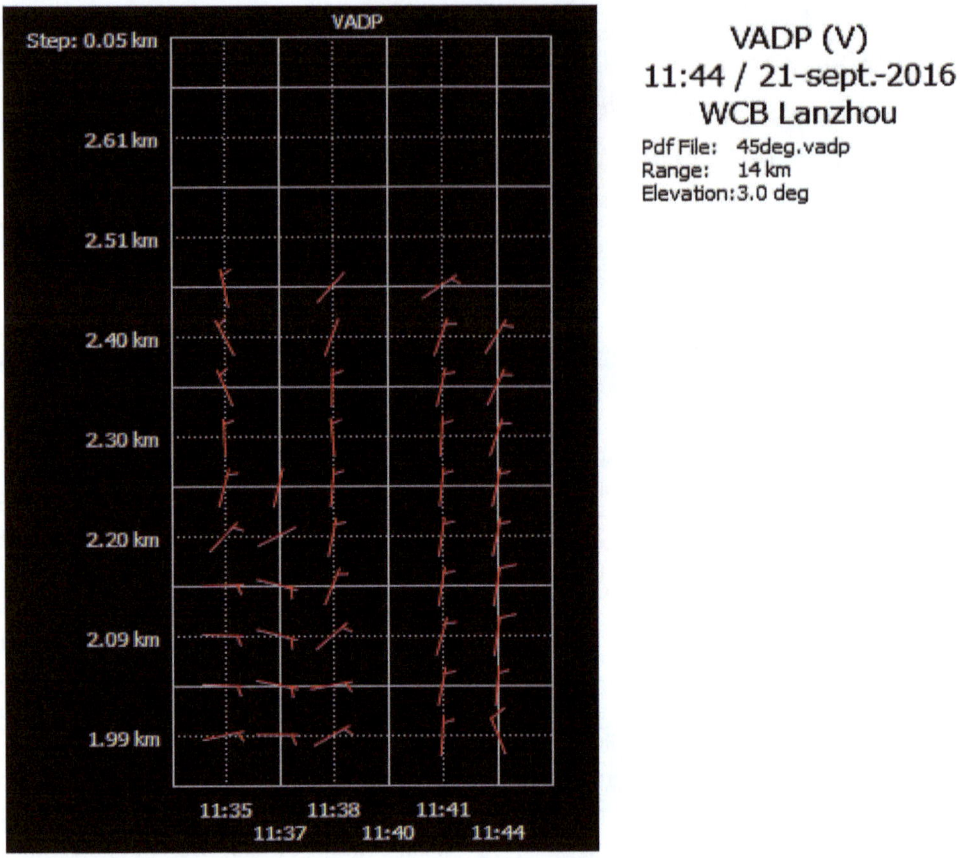

Figure 9
Example vertical wind profiles produced by Rainbow5 software from WindCube400S CDL data during a wind shear event on 21 September 2016

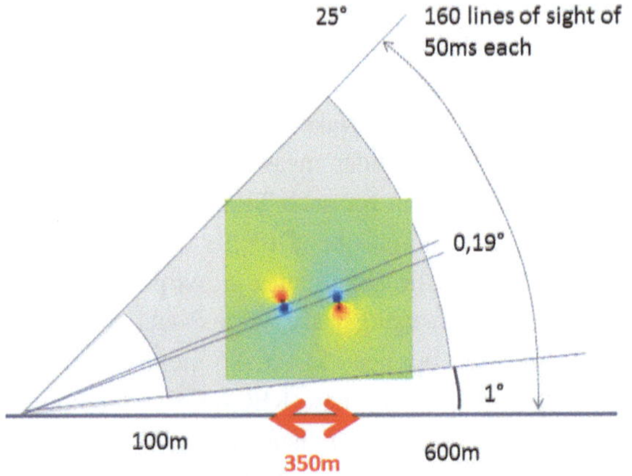

Figure 10
Scanning scenarios used by the WindCube200S lidar for collection of wake vortex data at Paris-Charles De Gaulle Airport

vortices very close to the ground. The vertical resolution of 2 m was achieved using an angular resolution of 0.19°, leading to an accumulation time per beam of 50 ms, while a range-gate overlapping technique was used to achieve the targeted range resolution of 5 m.

Overall, the results showed that approximately 25,000 wake vortex tracks were collected by the 200S lidar during the study period. Two-thirds of these tracks were found for medium-sized aircraft. Detailed analysis of the wake vortex data from the lidar was carried out based on the following three parameters:

1. Hit rate, defined as the probability of detecting a pair of wake vortices at least in the first scan of a track
2. Vortex span, defined as the initial distance between the two vortex cores (immediately after the merge of the marginal wake vortices) for one aircraft

3. Initial circulation, defined as the maximum circulation of the vortices in a wake vortex track

The results were evaluated for heavy and jumbo aircrafts using lidar observations for Vis > 10 km with no rain. The hit rate for the wake vortex data based on all aircraft was found to be about 94%. The accuracy results for the heavy and jumbo aircraft categories are collated in Table 4, revealing reasonable performance for detection of wake vortex. Using an earlier lidar system based on a wake vortex characterization algorithm (Rooseleer and Treve 2015), Thobois et al. (2018) suggested that the difference between the new analysis results and results from old data was about 2%.

4.9. Wake Vortex Generated by a Heavy Aircraft

The RHI scans with high spatial resolution by the scanning coherent Doppler lidar with short pulses allow observation of the complex flows of both vortices plus the background wind conditions (Chan and Shao 2007). The time evolution of a pair of wake vortices generated by a jumbo aircraft on 20 November 2015 during low crosswinds is shown in Fig. 11. This sequence of scans was collected by the Doppler lidar from 6:42:21 AM up to 6:45:33 AM. The single and first pair of wake vortices was observed from 6:42:21 AM to 6:43:53 AM. A second pair of wake vortices was observed above the runway at 6:44:16 AM, separated horizontally from the lidar by 365 m. This case could be characteristic of wake vortex encounters, since the crosswind was very low (~ 1 m/s) and the turbulence was weak. The high resolution of 5 m horizontally and 2 m vertically of the WindCube lidar were used to visualize the wake vortices and the vertical structure of the crosswind.

In Europe, the reference document for wake vortex regulations can be found in the RECAT-EU project (Rooseleer and Treve 2015), including an evaluation of the risk using various severity metrics.

5. Discussion

The concept of coherent Doppler lidar is based on the physics of light propagation. Atmospheric conditions such as precipitation, clouds, and fog can impact on the precision and accuracy of the obtained wind speed data. Unlike other remote-sensing technologies such as sodars, Doppler lidar can provide accurate wind components and derived turbulence parameters for both cloud-free conditions and in the presence of clouds with optical thickness < 2–3 (Gultepe and Heymsfield 2016). Therefore, detection of ice fog (composed of suspended ice crystals) is possible with Doppler lidar systems (Gultepe et al. 2014). The precision and accuracy of Doppler lidar results are ensured by calibration procedures applied by the manufacturer.

The results provided by the WindCube400S-AT lidar provide important insights to better characterize the local wind shear and wake vortices at airports, including typical duration, headwind changes, headwind to tailwind changes, crosswind components, and turbulence, as well as their impact on air-traffic delays. Forecasters can use the resulting wind shear alerts and wind data from the Doppler lidar to evaluate wind shear events, leading to an increase of ~ 25% in the forecasting scores for dry wind shear. This increase reached up to 50% when taking into account additional pilot reports related to crosswind and turbulence (Fig. 12).

Table 4

Accuracy of initial vortex span and initial circulation for heavy and jumbo aircraft

Criterion	Mean absolute error	Mean relative error (%)
Vortex span for heavy aircraft	1.49 m	3.16
Vortex span for jumbo aircraft	2.21 m	3.50
Initial circulation of left vortex for heavy aircraft	24.32 $m^2 s^{-1}$	5.52
Initial circulation of right vortex for heavy aircraft	34.53 $m^2 s^{-1}$	7.65
Initial circulation of left vortex for jumbo aircraft	46.18 $m^2 s^{-1}$	6.12
Initial circulation of right vortex for jumbo aircraft	86.40 $m^2 s^{-1}$	11.3

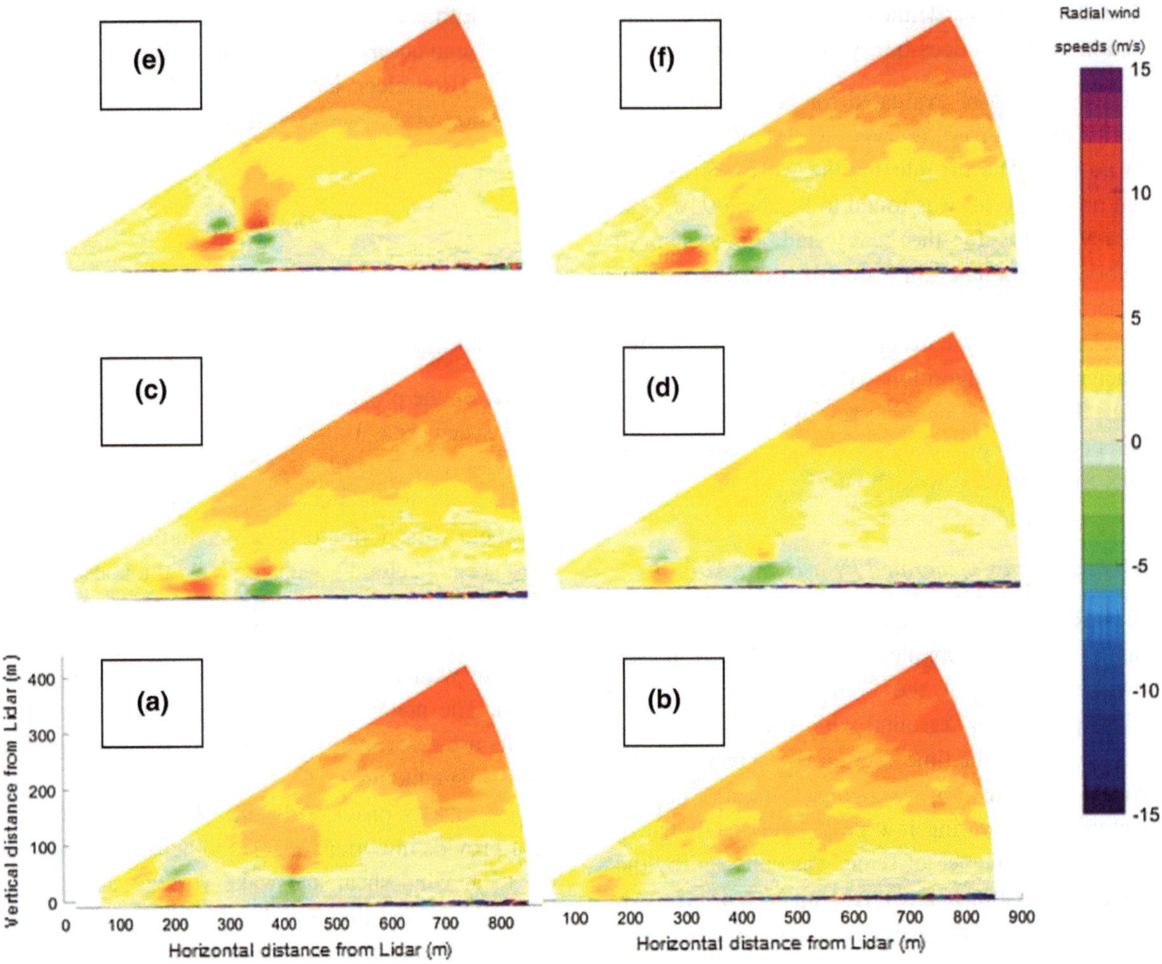

Figure 11
Radial wind measurements provided by the WindCube lidar, showing the evolution of a pair of wake vortices generated by a Jumbo aircraft on 20 November 2016 at 18:40:58 (**a**), 18:41:17 (**b**), 18:41:35 (**c**), 18:41:54 (**d**), 18:42:13 (**e**), and 18:43:00 (**f**)

Dry wind shear affects airports worldwide, being severe for those located close to shorelines, valleys, or mountains. Dry wind shear forms in a very complex way and depends on local weather conditions and topography. Doppler lidars can help improve evaluation of aviation meteorology concepts such as vortex, shear, and turbulence (Wieringa 1980). The use of fiber-based technology allows the lidar configuration to be adjusted flexibly to monitor dry wind shear. This technology has demonstrated its ability to achieve short time scales for operational purposes. The easy deployment of fiber-based lidars allows better and effective operational weather and aviation forecasts.

Lately, Doppler lidars have become commercially available instruments, such as those from Lockheed Martin Corporation, Leosphere, and Halo Photonics, opening their scope of application even to nonscientific end users, for example, to monitor flow around hills, and wind in the area of airstrips (Stawiarski et al. 2013; Tang et al. 2011a, b). Lidar can be a good tool for aviation operations for detecting wind-related parameters such as gust, wind shear, turbulence, and eddies, and clear-air turbulence, as well as particles which affect visible-light extinction. During flights, turbulence may be a major cause of high-level flight accidents. An ultraviolet lidar sensor developed by the German Aerospace Center (DLR) was tested

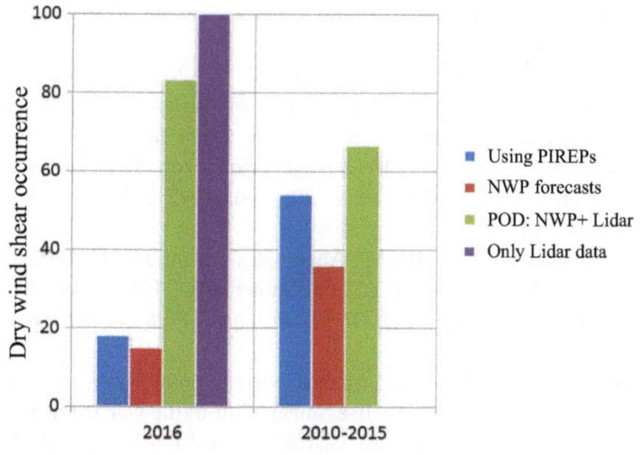

Figure 12
Percentage of wind shear events estimated using various methods, including pilot reports (PIREPs), NWP forecasts, and the POD based on both NWP predictions and lidar monitoring data. Using observations from the WindCube400S-AT lidar located at Lanzhou Airport, improved monitoring of wind shear conditions is clearly seen

during the Thales-led Delicat (Demonstration of LIdar based Clear Air Turbulence) program to reduce passenger and crew injuries by providing sufficient advance warning of an encounter (Vrancken et al. 2010). When installed onboard an aircraft, this new lidar system emits short-wave ultraviolet (UV) laser radiation along the direction of flight. The density of the air (and its fluctuations) can then be determined from the backscattered value measured for oxygen and nitrogen molecules in the air. These fluctuations provide information about clear-air turbulence ahead. Such advanced technology for aircraft protection against clear-air turbulence hazards requires validation by the aviation industry. Improving CAT monitoring in aircraft would increase both customer satisfaction and aviation safety. Based on traffic and accident statistics, use of UV lidar turbulence protection equipment could avoid 8–10 turbulence accidents per year and reduce by 15–20 (or 40%) the number of accidents due to turbulence per year when fully developed (http://www.delicat.inoe.ro/).

Using pulsed Doppler lidar systems, long-range high-resolution measurements of radial wind velocities are possible. With the development of commercial Doppler lidars and the reduction of their acquisition costs, dual-Doppler lidar systems will become increasingly accessible in coming years. Starwiarski et al. (2013) studied a highly synchronized long-range dual-Doppler lidar system, and extensively discussed the errors involved in this complex measurement technique. The sources of errors and interactions for this system relate to the retrieval algorithm, including errors from each single Doppler lidar and those occurring from the combination of instruments, such as relative beam angle, the temporal and spatial scales of the scan pattern, and atmospheric conditions. Certainly, deployment of two or even three Doppler lidars is necessary to measure two or three components of the wind vector in the atmospheric flow aloft. Details on the errors from a single and multiple lidars can be found in the work of Frehlich (2001), Frehlich et al. (1994, 1997, 1998), and Starwiarski et al. (2013).

6. Conclusions

Use of Doppler lidars in aviation weather applications is currently limited to about 10 pioneer projects being conducted worldwide, although some of these are using lidar data operationally (at airports in Japan, Hong Kong, and Italy). Their benefits for measuring wind shear, wind profiles, and turbulence and wake vortices are shown here using a state-of-the-art technology called Leosphere WindCube lidar.

The benefits of Doppler lidar to study the dynamics of weather systems are making them more popular for use in airport applications. Using Doppler lidar technology, three-dimensional wind components, covering various levels from the surface to higher levels, can also be obtained over and near airports, thereby improving flight dynamics. In addition to the Doppler velocity field, backscattered data obtained from the lidar return power have also recently been applied to calculate the cloud base, visibility, aerosol extinction, and volcanic ash hazards. This new capability allows the development of a multifunction meteorological observation system that could be integrated into supersites to improve nowcasting (Gultepe et al. 2019, current issue).

The most important result of this work is the suggestion that wind shear, turbulence, gust fronts, and wake vortices can be detected using Doppler lidars, and their time development can be studied in both horizontal and vertical directions. The uncertainty on the wind speed can be about 0.1 m s^{-1}. It was also found that inclusion of Doppler lidars in nowcasting systems, as well as in integrated observing systems, could improve forecasts of aviation weather by up to 50% when no precipitation occurs. Clearly, previous ICAO criteria should be improved for future aviation meteorology applications.

The final step of the use of lidars in airport applications is the integration of lidar data into airport forecasting systems for optimizing air-traffic operations; this will allow airport authorities to reduce the occurrence of accidents due to wind- and turbulence-related weather events, mitigating atmospheric hazards.

Acknowledgements

This work was mainly supported by Leosphere of France, and performed in collaboration with EUROCONTROL for the wake turbulence project at Paris-Charles de Gaulle Airport and with Norths West ATMB from Gansu, China for the wind shear project at Lanzhou Airport. This work is also partially supported through the Satellite Applications for Arctic Weather and SAR (Search and Rescue) Operations (SAAWSO) project by the SAR office of DND and ECCC of Canada.

Publisher's Note Springer Nature remains neutral with regard to jurisdictional claims in published maps and institutional affiliations.

References

Barbaresco, F. (2015). European FP7 UFO (Ultra-Fast wind sensors for wake-vortex hazards mitigation) Project: Technological Developments and Validation Campaign of New Generation Multifunction X-band and 1.5 μm LIDAR sensors on Airport for wind hazards monitoring on airport. *WAKENET-EU Workshop*, Bretigny, France, 13th and 14th May, 2014 **(Oral Presentation)**.

Cariou, J. P., Augere, B., & Valla, M. (2006). Laser source requirements for coherent Lidars based on fiber technology. *Comptes Rendus Physique, 7,* 213–223.

Chan, P. W. (2008). Applications of an infrared Doppler lidar. *Journal of Atmospheric and Oceanic Technology, 25,* 637–655. https://doi.org/10.1175/2007jtecha1057.1.

Chan, P. W., & Shao, A. M. (2007). Depiction of complex airflow near Hong Kong International Airport using a Doppler LIDAR with a two-dimensional wind retrieval technique. *Meteorologische Zeitschrift, 16,* 491–504.

Chan, P.W., Shun, C. M., & Wu, K. C. (2006). Operational LIDAR-based system for automatic windshear alerting at the Hong Kong International Airport. *12th Conference on Aviation, Range, & Aerospace Meteorology*, American Meteorological Society, Atlanta, GA, USA, 29 January—2 February 2006, Reprint 613.

Dolfi-Bouteyre, A., Valla, M., Augère, B., Cariou, J.P., Goular, D., Fleury, D., et al. (2007). 1.5 μm all fiber pulsed LIDAR for wake vortex monitoring. 14th *Coherent Laser Radar Conference*. pp. 27–31.

Eurocontrol. (2013). Performance review commission, performance review report. An assessment of air traffic management in Europe during the calendar year 2013, Final report, PRR 2013.

Frech, M., & Holzäpfel, F. (2008). Skill of an aircraft wake-vortex weather prediction and observation. *Journal of Aircraft, 45,* 461–470.

Frehlich, R. (2001). Estimation of velocity error for Doppler lidar measurements. *Journal of Atmospheric and Oceanic Technology, 18,* 1628–1639.

Frehlich, R., Hannon, S. M., & Henderson, S. (1994). Performance of a 2-mm coherent Doppler lidar for wind measurements. *Journal of Atmospheric and Oceanic Technology, 11,* 1517–1528.

Frehlich, R., Hannon, Stephen M., & Henderson, S. W. (1997). Coherent Doppler Lidar measurements of winds in the weak signal regime. *Applied Optics, 36,* 3491–3499.

Frehlich, R., Hannon, S. M., & Henderson, S. (1998). Coherent Doppler lidar measurements of wind field statistics. *Boundary-Layer Meteorology, 86,* 233–256.

Fuertes, F. C., Iungo, G. V., & Porté-Agel, F. (2014). 3D turbulence measurements using three synchronous wind lidars: Validation against sonic anemometry. *Journal of Atmospheric and Oceanic Technology, 31,* 1549–1556.

Gerz, T., Holzäpfel, F., Bryant, W., Köpp, F., Frech, M., Tafferner, A., et al. (2005). Research towards a wake-vortex advisory system for optimal aircraft spacing. *Comptes Rendus Physique, 6*(4/5), 501–523. **(Special issue on Aircraft trailing vortices)**.

Gultepe, I., Sharman, R., Williams, P.D., Zhou, B., Ellrod, G., Minnis, P., Trier, S., Griffin, S., Yum, S.S., Gharabaghi, B., Feltz, W., Temini M., Pu, Z., Storer, L.N., Kneringer, P., Weston, M.J., Chuang, H.Y., Thobois, L., Dimri, A.P., Dietz, S.J., Gutenberg, Almeida, M.V., Neto, F.L. A. (2019). A review of high impact weather for aviation meteorology. Pure and Applied Geophysics January. **(Accepted)**.

Gultepe, I., & Heymsfield, A. J. (2016). Ice fog, ice clouds, and remote sensing; Introduction. *Pure and Applied Geophysics, 173*(9), 2977–2982. https://doi.org/10.1007/s00024-016-1380-2.

Gultepe, I., Heymsfield, A. J., & Lenschow, D. H. (1990). A comparison of vertical velocity in cirrus obtained from aircraft and Lidar measurements. *Journal of Atmospheric and Oceanic Technology, 7,* 58–67.

Gultepe, I., Kuhn, T., Pavolonis, M., Calvert, C., Gurka, J., Isaac, G. A., et al. (2014). Ice fog in Arctic during FRAM-IF project: Aviation and nowcasting applications. *Bulletin of the American Meteorological Society, 95,* 211–226.

Gultepe, Müller, M. D., & Boybeyi, Z. (2006). A new warm fog parameterization scheme for numerical weather prediction models. *Journal of Applied Meteorology, 45,* 1469–1480.

Gultepe, I., Pearson, G., Milbrandt, J. A., Hansen, B., Platnick, S., Taylor, P., et al. (2009). The fog remote sensing and modeling (FRAM) field project. *Bulletin of the American Meteorological Society, 90,* 341–359.

Gultepe, I., Tardif, R., Michaelides, S. C., Cermak, J., Bott, A., Bendix, J., et al. (2007). Fog research: A review of past achievements and future perspectives. *Journal of Pure and Applied Geophysics, 164,* 1121–1159. **(Special issue on fog, edited by I. Gultepe).**

ICAO. (2005). *Manual on low-level wind shear and turbulence first edition—2005. Doc 9817, AN/449.* International Civil Aviation Organization (ICAO). DOC-09817-001-01-E-P.

ICAO-PANS-ATM. (2007a). *ICAO Current distance separations ICAO minima are detailed and illustrated in PANS-ATM Doc 4444.* Amendment No. 1 to the Procedures for Air Navigation Services (PANS), AIR TRAFFIC MANAGEMENT (ATM) (Doc 4444).

ICAO-PANS-ATM. (2007b). *Annex 3 to the convention on international civil aviation, meteorological service for international air navigation—Part I: Core SARPs—Part II appendices and attachments* (16th ed.). 8/07, E/P1/2000.

ISO 28902-2. (2017). *Air quality—Environmental meteorology—Part 1: Ground-based remote sensing of wind by heterodyne pulsed Doppler Lidar.* WMO ISO Standards.

Juge, P. & F. Barbaresco, 2014: Progress of European SESAR P12.2.2—Runway wake vortex detection, prediction and decision support tools—Project and presentation of 1 year Radar/LIDAR trials at Paris-CDG Airport planned in 2014–2015. *WAKENET-EU workshop,* Bretigny, France, 13th and 14th May, 2014 **(Oral Presentation).**

Klein, A., & Kavoussi, R. (2009). Weather forecast accuracy: Study of impact on airport capacity and estimation of avoidable costs. *Eighth USA/Europe Air Traffic Management Research and Development Seminar* (ATM2009), Fairfax, VA, USA.

Kongara, S., Calhoun, R., Choukulkar, A., & Boldi, M. (2012). Velocity retrieval for coherent Doppler lidar. *International Journal of Remote Sensing, 33,* 5306. https://doi.org/10.1080/01431161.2012.663524.

Krishnamurthy, R., Choukulkar, A., Calhoun, R., Fine, J., Oliver, A., & Barr, K. S. (2012). Coherent Doppler Lidar for wind farm characterization. *Wind Energy.* https://doi.org/10.1002/we.539.

Kulesa, G. (2003). Weather and aviation: How does weather affect the safety and operations of airports and aviation, and how does FAA work to manage weather-related effects? The potential impacts of climate change on transportation. DOT report. *Conference: The Potential Impacts of Climate Change on Transportation,* Washington, D.C., 2002-10-1 to 2002-10-2.

Leweke, T. (2005). *FAR-wake; Fundamental research on aircraft wake phenomena.* EU project Funded under: FP6-AEROSPACE Ref.: AST4-CT-2005-012238, Final Activity Report.

Luce, H., Nakamura, T., Yamamoto, M. K., Yamamoto, M., & Fukao, S. (2010). MU radar and lidar observations of clear-air turbulence underneath cirrus. *Monthly Weather Review, 138,* 438–452.

Matejka, T., & Srivastava, R. C. (1991). An improved version of the extended velocity-azimuth display analysis of single-Doppler radar data. *Journal of Atmospheric and Oceanic Technology, 8,* 453–466.

Mutuel, L. H., Barbaresco, F., Juge, P., Klein, M., Canal, D., Ricci, Y., et al. (2014). ATM decision support tool for wake vortex hazard management combining sensors and modeling. AIAA Paper n°2014-2332. *AIAA Aviation, 6th AIAA Atmospheric and Space Environments Conference,* 16–20 June, 2014, Atlanta, Georgia, USA AIAA 2014-2332, pp. 364–378.

Rooseleer F., & Treve V. (2015). *RECAT-EU European Wake Turbulence Categorization and Separation Minima on Approach and Departure,* EUROCONTROL technical report, 15/07/2015.V 1.1 available from EUROCONTROL Headquarters, 96 Rue de la Fusée B-1130 Brussels.

Sawyers, D. (2010). *AWIATOR (aircraft wing with advanced technology operation) passive flow control on civil aircraft flaps using sub-boundary layer vortex generators.* G4RD-CT-2002-00836, Funded under: FP5-GROWTH.

Singhal. J. (2014). From SESAR to implementation: Completing the TBS picture. *2014 Workshop of WakeNetEu network,* 13–14 May 2014, Bretigny, France, **(Oral Presentation)**.

Speijker, L. (2009). *Crosswind-reduced separations for departure operations (CREDOS) WP4-final report.* EU Project Program: FP6-AEROSPACE, Ref.: 30837.

Srivastava, R. C., Matejka, T. J., & Lorello, T. J. (1986). Doppler radar study of the trailing anvil region associated with a squall line. *Journal of Atmospheric Science, 43,* 356–377.

Stawiarski, C., Träumner, K., Knigge. C., Calhoun. R. (2013). Scopes and challenges of dual-Doppler lidar wind measurements—an error analysis. *Journal of Atmospheric and Oceanic Technology, 30*(9), 2044–2061.

Steen, M., Schonhals, S., Polvinen, J., Drake, P., Cariou, J. P., Bouteyre, A. D. et al. (2010). Candidate technologies survey of airport wind & wake-vortex monitoring sensors. Sensors for weather & wake-vortex hazards mitigation. *9th Innovative Research Workshop & Exhibition,* December 7–9, 2010.

Tang, W., Chan, P. W., & Haller, G. (2011a). Lagrangian coherent structure analysis of terminal winds detected by lidar. Part I: Turbulence structures. *Journal of Applied Meteorology and Climatology, 50,* 325–338.

Tang, W., Chan, P. W., & Haller, G. (2011b). Lagrangian coherent structure analysis of terminal winds detected by lidar. Part II: Structure evolution and comparison with flight data. *Journal of Applied Meteorology and Climatology, 50,* 2167–2183.

Thobois, L., Cariou, J.-P. (2017). Next generation scanning LIDAR systems for optimizing wake turbulence separation minima[J]. *Journal of Radars, 6*(6), 689-698.

Thobois, L., Cariou, J.-P., Cappellazzo, V., Musson, C., & Treve, V. (2018). Comparison and validation of wake vortex characteristics collected at different airports by different scanning Lidar sensors. *The European Physical Journal Conferences, 176*(2), 06002. https://doi.org/10.1051/epjconf/201817606002. **(ILRC 28)**.

Tian, L., Heymsfield, G. M., Didlake, A. C., Guimond, S., & Li, L. (2015). Velocity-azimuth display analysis of Doppler velocity for HIWRAP. *Journal of Applied Meteorology and Climatology, 54,* 1792–1808.

Vrancken, P., Wirth, M., Rempel, D., Ehret, G., Bouteyre, A. D., Lombard, L., et al. (2010). Clear air turbulence detection and characterization in the DELICAT airborne Lidar project. *The 25th International Laser Radar Conference (ILRC25), in Saint Petersburg, Russia, on 06/07/2010.* Presentation.

Wagner, R., Courtney, M. (2014). Scanning lidar verification study. DTU Report.

Wieringa, J. (1980). Representativeness of wind observations at airports, Royal Netherlands Meteorological Institute. *Bulletin of the American Meteorological Society, 61,* 962–971.

Winckelmans, G., Cocle, R., Dufresne, L., & Capart, R. (2005). Vortex methods and their application to trailing wake vortex simulations. *Comptes Rendus Physique, 6*(4/5), 467–486. **(Special issue on Aircraft trailing vortices).**

(Received July 13, 2018, revised November 6, 2018, accepted November 20, 2018, Published online December 17, 2018)

Pure Appl. Geophys. 176 (2019), 1977–2015

https://doi.org/10.1007/s00024-018-1880-3

Pure and Applied Geophysics

A Meteorological Supersite for Aviation and Cold Weather Applications

ISMAIL GULTEPE,[1,2] M. AGELIN-CHAAB,[1] J. KOMAR,[3] G. ELFSTROM,[3] F. BOUDALA,[2] and B. ZHOU[4]

Abstract—The goal of this study is to better understand atmospheric boundary layer processes and parameters, and to evaluate physical processes for aviation applications using data from a supersite observing site. Various meteorological sensors, including a weather and environmental unmanned aerial vehicle (WE-UAV), and a fog and snow tower (FSOS) observations are part of the project. The PanAm University of Ontario Institute of Technology (UOIT) Meteorological Supersite (PUMS) observations are being collected from April 2015 to date. The FSOS tower gathers observations related to rain, snow, fog, and visibility, aerosols, solar radiation, and wind and turbulence, as well as surface and sky temperature. The FSOSs are located at three locations at about 450–800 m away from the PUMS supersite. The WE-UAV measurements representing aerosol, wind speed and direction, as well as temperature (*T*) and relative humidity (RH) are provided during clear weather conditions. Other measurements at the PUMS site include cloud backscattering profiles from CL51 ceilometer, MWR observations of liquid water content (LWC), *T*, and RH, and Microwave Rain Radar (MRR) reflectivity profile, as well as the present weather type, snow water depth, icing rate, 3D-ultrasonic wind and turbulence, and conventional meteorological observations from compact weather stations, e.g., WXTs. The results based on important weather event studies, representing fog, snow, rain, blowing snow, wind gust, planetary boundary layer (PBL) wind research for UAV, and icing conditions are given. The microphysical parameterizations and analysis processes for each event are provided, but the results should not be generalized for all weather events and be used cautiously. Results suggested that integrated observing systems based on data from a supersite as well as satellite sites can provide better information applicable to aviation meteorology, including PBL weather research, validation of numerical weather model predictions, and remote-sensing retrievals. Overall, the results from the five cases are provided and challenges related to observations applicable to aviation meteorology are discussed.

[1] Faculty of Engineering and Applied Science, University of Ontario Institute of Technology (UOIT), Oshawa, ON, Canada. E-mail: Ismail.gultepe@canada.ca

[2] Environment and Climate Change of Canada, MRD, Toronto, ON M3H 5T4, Canada.

[3] UOIT, ACE, Oshawa, ON, Canada.

[4] IMSG and EMC/NCEP/NOAA, College Park, MD, USA.

1. Introduction

Atmospheric observations at the surface and boundary layer strongly impact aviation forecasts and nowcasting that are important to advance numerical weather prediction, climate, and cloud or fog model developments, as well as improving remote-sensing retrievals. Dines (1917) stated that "thus it appears that the demand of the airman on the meteorologist is that he shall be able to forecast wind and fog, and to a less extent clouds, on the route the airman is proposing to follow". Clearly, over years of scientific developments, improvements on fog and precipitation visibility simulated by numerical weather prediction (NWPs) improvements are still limited (Gultepe et al. 2017b, c; Rasmussen et al. 1999, 2012; Roquelaure et al. 2009; Wong et al. 2013). According to Zhou et al. (2012), the performance of operational visibility forecast at NCEP is much lower than that of precipitation. Without in situ observations, remote-sensing retrievals and NWP validations cannot be obtained accurately to develop algorithms properly. These observations can be very important when there are limited observations especially in the northern climates and mountainous regions (Fernando et al. 2015). To validate model simulations, supersite and satellite site observations around it are required because of low visibility conditions, light snow precipitation with low visibility, strong wind gust conditions, and variability of fog and precipitation over short distances and time-dependent processes (Gultepe et al. 2009, 2017a; Higgins et al. 2006).

Aviation businesses related to transportation in northern climates will increase extensively in the future because of the addition of the UAVs to market and that will create more issues related to aviation operations. Therefore, both safety and efficiency of aircraft and UAV operations are needed for better airport in situ observations, accurate predictions related to wind, turbulence, icing, as well as convection. Improvements in the prediction and observation systems are critical for reducing costs that likely happen due to in-flight injuries, structural damage, and flight delays and hazards (e.g., Sharman et al. 2012; Sharman and Lane 2016; Sharman and Pearson 2017; Gultepe et al. 2014). Earlier studies suggested that low visibility, precipitation during cold seasons, strong wind, and turbulence conditions are the most important parameters affecting aviation performance (Keller et al. 2015; Darby and Poulos 2006; Gultepe et al. 2016a, b, 2017b). A fine-scale NWP model for aviation applications has been developed by Wong et al. (2013) that suggested the importance of in situ observations for aviation operations. Hu et al. (2014) also emphasized the initial conditions' impact on NWP simulations using the WRF model. The impact of in situ observations on model simulations for aviation applications is also reported by Mandel (1975) in which an unmanned automated weather station was for the first time suggested for surface aviation operations.

A research done by Gultepe et al. (2016a) summarized the impact of weather-related accidents because of visibility (Vis) and ceiling, wind and wind gust, and icing, respectively, at 32, 40, and 7%. The work of Turcotte (2016) for Canadian Defense Forces suggested that weather requests by clients were mostly related to Vis/ceiling and wind at 25 and 50%, respectively. Clearly, the impact of these parameters on the economy can be very large (Chun et al. 2017; Kim and Chun 2011). Therefore, better observing systems are needed to improve NWPs and remote-sensing retrievals (RSR) on these issues. The latest issues on the subjects of cold precipitations are emphasized to be important for future research and recognized by AMS. A review book on ice phase in the atmosphere published by the AMS Monographs that includes topics like snow, and mixed-phase precipitation (McFarquhar et al. 2017; Gultepe et al. 2017b, c).

The UAVs developed for weather and environmental applications (WE-UAVs; Gultepe et al. 2017c; Bates et al. 2013) can be used for atmospheric profiling of the thermodynamical parameters that include wind speed and turbulence (w'), temperature (T), relative humidity (RH), and vapor mixing ratio (q_v). These parameters are similar to parameters collected by the radiosondes and provide continuous observations of meteorological parameters if needed. Uncertainty in the WE-UAV measurements can be significant for wind-related parameters because of the impact of the rotors or body of UAV on airflow. Therefore, additional work is needed to improve wind measurements from UAVs (Reineman et al. 2016). Various meteorological parameters from the UAVs are also being used by others (Jonassen et al. 2012) to improve BL research. These parameters include aerosol number concentrations, fog droplets, visibility, particle spectra, and air quality parameters (Boer et al. 2017; Gultepe et al. 2017c).

The supersites for operational applications related to aviation can help to validate and verify the accuracy of the numerical model predictions' (NWPs) simulations and remote-sensing platform-based retrievals. The supersites are commonly developed for a specific purpose such as aviation (Gultepe et al. 2014), hydrometeorological (Wood et al. 2015), and Arctic applications (Gultepe et al. 2017c; Uttal et al. 2016; Starkweather and Uttal, 2016; McFarquhar et al. 2011). Jacobs and Maat (2005) developed a numerical guidance method that is used for decision-making support for aviation meteorological forecasters. In their work, a new method is developed as a combination of statistical and physical post-processing of NWP model data and in situ observations and is used to derive forecasts for all aviation-related meteorological parameters at the airports. In this respect, integrated observing systems utilizing various microphysical, dynamical, thermodynamical in situ sensors, as well as data from remote-sensing platforms are needed for aviation and BL meteorology research and operations.

The low skill in visibility prediction is likely in part due to numerical model physics but most likely due to the visibility parameterization (Zhou and Du 2010; Zhou et al. 2012; Gultepe et al. 2007; Stoelinga and Warner 1999). Current operational visibility in post processors is obtained using simulations of WRF (Pu et al. 2016) and NAM NCEP (Rogers et al. 2017). In these NWP models, Vis is correspondingly based Stoelinga and Warner (1999) and GSD's improved scheme (Smirnova and Brown 2000). However, those parameters are based on limited evaluations and required additional calibrations. The aviation supersites can provide a unique opportunity for this purpose. Ceiling (cloud base height when total cloud cover ≥ 50%) is an important condition for aviation operations, but it is hard to verify in the current data availability (Boudala et al. 2018). Presently, a ceilometer and broadband radiative fluxes can provide cloud cover conditions accurately (Long and Ackerman 2000). The Planetary Boundary Layer (PBL) structure is critical for fog prediction (Gultepe et al. 2007); therefore, the vertical temperature and humidity data should be collected at the surface and higher levels for the aviation supersites, and represent vertical resolution in meters.

In the current work, the main goal in using case studies is to emphasize the importance of the aviation/weather supersites to develop microphysical/dynamical algorithms fog, cloud, precipitation microphysics, radiation, wind and turbulence, icing, blowing snows and wind gust, as well as to provide knowledge on how to improve observational and prediction research. The use of these case studies is to illustrate the potential usage of supersite data and the statistical relationships derived from data sets serve as blueprint for a possible future work. The cases used in the analysis are given in Table 1. In the analysis, integrated physical parameterization for visibility, icing, blowing snow, and wind gust conditions are first introduced based on instruments available at the PUMS meteorological supersite. Afterward, the results, discussion, and conclusions will follow.

2. *Observations*

The meteorological supersite, called PanAm UOIT (University of Ontario Institute of Technology) Meteorological Supersite (PUMS), was the PanAm project site collecting observations applicable to the BL research from April 2015 to date (Dec 2017). The PUMS is set up with the state of the art instruments and platforms (Table 2), including a Weather and Environmental UAV (WE-UAV), Fog and Snow Observing System (FSOS) tower, and Gondola-based icing research. The 3-satellite sites (Table 3) around the PUMS site (Fig. 1a) were also designed for measuring variability in the measurements over less than a km (see figure for distances between sites). The Gondola-based research with microphysical sensors (Table 4) is developed for fog and icing research, and tested at the UOIT Automobile Center of Excellence (ACE) facilities. The five case studies given in Table 1 are used to characterize various specific weather conditions related to fog, rain, snow, blowing snow, visibility, icing, and wind gust conditions. The following subsections will focus on details of the platforms.

2.1. *Meteorological Supersite (PUMS)*

The PUMS supersite includes various sensors and platforms given in (Table 2) and summarizes the sensor names and measurement ranges. Figure 1 shows some of the instruments and project site at the UOIT Campus, Oshawa, Ontario. Details on these

Table 1

Cases used during this PUMS project

Case	Day	Weather type	Synoptic conditions
1	February 22 2017	FFG	Warm front
2	January 10 2017	HSN	Cold front
3	June 23 2017	HRN	Warm ront
4	April 7 2017	BSN	Cold front
5	April 6 2017	GST	Warm front
6	July 10 2015	CLR	High-pressure system

The rain (RN) and snow (SN) for light, moderate, and heavy conditions are given as LRN, MRN, and HRN (that correspond to PR < 2.5, PR > 2.5 and < 10, and PR > 10 mm h^{-1}, respectively), and as LSN, MSN, and HSN (that correspond to PR < 1.0, PR > 1.0 and < 5.0, and PR > 5.0 mm h^{-1}), respectively. FFG signifies the freezing fog, BSN the blowing snow, GST the wind gust conditions, and CLR the clear weather conditions. See text for other definitions

Table 2

PUMS supersite instrument list and their measurements at the UOIT, Oshawa, Ontario

ID	Instrument	Measurement	Company
1	FMD fog monitor	Fog droplet spectra	DMT Inc
2	GCIP probe	Precip spectra	DMT Inc + EC
3	OTT ParSiVel	Precip spectra	METEK GmbH
4	MAP aerosol profiler	Aerosol spectra	METEK GmbH
5	GEONOR precip	PA and PR	Geonor Inc
6	Sentry Vis	Visibility	R.M. Young Company
7	PWD22	Visibility, PA, PR	Vaisala
8	CL51 Ceilometer	Cloud profiling	Vaisala
9	Heated tipping bucket	PA	Campbell Scientific
10	Young USA 81000	3D wind	R.M. Young Company
11	Young 2D	2D wind	R.M. Young Company
12	SPN-1	Direct and diffused solar rad	Delta-T Devices Ltd
13	RID	Icing amount	Campbell Inc
14	Pluvio2 (OTT)	PR and PA	METEK GmbH
15	CRN1	IR and SW BRF	Kipp&Zonen
16	HSS SWS-200	Visibility, PA, PR	METEK GmbH
17	LPM 5.4110	Precip spectra	METEK GmbH
18	Axis camera system	Pictures	AXIS Comm. Inc.
19	CampbellSci 5 × camera	Visibility/picture	Campbell Inc
20	Metek USA1 3d -heated	3D wind	METEK GmbH
21	Luft R2S-UMB	PR and PA	Lufft Inc
22	Young capacitor-100 cm^2	PA -Model 50203	Campbell Inc
23	3017 MWR	LWC, T, q_v profiles	Radiometeric Corp
24	MiniOFS	Visibility	Sten Lofving Optical Sensors
25	OSI vis sensor	Visibility, PA, PR	OSI Inc.
26	SWS Metek Vis sensor	Visibility, PA, PR	METEK GmbH
27	CR3000 Data logger	DAS	Campbell Inc
28	Camera Micro7	Pictures	Microseven Inc.
29	Vaisala WXT50 (2)	Wind, T, RH, precip	Vaisala
30	Airmar	Met parameters	Airmar Inc
31	MRR radar	Radar reflectivity	METEK GmbH
32	WE-UAV (Weather and Environmental UAV)	Wind, aerosol concentration, met parameters	Aeryon Inc. and EC3
33	Gondola (CDP&BCP)	Droplet spectra	DMT Inc.
34	CS725	SWE	Campbell Sci.

The PA and PR represent precipitation amount and precipitation rate, respectively. WE-UAV: Weather Unmanned Aerial Vehicle (developed by Gultepe et al. 2017a)

Table 3

Satellite sites (FSOS towers) with sensor list around PUMS supersite (~ 1 km apart from PUMS)

House site (satellite site)	Barn site (satellite site)	UOIT campus site (satellite site)
CR1000 data logger	CR1000 data logger	CR1000 data logger
PWD22	PWD22	PWD22
WXT520	WXT520	WXT520
Rotronic met sensor	Rotronic	Rotronic Instrument Corp.
Modem, brick computer	Modem, brick computer	Modem, brick computer
Camera Micro7	Camera Micro7	Camera Micro7

See Table 1 for sensor details. The FSOS signifies fog and snow observing system developed by Gultepe et al. (2016a, 2017c) sensors and platforms can be found in Gultepe et al. (2009, 2015, 2016b). The Pluvio in the Double Fence International Reference system (DFIR) setup for precipitation measurements is shown in Fig. 1a. The visibility sensors used in this work are the present weather sensors, e.g., PWD52, SWS200, OSI, and Sentry (Fig. 1b–e). These sensors measure precipitation type, precipitation rate (PR) and amount (PA),

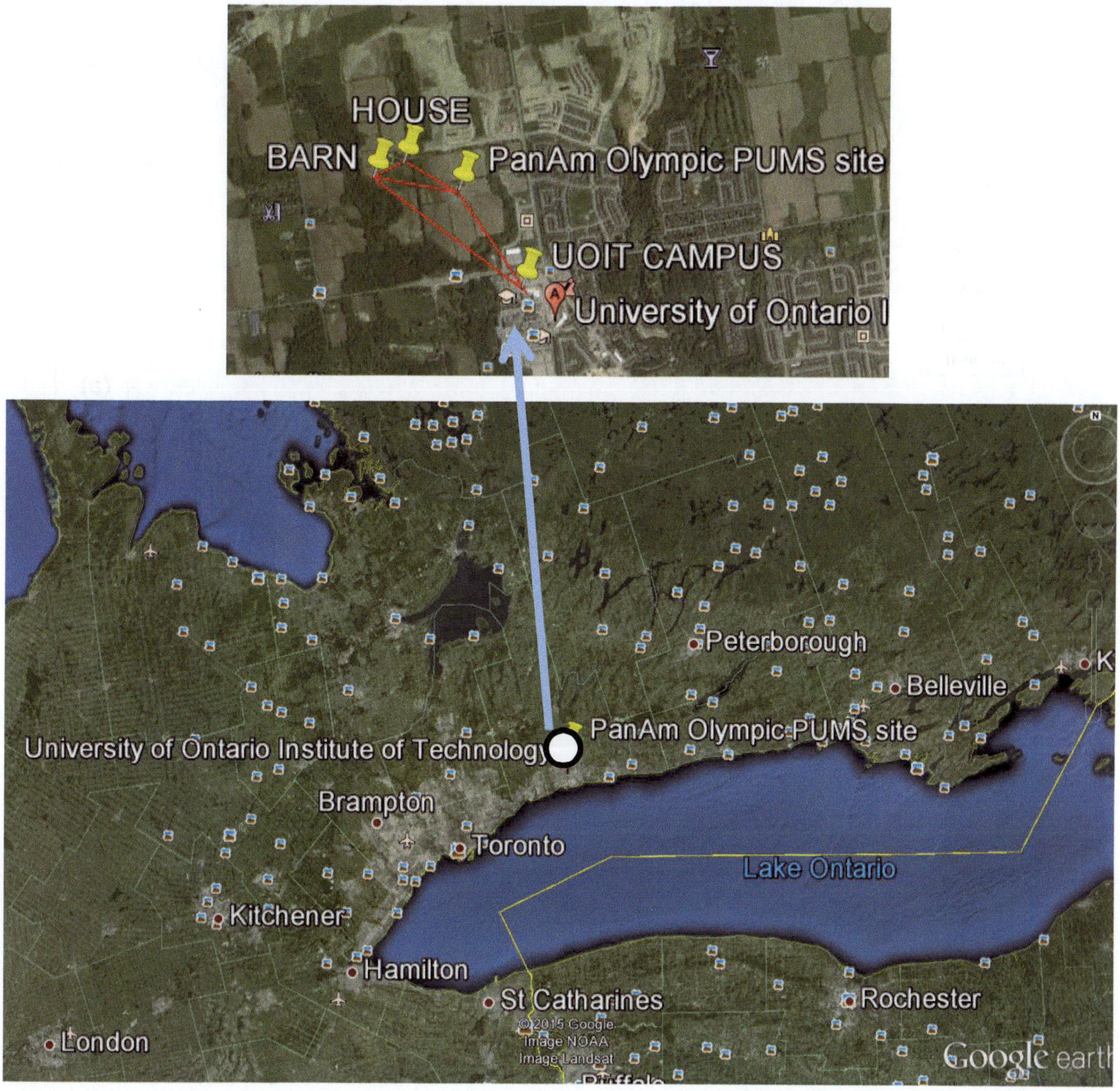

Figure 1
The bottom box is for the PUMS project site at the UOIT Campus nearby Lake Ontario. The white circle indicates the PUMS project site (P) (43°57′35.46″; 78°54′24.12″ at 168 m), and three satellite sites called Barn (B), House (H), and UOIT Campus (U) (top box) are shown with yellow pins. The enlarged area at the top shows all the sites from which data were used in the analysis. Distance between P–U is 800 m, P–H is 450 m, P–B is 600 m, and B–H is 200 m. **b–h**: **a** UOIT PUMS supersite for boundary layer research and aviation meteorology applications. The Data Acquisition System (DAS) is kept inside the trailer. Pluvio and 2D sonic wind measurements used for snow measurements within the Double Fenced Intercomparison Reference (DFIR) fence system is seen on the right. The rest of the sensors are over the trailer or attached to the tressel. The WE-UAV is given in the inlet at the right top corner of the picture. Visibility sensors are given in **b** PWD50, **c** SWS, **d** Sentry, and **e** OSI. **f** LPM is for precipitation. **g** FMD is for fog spectra (DMT Inc.). **h** is for other precipitation sensors without shelters, as well as the CL51 ceilometer

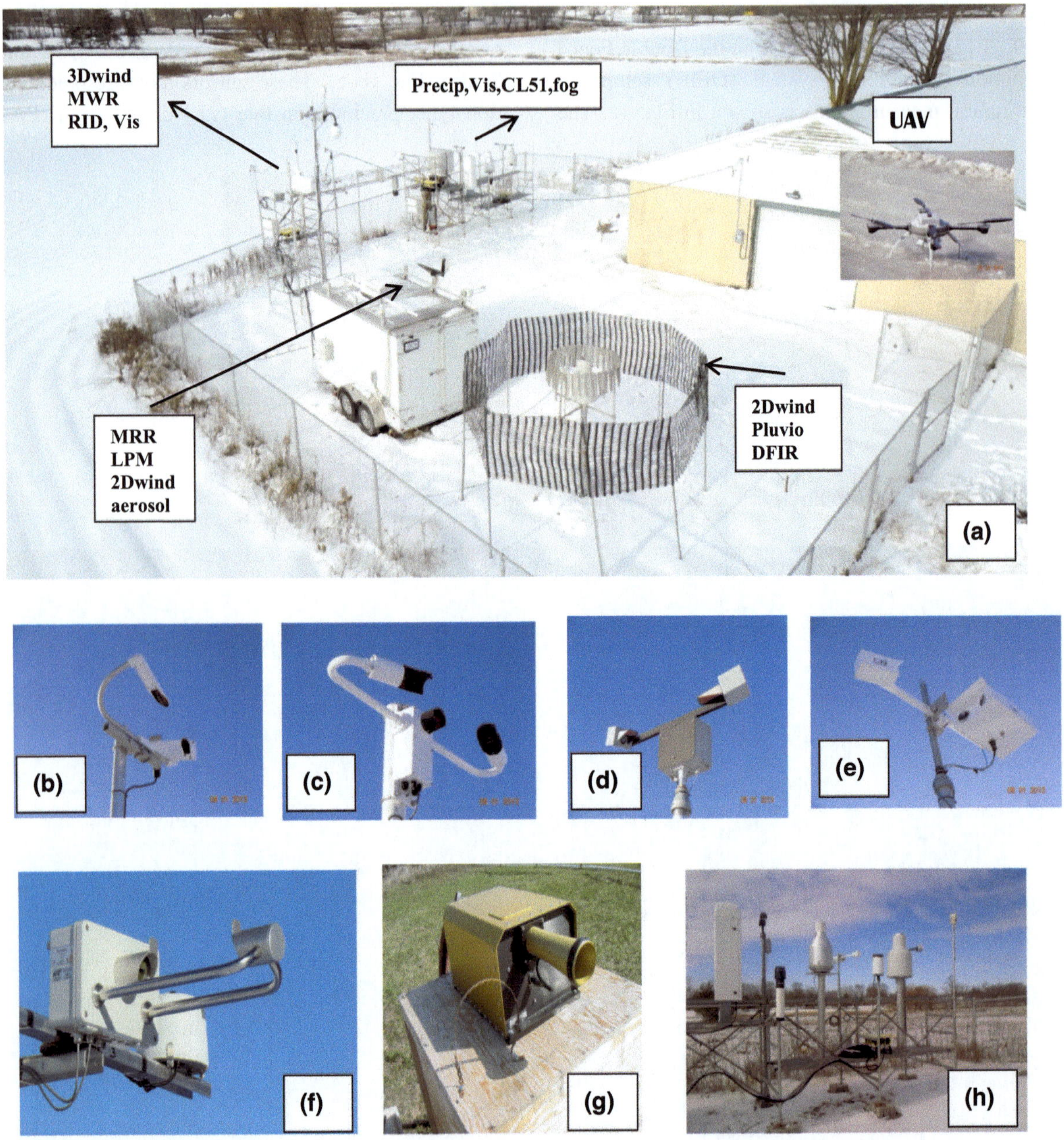

Figure 1
continued

and visibility (Vis). Particle spectra for aerosols, fog, snow, and rain (Figs. 1f, g) were also collected. Figure 1h contains a Geonor, Pluvio, Young tipping bucket, and Young Capacitor sensors, which also provides precipitation measurements.

2.2. *FSOS Tower*

Table 3 shows the Fog and Snow Observing System (FSOS) tower instruments located at three locations near the PUMS site. The FSOS tower (Fig. 2a) is designed for measuring all basic parameters related to fog and snow characterization that

Table 4

UOIT ACE-PUMS supersite sensors used in the ACE CWT (climate wind tunnel) for icing research

Sensor name	Measurements	Model/company info
Gondola	Platform for balloon	DMT-EC
CDP	1–50-μm spectra	DMT Inc.
BCP	5–100-μm spectra	DMT Inc.
GCIP	15–940-μm spectra	DMT Inc.
RID	Icing amount/indicator	DMT Inc.
MiniOFS	Visibility	SLOS
WXT520	U_h, T, RH, pressure, T_d	VAISALA Inc.
LPM	0.1-mm to ~ 1-cm size spectra	Metek Inc.

includes wind speed and direction, turbulence, aerosol, precipitation rate and amount, snow depth, and visibility. It also gathers incoming SW radiative fluxes and surface temperature. The three FSOS towers were located at about 1 km from the PUMS supersite to show variability in precipitation and visibility characteristics rather than depend on only single point measurements.

2.3. *WE-UAV Platform*

The WE-UAV (Fig. 1a inlet) is described in the work of Gultepe et al. (2017c) and its measurements were performed during the PUMS PanAm Olympic Project from July 1 to July 20 2015. The WE-UAV-based sensors are shown in Fig. 2b, c. Specifically designed MiniVis sensor is shown in Fig. 1b which is tested at the UOIT climate chamber. The WE-UAV typical measurements were air temperature (T_a), relative humidity (RH), pressure, horizontal wind and wind gust, aerosol number concentration (N_a) with sizes between 0.3 and 1 μm, CO_2, Vis, and UAV position angles.

The Radiometric Corporation Profiling Microwave Radiometer (PMWR) (Ware et al. 2003) is used for BL profiling of T_a, RH, and liquid water content (LWC). To specify cloud boundaries, Vaisala CL51 backscatter characteristics for the 10 July 2015 case (Table 1) are also used to analyze aerosol/cloud properties.

Figure 2d shows both Ground-Based Cloud Imaging Probe (GCIP) and W725 instruments used for calculations of snow particle spectra and Vis, and snow water equivalent (SWE), respectively. Figure 2e shows Kipp&Zonen IR and SW incoming and upgoing broadband radiative fluxes. The IRT black body temperature sensors for the surface and sky conditions are shown in Fig. 2f and are used for evaluating fog thermal processes. The global incoming SW radiative flux and diffused SW measured by the SN1 sensor is shown in Fig. 2g.

2.4. *UOIT ACE CWT (climate wind tunnel) and Gondola*

A Gondola with meteorological sensors is designed for a tethered balloon-based research and was supported by Droplet Measurement Technologies (DMT). The Gondola was also used to test the UAV and aircraft icing sensors in the ACE CWT (climate wind tunnel). The Gondola (Fig. 3a) includes DMT Cloud Droplet Probe (CDP) and backscatter cloud probe (BCP) spectrometers that were used to characterize fog/droplet and icing conditions.

During the ACE tests, FMD (FS100) and GCIP were used for fog and snow spectral measurements (Fig. 3b). The CDP and BCP provide particle spectra from 1 to 50 and 5 to 100-μm size ranges, respectively. Figure 3c shows a Rosemount Icing Detector for discriminating icing conditions at the PUMS site that was also used in the ACE CWT. Measurements from the Gondola were used to obtain microphysical parameters such as liquid water content (LWC), mean volume diameter (MVD), and droplet number concentration (N_d) that are important for the characterization of fog and icing conditions, as well as precipitation.

The above measurements were then used in the analysis of microphysical parameters for the five cases, including fog, rain, snow, blowing snow, wind gust, and icing conditions, and provided microphysical relationships and characterization weather events.

3. *Method*

In this section, the concept of visibility related to fog and precipitation, blowing snow, icing, and wind gust are provided for possible numerical weather prediction (NWP) modeling applications. A similar

(a)
(b)
(c)
03/30/2012 10:53
(d)
(e)
Kipp & Zonen
(f)
(g)

◀Figure 2
Fog and Snow Observing System (FSOS) tower (**a**), the miniOFS visibility sensor (**b**), RH, T, CO_2, and O_3 sensors (**c**), GCIP (right) and CS275 sensor for snow water equivalent (SWE) (left) (**d**), four piece K&Z BRF sensor for IR and SW (**e**), and IR T up and down sensors (**f**), and SPN1 for SW direct and diffuse radiative fluxes (**g**)

methodology can also be used for extreme weather analysis. The issues related to the analysis of observations are also discussed. In the analysis, the visibility (Vis) parameter obtained from the supersite present weather and microphysical optical instruments, e.g., FMD, GCIP, and CDP, are first provided, and then, Vis and LWC, or PR for various precipitation types are calculated from the particle spectra measurements. The correction of snow PR is provided after removing wind effects that are based on DFIR reference values. The blowing snow visibility is parameterized as a function of wind speed (U_h) after removing snow PR segments. The following sections provide the methods for each case study given in Table 1.

3.1. *Visibility Parameterizations for Fog, Rain, and Snow*

Visibility measurements are used for physical parameterizations for each case, e.g., fog, rain, and snow, that is based on extinction of forward scattering visible light (β_{ext}) obtained from the present weather sensors (such as PWD50, Table 2). Extinction parameter is also calculated directly from the distrometers, e.g., LPM or OTT ParSiVel, or from droplet (ice) spectrometers, e.g., CDP, BCP, and GCIP (Table 2). Extinction parameter is calculated as follows:

$$\beta_{\mathrm{ext}} = \sum_{k=0}^{n} Q_{\mathrm{ext}} n(r) \pi r^2 \mathrm{d}r, \qquad (1)$$

where the Q_{ext} and $n(r)$ are extinction coefficient and particle spectral density at the given size (r) interval, respectively. The meteorological observing range (MOR) visibility is based on Koshmider Law that is obtained as a function β_{ext} as follows:

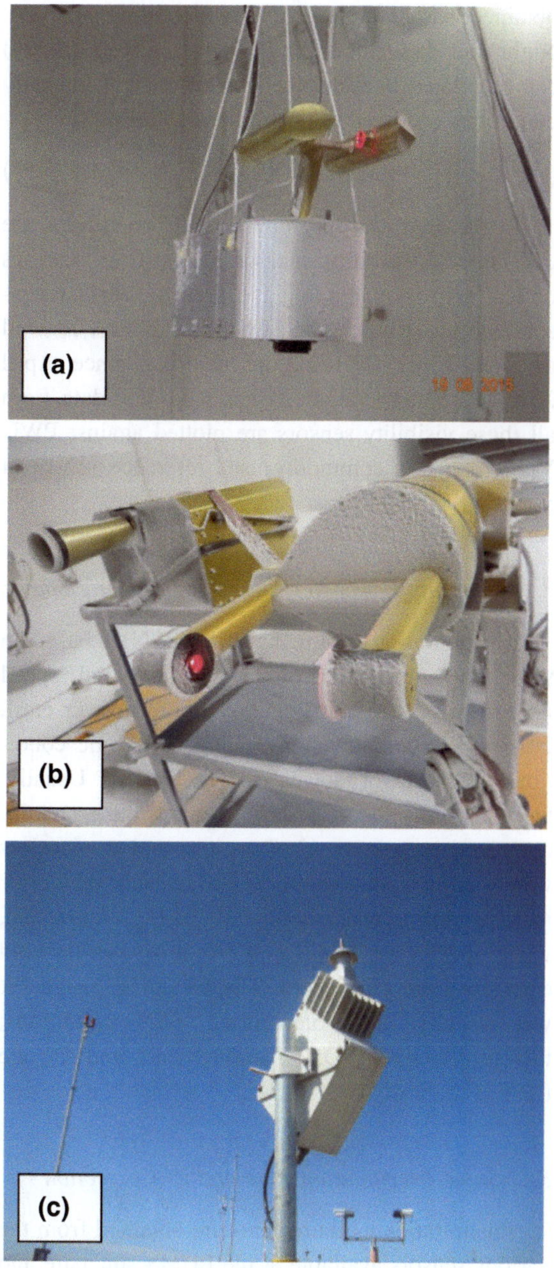

Figure 3
Icing test units used in the UOIT ACE CWT: Gondola-based CDP (forward looking) and BCP sensors (backscattering and down looking) (**a**), FMD and GCIP (CIP) sensors in the wind tunnel (**b**), and Rosemount icing detector (**c**). Table 4 provides sensor info used during various icing tests

$$\mathrm{Vis} = -\ln(0.05)/\beta_{\mathrm{ext}}. \qquad (2)$$

Thereafter, Vis–PR, Vis–LWC–N_d (or IWC), and Vis–U_h for blowing snow (BSN) conditions are obtained based on statistical curve fitting analysis. Night and day visibilities (V_n and V_d, respectively) can be related to each other for lower light intensity conditions observed at the airports as follows:

$$V_n = 1.8507 V_d^{0.814}. \quad (3)$$

In the analysis, only daytime Vis conditions are considered. Visibility measurements from various optical sensors, including the present weather sensors such as PWD, SWS, OSI, and Sentry, as well as from FMD (FS100), are gathered since April 2015. To better evaluate issues with Vis, data from all these visibility sensors are plotted against PWD Vis which is a commonly used reference sensor at airports.

3.2. Visibility Parameterizations for Blowing Snow

The definition of blowing snow (BSN) is based on when the magnitude of the 1-min wind speed $U_h > 5$ m s^{-1} (Li and Pomeroy 1997; Hoover et al. 2014). This definition may change in Arctic conditions because of lower snow density. Usually, blowing snow ice crystals in a snow flake are dry compared to warmer air T regions and carries less water equivalent amount. To reduce the error included in the analysis of the Vis–U_h relationship, winds speeds less than 3 m s^{-1} in this work for cold conditions are considered. The weighing gauge PR was usually negligible during BSN conditions; therefore, only PWD50 measurements and U_h are used in the analysis.

3.3. Snow Precipitation Rate and its Correction

Precipitation measurements are obtained from the various sensors available at the PUMS supersite and FSOS towers located at the project satellite sites (Fig. 1). Here, precipitation type is defined as a function of temperature and signal processing from the present weather sensors (see sensor manuals). When $T < 0$ °C, snow occurrence is assumed as the precipitation type, and then, a collection efficiency correction for PR measurements is applied that utilized DFIR reference data available from the PUMS site.

A major uncertainty in the PR for SN conditions is due to the assumed snow water equivalent (SWE) value in the algorithms used by the sensors. The present weather sensors (e.g., PWS100, Campbell Sci.) use snow density versus size relationships. On the other hand, snow depth sensor (e.g., SR50) measurements are converted to SWE values using a 10% ratio of snow depth. The PWD50 uses a capacitor to estimate melted snow amount. Therefore, PR and PA from the optical sensors are strong functions of the algorithms used for the conversion rates.

3.4. Wind Gust and Turbulence

Atmospheric turbulence and wind gust are important for aviation applications for both at the surface and at the higher altitudes, and their predictions from an NWP are related to how good their parameterizations are used in the NWPs. Turbulence is usually related to thermal and dynamical instabilities in the scale of a few mm up to km scales, depending on aircraft speed and observation sampling rate, e.g., 16–32 Hz (Cho et al. 2003; Gultepe and Starr 1995). The ratio of wind shear produced turbulence to the thermally generated turbulence is about 2:1 for weak turbulence and ~ 3:1 for strong turbulence ($\varepsilon_{dis} > 10^{-4}$ m^2 s^{-3}). The structure function (D_{it}) for isotropic turbulence can be related to the dissipation rate (ε_{dis}) as follows:

$$D_{it} = 4C(\varepsilon_{dis} L_s)^{\frac{2}{3}}/3, \quad (4)$$

from which ε_{dis} is derived by Gultepe and Starr (1995) as follows:

$$\varepsilon_{dis} = \frac{1}{L_s}\left[\frac{3D_{it}}{4C}\right]^{3/2}, \quad (5)$$

and was used for aircraft based eddy dissipation rate calculation. In the above equation, C is 2.05 and L_s is the length scale which is defined as the distance between consecutive data points. For 3D isotropic turbulence, structure function (Cho et al. 2003; Gultepe and Starr 1995) in Eq. 5 can be calculated using the wind fluctuations at high sampling rates as follows:

$$D_{\rm it}({\rm d}t) = \frac{1}{3}\left[\frac{8}{7}(D_{u'} + D_{v'} + D_{w'})\right], \quad (6)$$

where D for wind z (also for another axis) component is calculated by Gultepe and Starr (1995) as follows:

$$D_{w'} = [w(t + {\rm d}t) - w(t)]^2. \quad (7)$$

In the above equations, the subscript u', v', and w' are x, y, and z wind speed changes at the consecutive time intervals (dt). After that, the $L_s = {\rm d}t{\cdot}U_h$ relationship is used for transforming the time scale to L_s. In the analysis, ε_{dis} is calculated using Eq. 5.

The wind gust calculation is performed in various ways in the literature. Here, it is done based on NOAA NWS definition and calculated using the FAA ASOS method. The NOAA NWS (1996) describes wind gust as "a sudden, brief increase in speed of the wind and wind gusts are reported when the peak wind speed reaches at least 16 knots and the variation in wind speed between the peaks and lulls is at least 9 knots over 10-min intervals and the duration of a wind gust is usually less than 20 s. In METARS (Kantor et al. 1996), wind gusts are coded in two or three digits immediately following the wind speed. For example, a wind out of the west at 20 knots with wind gusts to 35 knots would be coded 27020G35KT".

In the analysis, 1-s data were averaged over 2- and 10-min intervals to show variability in wind gust conditions. Thereafter, wind speed maxima and minima at 1 Hz data over 10-min intervals are checked for d$U_h \geq 10$ knots where wind speed with a variation of 10 knots or more between peaks and lulls is satisfied for wind gust calculations. The satisfaction of peak winds is provided with a condition that max U_h at 1 Hz is at least 16 knots ($\sim$ 8 m s^{-1}) and min U_h is $\geq$ 6 knots over 10-min intervals. The wind gust velocities using 1 Hz measurements are calculated over 10-min intervals. In the ASOS system used by FAA, 5-s averages over 10 min are used in the current 2-min average U_h calculation (U_{h2m}), and then compares it with max 5-s mean U_{h1m} during past minute. If $U_{h2m} > 9$ knots and $U_{h1m} > 5$ knots, then the greatest 5-s maximum wind speed observed during the past minute is stored as a wind gust over 10 min. The weak and strong turbulence conditions during wind gust events can be calculated using $\varepsilon < 10^{-4}$ m^2 s^{-3} and $\varepsilon > 10^{-4}$ m^2 s^{-3}, respectively (Cho et al. 2003).

3.5. WE-UAV Thermodynamic Profiling

The measurements from both WE-UAV-based sensors and ground-based remote-sensing platforms are used to provide thermodynamic profiling in the boundary layer. The results of BL profiling for a cloud-free case occurred on the July 10 2015 are given qualitatively. For this day, time series and profiling data from WE-UAV are matched with CL51 and PMWR profiles for the same time period. Possible issues for comparisons are discussed and the new sensors related to both aerosol number concentration (N_a) and Vis for a future WE-UAV platform are provided.

3.6. Icing Envelopes and Precipitation Rate

Icing rates (ice amount over unit time) are important for aviation applications and the intensities of icing amount have been provided by earlier works such as Newton (1978), Politovich (1996). Icing intensities as described as trace for > 1 h), light (15–60 min), moderate (5–10 min), and intense (< 5 min) based on the time required for icing accumulation of ¼ inch (6 mm) over an icing rode (Jeck 2002). The latest work also converted the rates to represent distances. The Newton results suggested the envelopes for icing rates and commonly used by the FAA and aviation community. These intensities were set up for PR = 120, 60, and 10 mm h^{-1}. The liquid water content (LWC) as a function of mean volume diameter (MVD) is provided for the PR rates described above. In the present work, LWC is calculated from CDP and GCIP probes used in the UOIT ACE CWT, and that is obtained as a function of droplet number concentration (N_d), MVD, and PR. These parameters are calculated using particle spectra obtained from CDP and GCIP probes (Gultepe et al. 2017b).

The theoretical values of PR are calculated to evaluate icing conditions and its derivation is given below. The measured and calculated values of microphysical parameters are then used to study present icing envelopes.

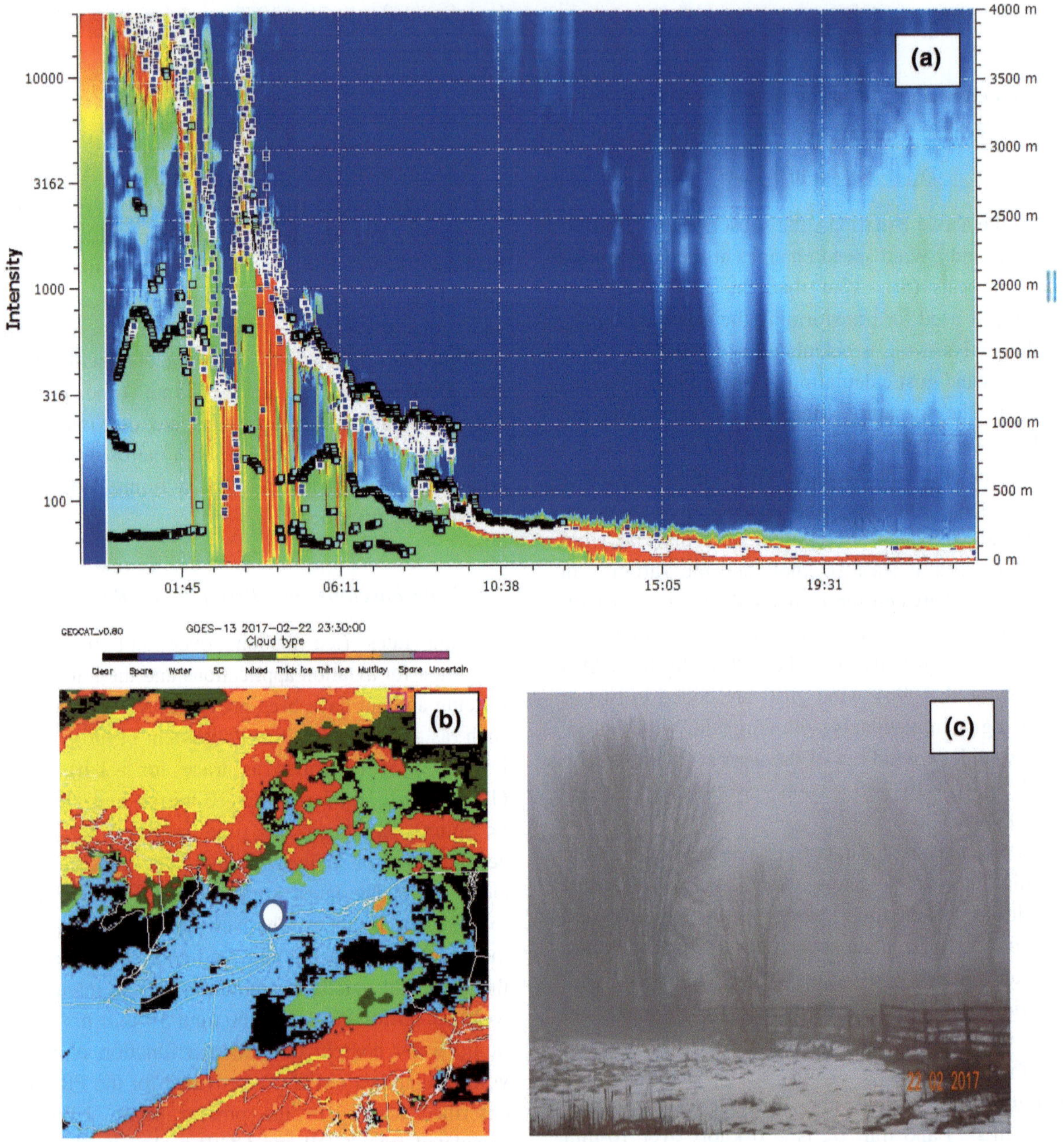

Figure 4
Cloud base height from CL51 on February 22 2017 for freezing fog case (box **a**). Fog started at about 15 UTC and lasted more than 10 h. Using the analysis of cloud type (Pavolonis 2010) based on GOES-13 data (available at http://cimss.ssec.wisc.edu/geocat/). Box **b** shows the fog occurred (blue region) at about 0 °C behind the cold front system. Box **c** indicates melting snow at the surface and low Vis values less than a few hundred meters

Using the definition of LWC as

$$\mathrm{LWC} = \mathrm{MVD}^3 N_{\mathrm{dt}} \tag{8}$$

and precipitation rate (or icing rate) (PR) as

$$\mathrm{PR} = \varepsilon(\mathrm{LWC})(\mathrm{TAS}), \tag{9}$$

where ε and TAS are the collection efficiency and true air speed (or horizontal wind speed in this work), respectively. LWC can be obtained as a function of ε,

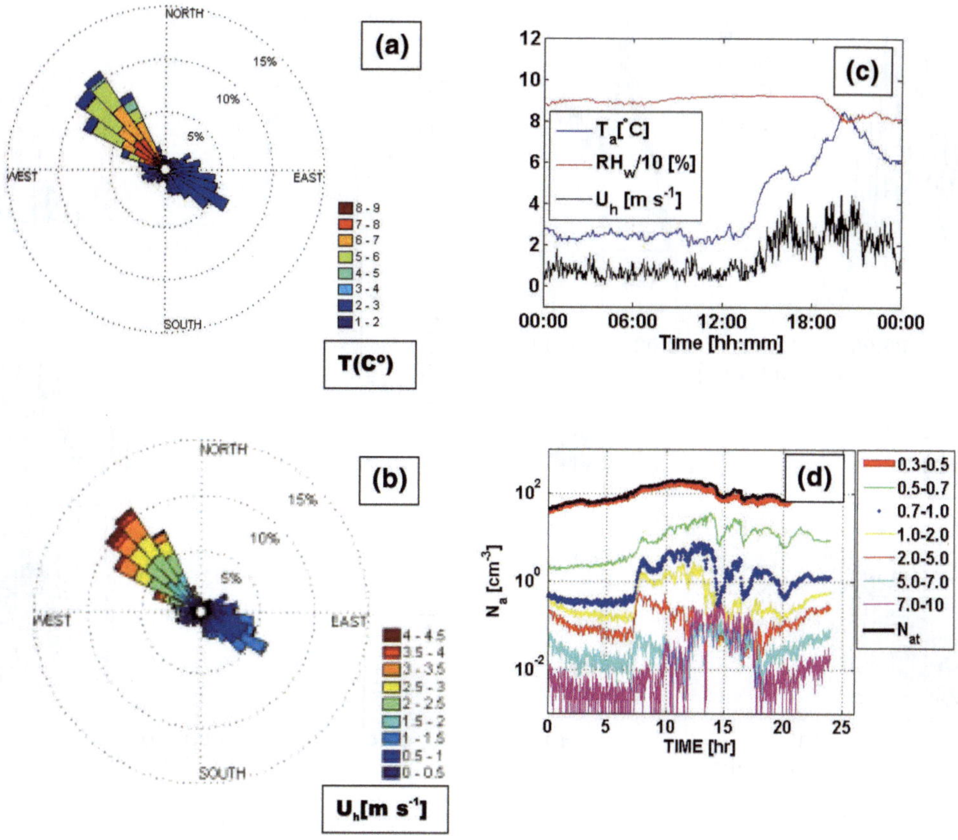

Figure 5

Temperature and wind roses are shown in **a**, **b**, respectively. Fog formation started at about 12 UTC (9 AM LST) (**c**) in which RH was about 95% and T started increasing from 2 to 8 °C (presented by the color squares in box **a**). Occurrences are shown in %. Horizontal wind (U_h) changed from 1.5 to 4 m s^{-1} (presented by the color squares in box **b**). Fog dissipated at about 18 UTC (3 PM LST) with U_h reaching 4 m s^{-1}. During dissipation, warm air advection was with max U_h and T from NW. Aerosol number concentration increases during fog events because of CN becoming CCN, resulting in total N_a (N_{at}) at about 200 cm^{-3} (**d**)

MVD, integrated total droplet number concentration (N_{dt}), TAS, and PR as follows:

$$\mathrm{LWC} = \left(\frac{\mathrm{PR\ MVD}^3 N_{dt}}{\varepsilon \mathrm{TAS}}\right)^{1/2}. \quad (10)$$

Equation 10 is used to show the impact of N_d, PR, TAS, and MVD on LWC values. In Eq. 10, PR (precipitation rate) is taken as 10, 60, and 120 mm h^{-1}, representing light, moderate, and severe icing conditions (Newton 1978). The ε is about 0.9 as peak collection efficiency for larger droplets but can decrease quickly for smaller droplets. The icing definitions related to the time duration (dt) are also obtained for 6-mm accumulation of ice over a 3-inch diameter cylinder (Jeck 2002). Equation 10 is used to obtain LWC versus MVD plots, assuming fixed N_d and PR with TAS of 90 m s^{-1} for a typical aircraft for various microphysical conditions.

4. Results

In the results, visibility related to various hydrometeors, e.g., fog, rain, snow, and blowing snow, is discussed for the cases given in Table 1. Then, the wind gust calculations based on turbulence intensity are provided that can be used for a new wind gust parameterization in the future. Afterward, the WE-UAV results are summarized for future BL research.

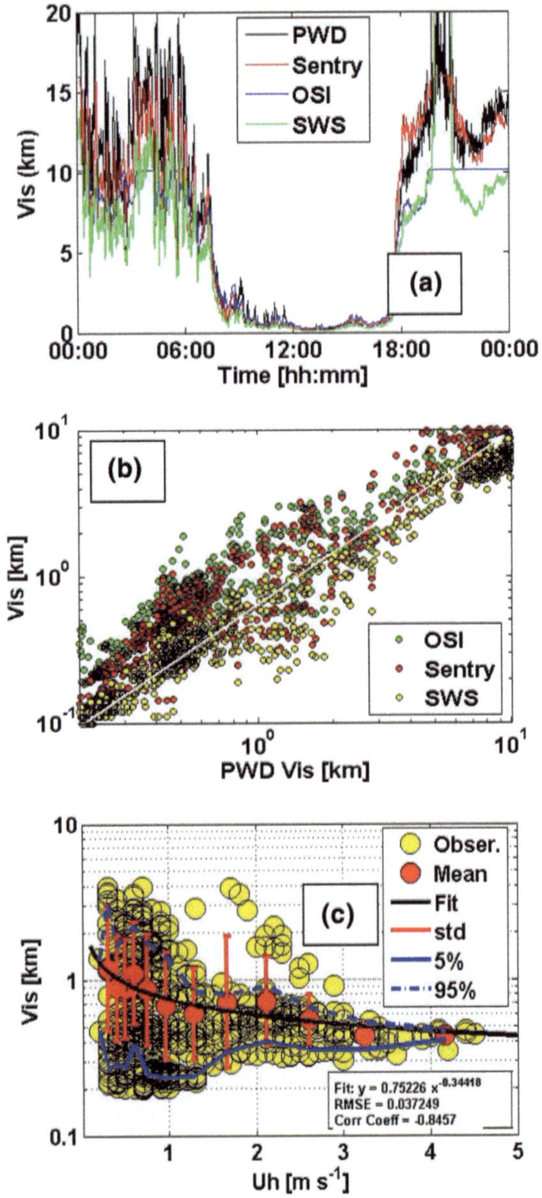

Figure 6
Visibility from the four present weather sensors (Table 2) is shown in **a**. Significant Vis differences are seen in high Vis conditions. To see variability in Vis, Vis values are plotted against PWD Vis (**c**). Clearly, other than SWS Vis, other Vis values are larger than PWD Vis at about 50%. Vis versus U_h is plotted to show wind impact on fog dissipation (**c**). This figure shows that when $U_h > 1$ m s^{-1}, fog losses its intensity quickly with increasing U_h

4.1. Vis for Freezing Fog Case (February 22 2017)

The fog occurred on February 22 2017 due to mainly moist air advection with southeasterly surface

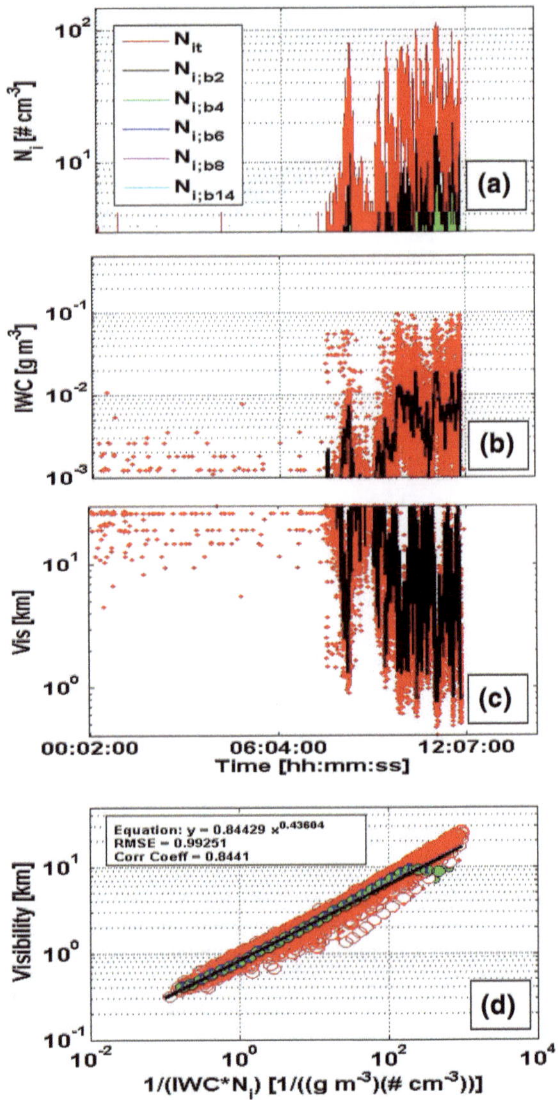

Figure 7
Microphysical characteristics of freezing fog FZFG are shown in **a–d**, respectively, for the time series of FZFG total number concentration (N_i), ice water content (IWC), Vis, and Vis parameterization (Vis versus fog factor, FF). The FF is 1/ (IWC × N_i). A fit to mean values is also given in the plot

Figure 8 ▶
Surface pressure heights (contours, yellow lines) together with radar reflectivity factor from NOAA archives are shown in **a** to explain large-scale conditions for snow formation (white circle shows the PUMS project site). The cloud top temperature image which is available from University of Wisconsin, Madison, is shown in **b**. A strong upper air flow with a low-pressure system moved over the PUMS project region on January 10 2017. The CL51 backscatter coefficient representing cloud base conditions is shown in **c** where snow event lasted about 7 h from 1000 UTC to 1700 UTC

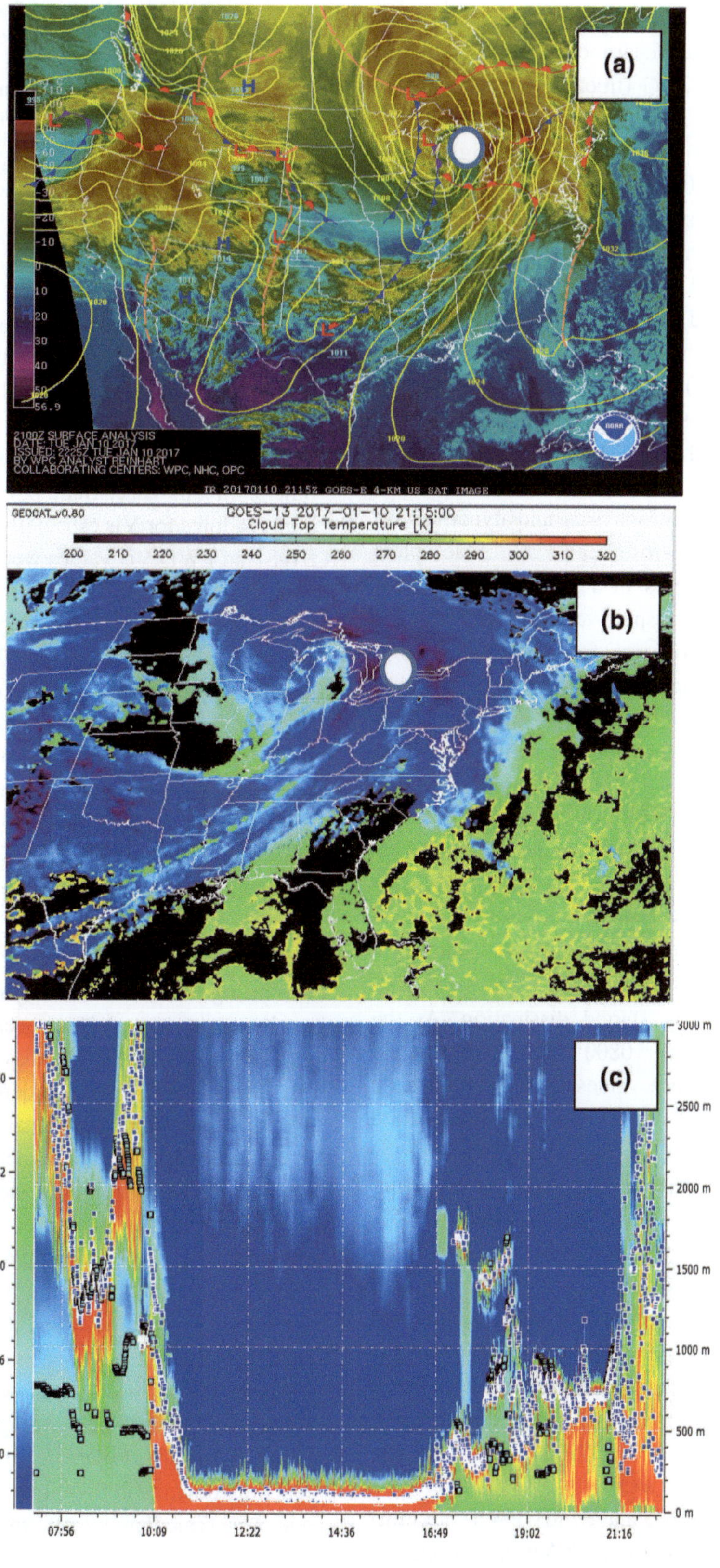
(a)
2100Z SURFACE ANALYSIS
DATE: TUE JAN 10 2017
ISSUED: 2235Z TUE JAN 10 2017
BY WPC ANALYST REINHART
COLLABORATING CENTERS: WPC, NHC, OPC
IR 20170110 2115Z GOES-E 4-KM US SAT IMAGE
GEOCAT_v0.80
GOES-13 2017-01-10 21:15:00
Cloud Top Temperature [K]
200 210 220 230 240 250 260 270 280 290 300 310 320
(b)
(c)
Intensity
10000
3162
1000
316
100
3000 m
2500 m
2000 m
1500 m
1000 m
500 m
0 m
07:56
10:09
12:22
14:36
16:49
19:02
21:16

winds $U_h \sim 1$ m s^{-1}. The snow melting contributed to the fog formation at the surface where T increased up to 3 °C prior to FZFG (freezing fog) formation. Figure 4a shows the CL51 backscatter image for 24-h time period. Prior to the fog formation, light snow occurred at about 0100-0611 UTC. There were BL clouds with their base at about 1 km. After 1000 UTC, cloud base decreased close to the ground till 1500 UTC (10 AM LST). The fog intensified after 1500 UTC till the end of the day. GOES-13 image at 2330 UTC (Fig. 4b) suggested low-level water/fog cloud (fog, blue color). The cloud-type algorithm can be found in Pavolonis (2010). Figure 4c shows the fog picture at about 1500 UTC, indicating very low Vis and melting snow at the project site.

To evaluate fog microphysics and dynamics on this day, temperature-rose and wind-rose plots together with time series of T_a, RH_w, and U_h are provided in Fig. 5a–c, respectively. Figure 5a shows that colder T advection was from SE and warmer T advection with stronger winds (up to 4 m s^{-1}) was from NW. Based on T, RH_w, and U_h time series (Fig. 5c), it is found that fog intensity (e.g., Vis) decreased after 1800 UTC (1 PM LST) during which turbulence intensity increased significantly (indicated by fluctuations). Prior to fog formation, winds were calm at about 1 m s^{-1}. Figure 5d shows the time series of MetOne aerosol number concentration (N_a) time series at eight channels from > 0.3 mm to > 10 mm. At about 1800 UTC, CCN decreased significantly where fog started dissipation. At the start of fog occurrence ($\sim$ 0800 UTC), CCN reached to 100–200 cm^{-3}. These values are closer to those seen from FMD fog device and explained more below.

Visibility measurements from various Vis sensors for the FZFG case are plotted that includes PWD52, Sentry, OSI, and SWS present weather sensors (PWS) (Table 2). The Vis values from these sensors are usually different for clear air conditions, meaning that their sensitivity can be different for higher Vis conditions. Figure 6a shows time series of Vis from these sensors. FZFG duration is indicated in this plot from about 0800 UTC to 1800 UTC. Over the fog duration, SWS PWS seems provide lower Vis compared to others. PWD and Sentry PWSs are most consistent with each other at high Vis conditions. To see details in Vis from all sensors, Fig. 6b is provided for PWD Vis (x axis) versus other Vis values (y axis). A 1:1 line is also shown on the plot. During FZFG conditions, OSI PWS usually performs worst (Gultepe et al. 2016c) at high Vis values compared to others. In this case, Sentry Vis was also providing high readings for Vis especially for Vis < 1 km. Both SWS and PWD Vis values were more consistent with each other along to 1:1 line.

To evaluate wind effects on Vis, PWD Vis is plotted against U_h in Fig. 6c. This plots shows mean, standard deviation (SD), and percentile values at 5 and 95% of measured Vis and a fit to scatter data points. Results showed that 5% Vis gets much lower when winds are less than about 1.5 m s^{-1}, but variability is usually also large (0.2–5 km). On the other hand, increasing wind and fluctuations makes the mixing processes stronger, leading to higher 2% Vis values (for $U_h > 2$ m s^{-1}), but overall, Vis mean values along the fit (black line) decrease with increasing wind. This suggests that in advective environments, e.g., lake or ocean breezes, winds play an important role for fog dissipation and formation, but this should be considered as a function of advective air volume physical characteristics

Characteristics of fog microphysical parameters, including N_i (freezing fog ice crystal number concentration), IWC (Ice Water Content), and Vis are shown in Fig. 7a–c, respectively. Because of sensor issues, data are shown only up to 1200 UTC where the heavy fog was observed. Figure 7a indicates that total N_i (N_{it}) reaches to about 100 cm^{-3}, and sizes less than about 10 μm; meanwhile, 1-s IWC is about 0.1 g m^{-3}. With 1-min averages, IWC is about less than 0.02 g m^{-3}. Figure 7c shows the 1-s Vis values reaching down to 200–300 m. The values of N_i, IWC, and Vis (calculated from Eq. 2) are then used to

Figure 9 ▶
LPM-based particle counts (color squares) are shown in fall velocity (V_f)-D (diameter) coordinate system (**a**). This plot is obtained at about 1130 UTC shown in **b** where PR_{SN} time series during storm is given. Vis time series are shown in **c** and T_a, RH_w, and U_h are shown in **d**. Vis variability is shown in **e** where PWD50 Vis is given in x axis compared to others in y axis. Vis versus PR_{SN} scatter plot is shown in **f**

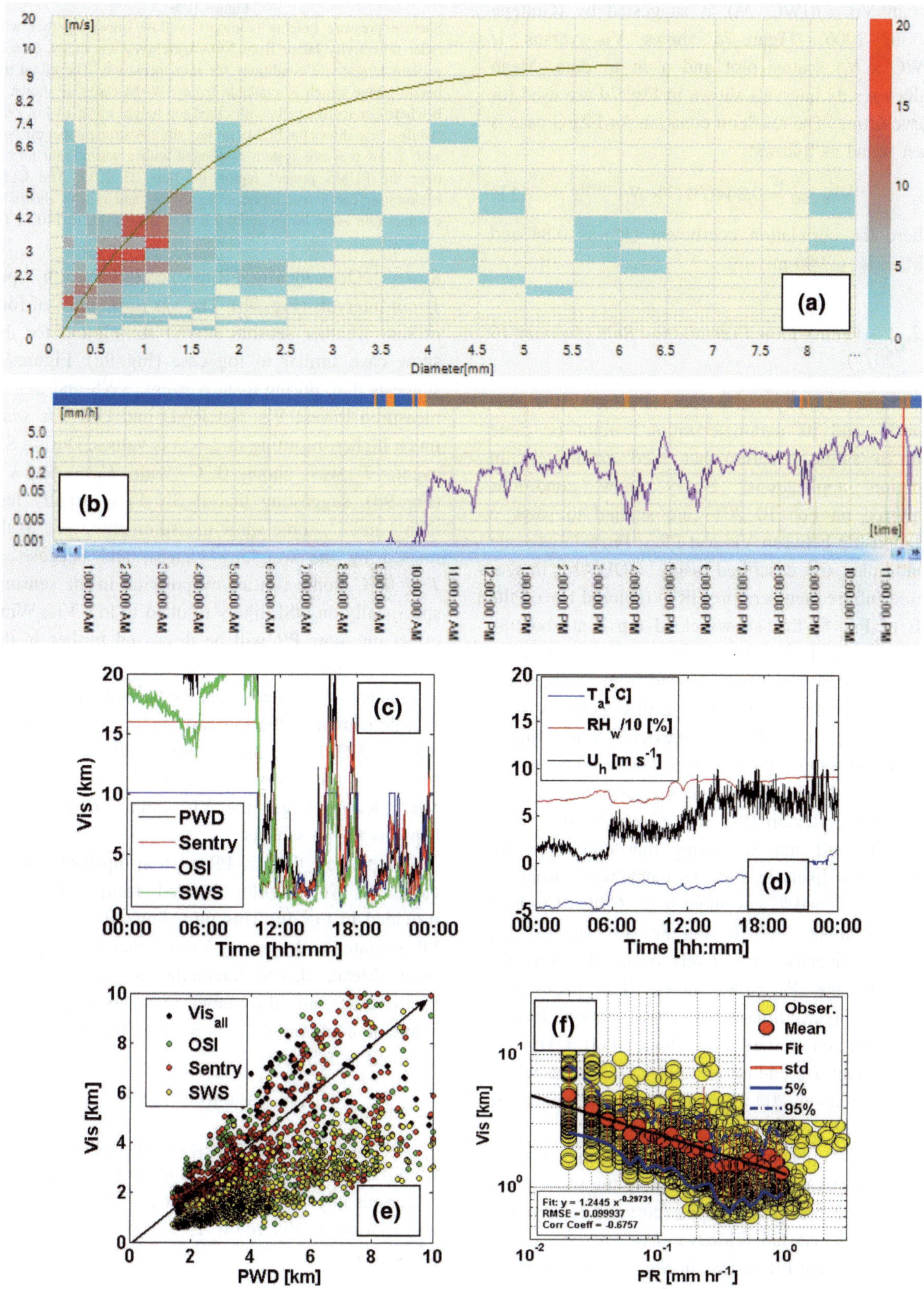
(a)
[m/s]
Diameter[mm]
(b)
[mm/h]
[time]
(c)
Vis (km)
PWD
Sentry
OSI
SWS
Time [hh:mm]
(d)
T_a[°C]
RH_w/10 [%]
U_h [m s^-1]
Time [hh:mm]
(e)
Vis_all
OSI
Sentry
SWS
Vis [km]
PWD [km]
(f)
Obser.
Mean
Fit
std
5%
95%
Vis [km]
PR [mm hr^-1]
Fit: y = 1.2445 x^-0.29731
RMSE = 0.099937
Corr Coeff = -0.6757

obtain Vis = f(IWC, N_i) as suggested by (Gultepe et al. 2006). Figure 7d shows Vis versus 1/(IWC × N_i) scatter plot and a fit to data. Mean values for dx intervals shown in Fig. 7d are used for curve fitting. The resultant equation for FZFG case is then found as follows:

$$\mathrm{Vis_{FZFG}} = 0.844(IWC \times N_i)^{-0.44}, \qquad (11)$$

where the correlation coefficient (R^2) is 0.84 and RMSE is ~ 900 m.

4.2. *Vis Versus Snow Precipitation Rate (January 10 2017)*

Snow microphysical properties affect Vis significantly, and its parameterization cannot be done with a single equation but used extensively in aviation applications. Heavy snow conditions occurred on Feb 10 2017 are studied to show a relationship between Vis and PR_{sn}. First, large-scale conditions are described using GOES-13 image-based infrared temperature (IRT) (colored bar on the left of Fig. 8a that shows cloud top T at about − 60 °C above the PUMS project site. A low-pressure system with a warm front moved from southern US and a cold front was just behind this occluded front. Upper air jet from Gulf of Mexico.lf was bringing lots of moisture at the higher levels (atmospheric river) which fed the convective system over the Great Lake Region (Fig. 8b). Closed isobars were down to 981 mb, indicating that conditions for heavy snow likely existed and winds were strong up to 20 m s^{-1} and T was about 0 °C. GOES-13 cloud top T at about − 60 °C (Fig. 8b) also showed that intense convective cells likely occurred during the front passage. Figure 8c shows CL51 backscatter extinction (β_{ext}) time-height cross section. It suggests that snow started at about 1000 UTC and lasted about 7 h ending at 2200 UTC during which β_{ext} below about 200 m has a strong attenuation due to large LWC.

Figure 9a shows laser precipitation monitor (LPM) observations collected during the snow storm that include precipitation type (curve is for droplets fall velocity, V_f), fall velocity (y axis), particle counts (color bar), and PR (y axis in Fig. 9b). The sizes of particles were up to 1 cm and max V_f was about 8 m s^{-1}. Occasionally, PR reached to 10 mm h^{-1} per 1-min intervals (Fig. 9b). The Vis time series of four present weather sensors (PWS) are plotted for the snow case similar to fog case (Fig. 9c). Figure 9c suggests that, during a snow event, SWS and Sentry measured lowest Vis, but PWD and OSI Vis were much higher, resulting in 2 × Vis values. During SN event, T was about 0 °C with U_h ~ 5 m s^{-1} (Fig. 9d), suggesting that if SN V_f is usually less than 1 m s^{-1}, some snow measurements were likely missed by the weighing gauges, and because of T ~ 0 °C, some wetting/evaporation in the sensors, specifically on OSI, likely resulted in low Vis. Wind effect on snow PR will be discussed further in the Discussion section.

Figure 10 ▶
Surface pressure heights (contours, yellow lines) together with radar reflectivity factor from NOAA archives are shown in **a** to explain large-scale conditions for rain formation. The cloud top height image which is available from UW, Madison, is shown in **b** where convective cells with cloud top height up to 14-km exist (white circle shows the PUMS project site). A strong upper air flow with a low-pressure system associated with a warm front moved over the PUMS project region on June 23 2017. The CL51 backscatter coefficient representing clouds and rain is shown in **c** where rain event lasted about 9 h from 0500 UTC to 1400 UTC

Scatter Vis values for PWSs are provided in Fig. 9e similar to the fog case. Figure 9e shows that SWS and OSI Vis values were usually smaller than PWD Vis, but sentry Vis was comparable to PWD Vis. Clearly, larger the Vis values, more scatter happens for all sensors.

To evaluate the Vis–PR relationship for the snow case, Vis versus PR obtained from PWD50 is provided in Fig. 9f. This plot shows that increasing PR is related to decreasing Vis and that is an expected result. Mean, sd, and percentiles, as well as a fit to mean values, are also shown in the same plot. The resultant equation for Vis = f(PR) for this case is obtained as follows:

$$\mathrm{Vis_{SN}} = 0.637\mathrm{PR_{SN}^{-0.55}}. \qquad (12)$$

Note that for a given PR, Vis can span an interval from 0.5 km up to 2 km, and this indicates the effect of various microphysical properties of the snowflakes on variations in Vis. In Eq. 12, RMSE is 0.02 km and R^2 is − 0.93.

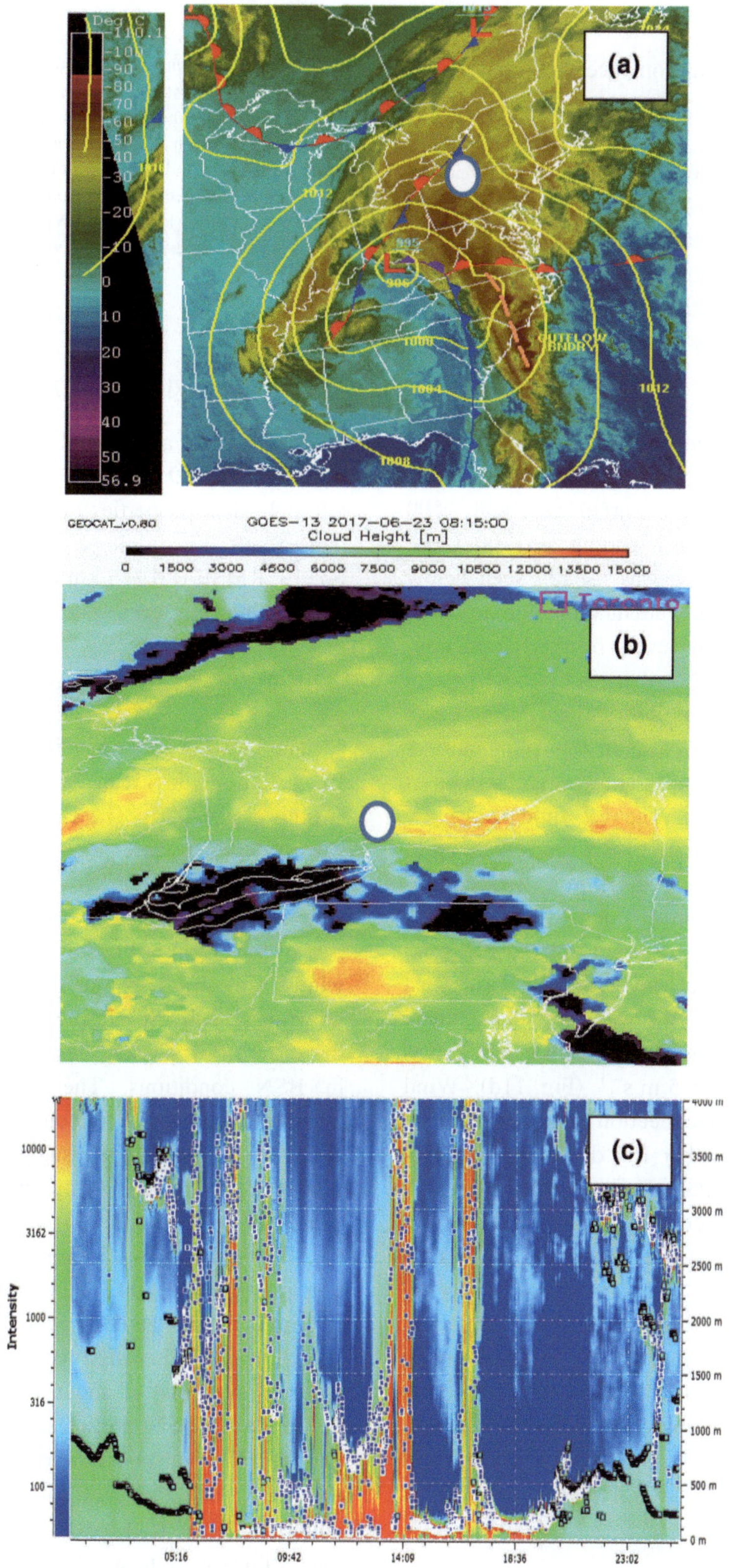
(a)
Deg C
-110.1
-100
-90
-80
-70
-60
-50
-40
-30
-20
-10
0
10
20
30
40
50
56.9
1012
1000
1004
1008
GEOCAT_v0.80
GOES-13 2017-06-23 08:15:00
Cloud Height [m]
0 1500 3000 4500 6000 7500 9000 10500 12000 13500 15000
(b)
Toronto
(c)
Intensity
10000
3162
1000
316
100
4000 m
3500 m
3000 m
2500 m
2000 m
1500 m
1000 m
500 m
0 m
05:16
09:42
14:09
18:36
23:02

4.3. Vis Versus the Rain Case (June 23 2017)

Heavy rain (RN) conditions occurred on June 23 2017 (Table 1) are analyzed to show the impact of rain PR (PR_{RN}) on Vis. Similar to SN case, large-scale conditions are first described using GOES-13 IRT image (colored bar on the left of Fig. 10a) that shows cloud top T was at − 55 °C over the PUMS project site. A low-pressure system with its center at 996 mb moved from the southern US over the project site; a stationary front moved in first over the project area slowly. GOES-13 cloud top T at about − 55 °C (Fig. 10b) also show that intense convective cells occurred during the front passage. Figure 10c shows CL51 backscatter extinction time-height cross section. Rain started at about 0500 UTC and lasted about 14 h ending at 1900 UTC. During the RN event, β_{ext} was usually seen below about 200 m with strong attenuation due to large water amounts at higher levels resulted from convective motions.

Figure 11a shows LPM observations during a rain storm similar to the SN case. In this plot, number concentration of RN drops is shown as functions of particle diameter and V_f. The PR time series are shown in Fig. 11b. The sizes of RN drops were up to 0.5 cm and V_f was more than 10 m s^{-1}. Occasionally, PR reached to 40 mm h^{-1} per 1-min intervals. The time series of 4 PWS (Present Weather Sensors) Vis are shown in Fig. 11c. The Sentry Vis was lowest among all Vis, but OSI Vis was more than others during low Vis conditions. During RN event, T was about 20 °C with $U_h < 5$ m s^{-1} (Fig. 11d). Wind effect on precipitation collection in nature can be minimal because of higher RN drop fall velocities (V_f) compared to snowflake V_f. Interestingly, the Vis variations in Fig. 11e were very large during this rain event, and it is likely due to turbulence and sensor wetting issues.

To evaluate the Vis–PR relationship for rain case, Vis versus PR obtained from the PWD50 is shown in Fig. 11f. Similar to the snow case, Vis decreases with increasing PR. Mean, sd, and percentiles, as well as a fit to mean values, are also shown on the plot and a fit to the mean values for RN case is obtained as follows:

Figure 11 ▶
LPM-based particle counts (color squares) are shown in fall velocity (V_f)-D (Diameter) coordinate system (the black solid line is for droplet fall velocities) (**a**). This plot is obtained at about 0545 UTC shown in **b** where PR_{RN} (max at 50 mm h^{-1}) time series during storm is given. Vis time series are shown in **c** and T_a, RH_w, and U_h are shown in **d**. The impact of rain on Vis sensors is shown in **e** where PWD Vis is given in x axis compared to others in y axis. Sentry Vis is usually lower than PWD Vis at about 50%. Scattering increases with increasing Vis. Vis versus PR_{RN} for rain shown in **f**

$$\mathrm{Vis_{RN}} = 4.379\mathrm{PR}_{\mathrm{RN}}^{-0.37}. \quad (13)$$

Note that for a given PR, Vis can span an interval from 1 km to 20 km, suggesting how important the microphysical properties of the RN drop such as N_d that affects Vis (Gultepe and Milbrandt 2010). In Eq. 13, RMSE is 0.245 km and R is − 0.72 when the mean values are used.

4.4. Vis Versus U_h for BSN Case (April 7 2017)

The blowing snow conditions are described when Vis goes down less than 10 km m and $U_h > 4$ m s^{-1} (Knuth et al. 2010). The case of April 07 2017 case was related to a winter snow storm. Figure 12a shows the reflectivity factor (Z) in dBZ at 1200 UTC, in which $Z > 30$ dBZ. The closed isobars in the west part of the storm at the surface indicate that stronger winds are seen behind the low-pressure system movement with 986-mb pressure at the center. After the snow storm passes, these stronger winds resulted in BSN conditions. The GCIP particle images obtained during BSN conditions are shown in Fig. 12b. These images suggest that BSN crystal sizes were less than 300 mμ and had various shapes. Time-height cross section of β_{ext} (Fig. 12c) from CL51 suggests that, before the end of the heavy precipitation (at 1200 UTC), PR_{SN} was about 3 mm h^{-1} and V_f was about 9 m s^{-1} when $D < 0.5$ mm (Fig. 13a, b). The high counts of LPM particles (colored symbols) at about 1000 also indicated the existence of BSN particles. The Vis values from various PWS sensors showed highly variable results, suggesting that they are not calibrated at the BSN conditions (Fig. 13c). After the end of the storm, U_h max values were near 20 m s^{-1} and stayed between

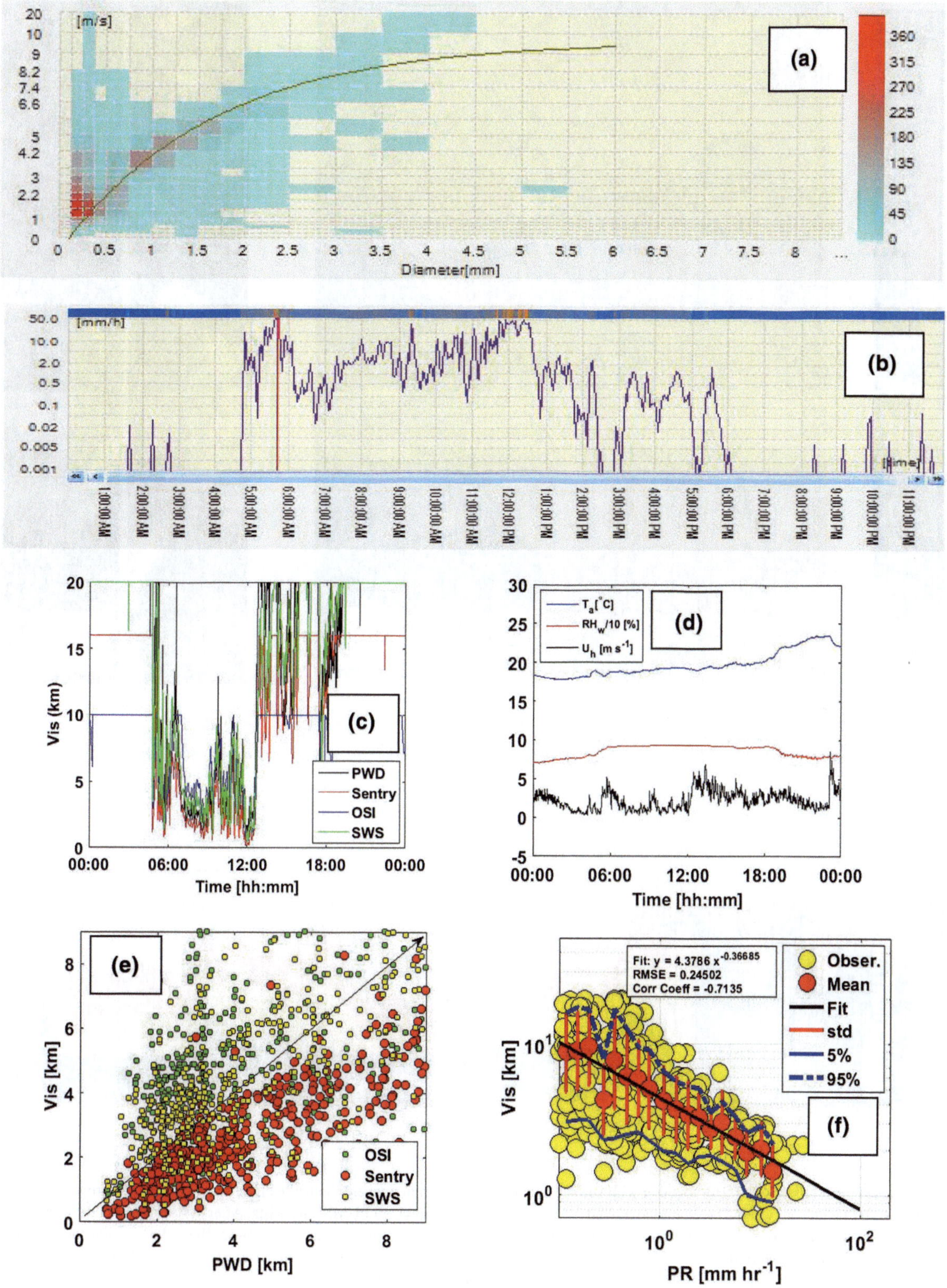
(a)
[m/s]
Diameter[mm]
(b)
[mm/h]
(c)
Vis (km)
PWD
Sentry
OSI
SWS
Time [hh:mm]
(d)
T_a[°C]
RH_w/10 [%]
U_h [m s^{-1}]
Time [hh:mm]
(e)
Vis [km]
PWD [km]
OSI
Sentry
SWS
(f)
Fit: y = 4.3786 x$^{-0.36685}$
RMSE = 0.24502
Corr Coeff = -0.7135
Obser.
Mean
Fit
std
5%
95%
Vis [km]
PR [mm hr^{-1}]

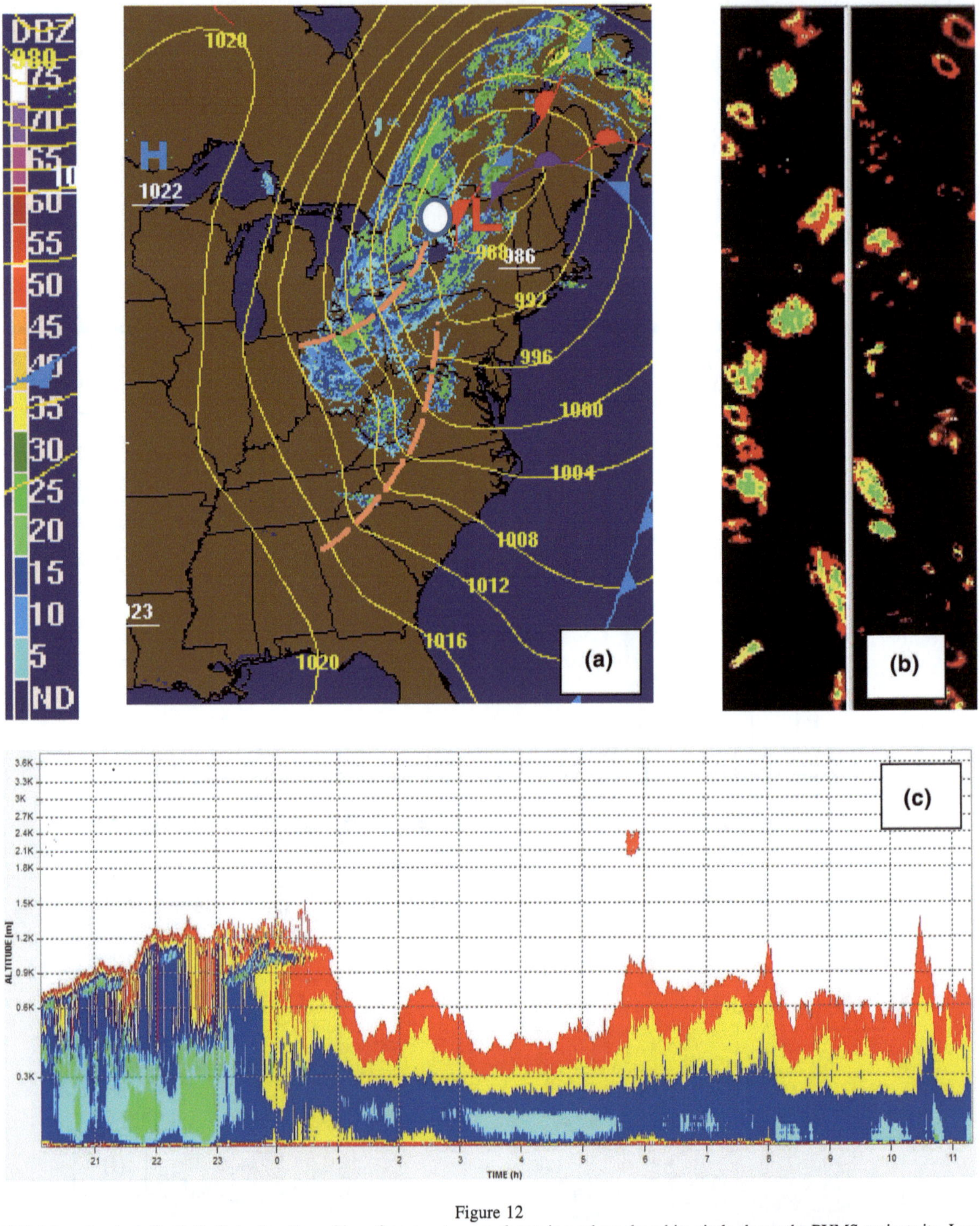

Figure 12
NOAA-based radar reflectivity factor together with surface counters are shown in **a** where the white circle shows the PUMS project site. Low-pressure system with heavy snow precipitation moved over the Toronto region where PUMS site exists. After system moves out of the region back of the system had closed isobars indicating intensity of the winds, radar reflectivity Ze was about 30–35 dBZ. The blowing snow particles from GCIP probe are shown in **b**. The CL51 shows backscatter coefficient that indicates snow particles existing up to 800 m during which U_h was usually larger than 8 m s^{-1} (Fig. 13)

10 and 15 m s^{-1} with wind gusts at about 5 m s^{-1} (Fig. 13d). Before the storm ends, T was about 0 °C and reached about 5 °C after the storm (after 1300 UTC). RH_w was about 70% during the BSN conditions.

Vis from the PWS sensors versus the PWD Vis is shown in Fig. 13e. Clearly, OSI and SWS Vis were more than the PWD Vis for Vis < 1 km. Both the OSI and Sentry Vis were less than the PWD Vis when high Vis values are considered. When Vis is plotted against U_h for PR less than 0.5 mm h^{-1} (Fig. 13f), Vis decreased from 10 km at 2–3 m s^{-1} down to 600 m at 8 m s^{-1} (with 5% line). The mean values and a fit to means suggest that when $U_h \sim 30$ m s^{-1}, Vis becomes about 100 m. However, occasionally, Vis is about 100 m when $U_h = 7$ m s^{-1}. Therefore, any physical parameterization used for the BSN prediction should consider variability in the observations. The best fit for the BSN conditions for this event is obtained as follows:

$$Vis_{BSN} = 127.31 U_h^{-1.62}, \quad (14)$$

where the correlation coefficient (R) is − 0.86 and RMSE is 0.19 km.

4.5. Wind Gust and Dissipation Rate (April 6 2017)

Wind gust and dissipation rates are usually defined based on their applications. Here, GST is described as in the method section; time interval (dt) = 10-min segments were used to calculate 16 knots' line which is required for the mean wind values. Thereafter, 2-min intervals are used with lulls and peaks that have a variation of 10 knots. Figure 14a shows a strong low-pressure system with some precipitation (Z between 25 and 35 dBZ). Surface-level isobars were closer near the north-west corner of the system, an indication of possible wind gust occurrence. This was also consistent with strong up and downdrafts occurring along the upper level jet (Fig. 14b). This figure shows that 850 mb wind speed was at 50–55 knots at the NW quadrant and 70 knots at the SE quadrant. Both regions were highly susceptible to wind gust occurrence. Figure 14c shows CL51 BL cloud top conditions with light snow occurring during wind gust conditions. The dry air above the cloud top sinks down to 300-m level from about 700 m. The red-colored areas in Fig. 14c are due to falling light snow particles.

For the gust (GST) analysis, the time series of Young 3D sonic anemometer measurements used to calculate 1-s, 2-min, and 10-min averages of U_h, and a 16 knots' line representing wind gust conditions are shown (Fig. 15a). At about 1000 UTC, U_h changed from about 5 m s^{-1} to 35 m s^{-1}. The 2-min averages were above the 7 m s^{-1} (14 knots) with $U_h' > 5$ m s^{-1} (10 knots) and that suggests the wind gust was almost occurring for the entire day. The maximum intensity as 30 m s^{-1} is seen at 1000 UTC. Figure 15b shows 1 and 2 s fluctuations of U_h after removing 2-min filtered values from the observations. It is seen that U_h' for 1 s was about − 5 and + 5 m s^{-1}. After the strong shear happened at 1000 UTC, U_h' was between − 10 and 10 m s^{-1}. In addition, dissipation rate is calculated using Eq. 5, while data are averaged over 60-s intervals. Results are shown in Fig. 15c. Based on the work of Cho et al. (2003), dissipation rate more than 20 m^2 s^{-3} can be assigned as severe turbulence conditions (Fig. 15c).

4.6. WE-UAV Thermodynamic Profiling

The WE-UAV is developed based on Aeryon UAV that utilizes position angles and resistance of airflow around it to calculate horizontal wind, wind gust, and direction. The data is being recorded at 1 Hz in WE-UAV as described earlier. Other observations such as aerosol total number concentration (N_a) (size > 0.5 and 1.0 μm), T, and RH_w were also collected during flight patterns defined as step-ups and lifting up/down directly at constant flight speeds up to 15 m s^{-1}, although it can fly in the much higher wind conditions (e.g., 30 m s^{-1}).

Figure 16a shows the ceilometer backscattering time-height cross section (β_{ext}) during which WE-UAV flights (at 1620 UTC) are performed. During this clear air flight case, BL height was shallow at about 200 m, and a thin cloud base was decreasing at $z = 3$ km level at about 1300 UTC. At the same time, the PMWR-based T, RH_w, and q_v were about 25 °C, 80%, and 10 g m^{-3}, respectively (Fig. 16b–d). Because of clear air conditions, heating at the surface around noon LST (1700UTC) suggests some thermal

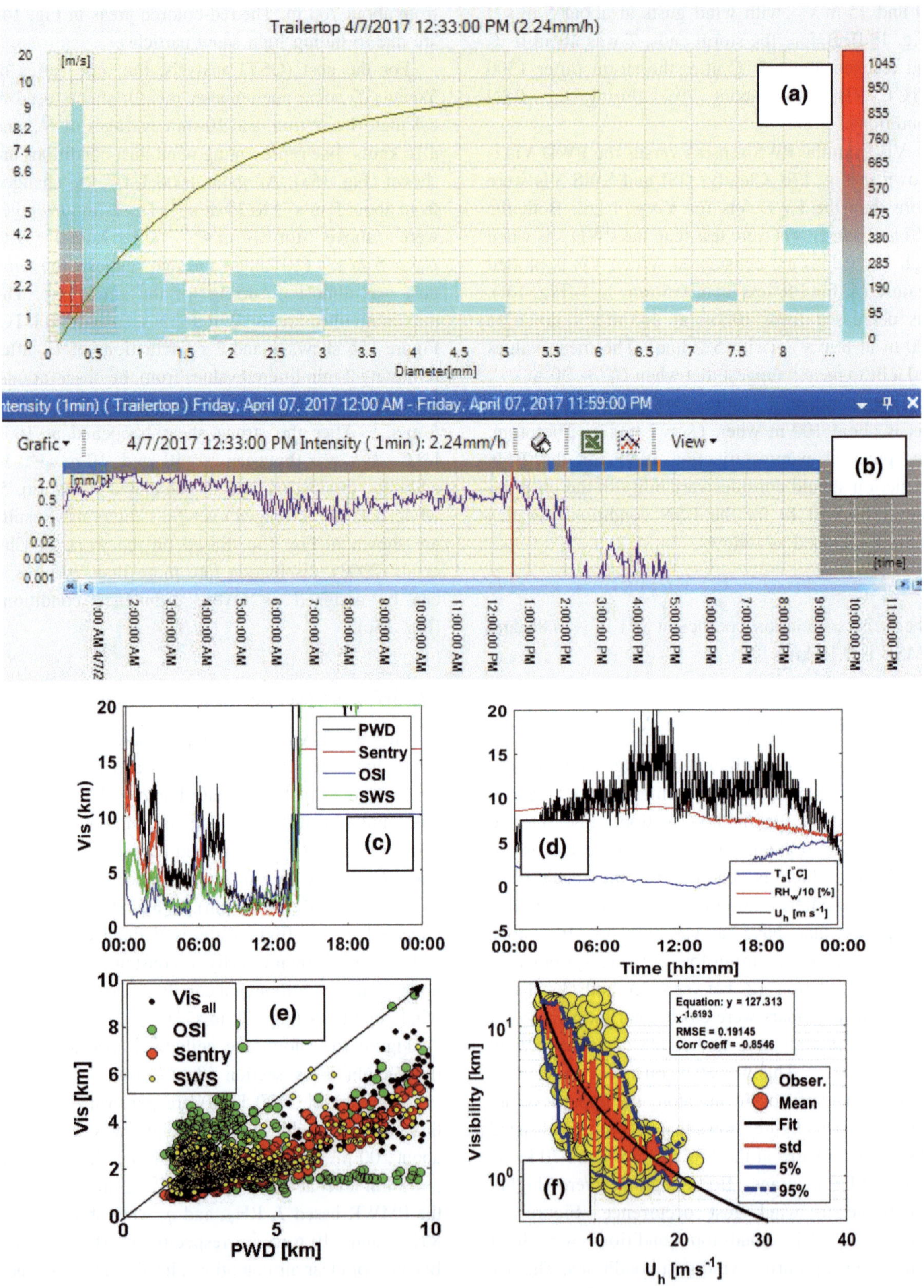
Trailertop 4/7/2017 12:33:00 PM (2.24mm/h)
[m/s]
Diameter[mm]
(a)
Intensity (1min) (Trailertop) Friday, April 07, 2017 12:00 AM - Friday, April 07, 2017 11:59:00 PM
Grafic
4/7/2017 12:33:00 PM Intensity (1min): 2.24mm/h
View
[mm/h]
[time]
(b)
Vis (km)
PWD
Sentry
OSI
SWS
(c)
(d)
Ta [°C]
RHw/10 [%]
Uh [m s-1]
Time [hh:mm]
Vis [km]
Visall
OSI
Sentry
SWS
(e)
PWD [km]
Visibility [km]
Equation: y = 127.313 x-1.6193
RMSE = 0.19145
Corr Coeff = -0.8546
Obser.
Mean
Fit
std
5%
95%
(f)
Uh [m s-1]

◀Figure 13
LPM-based particle counts (color squares) are shown in fall velocity (V_f)-D (diameter) coordinate system (the black solid line is for droplet fall velocities) (**a**). This plot is obtained at about 1230 UTC shown in **b** where PR_{SN} (max at 3 mm h^{-1}) time series during storm is given. Vis time series are shown in **c** and T_a, RH_w, and U_h time series are shown in **d**. Vis values from various sensors (*y* axis) versus PWD Vis are shown in **e** where scattering decreases with increasing Vis. The impact of blowing snow on Vis sensors is shown in **f** where PWD Vis is given in *y* axis and a fit for BSN conditions is shown in the plot

instability and that likely resulted in wind gusty and turbulent conditions. In this subsection, the goal is to indicate the capability of WE-UAV for BL research and to make some qualitative comparisons of UAV measurements with these other platforms (e.g., PMWR or CL51). In the future, comparisons of the UAV-based measurements with tall tower meteorological measurements will be performed quantitatively.

During the PanAm Olympic Games in the summer of 2015, WE-UAV measurements were provided for clear air conditions because of the line of sight issues and permissions from Transport Canada. Using WE-UAV 1 Hz $N_{a0.5}$ and $N_{a1.0}$, aerosol measurements (> 0.5 and > 1.0 μm), as well as horizontal wind (U_h) and wind gust (w_g), collected along the constant altitude flight patterns are shown in Fig. 17a–c, respectively, for July 10 2015 in Oshawa, Ontario. The time series of horizontal wind and wind gust velocities (1 s) are obtained directly from the inertial navigation system of the UAV. The wind gust velocities are obtained from smoothed values of horizontal wind over 2-min intervals. It is seen that wind gust velocities were up to 7 m s^{-1} at 16:14 and 16:21 LST.

The smoothed lines in Fig. 17 are for 1-min running averages. Two vertical lines on the left (vertical lines with black color, 1614 UTC) and on the right (vertical lines with green color, 1621 UTC) are for wind gusts up to 7.0 m s^{-1}. During these segments, $N_{a0.5}$ increases and decreases more than 200 cm^{-3} which are likely resulted from vertical air velocity fluctuations that are not measured.

Comparisons of T and RH_w values from the WE-UAV at the first level (~ 30 m) of PMWR indicated that T was about 25 °C and RH_w = 80% (not shown). These values were comparable to PMWR values, and differences between T and RH_w were about a few °C and 10%, respectively. Note that height resolution of PMWR compared to WE-UAV was much coarse; therefore, additional work in this area is needed and this kind of experiment likely requires tower based in situ observations. The measurements of WE-UAV N_a are found to be comparable with these surface-based MetOne N_a observations at 3 m level, and its uncertainty was usually about less than 10–20% over a time period of 30 s.

The results suggested that preliminary measurements from the WE-UAV that included aerosol number concentrations at the channels for diameter (d) > 0.5 and d > 10 μm, wind speed, as well as T, RH, (including future Vis measurements) can be useful in the future for aviation meteorological applications within the BL. Here, only a limited number of measurements are provided and more tests are required to improve WE-UAV operations for aviation applications.

4.7. *Icing Envelopes and Gondola Measurements*

The Gondola-based microphysical observations, e.g., using DMT CDP and GCIP, collected in the UOIT ACE facilities are provided for better understanding of fog and drizzle microphysics effects on icing conditions at T = – 20 °C with using U_h of 20, 40, and 60 m s^{-1} These data sets can also be used for developing fog visibility and icing/freezing parameterizations but were not performed here. Figure 18a is obtained from a Gondola-based CDP probe measurements obtained during the ACE icing test performed on 26 August 2016 at T = – 2 °C with U_h < 20 m s^{-1} and U_h > 45 m s^{-1}. In this figure, the LWC is found higher with U_h > 45 m s^{-1} compared to lower wind speed. For a given MVD at high values, the LWC is usually larger compared to the low wind speed values. Clearly, the icing rate is a function of flow rate, the pressure at the nozzle, water temperature, as well as the nozzle orifice size. Icing that occurred on the Gondola box during a 10-min experiment in the ACE climate facilities is shown in Fig. 18b. This figure suggests that icing was not glaze icing but rather rime icing.

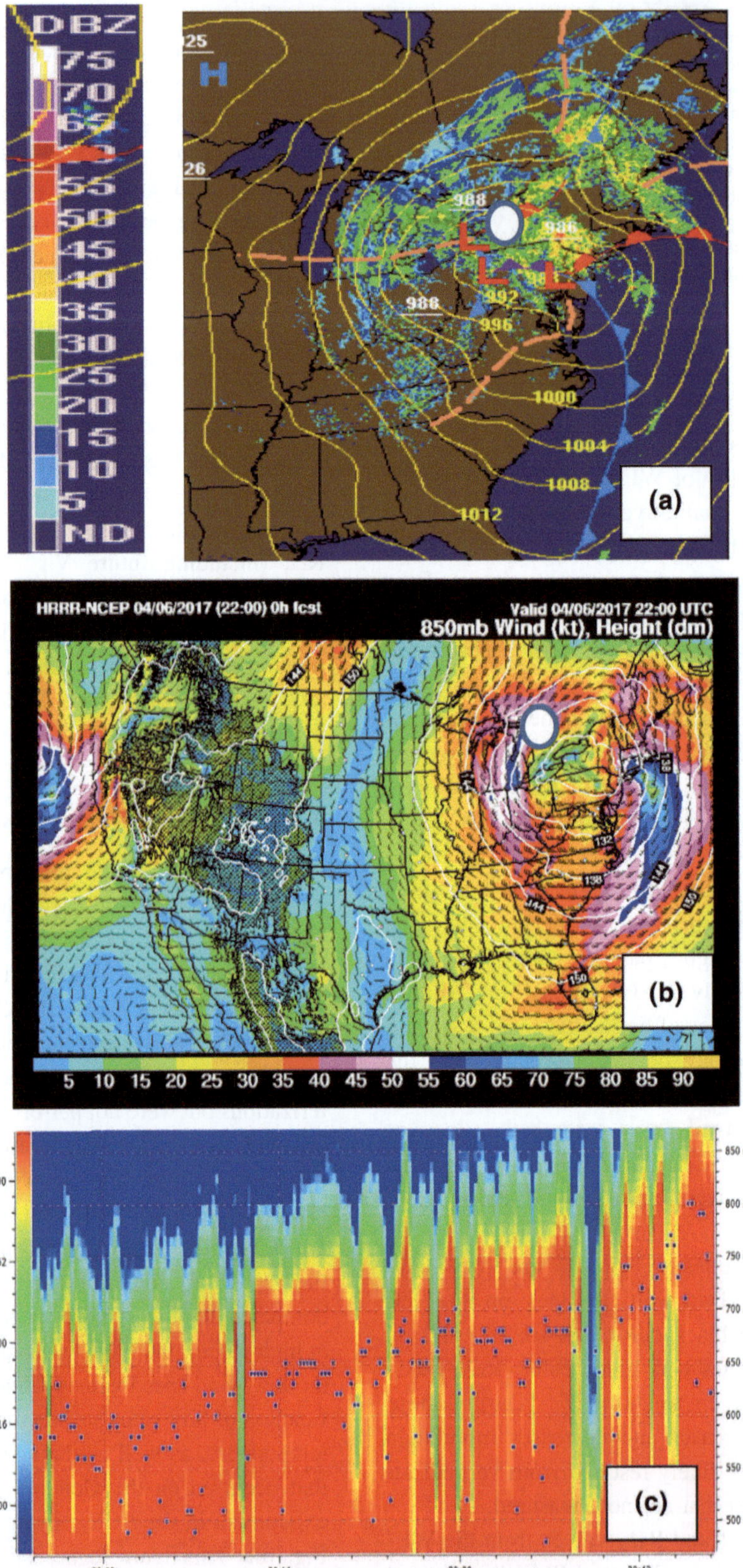
DBZ
75
70
55
50
45
40
35
30
25
20
15
10
5
ND
988
986
992
996
1000
1004
1008
1012
(a)
HRRR-NCEP 04/06/2017 (22:00) 0h fcst
Valid 04/06/2017 22:00 UTC
850mb Wind (kt), Height (dm)
(b)
5 10 15 20 25 30 35 40 45 50 55 60 65 70 75 80 85 90
Intensity
10000
3162
1000
316
100
850 m
800 m
750 m
700 m
650 m
600 m
550 m
500 m
20:03
20:16
20:30
20:43
(c)

◀Figure 14
Wind gust conditions occurred during a precipitation event is shown in **a** where a low-pressure system together with a warm front located nearby the PUMS project site (**a**). During storm pass, isobars were very close to each other, indicating of strong channeling flows. Reflectivity factor (*Z*) was about 35 dBZ. The 850-mb winds (in knots) and heights (dm) from NOAA HRRR-NCEP model runs at 2200 UTC on April 6 2017 indicate that stronger winds just above the surface was east and west part of the system (**b**). The CL51 backscattering intensity after the storm (**c**) shows clearly incursion of dry air from high levels (850 m) down to 500-m level (red color shows light precipitation), meaning that strong eddies were occurring

Based on Eq. 8, the LWC versus MVD is obtained theoretically as a function of N_d that changed from 0.01 up to 500 cm^{-3} (Fig. 18c). In this figure, high LWC values are observed with MVD > 10 mμ and N_d between 10 and 500 cm^{-3}. This also suggests that icing due to freezing fog particles can be significant for aviation operations.

Observations from the present work obtained using Gondola CDP and GCIP probes located at the UOIT ACE CWT and earlier work results are shown in Fig. 18d. The CDP data (blue filled circles) with $U_h > 45$ m s^{-1} at $T = -20$ °C and GCIP data (green filled circles) indicate that icing occurred at MVD from 5 mμ up to more than 100 mμ (with supercooled large droplets, SLD). The various N_d and PR are also plotted on the same figure (colored solid lines, see inlet panel). The highest icing rates occurred when $N_d = 100$ cm^{-3} (red lines, Fig. 18d). However, when MVD increases, severe icing rates occur at the even very low N_d values. Note that PR ~ 120 mm h^{-1} represents severe icing conditions based on Newton's definitions (Lewis 1947; Newton 1978). The yellow-colored boxes in Fig. 18d are the FAA icing envelopes based on Newton's work (1978). The purple-colored surfaces are for wind tunnel tests based on Steen et al. (2015). The dashed black lines are for icing rates used in the FAA Appendix C requirements for various icing intensities from light, moderate, and severe icing conditions. The results of Politovich (1996) also suggested an upper envelope for severe icing conditions (solid thick black line), as shown in Fig. 18d. Figure 18d suggests that icing conditions and classifications are strong functions of MVD, N_d, and PR (or flow rate in wind tunnel). Severe icing conditions can occur anywhere along the PR = 120 mm h^{-1}. Note that in the wind tunnels, accuracy of MVD and PR can be a strong function of the accuracy of water pressure/flow rate, wind speed, and turbulence, as well as the collision of droplets beyond the controlled experiment.

These results of icing tests are strong functions of extreme temperatures, dynamical effects, and microphysical conditions, as well as nozzle physical and dynamical characteristics. Therefore, to generate supercooled droplets for icing applications in a wind tunnel, nozzle characteristics need to be further researched.

5. Discussion

In this work, various meteorological parameters and new platforms related to aviation operations are evaluated and issues related to their use for operations are discussed below.

5.1. Scale Issues

In the analysis, issues related to variability over short time and space scales can be important for the model and remote-sensing validations, and considered significant for presenting the results. To emphasize the variability issues, precipitation rate (PR), visibility (Vis), and wind speed (U_h) are provided from the three sites (PUMS supersite, Barn site, and House site) that were apart from each other by only less than 1 km. Figure 19 shows the time series and scatter plots of PR_{SN} in box a and box b, Vis in box c and box d, and U_h in box e and box f, respectively, on January 10 2017 snow case. This figure shows that variations in PR_{SN}, Vis, and U_h can be more than 5 mm h^{-1}, 2 km, and 4 m s^{-1}, respectively. For given *x* axis values of PR_{SN}, Vis, and U_h representing the PUMS supersite, Barn satellite site, and House site, the changes on these values reach to 100, 50, and more than 100%, respectively. Therefore, any NWP model run validations using observations should cover the spatial coverage of related parameters, not only a single station's point measurements.

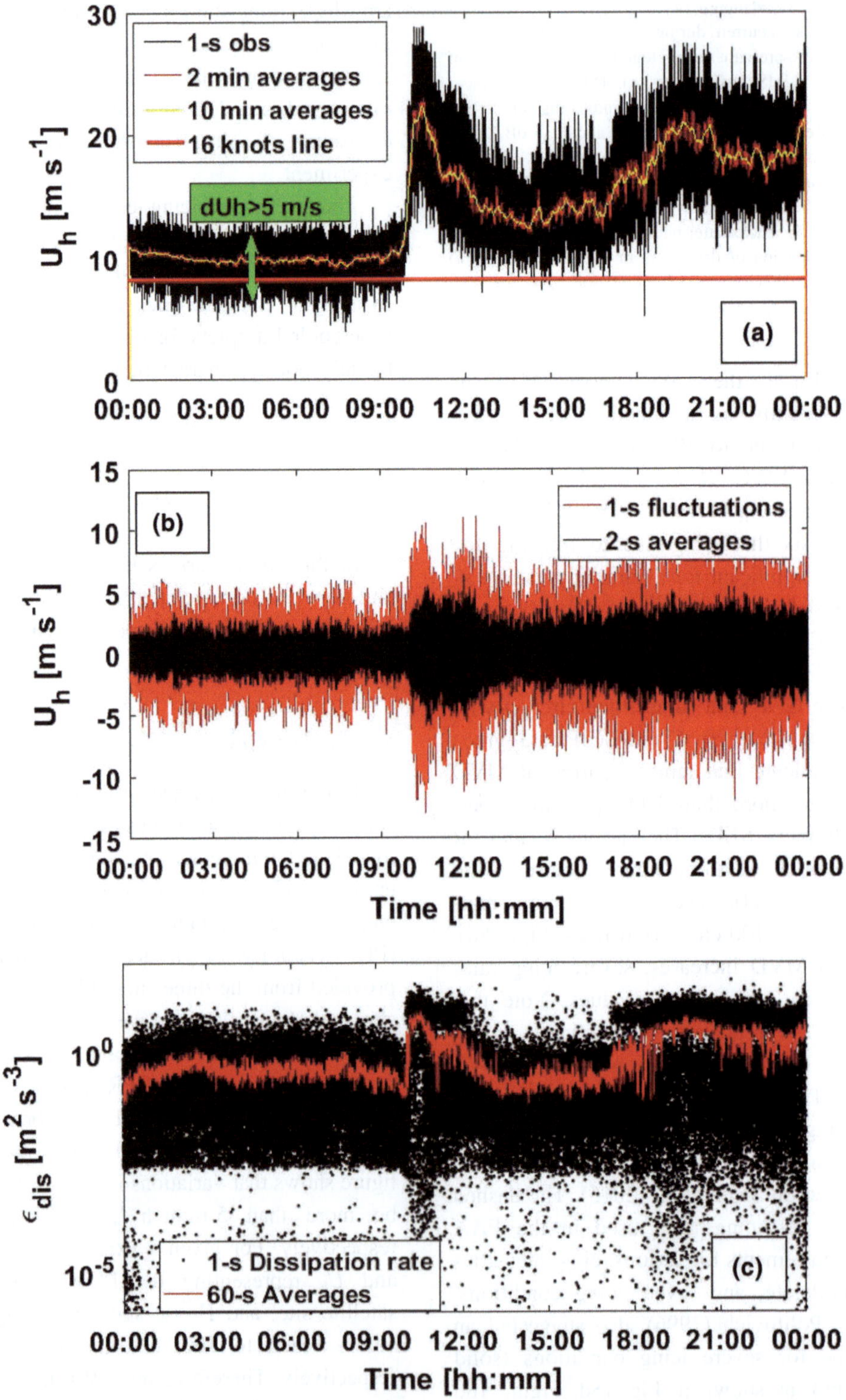

Figure 15

Horizontal wind component (U_h) obtained from 1-Hz data using WXT520, and 2 min, and 10-min filtered observations together with wind gust occurrence line (horizontal solid line) are shown in **a**. Time series of 1-s and 2-s smoothed data of U_h are shown in **b**. Eddy dissipation rate calculated using Eq. 5 for 60-s filtered data and 1-s eddy dissipation rates are shown in **c**

5.2. Variability

To show time variations' wind speed (U_h) parameters, observations are averaged over 1-, 5-, 10-, 20-, 30-, and 60-min intervals for the PUMS supersite (Fig. 20a). Based on Fig. 20a, maximum U_h versus time for 1-h time period is shown in Fig. 20b. This is done by zooming in the time period between 1300 and 1360 UTC in Fig. 20a. This figure shows that as the averaging time increases, maximum wind speed decreases and variability also decreases. To show wind speed versus distance relationship, time is converted to distance scale (L_s) using the measured wind speeds, and the results are shown in Fig. 20c. This figure indicates that representation of max wind speed in an NWP model is strongly related to averaging scales; U_h decreases from 19 m s^{-1} (at 20 m) to 8 m s^{-1} (at 500 m). A fit applied to scatter plot in Fig. 20c is obtained as follows:

$$U_{\mathrm{hmax}}(\mathrm{m/s}) = -2.85\ln(L_s(\mathrm{m})) + 26.14, \qquad (15)$$

where R^2 is 0.96 and this relationship shows that increasing L_s results in decreasing max U_h. Overall, the results suggest that variations in sub-grid and averaging scales need to be considered in the NWP model simulations.

5.3. 3D Versus 2D Wind for Turbulence and Wind Gust Estimation

The WE-UAV wind measurements can be highly susceptible to variation of aircraft position angles and propeller generated downwind flushes. Therefore, an additional sensor to analyze the wind/wind gust conditions is needed to be located above or in front of the forward flying WE-UAV. A UAV's high sampling rate measurements within the BL can be very powerful for BL and aerosol research on ice nucleation or CCN, turbulence flux, and atmospheric profiling, as well as emergency applications. The 1-Hz wind observations and eddy dissipation rate estimations can be as much as two times more than 1-min average values (Fig. 21). This suggests that an aircraft flying over 100 m s^{-1} can be severely affected by wind gust velocities. For wind gust estimation, the present observations use only wind x and y components but vertical air wind gust values (Fig. 21 green lines) in non-isentropic turbulence conditions, e.g., changing wind components at all three axes, (Fig. 21) can also severely impact aircraft position angles. Therefore, the inclusion of vertical air velocity in the analysis increases possible accident occurrence. The smoothed lines over 1-min intervals in Fig. 21 are shown in bluish colors. A strong wind gust formation is seen at about 1000 AM in the 3D axes.

Duration of wind gusts usually happen in short time scales, e.g., less than 20 s; therefore, fast response wind measurements are needed for wind gust and turbulence estimates. The method using various parameters from the ultrasonic sensors here showed that eddy dissipation rates and wind gust velocities can be obtained accurately, and these can be used for warnings during aircraft landing and taking off. Wind gust can be related to wind shifts, which is so-called when the wind direction changes more than 45° in less than 15 min and sustained winds > 10 knots. In addition, the wind direction is specified as variable if, during the 2-min time period, it varies by 60° or more during U_h that is greater than six knots (NWS definition; Kantor et al. 1996).

One of the important aspects of wind speed distribution over airport areas that affects aircraft landing and takeoff is related to wind shear and turbulence (Thobois et al. 2018). At the PUMS supersite, although a Halo Doppler Lidar was available, it was not allowed to be used during the project, but in EC3, it was used, for the first time, during an Arctic project called SAAWSO (satellite And Arctic Weather for Search and Rescue (SAR) Operations) to study turbulence and wind gust in Goose Bay, NFL, Canada, and during a fog project called MATERHORN to study wind effects on ice fog physics. In this respect, readers should see the paper submitted by Thobois et al. (2018), and Gultepe et al. (2015) and Fernando et al. (2015) for Doppler Lidar use for wind research and applications.

5.4. Wind Effect on Snow Measurements

Wind effects on snow collection efficiency are an important issue for both aviation and water budget applications. For example, Rasmussen et al. (2012) and Gultepe et al. (2017a) showed that wind speeds

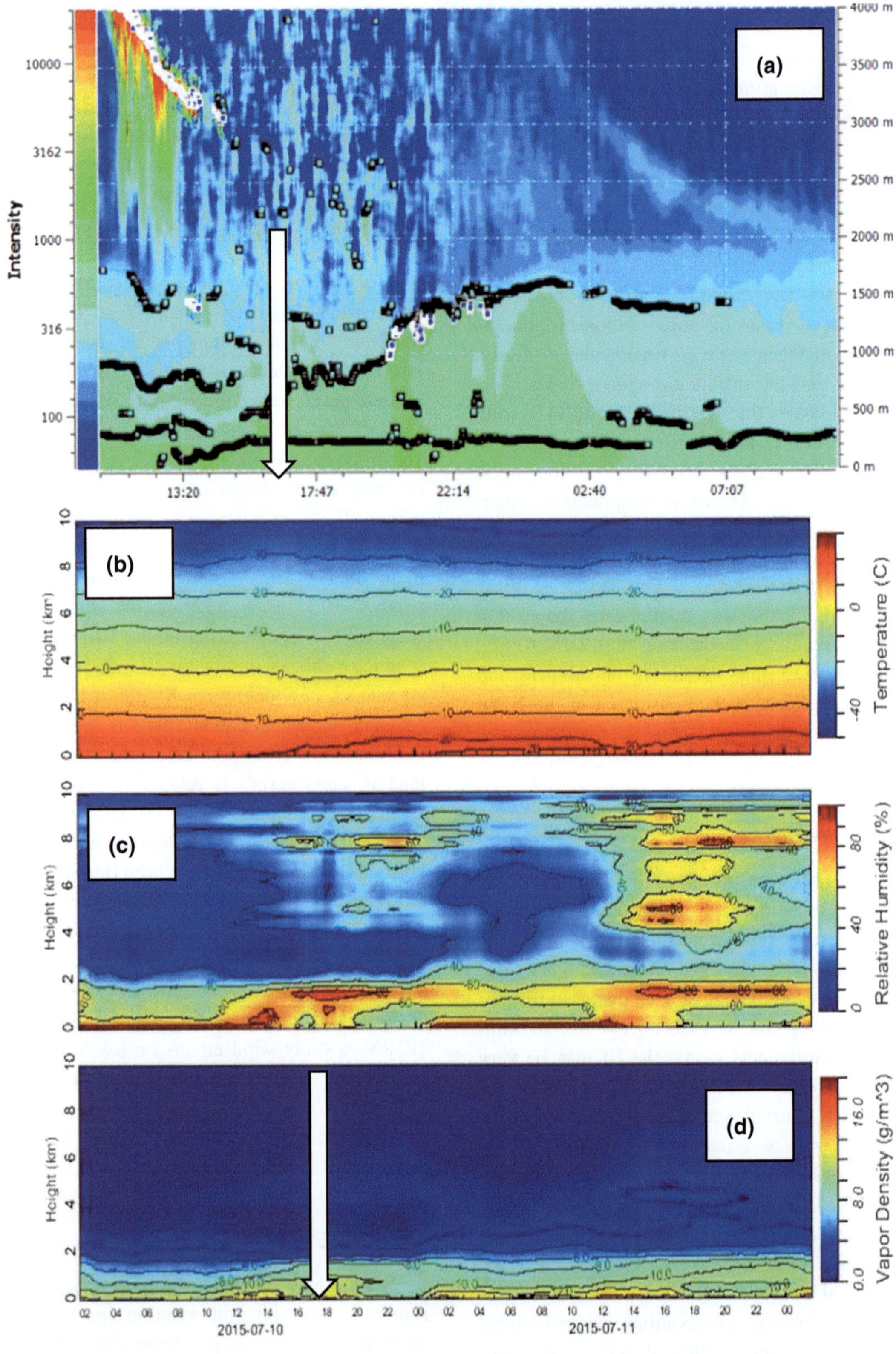
(a)
(b)
(c)
(d)
Intensity
Height (km)
Temperature (C)
Relative Humidity (%)
Vapor Density (g/m^3)
13:20
17:47
22:14
02:40
07:07
2015-07-10
2015-07-11

◀Figure 16
CL51 backscatter time-height cross section together with approximate location of UAV flight happened (white vertical arrow) is shown in **a**. Temperature, RH_w, and vapor mixing ratio are shown in **b–d**, respectively

more than 5 m s^{-1} can severely affect PR_{SN}, resulting in decreasing its accumulation at about 50%. For the snow case used here, Fig. 22a shows PA for various snow measuring sensors plus a DFIR based Pluvio PA_D and wind speed (U_h). Increasing wind speed results in large discrepancy among PA values and PA_D. With modified DFIR, PA_D shows large snow accumulation amounts (red line). Geonor- and Pluvio-based PA without a shelter is found very similar to each other. PWD50 based PA_P is smaller than other PR values during heavy snow that likely happened due to the internal algorithm used. Figure 22b shows DFIR Pluvio PA_D and Pluvio PA_P versus U_h scatter plot. Increasing U_h results in increase in both PA values, but PA_D is usually larger

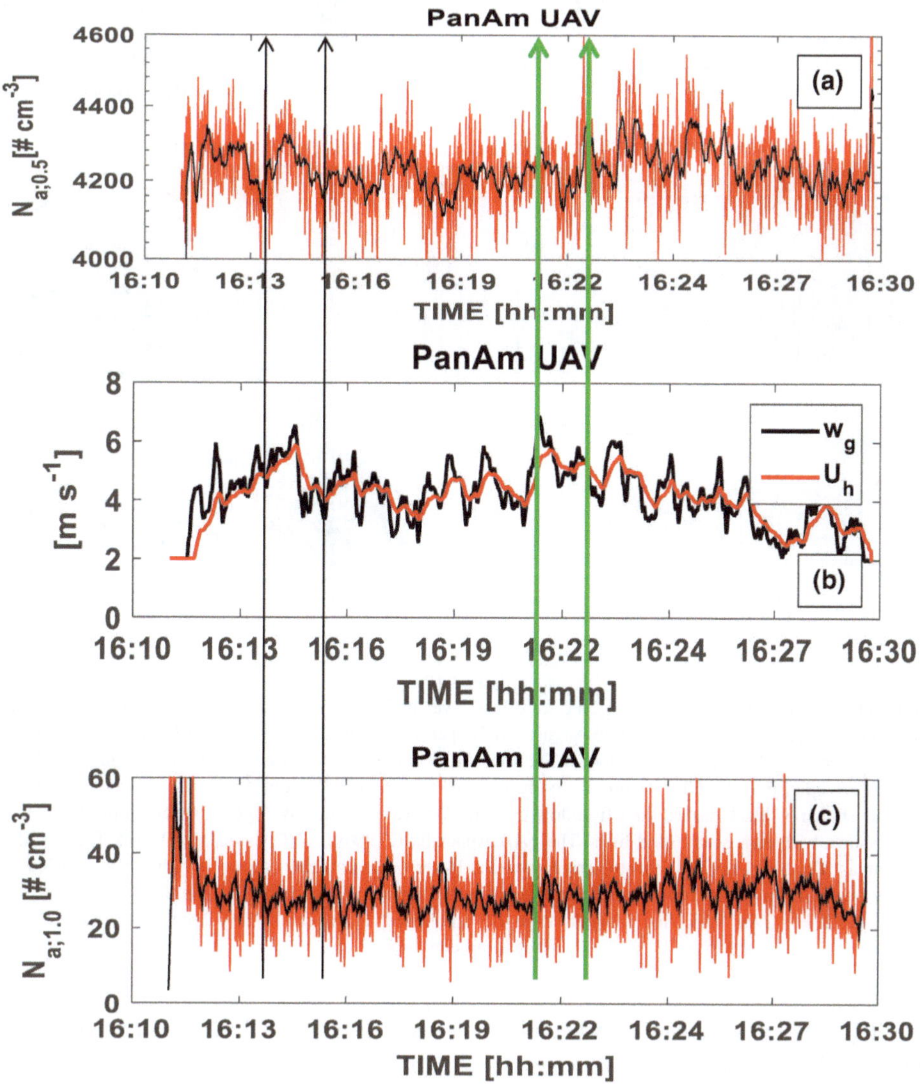

Figure 17
WE-UAV time series of $N_{a0.5}$ (**a**), U_h (smoothed over 1 min), and vertical wind gust w_g (1-s raw data) from WE-UAV GPS system (**b**), and $N_{a1.0}$ (**c**). The subscripts 0.5 and 1.0 mean aerosols sizes larger than these thresholds. Smoothed black solid lines are for 1-min intervals and red lines are for 1-s measurements

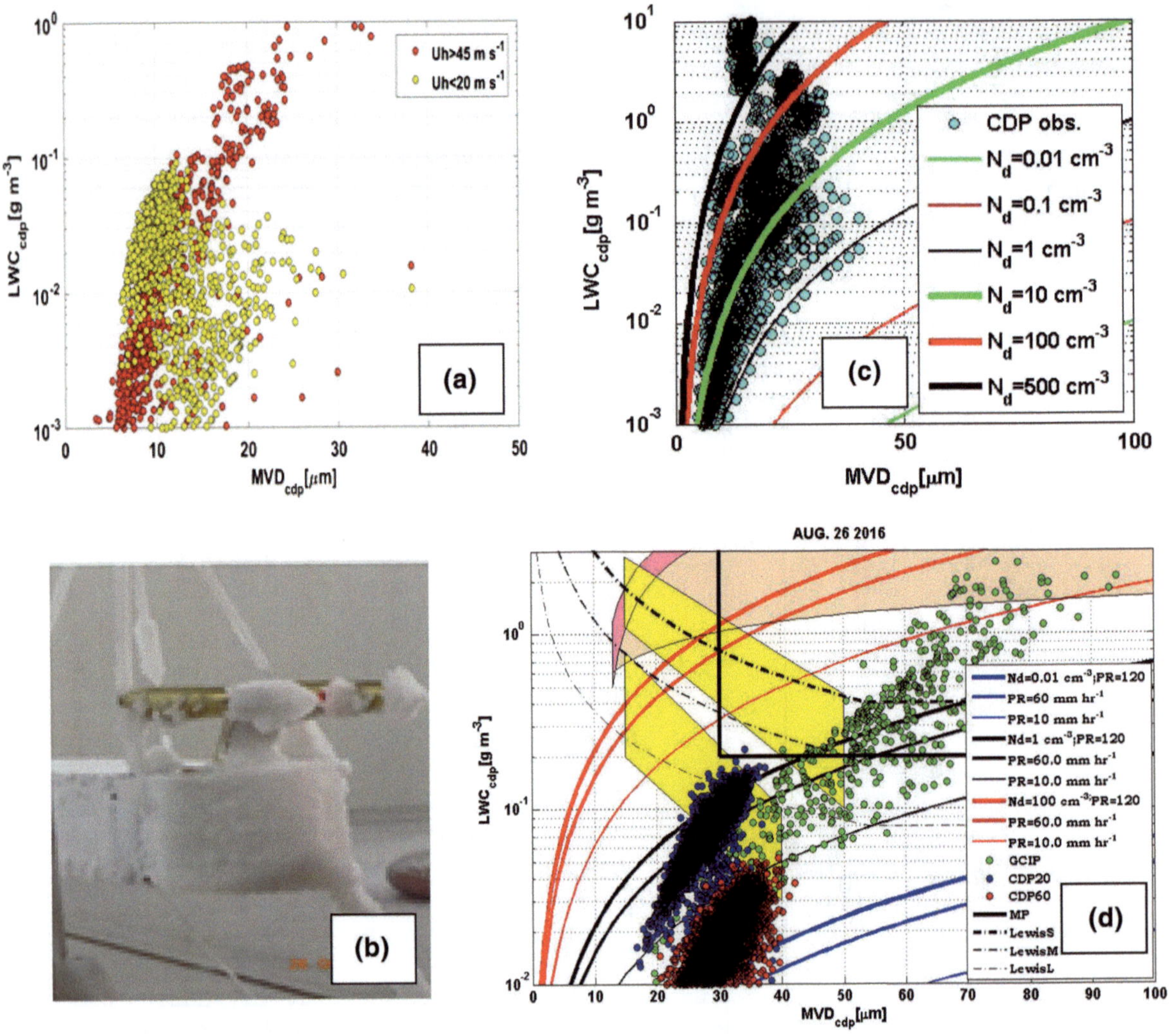

Figure 18

The LWC from CDP measurements for $U_h > 45$ m s^{-1} and < 20 m s^{-1} versus MVD is shown in **a**. The Gondola that includes CDP and BCP sensors for intense icing conditions is shown in **b** on 26 August 2017 experiment. The CDP LWC versus MVD is shown in **c** for various fixed N_d conditions (colored lines; Eq. 8), representing clouds. The blue filled circles are for wind tunnel observations. LWC versus MVD for $N_d = 0.01$, 1, 100 cm^{-3} while PR = 10, 60, 120 mm h^{-1} corresponding to 1, 6, and 12 g cm^{-2} h^{-1} (**d**). Newton (1978) envelopes for 1, 6, and 12 g cm^{-2} h^{-1} are shown by black dashed lines from low to high LWC values. The envelope for severe icing (LWC > 0.2 g m^{-3}) from Politovich (1996) is shown with dark heavy solid lines. The FAA Appendix C (2009) LWC versus MVD for a TAS = 225 knots is shown with yellow shaded areas. The Mod1 nozzles (air-atomizing nozzle at lower water flow rate) shown in light-brown shaded area and standard nozzles (higher water flow rates) in darker brown shaded area are shown at the top of the figure (Steen et al 2015). Note that icing rate can be estimated using IR = dR/dt = ε LWC TAS, then, icing duration (dt) is obtained. The dt can be very large for small LWC values representing trace icing conditions

than PA_P. After $U_h > 4$–5 m s^{-1}, wind effects on PA seem very chaotic. To see how wind affects PA, collection efficiency $\varepsilon_{ce} = PA_P/PA_D$ is plotted against U_h (Fig. 22c) in which a line shows the ε_{ce} from Rasmussen et al. (2012). This figure suggests that ε_{ce} can be highly variable when 1-min observations are used. It is found that when T is close to 0 °C, ε_{ce} can change significantly because of the water content in the snowflakes. Therefore, these ε_{ce} versus wind relationships need to be tested at various T and show water content conditions.

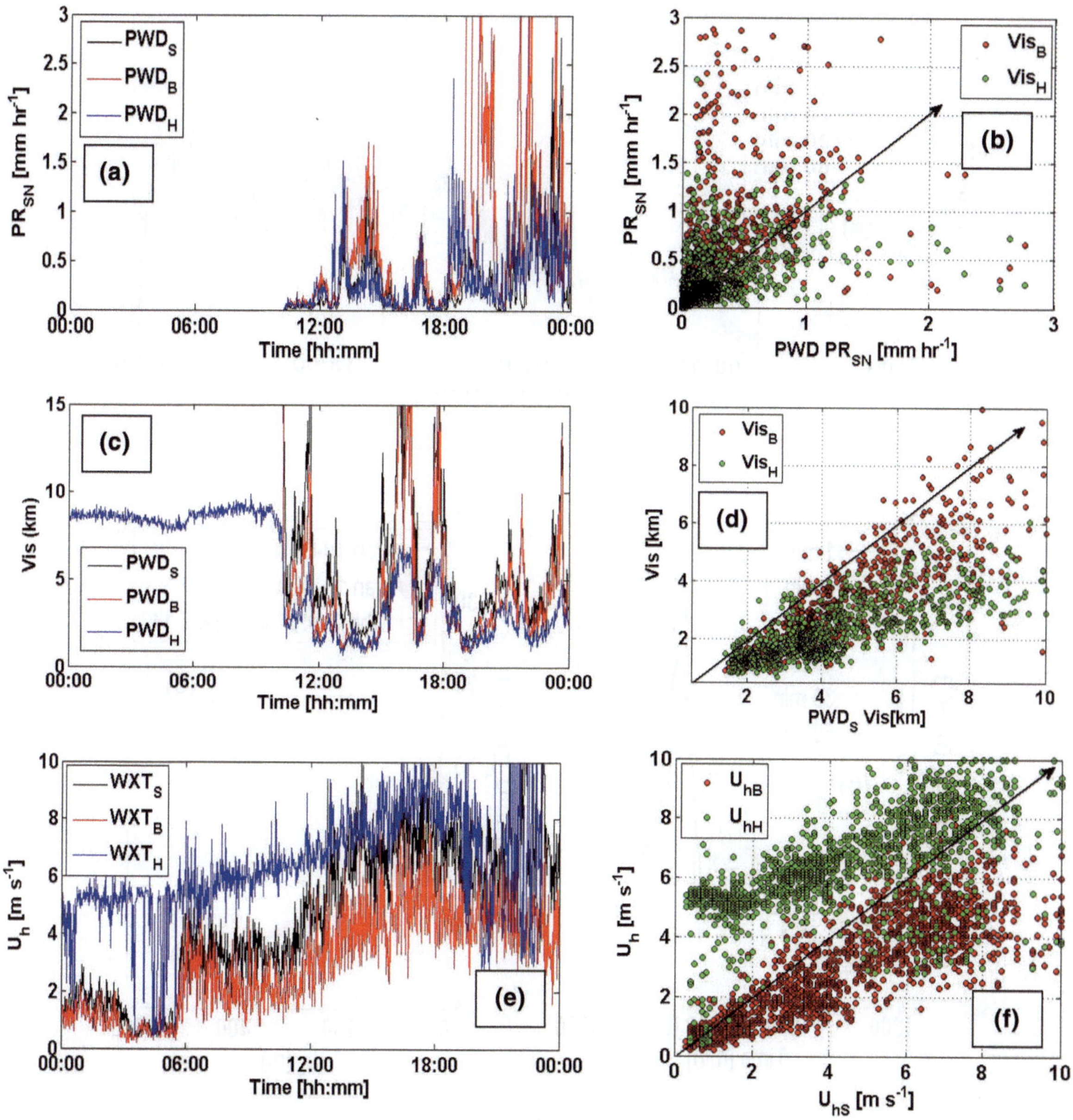

Figure 19

Time series of PR_{SN} obtained from PWD sensors located at the Supersite (PWD_S), Barn site (PWD_B), and Home site (PWD_H) are shown in **a** and scatter plots using *x* axis as supersite observations in **b**. Vis time series from the PWD are shown in **c** and scatter plots in **d**. U_h time series from the WXT520 are shown in **e** and scatter plots in **f**

5.5. Icing Environments

Aircraft icing conditions because of supercooled droplets at various sizes can be significant for in-cloud flight conditions dependent on flight duration, flight dynamics, and aircraft type. Therefore, the LWC and MVD relationships should be considered together with precipitation rates and droplet number concentrations (Fig. 18). Wind tunnel tests suggest that nozzle characteristics, e.g., size and pressure at the orifice are also important for particle microphysics and RH_w in the wind tunnel, as well. Therefore, further analysis related to droplet

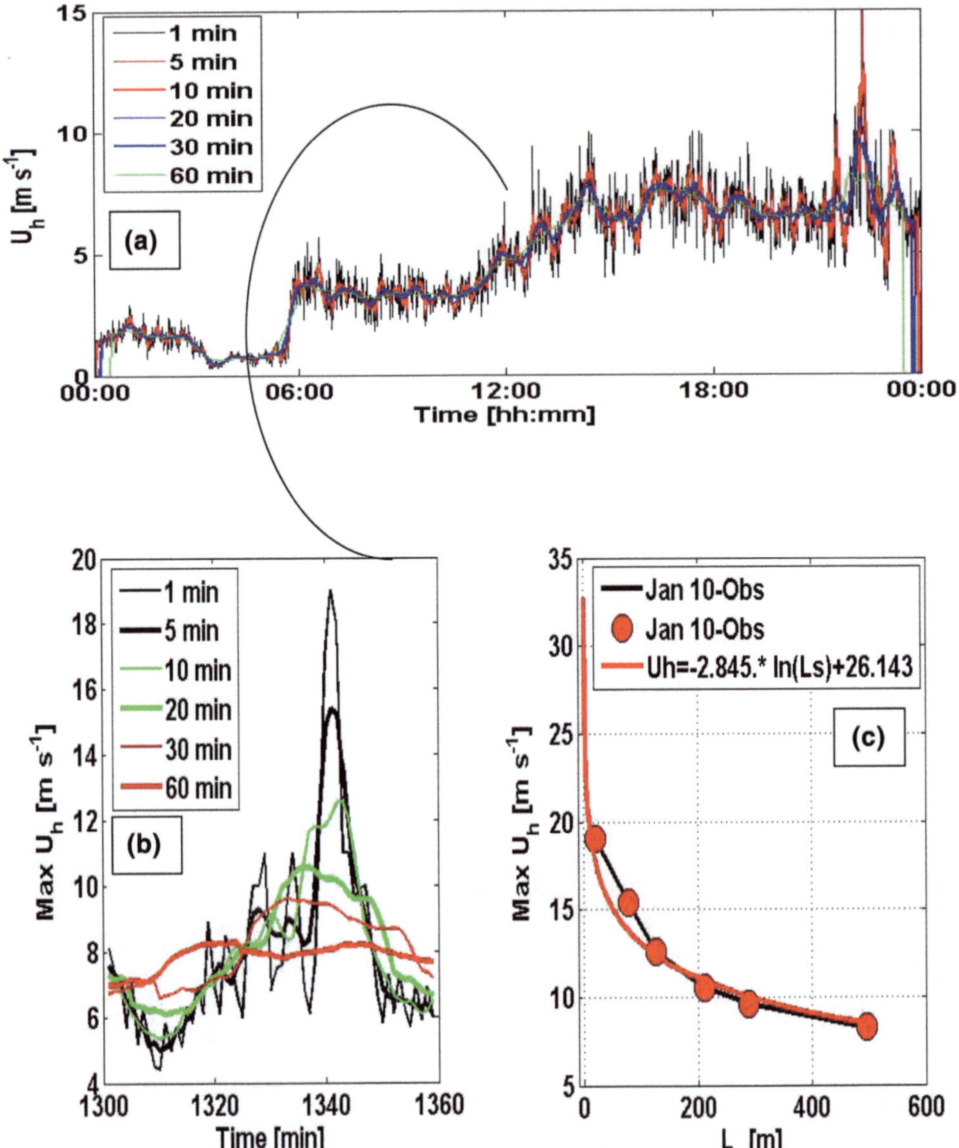

Figure 20
Time series of U_h smoothed at 1-, 5-, 10-, 20-, 30-, and 60-min intervals from the PUMS supersite WXT520 measurements are shown in **a**. 1-h time period zoomed in at about 1200 UTC is shown in **b**, and max U_h versus averaged scales is shown in **c**

microphysics and wind tunnel dynamical environments should be performed for generating stable microphysical conditions that can be utilized to define the icing environments. Note that wind tunnel tests should also be evaluated based on real flight conditions.

Overall, issues related to sensor calibrations and to both physical and dynamical environmental conditions affect the quality of measurements, and therefore, microphysical and dynamical parameterizations of icing conditions should be evaluated carefully. In this respect, because of the complexity of icing research environments, further in situ field and climate/cloud chambers projects are needed. In line with more evaluation and understanding of the physical processes, improved replication of real environmental

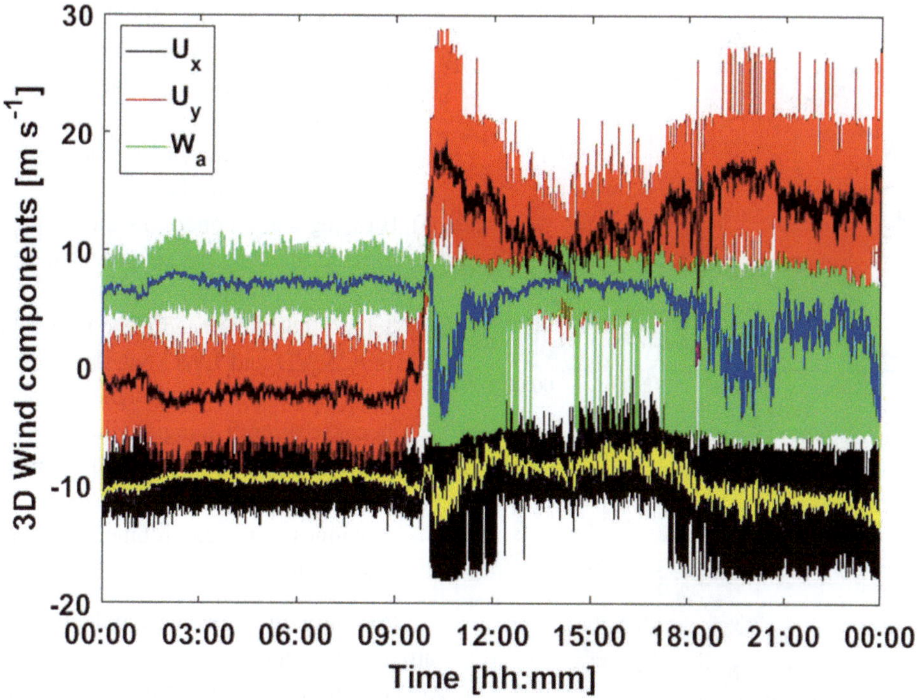

Figure 21
Time series of 3D-wind (U_x, U_y, w_a) components from Young sonic anemometer are shown. The bluish colors represent 1-min averages of the data. The quick jump at about 1000 UTC is due to strong wind gust occurrence

conditions using the UOIT Climatic Wind Tunnel can provide specific applications for aviation and other Industrial applications; therefore, developments in environmental processes, in turn, can lead to advances for the regulatory testing and safety.

6. Conclusions

This work summarizes important weather processes based on in situ observations that affect an aircraft flying environment. The various case studies on visibility, precipitation, blowing snow, wind gust, eddy dissipation, icing, and fog are studied. In addition, a new flight platform known as WE-UAV was introduced for boundary layer weather research. Most of the issues related to measurements and parameterizations of meteorological processes are given to be considered in a future aviation meteorology research and operations. The parameterizations given here should be considered cautiously, because they represent specific conditions and more cases are needed to make concrete conclusions.

The following conclusions can be drawn from this work:

(a) The measurements of small fog particles with sizes < 30 μm at cold temperatures are difficult due to instrument related issues and low-level flying restrictions that prevent aircraft flying within the surface boundary layer. However, unmanned aerial vehicles (UAVs) such as WE-UAV can be operated safely to measure ice nuclei (IN) number concentration that can be used for ice fog visibility parameterizations.
(b) The current supersite can help to develop integrated measurement systems for aviation applications and provide physical parameterizations.
(c) The use of WE-UAV together with fog and snow observing system (FSOS) towers located at satellite sites for fog and light snow research can improve the variation analysis of precipitation.
(d) The FSOS tower developed can be used for precipitation and visibility measurements over

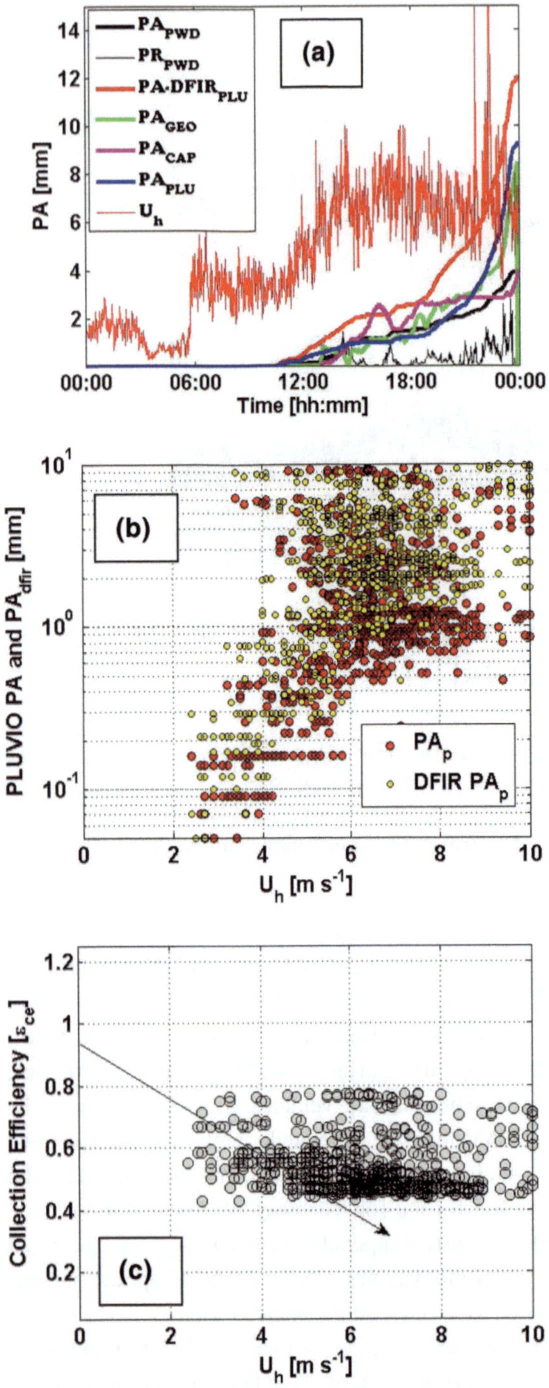

Figure 22

Time series of precipitation amount (PA) from various sensors that includes PWD, Geonor (Geo), Climatronics Capacitor (CAP), Pluvio (PLU), Double Fenced Intercomparison Reference (DFIR) sheltered Pluvio, and horizontal wind speed (U_h) (**a**). Precipitation rate (PR) is from PWD sensor. Scatter plot of Pluvio PA versus U_h is shown in **b**. Collection efficiency versus U_h is shown in **c**. The solid line is from Rasmussen et al. (2012)

various scales that allow us to validate BL processes derived from an NWP model.

(e) The new wind gust calculation and eddy dissipation rate classification using 1 Hz data, as well as vertical air velocity, are needed for aircraft safety during takeoff and landing.
(f) Blowing snow conditions are strongly a function of wind intensity but also particle density and that needs to be further researched
(g) Icing algorithms need to be researched using not only MVD and LWC but also with precipitation rates based on well-designed wind tunnel tests, because aircraft icing studies in a lab are strong functions of nozzle physical characteristics such as pressure, flow rate, nozzle size, and the dynamics of the environment such as wind speed and turbulence.

Presently, the supersite and the three FSOS towers around it represent observations collected over 3 years. In the future, the processes studied here will also be supported using similar additional cases and provide more meaningful statistics that represent various extreme conditions related to temperature, heavy snow, icing, visibility, and wind gust conditions. The FSOS towers are highly portable platforms for cold and extreme weather research, and its measurements can be valuable for areal validations of NWP simulation outputs.

Acknowledgements

The authors gratefully acknowledge the major funding for this work provided by Environment and Climate Change of Canada, Toronto, Ontario, and Canadian Defense Forces Search and Rescue (SAR) Office. The technical support provided by the staff of the Automobile Center of Excellence (ACE) at the University of Ontario Institute of Technology (UOIT) is also acknowledged.

References

Bates, T. S., Quinn, P. K., Johnson, J. E., Corless, A., Brechtel, F. J., Stalin, S. E., et al. (2013). Measurements of atmospheric aerosol vertical distributions above Svalbard, Norway, using unmanned aerial systems (UAS). *Atmosphere Measurements Technique, 6,* 2115–2120.

Boer, G. D., Palo, S., Argrow, B., LoDolce, G., Mack, J., Gao, R.-S., et al. (2017). The pilatus unmanned aircraft system for lower atmospheric research. *Atmosphere Measurements Technique Discussion, 8,* 11987–12023.

Boudala, F.S., Gultepe, I., & Wu, D. 2018. The performance of commonly used surface based instruments for measuring visibility, cloud ceiling, and humidity at Cold Lake, Alberta. *Pure And Applied Geophysics* **(submitted)**.

Cho, J. Y. N., Newell, R. E., Anderson, B. E., Barrick, J. D. W., & Lee Thornhill, K. (2003). Characterizations of tropospheric turbulence and stability layers from aircraft observations. *Journal of Geophysics Research, 108*(D20), 8784–8793.

Chun, H., Kim, J., Lee, D., Kim, S., Strahan, M., Pettegrew, B., et al. (2017). Research collaborations for better predictions of aviation weather hazards. *Bulletin of the American Meteorological Society, 98,* ES103–ES107. https://doi.org/10.1175/BAMS-D-17-0010.1.

Darby, L. S., & Poulos, G. S. (2006). The evolution of lee-wave–rotor activity in the lee of Pike's Peak under the influence of a cold frontal passage: Implications for aircraft safety. *Monthly Weather Review, 134,* 2857–2876.

Dines, W. H. (1917). Meteorology and aviation. *Monthly Weather Review, 45,* 401.

FAA Appendix C. 2009. *Atmospheric icing conditions*. Doc. No. 4080, 29 FR 17955, Dec. 18, 1964, Amdt. 25-129, 74 FR 38340, Aug. 3, 2009, pp. 536–543.

Fernando, H. J., Pardyjak, E. R., Di Sabatino, S., Chow, F. K., De Wekker, S. F., Hoch, S. W., et al. (2015). The MATERHORN: Unraveling the intricacies of mountain weather. *Bulletin of the American Meteorological Society, 96,* 1945–1967.

Gultepe, I., Burrows, W., Pavolonis, M., Milbrandt, J., Hansen, B., & Victor, C. (2016a) Marine fog visibility: measurements and forecasting during SAR projects. Session 20300 - fog or low visibility in Atlantic Canada, CMOS 2016 annual meeting, Fredericton, NS, Canada.

Gultepe, I., Fernando, H. J. S., Pardyjak, E. R., Hoch, S. W., Silver, Z., Creegan, E., et al. (2016b) An overview of the MATERHORN fog project: observations and predictability. *Pure and Applied Geophysics, 173*(9), 2983–3010. https://doi.org/10.1007/s00024-016-1374-0.

Gultepe, I., Heymsfield, A. J., Field, P. R., & Axisa, D. (2017a) Ice-phase precipitation. *Meteorological Monographs, 58,* 6.1–6.36. https://doi.org/10.1175/AMSMONOGRAPHS-D-16-0013.1.

Gultepe, I., Heymsfield, A. J., Gallagher, M., Ickes, L., & Baumgardner, D. (2017b) Ice fog: the current state of knowledge and future challenges. *Meteorological Monographs, 58,* 4.1–4.24, https://doi.org/10.1175/AMSMONOGRAPHS-D-17-0002.1.

Gultepe, I., Heymsfield, A. J., Fernando, H. J. S., Hoch, S. W., Pardyjak, E., Boudala, F., et al. (2017c) UAV applications for atmospheric profiling and ice fog. *EGU Annual Meeting*, Vienna, Austria. Oral presentation. AS1.5.

Gultepe, I., Kuhn, T., Pavolonis, M., Calvert, C., Gurka, J., Heymsfield, A. J., et al. (2014) Ice fog in arctic during FRAM–ice fog project: aviation and nowcasting applications. *Bulletin of the American Meteorological Society, 95,* 211–226.

Gultepe, I., & Milbrandt, J. A. (2010) Probabilistic parameterizations of visibility using observations of rain precipitation rate, relative humidity, and visibility. *Journal of Applied Meteorology and Climatology, 49,* 36–46.

Gultepe, I., Müller, M. D., & Boybeyi, Z. (2006) A new warm fog parameterization scheme for numerical weather prediction models. *Journal of Applied Meteorology, 45,* 1469–1480.

Gultepe, I., Pearson G., Milbrandt, J. A., Hansen, B., Platnick, S., Taylor, P., et al. (2009) The fog remote sensing and modeling (FRAM) field project. *Bulletin of the American Meteorological Society, 90,* 341–359.

Gultepe, I., Rabin, R., Ware, R., & Pavolonis, M. (2016c) Light snow precipitation and effects on weather and climate. *Advances in Geophysics, 57,* 147–210.

Gultepe, I., & Starr, D. O. (1995) Dynamical structure and turbulence in cirrus clouds: aircraft observations during FIRE. *Journal of the Atmospheric Sciences, 52,* 4159–4182.

Gultepe, I., Tardif, R., Michaelides, S. C., Cermak, J., Bott, A., Bendix, J., et al. (2007) Fog research: a review of past achievements and future perspectives. *Pure and Applied Geophysics, 164,* 1121–1159.

Gultepe, I., Zhou, B., Milbrandt, J., Bott, A., Li, Y., Heymsfield, A. J., et al. (2015) A review on ice fog measurements and modeling. *Atmospheric Research, 151,* 2–19.

Higgins, W., Ahijevych, D., Amador, J., Barros, A., Berbery, E. H., Caetano, E., et al. (2006). The NAME 2004 field campaign and modeling strategy. *Bulletin of the American Meteorological Society, 87,* 79–94.

Hoover, J. D., Doesken, N., Elder, K., Laituri, M., & Liston, G. E. (2014). Temporal trend analyses of alpine data using north american regional reanalysis and in situ data: Temperature, wind speed, precipitation, and derived blowing snow. *Journal of Applied Meteorology and Climatology, 53,* 676–693. https://doi.org/10.1175/JAMC-D-13-092.1.

Hu, H., Zhang, Q., Xie, B., Ying, Y., Zhang, J., & Wang, X. (2014). Predictability of an advection fog event over North China. Part I: Sensitivity to initial condition differences. *Monthly Weather Review, 142,* 1803–1822.

Jacobs, A. J., & Maat, N. (2005). Numerical guidance methods for decision support in aviation meteorological forecasting. *Weather And Forecasting, 20,* 82–100.

Jeck, K. R. 2002: *Icing design envelopes (14 CFR Parts 25 and 29, Appendix C) converted to a distance-based format.* DOT/FAA/AR-00/30, Office of Aviation Research, Washington DC 20591. US Department of Transportation FAA, p. 48.

Jonassen, M. O., Ólafsson, H., Ágústsson, H., Rögnvaldsson, Ó., & Reuder, J. (2012). Improving high-resolution numerical weather simulations by assimilating data from an unmanned aerial system. *Monthly Weather Review, 140,* 3734–3756.

Kantor, M., Baker, J., & Friday, E. W. 1996: *AVIATION WEATHER OBSERVATIONS for Supplementary Aviation Weather Reporting Stations (SAWRS). Manual Observations.* National Weather Service (NWS) Observing Handbook No 8, p. 106.

Keller, T. L., Trier, S. B., Hall, W. D., Sharman, R. D., Xu, M., & Liu, Y. (2015). Lee waves associated with a commercial jetliner accident at Denver international airport. *Journal of Applied Meteorology and Climatology, 54,* 1373–1392.

Kim, J., & Chun, H. (2011). Statistics and possible sources of aviation turbulence over South Korea. *Journal of Applied Meteorology and Climatology, 50,* 311–324. https://doi.org/10.1175/2010JAMC2492.

Knuth, S. L., Tripoli, G. J., Thom, J. E., & Weidner, G. A. (2010). The influence of blowing snow and precipitation on snow depth change across the Ross ice shelf and Ross sea regions of Antarctica. *Journal of Applied Meteorology and Climatology, 49,* 1306–1321.

Lewis, W. 1947. *A flight investigation of the meteorological conditions conducive to the formation of ice on airplanes.* NACA TN 1393.

Li, L., & Pomeroy, J. W. (1997). Estimates of threshold wind speeds for snow transport using meteorological data. *Journal of Applied Meteorology and Climatology, 36,* 205–213.

Long, C. N., & Ackerman, T. P. (2000). Identification of clear skies from broadband pyranometer measurements and calculation of downwelling shortwave cloud effects. *Journal of Geophysical Research, 105,* 15609–15626.

Mandel, E. (1975). An early look at the development of an unmanned automated surface aviation weather observation system. *Bulletin of the American Meteorological Society, 56,* 979–982.

McFarquhar, G. M., Baumgardner, D., Bansemer, A., Abel, S. J., Crosier, J., French, J., et al. (2017). Processing of ice cloud in situ data collected by bulk water, scattering, and imaging probes: fundamentals, uncertainties, and efforts toward consistency. *Meteorological Monographs, 58,* 11.1–11.33. https://doi.org/10.1175/AMSMONOGRAPHS-D-16-0007.1.

McFarquhar, G. M., Ghan, S., Verlinde, J., Korolev, A., Strapp, J. W., Schmid, B., et al. (2011). Indirect and semi-direct aerosol campaign. *Bulletin of the American Meteorological Society, 92,* 183–201.

Newton, D. W. (1978). An integrated approach to the problem of aircraft icing. *Journal of Aircraft, 15,* 374–381.

NWS. 1996. *Aviation weather observations.* October 1996. NWS Observing Handbook, No. 8. US Department of Commerce, p. 15.

Pavolonis, M. J. 2010. *GOES-R Advanced Baseline Imager (ABI) algorithm theoretical basis document for cloud type and cloud phase, version 2.0.* NOAA/NESDIS/STAR, 86 pp., Available from University of Wisconsin, Madison, WI, http://www.goes-r.gov/products/ATBDs/baseline/Cloud_CldType_v2.0_no_color.pdf.

Politovich, M. K. (1996). Response of a research aircraft to icing and evaluation of severity indices. *Journal of Aircraft, 33,* 291–297.

Pu, Z., Chachere, C., Pardyjak, E., Hoch, S., & Gultepe, I. (2016). Prediction of ice fog with WRF model: performance and early evaluation during MATERHORN-Fog. *Pure and Applied Geophysics, 173,* 3165–3186.

Rasmussen, R., Baker, B., Kochendorfer, J., Meyers, T., Landolt, S., Fischer, A. P., et al. (2012). How well are we measuring snow: The NOAA/FAA/NCAR winter precipitation test bed. *Bulletin of the American Meteorological Society, 93,* 811–829.

Rasmussen, R. M., Vivekanandan, J., Cole, J., Myers, B., & Masters, C. (1999). The estimation of snowfall rate using visibility. *Journal of Applied Meteorology, 38,* 1542–1563.

Reineman, B. D., Lenain, L., & Melville, W. K. (2016). The use of ship-launched fixed-wing UAVs for measuring the marine atmospheric boundary layer and ocean surface processes. *Journal of Atmosphere Oceanic Technology, 33,* 2029–2052.

Rogers, E., Carley, J. R., Ferrier, B. S., Aligo, E., Gayno, G., Janjic, Z., Lin, Y., Liu, S., Lou, G. P., Pyle, M., Wu, W. S., Wu, Y., & DiMego, G. J. 2017. Mesoscale modeling development at the national centers for environmental prediction: version 4 of the NAM forecast system and scenarios for the evolution to a high-resolution ensemble forecast system, 3B.4. *28th Conf. on Numerical Weather Prediction, Seattle, WA,* January 2017.

Roquelaure, S., Tardif, R., Remy, S., & Bergot, T. (2009). Skill of a ceiling and visibility local ensemble prediction system (LEPS) according to fog-type prediction at Paris-Charles de Gaulle Airport. *Weather Forecasting, 24,* 1511–1523.

Sharman, R., & Lane, T. (2016). *Aviation turbulence: Processes, detection, prediction* (p. 523). Berlin: Springer.

Sharman, R., & Pearson, J. M. (2017). Prediction of energy dissipation rates for aviation turbulence. Part I: Forecasting nonconvective turbulence. *Journal of Applied Meteorology and Climatology, 56,* 317–337.

Sharman, R., Trier, S. B., Lane, T. P., & Doyle, J. D. (2012). Sources and dynamics of turbulence in the upper troposphere and lower stratosphere: A review. *Geophysical Research Letters, 39,* L12803. https://doi.org/10.1029/2012GL051996.

Smirnova, T. G., Benjaminm S. G., & Brown, J. M., et al. 2000. Case study verification of RUC/MAPS fog and visibility forecasts. Preprints, *9th Conference on Aviation, Range, and Aerospace Meteorology, American Meteorology Society,* Orlando, FL, 2000.

Starkweather, S., & Uttal, T. (2016). Cyberinfrastructure and collaboratory support for the integration of arctic atmospheric research. *Bulletin of the American Meteorological Society, 97,* 917–922. https://doi.org/10.1175/BAMS-D-14-00144.1.

Steen, L. E., Ide, R. F., Zante, J. F. V., & Acosta, J. F. V. 2015. *NASA Glenn icing Research Tunnel: 2014 and 2015 cloud calibration procedures and results.* NASA/TM-2015-218758, p. 22.

Stoelinga, M. T., & Warner, T. T. (1999). Nonhydrostatic, mesobeta-scale model simulations of cloud ceiling and visibility for an east coast winter precipitation event. *Journal of Applied Meteorology, 38,* 385–404.

Thobois, L, Cariou, J. P., & Gultepe, I. 2018. Review of lidar based applications for aviation weather. *Pure and Applied Geophysics* **(submitted)**.

Turcotte, M.F. 2016. New concepts using numerical environmental Prediction to assist the Canadian armed forces, military meteorology and oceanography session 110500–CMOS 2016 Annual Meeting, Fredericton, NS, Canada.

Uttal, T., Starkweather, S., Drummond, J. R., Vihma, T., Makshtas, A. P., Darby, L. S., et al. (2016). International arctic systems for observing the atmosphere: An international polar year legacy consortium. *Bulletin of the American Meteorological Society, 97,* 1033–1056. https://doi.org/10.1175/BAMS-D-14-00145.1.

Ware, R., et al. (2003). A multichannel radiometric profiler of temperature, humidity, and cloud liquid. *Radio Science, 38*(4), 8079. https://doi.org/10.1029/2002RS002856.

Wong, W.-K., Lau, C.-S., & Chan, P.-W. (2013). Aviation model: A fine-scale numerical weather prediction system for aviation applications at the Hong Kong International Airport. *Advances in Meteorology, 2013,* 532475. https://doi.org/10.1155/2013/532475.

Wood, R., Wyant, M., Bretherton, C. S., Rémillard, J., Kollias, P., Fletcher, J., et al. (2015). Clouds, aerosols, and precipitation in the marine boundary layer: An arm mobile facility deployment. *Bulletin of the American Meteorological Society, 96,* 419–440.

Zhou, B., & Du, J. (2010). Fog prediction from a multimodel mesoscale ensemble prediction system. *Weather Forecasting, 25,* 303–322.

Zhou, B., Du, J., Gultepe, I., & Dimego, G. (2012). Forecast of low visibility and fog from NCEP: Current status and efforts. *Pure and Applied Geophysics, 169,* 895–909.

(Received January 18, 2018, revised March 29, 2018, accepted April 25, 2018, Published online May 18, 2018)

Pure Appl. Geophys. 176 (2019), 2017–2043

https://doi.org/10.1007/s00024-018-1821-1

Pure and Applied Geophysics

Applications of Geostationary Satellite Data to Aviation

GARY P. ELLROD[1] and KENNETH PRYOR[2]

Abstract—Weather is by far the most important factor in air traffic delays in the United States' National Airspace System (NAS) according to the Federal Aviation Administration (FAA). Geostationary satellites have been an effective tool for the monitoring of meteorological conditions that affect aviation operations since the launch of the first Synchronous Meteorological Satellite (SMS) in the United States in 1974. This paper will review the global use of geostationary satellites in support of aviation weather since their inception, with an emphasis on the latest generation of satellites, such as Geostationary Operational Environmental Satellite (GOES)-R (16) with its Advanced Baseline Imager (ABI) and Geostationary Lightning Mapper (GLM). Specific applications discussed in this paper include monitoring of convective storms and their associated hazards, fog and low stratus, turbulence, volcanic hazards, and aircraft icing.

Key words: Aviation weather, geostationary satellites, GOES-R, Himiwari, Meteosat, Fengyun, ABI, AHI, SEVIRI, thunderstorms, convective initiation, fog, low stratus, volcanic ash, volcanic SO_2, aircraft icing, microbursts, geostationary lightning mapper.

1. *Introduction*

Weather is the most important factor in air traffic delays in the United States' National Airspace System (NAS), accounting for more than one half of all delays lasting > 15 min (Federal Aviation Administration 2017). Geostationary (hereafter, GEO) satellites have been an effective tool for the monitoring of meteorological conditions that affect aviation operations since the launch of the operational Synchronous Meteorological Satellite (SMS) in the United States in 1974 (and the first of the subsequent series designated as the National Oceanic and Atmospheric Administration's (NOAA) Geostationary Operational Environmental Satellite (GOES)-A launched in 1975). (A series of experimental Applications Technology Satellites (ATS) had been launched by the National Aeronautics and Space Administration (NASA) between 1966 and 1974 to serve as a proof of concept.) The advantage of GEO satellites is that they frequently view the same large area of the earth's atmosphere from their equatorial vantage point, providing temporal continuity in the observation of evolving storm systems and other environmental hazards over land and ocean areas. There is an excellent coverage of tropical and mid-latitude weather from GEO satellites, but the imagery and derived products are degraded in Polar Regions due to the oblique viewing angle, requiring augmentation with polar-orbiting spacecraft such as the U.S. NOAA and the European Organization for the Exploitation of Meteorological Satellite (EUMETSAT) agency's MetOp series.

Prior to the GEO observations from space, aviation weather conditions and hazards could only be monitored using the surface and upper air observation network, land-based weather radar, and sporadic pilot reports (PIREPS). GEO satellites helped to create a more complete picture of ongoing weather throughout the airspace system. This paper will summarize the use of GEO observations for aviation applications from the earliest basic imaging spacecraft to the more advanced technology such as GOES-R.

2. *Brief History of Geostationary Satellite Technology*

Internationally, SMS and GOES in the U. S. were followed by the European METEOSAT and Japan's Geostationary Meteorological Satellite (GMS) in 1977. Coverage of the Indian sub-continent began

[1] EWxC, LLC, Granby, CT, USA. E-mail: gary.ellrod@gmail.com
[2] NOAA/NESDIS Center for Satellite Applications and Research, College Park, MD, USA.

with the Indian National Satellite (INSAT)-1B launched in 1983. China's Fengyun (FY) series of satellites began with FY-2 in 1997. The early satellites (SMS, GOES) had a simple two-band imager known as the Visible and Infrared Spin Scan Radiometer (VISSR) (Suomi and Krauss 1978) that viewed the earth in 2-km resolution visible (0.5 μm), and 8-km thermal infrared (IR 10–11 μm) spectral bands. Since the satellites were spin-stabilized, they only viewed the earth for 5% of each rotation, which reduced the potential for rapid scan imaging. The visible band was used for cloud, surface, and aerosol imaging during daytime. The lower resolution IR band observed clouds day and night but could not see aerosols unless they were exceedingly dense.

A water vapor (WV) absorption band centered near 6.5 μm was first introduced on the METEOSAT-1 launched in 1977 (Morel et al. 1978). Images from the WV band were able to observe changes in moisture patterns in the middle and upper troposphere associated with features such as jet streams, gravity waves, and upper troughs even in the absence of clouds, which was not previously possible. Beginning in the 1980s, WV band imagery became standard on all GEO satellites.

Each successive generation of GEO satellites has resulted in significant upgrades to their capabilities. A 12-band sounder, the VISSR Atmospheric Sounder (VAS), was introduced on GOES-4 in 1981. The GOES I-M series first launched in 1994 (Menzel and Purdom 1994) had a greatly improved, 19-band VAS with significant improvements in IR resolution from 7 or 14 km (depending on the detector used) to 4 and 8 km. A five-channel imager incorporated into GOES I-M (designated GOES 8–12 once in operational status) was capable of frequent (every 5–15 min), simultaneous observations at 4-km resolution in short-wave IR (3.9 μm), WV IR (6.7 μm), long-wave window (11.0 μm) and "split-window" IR (12.0 μm) bands for the first time. The three-axis stabilized GOES I-M series was capable of more efficient imaging and better quality signal to noise ratios (Menzel et al. 1998).

GOES-VAS also produced vertical thermal and moisture profiles and horizontal pressure gradient winds for input to numerical weather prediction (NWP) models and microburst prediction, as well as 8-km resolution-derived image products applicable to aviation, such as Lifted Index (LI) or cloud top pressure (CTP).

In the Eastern Asia field-of-view, Korea launched its first GEO satellite [Communication, Ocean and Meteorological Satellite (COMS)] in 2010 (Kim and Ahn 2014), with an imager that is similar to GOES-VAS and Japan's Multi-functional Transport Satellite (MTSAT) series of satellites.

While this paper will review aviation applications of prior GEO spacecraft, it will focus on advanced capabilities of the newest generations of GEO satellites: the GOES-R series in the U. S. (Schmit et al. 2005, 2017), METEOSAT Second Generation (MSG) in Europe (Schmetz et al. 2002), the advanced Himiwari in Japan (Bessho et al. 2016), and the FY-4 series in China (Yang et al. 2017). The MSG was launched in 2002 and became operational by 2004, while Himiwari and GOES-16 became operational in 2015 and December 2017, respectively. The first of China's advanced FY-4 series of GEO satellites was launched in late 2016, with several follow-on spacecraft planned through 2025. The METEOSAT Third Generation (MTG), which will be similar to GOES-R, is planned for launch by late 2021 (Bézy et al. 2005; Mohr 2014). Both MTG and FY-4 have advanced multispectral sounders onboard.

3. *GOES-16 Sensors and Applications*

3.1. *Advanced Baseline Imager (ABI)*

The Advanced Baseline Imager (ABI) on GOES-16 is a state-of-the-art instrument that is based on capabilities of several previous spacecraft including: prior GOES imagers and sounders, the Visible Infrared Imaging Radiometer Suite (VIIRS) (Hillger et al. 2013) on the polar-orbiting Suomi National Polar-orbiting Partnership (NPP), the High Resolution Infrared Sounder (HIRS) on the NOAA polar-orbiting series, and the Moderate Resolution Imaging Spectroradiometer (MODIS) onboard the NASA polar-orbiting Terra and Aqua spacecraft (Schmit et al. 2005, 2017).

Table 1 shows the spectral bands on the advanced geostationary satellites, their central wavelength

Table 1

Description of spectral bands on the advanced geostationary satellites including:, central wavelength, instantaneous geometric field-of-view (IGFOV) (km), uses in aviation forecasting and hazard detection, and availability on the GOES-16 ABI, Himiwari AHI and MSG SEVIRI

Central wavelength (μm)	Band type	IGFOV (km)	Aviation Uses	GOES ABI	Himiwari AHI	Meteosat/MSG SEVIRI	FY-4
0.47	Visible (blue)	1	Daytime clouds, fog, smoke, haze, volcanic ash	✓	✓		✓
0.51	Visible (green)	1	" " "		✓		
0.64	Visible (red)	0.5	" " "	✓	✓	✓	✓
0.86	Near-IR	1	" " "	✓	✓	✓	✓
1.37	Near-IR	2	Daytime thin cirrus, contrails	✓			✓
1.6	Near-IR	1	Daytime cloud phase discrimination (aircraft icing)	✓	✓	✓	✓
2.2	Near-IR	2	Cloud particle size, volcano "hot spots"	✓	✓		✓
3.9	IR	2	Fog and low clouds at night, volcanic ash	✓	✓	✓	✓
6.2	IR	2	Upper level water vapor, jet streams, clear air turbulence (CAT)	✓	✓	✓	✓

Table 1

continued

6.9	IR	2	Mid-level water vapor, jet streams, CAT, mountain waves, microbursts	✓	✓		✓[a]
7.3	IR	2	Lower level water vapor, jet streams, CAT, mountain waves, volcanic SO2, severe storms	✓	✓	✓	
8.4	IR	2	Cloud top phase, volcanic ash	✓	✓	✓	✓
9.6	IR	2	Ozone detection, CAT	✓	✓	✓	
10.3	IR	2	Cloud tops	✓	✓		✓
11.2	IR	2	Cloud tops, volcanic ash, fog	✓	✓	✓	✓
12.3	IR	2	Cloud tops, volcanic ash	✓	✓	✓	✓
13.3	IR	2	Cloud motion tracers, volcanic ash, cloud heights for Automated Surface Observing System (ASOS)	✓	✓	✓	✓

Slight differences may exist in wavelengths of equivalent bands between the spacecraft

[a]FY-4 Midwave IR spans the 6.9–7.3 μm wavelengths

(μm), nominal satellite instantaneous geometric field-of-view (IGFOV, km), sample aviation uses, and availability of each band on ABI, the Advanced Himiwari Imager (AHI), MSG Spinning Enhanced Visible and Infrared Imager (SEVIRI) and the FY-4 Advanced Geostationary Radiation Imagery (AGRI). The AHI is identical to the ABI except for the inclusion of a "green" visible band at 0.51 μm, which, along with the "blue" (0.47 μm) and "red" (0.64 μm) bands, allows the creation of "true color" imagery for easier interpretation of surface features (e.g., Miller et al. 2016). A simulated true color product is created using ABI by incorporating the "blue" and "red" visible image data with corrected "green" channel data from AHI to create a lookup table (Miller et al. 2016). Both the ABI and AHI have IGFOVs of 0.5 km for the "red" visible band (0.64 μm), 1.0 km for the "blue" visible band (0.47 μm) and two near-IR bands, and 2 km for two additional near-IR and ten IR bands.

The current MSG SEVIRI features 12 spectral bands with resolutions of 1 km for visible and 3 km for IR. The MTG series will have advanced capabilities, including a 16-band Flexible Combined Imager (FCI) similar to ABI and AHI, and a Lightning Imager (LI) on one platform, and separate IR and UltraViolet (UV) sounders on another spacecraft (Bézy et al. 2005; Mohr 2014). Specific applications of some of the spectral bands will be described in more detail in later sections.

In addition to more spectral bands and higher resolution, GOES ABI will be able to sample the earth at a much higher rate (Schmit et al. 2017). The full disk image, previously only sampled once every 3 h, will now be possible every 15 min. The Continental U. S. scan (and its equivalent for the west GOES spacecraft) will be 5-min interval, and two so-called mesoscale sectors (about 1000 × 1000 km) will be available once per minute (Schmit et al. 2015). This is especially important for aviation weather support, to detect rapidly changing conditions and provide timely warnings to aircraft.

Even more significantly, level 1b radiances will be directly assimilated into numerical weather prediction (NWP) models and used for model verification (Schmit et al. 2017). In addition, Schmit et al. (2008) suggest that data from the Cross-track Infrared Sounder (CrIS) and other polar-orbiting IR sounders in conjunction with the finer spatial resolution ABI data will provide a useful substitute for current sounder information for radiance uses within NWP. In effect, the improvements in spatial and temporal resolution of ABI radiances should positively impact NWP model data assimilation through the process of noise and error reduction.

3.2. Geostationary Lightning Mapper (GLM)

A Geostationary Lightning Mapper (GLM) was incorporated into GOES-R, the first time that an instrument of this type was placed into geostationary orbit. The GLM is based on the NASA Lightning Imaging Sensor (LIS), previously flown on the polar-orbiting Tropical Rainfall Measurement Mission (TRMM), and the Optical Transient Detector (OTD), flown on an independent low earth orbit spacecraft (Goodman et al. 2013). Lightning mappers are also planned for MTG (Bézy et al. 2005; Mohr 2014) and FY-4 (Yang et al. 2017).

The GLM has a sub-point resolution of 8 km, degrading to 14 km near the Earth's limb. The sensor unit observes each pixel in the field of view for a period of 2 ms (ms) at a visible wavelength of 777 nm (0.777 μm), which is an oxygen emission band. A lightning flash detection efficiency (DE) (or probability of detection) of 70% is required by users and it was estimated that an actual DE of close to 90% may be attainable. Detection efficiency [DE (%)] is defined as:

$$\mathrm{DE}\ (\%) = N_{\mathrm{det}}/N \times 100 \tag{1}$$

where N is the total number of lightning flashes, and N_{det} is the number of flashes detected by the instrument.

The false alarm rate (FAR) required for the GLM was 5%, where FAR is defined as:

$$\mathrm{FAR}\ (\%) = N_{\mathrm{f}}/N \times 100 \tag{2}$$

where N_{f} is the number of false detects by the instrument. FARs may be caused by noise in the data or sun glint. Since lightning flashes may be observed in adjacent pixels, groups of 2 × 2 pixels (an area of 16 × 16 km) are used to produce a "flash".

The GLM currently cannot distinguish between in-cloud (IC) and cloud to ground (CG) lightning, so it returns total lightning at each location. The main uses of GLM in aviation will be to help identify the most intense areas of convection in oceanic regions where surface-based radar data are sparse, or to help diagnose potential severe storms over both land and sea. Research using ground-based lightning detection systems has discovered that rapid increases in total lightning activity (referred to as lightning "jumps") often precede severe weather from thunderstorms (Schultz et al. 2009).

3.3. Space Weather Instruments

Among the suite of sun-pointing sensors on GOES-R are two that are used to monitor activity from solar storms that could disrupt terrestrial power grids and communication systems. Of particular importance to aviation are the Extreme Ultraviolet and X-Ray Irradiance Sensor (EXIS). Both EUV and X-ray spikes from solar flares bombard the ionosphere and an excess can lead to high-frequency (HF) blackouts that disrupt communication between aircraft and ground controllers. In the U.S., the NOAA Space Weather Prediction Center (SWPC) issues warnings of radio blackouts at low latitudes. There is also the risk of radiation exposure for aircrews that frequently fly on polar routes. Aircrews are now considered "radiation workers" by European and international agencies.

4. Specific Applications of Geostationary Satellite Data to Aviation

4.1. Convective Initiation

Organized, deep thunderstorms can be extremely disruptive to aircraft by producing enroute hazards such as hail and strong turbulence, and by forcing flights to reroute, leading to late arrivals, ground delays and airport gridlock. Identification of developing thunderstorms, especially near approaches to major airline hubs, is critical in managing air traffic flow and reducing costly inflight diversions. The earlier that developing storms are identified, the earlier controllers can make plans to accommodate traffic volume. For short duration flights, 2–6 h advanced notice is needed, with long haul flights requiring 4–8 h (Federal Aviation Administration 2017).

High-resolution visible GOES imagery, combined with IR bands can help anticipate developing storms and update short-range convective forecasts. Subjectively, when convection develops in a destabilizing air mass, there is a rapid increase in the area and brightness of cumulus clouds in visible images, along with a rapid cooling in IR imagery. Initiation of convective storms was observed to occur in GOES images along sea breeze fronts associated with land–water interfaces, or with mergers and intersections of cumulus cloud lines which may precede the first radar echoes by several hours (Purdom 1976). Once storms have formed, the leading edge of the storm-initiated mesohigh (associated with the surface gust front) is often marked by a narrow line of cumulus known as an arc line (Purdom 1973). When arc lines merge or intersect other unstable regions, the resulting surface moisture convergence may initiate new convection which can become severe (Purdom 1976).

An objective approach to predicting convective initiation (CI) (defined as cumulus cells that reach a radar reflectivity of $\geq$ 35 dBZ) was developed in the early 2000s (Mecikalski and Bedka 2006) using GOES-11 and GOES-12 data. It was later improved in a system known as Satellite Convection Analysis and Tracking System, version 2 (SATCASTv2) (Walker et al. 2012) using GOES-13. While partially successful based on evaluations in four regions east of the Rockies (combined Probability of Detection (POD) was 72% (with a high of 85% in the central U.S.) for cloud objects that were tracked), evaluations indicated that false alarm rates (FAR) were unacceptably high (48–60%). [A CI system based partially on SATCAST was also developed in Europe using data from MSG SEVIRI (Merk and Zinner 2013)]. The under-detection and false alarms for the SATCAST CI were mainly due to: (1) a mismatch of cloud object tracking and (2) sub-resolution initial cloud development that led to false indication of rapid cloud growth, and (3) cirrus cloud contamination leading to undetected CI. Also, rapid initial growth of cumulus clouds observed by satellite can be misleading, since atmospheric conditions such as a mid-layer capping inversion may not be favorable for continued development into convective storms. As expected, the performance of CI was worse at night due to IR-only data, and there were frequent instances of cloud objects being lost in the transition from daytime visible/IR combination to only IR at night.

A GOES-R version of the CI system was significantly improved when the satellite-only version was enhanced by numerical weather prediction (NWP)

Table 2

List of satellite parameters used in the GOES-R convective initiation (CI) algorithm (still under evaluation, specific values may change in future versions)

Satellite CI predictors
10.7 μm T_b (K)
13.3–10.7 μm T_b (K)
6.5–10.7 μm T_b (K)
15-min 10.7-μm cloud top cooling rate (K)
15-min change of 13.3–10.7 μm T_b (K)
Dominant cloud type at time 2
Dominant cloud type at time 1
Object size at time 1 (1 km pixels)
Object size at time 2 (1 km pixels)

data (Mecikalski et al. 2015). A total of 25 predictors were used, nine of which were satellite-based (Table 2). Two enhanced regression techniques, logistic regression (LR) and random forest (RF) were used to predict the likelihood of CI for each successive time interval. The FAR for both methods was lowered to 25–35%. A similar system has been developed in Korea based on the COMS data, with similar results (Han et al. 2015).

The GOES-R algorithm can be summarized as follows: first of all, a cloud "mask" is produced that identifies only cumulus clouds within an area of interest. The techniques involved in creating the cloud mask use visible brightness gradients, visible brightness thresholding, 10.7 μm IR brightness temperature (T_b) time-differences to screen out immature cumulus and non-cumulus clouds, and textural analysis of visible brightness values to remove low, thick stratus clouds from being incorrectly classified as cumulus. Then, low-level atmospheric motion vectors (AMVs) were determined to predict where each cumulus cloud would be located in successive images. Finally, nine satellite parameters and 16 NWP parameters are used for each time interval. The satellite parameters (shown in Table 2) are based on IR T_b values, trends of T_b, and multi-spectral T_b differences. (The original satellite-only CI algorithm had eight threshold tests). The NWP variables were derived from the 13-km rapid update (RAP) models. The output is a probabilistic value from 0 to 1 for the cloud object to reach the CI state, in which ground-based radar reflectivity reached $\geq$ 35 dBZ. Note that the GOES CI is not an operational "baseline" product, thus the procedures used to create CI such as the predictors shown in Table 2 may be further modified.

The CI "nowcasting" system is in experimental use by forecasters at the U.S. FAA, National Weather Service (NWS) Forecast Offices, the Aviation Weather Center (AWC) Testbed, and the NOAA Hazardous Weather Testbed (Mecikalski et al. 2015). It is expected that the current version of CI can provide advanced warning of developing convective storms with greater than 50% probability in the 30–50 min range on average. Recent examples of the use of CI available to operational forecasters can be seen in Gravelle et al. 2016.

The advantages of GOES-16 ABI imagery with the CI system are that both IR and visible data will have greater resolution, and the images will be more frequent (maximum 1–5 min versus the current 5–15 min). The additional spectral bands will also help in the identification of cloud top glaciation and cloud top heights.

4.2. Locating Hazardous Thunderstorms Over Land and Oceanic Areas

4.2.1 Global Convective Diagnostic (GCD)

Long overwater flights often traverse multi-layered mid-latitude frontal systems with embedded convection or tropical weather systems near the Inter-Tropical Convergence Zone (ITCZ). Avoidance of thunderstorm activity in these systems is a priority to prevent encounters with severe turbulence or icing. For pre-flight planning, international high-level Significant Weather (SIGWX) charts are issued by World Area Forecast Centers (WAFC) in London and Washington, DC, and for inflight avoidance Significant Meteorological (SIGMET) advisories are issued by numerous Meteorological Watch Offices (MWOs) and transmitted to airlines and other end-users.

Identification of oceanic deep convection uses primarily geostationary satellite imagery (IR and visible) and lightning data along with a few isolated island radar stations. Coverage using satellite data is excellent with the current network of Himiwari, GOES, and MSG. However, it is sometimes difficult to identify regions of the most active storm cells within multi-layered cold cloud systems.

Subjective methods to identify strong convection in cold cloud systems with satellite IR include the use of a threshold cloud top temperature (CTT) (i.e. 215 K) or identification of CTT gradients. An objective method using the T_b difference between 10.7 μm IR and 6.7 μm water vapor (WV) bands on GOES-12 was developed by Mosher (2001, 2002) for use at the U. S. NWS Aviation Weather Center (AWC) in Kansas City. The Global Convective Diagnostic (GCD) flagged a cloud as a possible thunderstorm if the WV–IR T_b difference was 1 °C or less. Verification of the GCD using storm height measurements with the Precipitation Radar on the Tropical Rainfall Measuring Mission (TRMM) spacecraft (Martin et al. 2008) showed that GCD significantly over-predicted the coverage of deep convection. However, performance improved when the Brightness Temperature Difference (BTD) was lowered to near 0 °C (no specific value was recommended). Also, GCD significantly out-performed the use of a single IR threshold of 215 K, except for a single case involving a Great Plains squall line.

The GCD is now used in combination with the Naval Research Laboratory's (NRL) satellite-derived Cloud Top Height (CTOP) (Donovan et al. 2008) and Cloud Classification (CC) algorithms (Tag et al. 2000), and the Earth Networks global ground-based lightning strike data (Liu and Heckman 2011) by the University Center for Atmospheric Research (UCAR) Research Applications Laboratory (RAL) to form a product known as the Convective Diagnostic Oceanic (CDO) product. A CDO index value (0–6) is created using a weighted average of the CTH, GCD, GOES-R Overshooting Tops (OT) (Bedka et al. 2010), and the Earth Networks lightning accumulation algorithm. The experimental CDO is available online at: http://www.rap.ucar.edu/projects/ocn/realtime_sys/wxHazards/. An example is shown in Fig. 1.

The description and evaluation of CDO has been completed and published by Donovan et al. (2008). The use of instruments on the Tropical Rainfall Measuring Mission (TRMM) to: (1) observe lightning (lightning imaging system), (2) precipitation intensity (precipitation radar), and (3) cloud properties [visible and infrared radiometer system (VIRS)] determined that both convective clouds with lightning (TRW) and without lightning (CLW) could be detected with a POD as high as 90% when lightning is used as a criterion for "hazardous" convective clouds. There is an ongoing effort to use these products to create the Ensemble Prediction of Oceanic Convective Hazards (EPOCH) system (Bass 2017). EPOCH real-time output of convective forecasts from 12 to 48 h in the future is viewed at the AWC but is not available to the public.

4.2.2 *Lightning Activity with Oceanic Convection*

Another important tool for thunderstorm identification in remote areas is lightning data from both time of arrival (TOA) and direction finding (DF), ground-based lightning systems first developed in the late 20th century. These systems observe electromagnetic pulses emitted by lightning with detection efficiency (DE%) of about 90% but with coverage limited to mostly continental areas. A World Wide Lightning Location Network (WWLLN) introduced in the early 2000s (Lay et al. 2004) uses the emission of very low-frequency (VLF) (1–24 kHz) radio pulses from lightning that can travel many thousands of km to expand detection coverage to oceanic areas. Location accuracy with WWLLN is good (15 km) but DE is estimated to be very low (10% for CG, 6% for total lightning). The GLM thus offers an excellent observational tool for frequent monitoring of oceanic convection in support of aviation with its near 90% DE and low FAR (Goodman et al. 2013).

An example of how GLM can identify oceanic thunderstorm activity associated with Hurricane Harvey over the western Gulf of Mexico is shown in Fig. 2. Parallax-corrected GLM lightning flashes for a 5-min period prior to a GOES visible image at 1317 UTC on 25 August 2017 is color-coded, with red being the oldest and yellow the newest lightning. Note the absence of lightning near the center of Harvey and westward toward the Texas coast. Lightning near the center of tropical cyclones (< 100 km) is typically episodic (DeMaria et al. 2012). Little or no lightning usually indicates a steady-state storm, while large lightning densities occur with either rapidly strengthening or weakening storms (DeMaria et al. 2012). Since overshooting

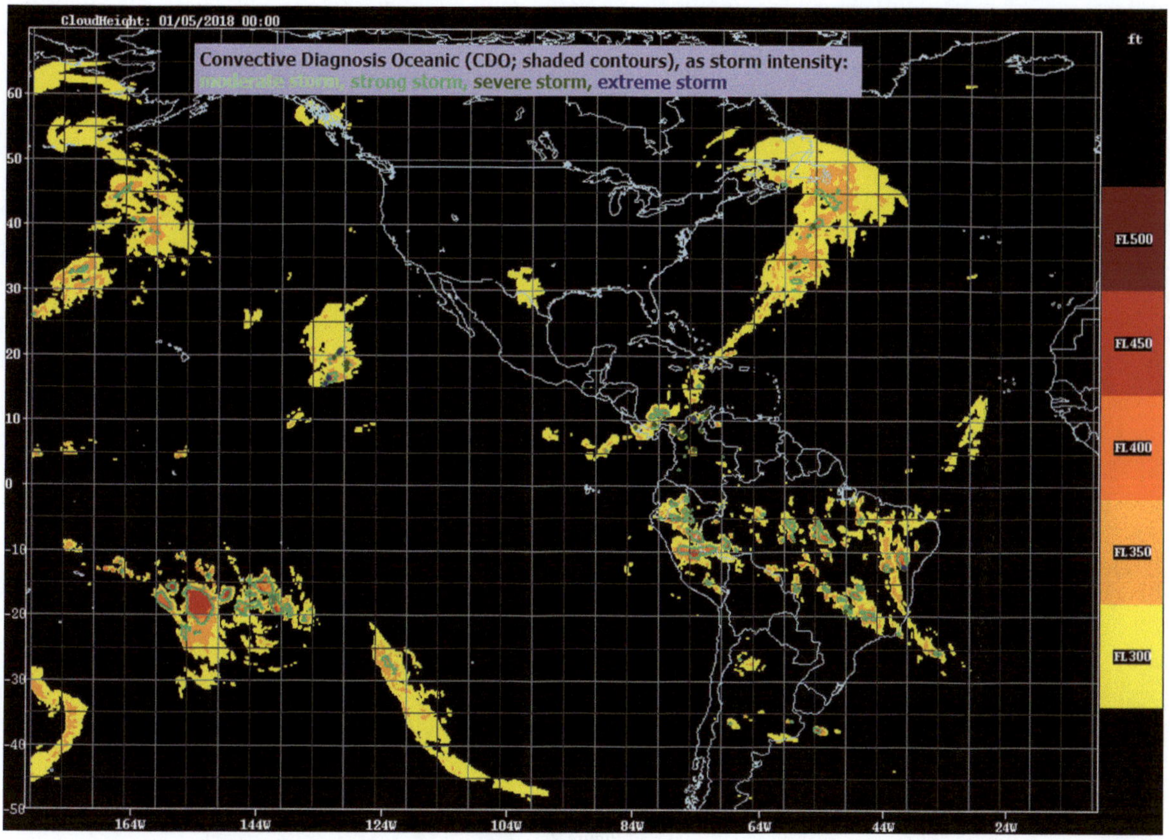

Figure 1
Convective Diagnosis Oceanic (CDO) product derived from GOES-EAST (16) and GOES-WEST (13) valid 0000 UTC, 5 January 2018. Cloud top heights are shown by the color scale at right as flight levels (100′s of feet). Storm intensity is shown by color scale in legend at the top. Only CDO values of 2 (moderate storm) through 5 (extreme storm) are displayed

cloud tops can be seen near the eye, they nevertheless represent a hazard to high-altitude flights through this area.

4.2.3 Severe Storm Identification

The identification of potentially severe convective storms over land has traditionally involved measurements of the time rate of change of CTT (K) or the growth (expansion of the area of the cirrus anvil) using both visible and IR data (Sikdar et al. 1970; Adler and Fenn 1979). Based on analysis of 5-min rapid scan data from Synchronous Meteorological Satellite (SMS)-2, storms with colder IR tops that grew more rapidly were more likely to be severe. Mean vertical velocity profiles for the severe elements were twice those of non-severe cases (Adler and Fenn 1979). Reports of tornadic activity on the ground occurred during, or immediately following, a rapid expansion of cold cloud top areas, indicating rapid ascent of the cloud. A lead time of as much as 30 min was observed. More recent work using GOES data, such as the University of Wisconsin Cloud Top Cooling (UW-CTC) algorithm (Hartung et al. 2013), shows good correlation between high rates of CTC and strong radar reflectivity (as measured by the WSR-88D in the Central U. S.), vertically integrated liquid (VIL) and maximum estimated size of hail (MESH). Lead times of 0–60 min were observed.

Features observed in individual IR images may also indicate the presence of a strong or severe convective storm. One of these is the “overshooting top” (OT) in which single or multi-pixel clusters of CTTs are significantly colder than the surrounding

Figure 2

A GOES-16 ABI Band 1 (visible) image at 1317 UTC on 25 August 2017 that shows Hurricane Harvey near the Texas coast, overlaid by GLM lightning data (non-operational, still undergoing beta-testing at the time of this writing) for a 5-min period prior to the image time. Lightning data are color-coded according to time (red is oldest, yellow is most recent). Note that the GLM data were parallax-corrected while the GOES-16 imagery was not (Source: Lindstrom 2017)

anvil cirrus. An OT by itself is most often associated with heavy rainfall (Negri and Adler 1981: Dworak et al. 2012). An example of the ability of the GOES-R (16) ABI to better detect OTs is shown in Fig. 3 for a Mesoscale Convective System (MCS) in northern Wisconsin on 16 May 2017. This feature sometimes has a V-shape, with warmer IR temperatures within the V downshear from the cold core (McCann 1983; Heymsfield and Blackmer 1988). The temperature difference in the cold-warm "couplet" varied from 7 to 17 K for severe storms. The cause of the downstream warm spot is believed to be subsidence of stratospheric air over the cloud top into the cirrus anvil, perhaps as a wave-like feature. A median lead time of 30 min was observed from identification of the enhanced-V to the occurrence of severe weather, although the low probability of detection ($\sim$ 40%) indicated that many severe storms do not have this feature (McCann 1983).

An objective, satellite-based OT detection algorithm has been developed for GOES-R using 5–30 min 11 μm IR data from GOES-12 for a 6-year period from January 2004 to December 2009 (Dworak et al. 2012). The best results (false alarms $\sim$ 15%) occurred when minimum CTT was < 200 K and when the satellite was in rapid scan (5 min interval imaging). This emphasized the importance of high-resolution, high-frequency imaging to observe rapidly evolving cloud top features. However, the GOES-R OT algorithm is not operational in the baseline product suite.

Dark bands observed along the upstream edge of strong convective storms in GOES 6.7-μm WV imagery are also good indicators of possible severe

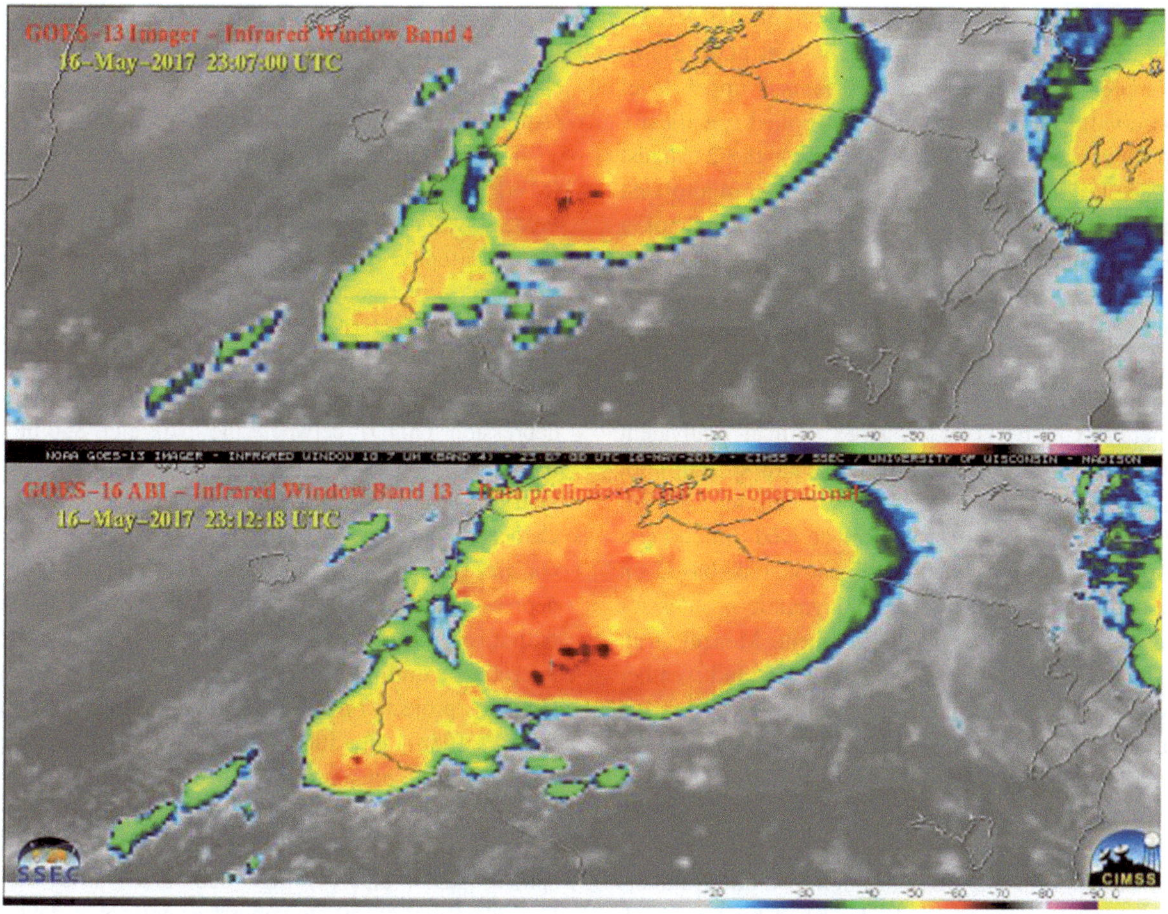

Figure 3
A comparison of GOES-13 IR (top) versus a non-operational GOES-16 IR mage (bottom) for strong storms over northern Wisconsin, USA, on 16 May 2017. The GOES-16 image is approximately 5 min later. Overshooting tops are the dark red to black circular clusters of pixels with temperatures near − 70 °C (203 K) (Source: Bachmeier 2017a)

weather (Ellrod 1990). The feature is likely caused by subsidence as strong upper level winds encounter a convective cloud with strong updrafts that act as a barrier to the flow. Of 147 incidences of this feature during the spring and early summer of 1989, 82% of the storms were severe. The predominant type of severe weather reported (83%) when this feature was present was large hail ($\geq$ 3/4″ diameter). The occurrence of the upstream dark band was most often observed in the central and northern Plains of the U. S.

The addition of the GLM on GOES provides a powerful additional tool for monitoring en route convection for severity. Rapid increases in total lightning flashes that often precede severe weather activity, known as lightning "jumps," have been observed in data from land-based Lightning Mapper Array (LMA) systems (Schultz et al. 2009). An algorithm that determined whether total lightning flash rates exceeded the average for non-severe storms by two standard deviations (2σ) or more performed the best, with POD of 87% and acceptable FARs (23–33%). This algorithm and others are in the process of evaluation for use with the GOES-R series.

4.2.4 Convective microbursts

The use of data from the GOES sounders for the nowcasting of convective storm potential commenced in the 1980s with the application of the VISSR Atmospheric Sounder (VAS) (Smith et al. 1981) to complement conventional surface and upper level observations. The VAS was first applied specifically to downburst potential assessment for the 2 August 1985 Dallas-Ft. Worth (DFW), Texas microburst storm (Ellrod 1989), which led to the crash of a commercial jetliner, killing 133. The following parameters were found to be most effective for prediction of the microburst that occurred with this case: large temperature lapse rates (°C km^{-1} from 850 to 700 hPa), high-convective available potential energy (CAPE) (J kg^{-1}), large vertical equivalent potential temperature (θe) difference (850–700 hPa), and low mid-tropospheric relative humidity (RH%). The existing GOES sounder-derived microburst products were designed to diagnose risk based on favorable environmental thermodynamic profiles for severe convective storm development. Menzel et al. (1998) describes the role, performance, and weather forecasting applications of the most recent generation of GOES sounders, as well as the thermodynamic profile generation process. (Although the GOES-R (16) instrument suite does not include a sounder, vertical temperature and humidity profiles can be generated with the ABI and high-resolution numerical models using methods described by Lee et al. (2014).)

The typical horizontal scale of a single-cell convective storm and larger downburst (or macroburst) is near 10 km (Byers and Braham 1949). Considering the 10-km spacing of sounding retrievals, the GOES-VAS was well-suited to observe horizontal variations in environmental conditions and associated parameters that indicate risk of strong winds produced by downbursts. In addition, a new version of the sounding physical retrieval algorithm (Li et al. 2008) was implemented into GOES operations in 2011.

Pryor and Ellrod (2004a, b) outlined the development of a suite of GOES sounder-derived products to assess the presence of conditions favorable for dry and wet microbursts. That study introduced the wet microburst severity index (WMSI) product to calculate potential magnitude of convective downbursts in humid environments over the eastern United States. The WMSI incorporated (surface-based) convective available potential energy (CAPE) as well as the vertical equivalent potential temperature difference ($\Delta\theta e$) between the surface and mid-troposphere (Atkins and Wakimoto 1991).

The Microburst Windspeed Potential Index (MWPI) (Pryor 2010, 2012, 2014), compared to the WMSI, was designed to quantify the most relevant factors in convective downburst generation in intermediate thermodynamic environments by incorporating: (1) surface-based CAPE, (2) temperature lapse rate between the 670 and 850 hPa levels ($|\Gamma|$), and (3) differences in dewpoint-temperature depression (ΔDD) between the 670 and 850 hPa levels. The MWPI was incorporated into a predictive linear model developed in the manner exemplified in Caracena and Flueck (1988). The MWPI formula consists of a set of predictor variables (i.e. dewpoint depression, temperature lapse rate) that generates output of expected microburst risk. Analysis of microbursts during the Joint Airport Weather Studies (JAWS) project (Wakimoto 1985; Caracena and Flueck 1988) identified the following favorable environmental characteristics for high-plain dry microbursts: (1) low-surface dewpoint temperatures, (2) high-convective cloud base, (3) small mid-tropospheric ΔDD, and (4) high sub-cloud lapse rate. Consideration of 50 downburst events over Oklahoma and Texas during the summer of 2009 revealed that for the majority of downburst events, CAPE was greater than 1000 J kg^{-1} and ΔDD was greater than 5 °C. Srivastava (1987) also noted that a minimum lapse rate of 5 °C km^{-1}, associated with heavy precipitation and a high reflectivity factor (> 50 dBZ), was necessary for intense downdraft generation. Thus, the scaling factors of 1000 J kg^{-1}, 5 °C km^{-1}, and 5 °C, respectively, are applied to the MWPI algorithm to yield a non-dimensional MWPI value that expresses wind gust potential on a scale from one to five (Pryor 2015):

$$\begin{aligned}\mathrm{MWPI} \equiv & \left\{\left(\mathrm{CAPE}/1000\mathrm{Jkg}^{-1}\right)\right\} \\ & + \left\{\Gamma/5\ {}^{\circ}\mathrm{C\,km}^{-1} + \left((T - T_{\mathrm{d}})_{850}\right.\right. \\ & \left.\left. -(T - T_{\mathrm{d}})_{670}\right)/5\ {}^{\circ}\mathrm{C}\right\}\end{aligned} \quad (3)$$

The MWPI algorithm is expected to be most effective in assessing downburst wind gust potential associated with ordinary cell and multi-cell

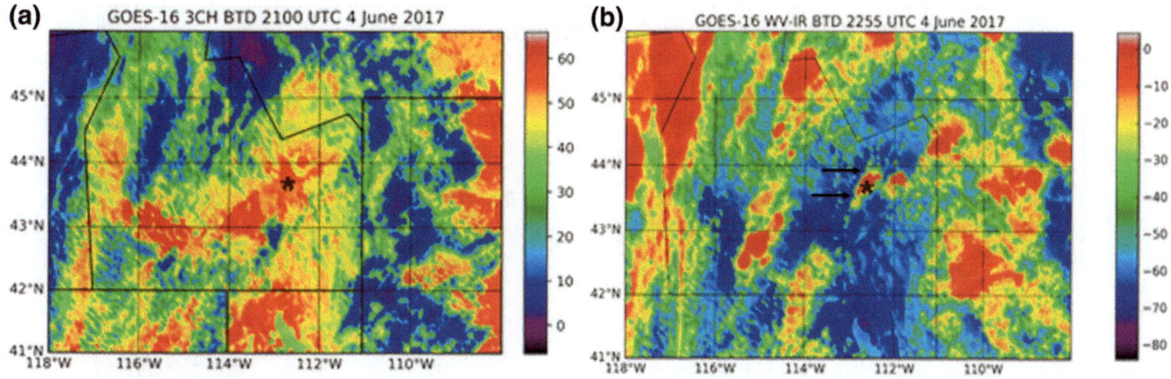

Figure 4
a GOES-16 ABI three-band BTD product image (based on bands 10 (7.3 μm), 14 (11.2 μm), and 15 (12.3 μm) at 2100 UTC, 4 June 2017 and **b** GOES-16 WV-IR BTD product image at 2255 UTC, 4 June 2017. An 'asterisk' marks the location of the Idaho National Laboratory, where a wind gust of 54 knots was recorded. Black arrows in **b** mark the location of dry-air notches (from Pryor 2017)

convective storms in weak wind shear environments. Employing 13-km resolution rapid refresh (RAP) model data as a proxy for GOES-16 ABI vertical sounding profile datasets, the MWPI and associated wind gust potential was effectively displayed for a severe downburst event over Idaho in June 2017. The reader is referred to Pryor (2017) for a detailed discussion of the convective environment and evolution of this downburst event.

More recently, the launch of GOES-16 and subsequent activation of the ABI has resulted in the increased accessibility to high spatial and temporal satellite image datasets for convective storm monitoring and nowcasting. Traditional meteorological satellite techniques for deep convective storm monitoring, including the WV–IR BTD, (Schmetz et al. (1997), have recently been extended by multi-band techniques for diagnosing attributes of a favorable downburst environment. The WV-IR BTD has demonstrated the ability to infer convective storm structure signatures (i.e. overshooting tops, lateral dry-air notches) that signify the occurrence potentially damaging straight-line winds resulting from downbursts, and, with 2-km resolution GOES-16 ABI data, the BTD product is able to detect fine detail signatures more effectively when compared to 4-km resolution GOES-13-15 imagery (Pryor 2017). Accordingly, a three-band BTD algorithm incorporating split window IR bands (11, 12 μm) and a WV band (7.3 μm) produces an output BTD that is proportional to downburst potential and supplements the sounder-derived MWPI. Figure 4 demonstrates the ability of two ABI-derived BTD products in diagnosing favorable storm structural patterns and environmental conditions for downburst occurrence in the upper Great Basin region on 4 June 2017. High thunderstorm-induced winds up to 28 ms^{-1} (54 kt) were observed at the Idaho National Laboratory (shown by an '*' in the images), where three-band BTD values of 50–60 K indicated wind gust potential of 26–31 ms^{-1} (50–60 kt).

4.3. Fog and Low Cloud Detection

4.3.1 Area coverage

Observing low clouds and fog with geostationary satellite imagery initially involved subjective pattern recognition that required considerable skill by the analyst/forecaster. The distinctive smooth, bright appearance of low clouds in daytime visible imagery makes them relatively easy to identify. At night, the problem became more difficult due to the formation of fog and low clouds beneath thermal inversions. In IR imagery, the tops of these clouds can be warmer, the same temperature, or even cooler than surface temperatures depending on the depth of the thermally mixed boundary layer.

Based on a bi-spectral approach used with AVHRR image data (Eyre et al. 1984), GOES VAS imagery was able to provide better detection of fog at night (Ellrod et al. 1989) over the Continental U. S. With the advent

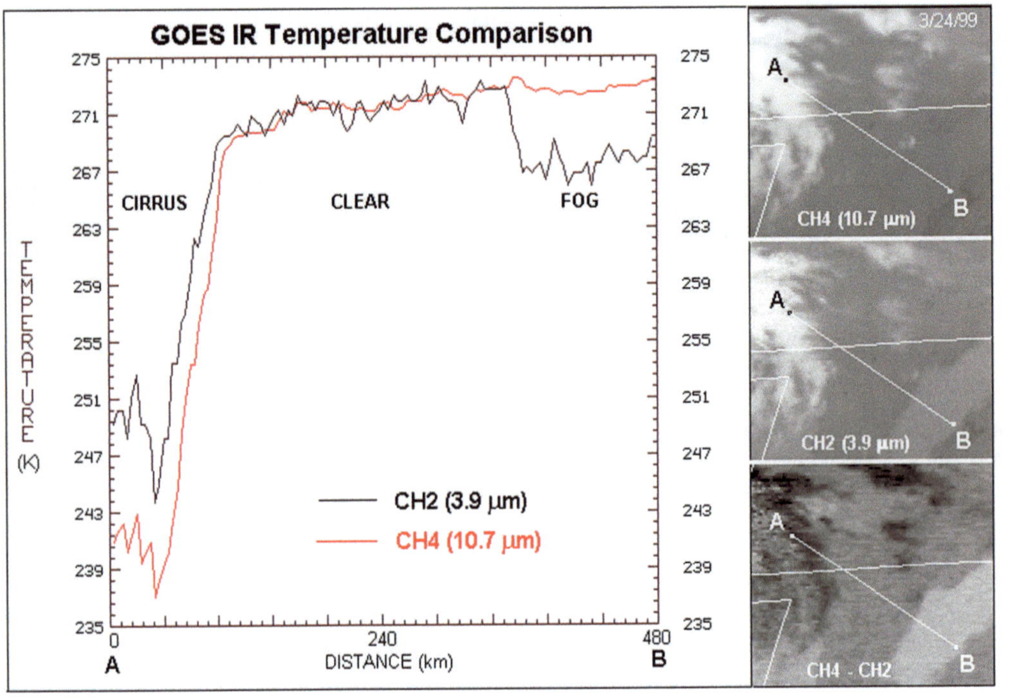

Figure 5
Comparison of IR Tb values across the transect A–B in CH2 (3.9 μm) and CH4 (10.7 μm) on 24 March 1999 showing the basis for nighttime fog imagery using GOES

of multi-spectral imager bands at 30-min intervals on GOES-8 in 1995, the use of a bi-spectral BTD between the short-wave IR (SWIR 3.9 μm) and window IR (10.7 μm) bands was shown to be far more effective in observing fog and stratus at night than single IR bands alone (e.g., Ellrod 1995; Lee et al. 1997). Image animations of the "fog product" could observe development, spread, and even dissipation of disruptive low clouds. The physical basis of the technique is that the radiative emissivity in the SWIR is lower than that of the LWIR, thus resulting in detection of cooler sub-cloud conditions with SWIR, and a BTD of ~ 1 to 4 °C. Figure 5 shows an area of low clouds and cirrus in Oklahoma in IR, and SWIR bands, along with a temperature comparison in the two IR bands along the line A–B.

One of the shortcomings of earlier geostationary systems is that the limited resolution (4–8 km) in the IR bands resulted in poor detection of valley fog. The GOES-R ABI and Himiwari AHI provide further upgrades for this product, with 2-km IR resolution and faster repeat scans (1–5 min). Figure 6 shows a comparison of the GOES-16 fog product versus one from GOES-13 for the same area of the Eastern U.S. on 17 October 2017. The improved quality of GOES-16 is evident due to the improved resolution and better precision (lower instrument noise).

4.3.2 Estimating Time of Clearing

The time required for the clearing or lifting of fog and low clouds is an important factor in aviation forecasting. Certain airports (especially San Francisco, Seattle, and Los Angeles in the U. S.) are especially prone to weather delays caused by fog. Early attempts to forecast dissipation of fog using GOES were based on the digital value of the brightness difference between the fog and adjacent cloud-free areas, a measure of fog thickness (Gurka 1978). While this method was effective, it could not be used until about 1 h after sunrise.

An IR enhancement technique developed using GOES sounder data from 1989 to 1992 based on SWIR-IR BTD values (Ellrod 1995) resulted in pre-dawn estimates of fog or stratus thickness which could then be used to determine approximate fog

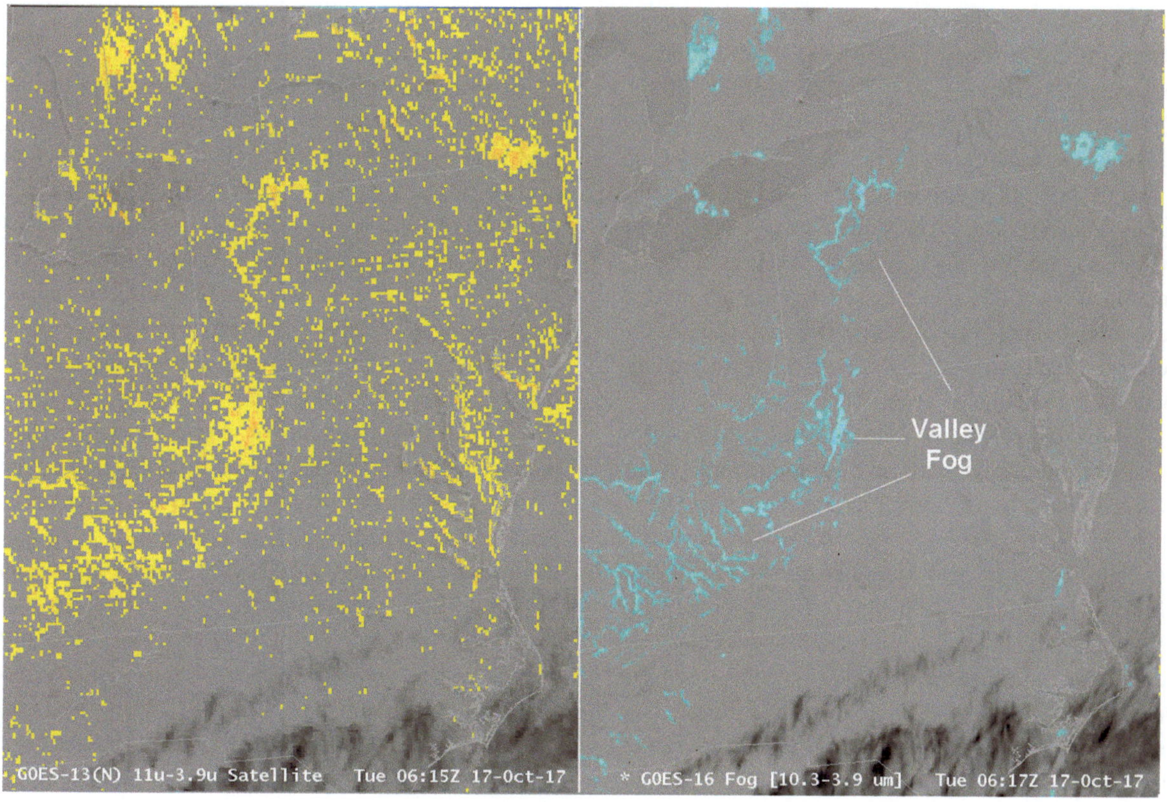

Figure 6
Nighttime fog imagery for GOES-13 (left) versus the same product for GOES-16 (right) at 0615 UTC (0215 EDT) on 17 October 2017. Valley fog is shown as yellow or cyan, cloud-free areas as gray, and cirrus is black (Source: University of Wisconsin-CIMSS 2017)

dissipation time after sunrise. Verification of this method using higher resolution GOES IR imager data from 1997 to 2001 and aircraft cloud top reports revealed a correlation of 0.625.

A case that demonstrated the potential improvements in forecasting the time of clearing of low stratus for airport operations that can be achieved using GOES-16 has been documented at San Francisco International Airport (SFO) (Eckert 2017). On the morning of March 3, 2017, a 3-h long Ground Delay Program (GDP) was initiated due to developing stratus near SFO. Based on GOES-16 time lapse visible imagery at 5-min intervals, clearing was observed to occur more rapidly than with the operational GOES-West images at 15-min intervals. This resulted in the lifting of the GDP about an hour earlier than possible using legacy GOES-West images alone, leading to a net cost saving of \$50 K to the airlines and passengers.

4.3.3 Estimating areas of IFR conditions

One of the shortcomings of satellite data in the detection of low clouds and fog is the uncertainty of the cloud base heights to determine the presence of instrument flight rule (IFR) (ceilings $<$ 1000 ft, visibilities $<$ 3 statute miles) or marginal visual flight rule (MVFR) conditions. There are pattern recognition techniques that provide clues, such as the texture of the cloud tops in IR and visible data, and cloud shadows. A low cloud base (LCB) product that combines the IR fog imagery with the surface temperature from ground observation sites was obtained to estimate areas where IFR conditions may exist (Ellrod 2002; Ellrod and Gultepe 2007). If the GOES IR cloud top minus surface shelter temperatures were $\leq$ 3 K and a low cloud was present in the GOES product, IFR ceilings ($<$ 1000 ft) were considered likely. This product was verified for all regions of the U.S. between May and August 2000 in comparison with > 1500 Meteorological Aeronautical Reports (METARs) (Ellrod 2002). Cases where the low clouds were obscured by higher cloud layers were not verified. The POD ranged from a low of 58% in the Northeast to a high of 77% in the upper Midwest. The FAR varied between 15% in the Northeast to 4% in the South Central U.S. A modification of this scheme used operational numerical model data over southern Ontario, Canada, to remove the effects of mid and high clouds to produce better results (Gultepe et al. 2007).

GOES bi-spectral IR fog and low cloud imagery has been combined with surface temperatures and Rapid Refresh (RAP) model data to obtain probabilities of IFR (ceilings 500–1000 ft), low IFR (ceilings < 500 ft), and Marginal Visual Flight Rule (MVFR) (ceilings 1000–3000 ft) (Calvert and Pavolonis 2011). (At this time, there is no capability to provide information on surface visibilities from satellites.) Fog depth (km) is also determined. The requirements for these products are a fog/low clouds detection rate of 70%, and a cloud layer thickness accuracy of 500 m. The model data help to provide details on sub-cloud moisture and temperature conditions that reduce false alarms and improve detection in areas where cirrus obscures lower cloud layers. Instead of BTD, the GOES-R algorithms use a parameter called pseudo-emissivity (EMS) for the 3.9-μm data. EMS is the ratio of the observed 3.9-μm radiance (numerator) and the 3.9-μm blackbody radiance calculated using the 11-μm brightness temperature (denominator). The advantage of using EMS over BTD is that it results in better skill scores because it is less sensitive to scene temperature. These products have been generated experimentally and began using GOES-R (16) data in late 2017.

4.4. *Aircraft Turbulence*

4.4.1 *Upper Level Clear Air Turbulence (CAT)*

Clear air turbulence (CAT) occurs at high altitudes (6–15 km) in a nearly cloud-free atmosphere, and is usually associated with vertical wind shears near the jet stream and upper level fronts. It is one of the most common causes of accidents involving large commercial jet airliners, resulting in injuries to air crews and passengers (e.g. NTSB).

In satellite imagery, CAT has been observed in the vicinity of cirrus or moisture boundaries associated with the jet stream and frontal zones. The nature of the cirrus itself often has clues regarding the presence of CAT in the form of striations or bands that are oriented transverse to the flow (e.g. Knox et al. 2010). Wider, thicker bands have been associated with a higher risk of moderate–severe turbulence. These bands have also been observed on the periphery of mesoscale convective systems (MCS) and tropical cyclones, especially where they interact with mid-latitude or sub-tropical jet streams.

Transverse cirrus bands tend to be irregular in shape and orientation and can be seen in both IR and visible data. Evenly-spaced wave clouds known as "billows" that are caused by gravity waves and Kelvin–Helmholtz Instability (KHI) have typically been observed only in high-resolution visible imagery, but modern imagers such as AHI and ABI can be used to observe them in the IR and WV bands also. A high-pass filtering technique has been able to identify small-scale gravity waves associated with aircraft turbulence over the western Pacific Ocean using 2-km resolution WV imagery from AHI (Wimmers et al. 2018).

On the poleward side of the jet stream, the air is often clear, but 6.7-μm IR water vapor (WV) images can indicate the presence of CAT if there is progressive warming with time (usually displayed as darkening gray shades). An evaluation of more than 100 cases of WV darkening from 1983 to 1985 indicated that moderate or greater CAT was present 80% of the time (Ellrod 1985). The image darkening has been associated with strong middle and upper tropospheric subsidence that occurs in conjunction with frontogenetic processes and possibly intrusion of stratospheric air via tropopause folds.

An objective technique has been developed to identify tropopause folds using the strength of gradients in GOES Layer Average Specific Humidity (GLASH) derived from 6.7-μm WV data (Wimmers and Moody 2001, 2004a, b). The likelihood of CAT with the tropopause folds is dependent on the orientation, slope and width of the dark bands. Empirical rules have been developed to identify those tropopause folds that are likely to contain CAT (Wimmers and Feltz 2010). Verification using GOES WV imagery and automated aircraft Eddy Dissipation Rate (EDR) data was completed from Nov 2005 to Dec 2007. An accuracy of 53% for detection of moderate or greater CAT was achieved (the product requirement is at least 50%). The Tropopause Folding Turbulence Product (TFTP) is planned for future implementation with the GOES-R series.

4.4.2 Mountain waves

When moderate to strong winds blow across mountain ridges in the presence of a stable layer near the mountain tops, mountain waves occur. When moisture is present near the base of the inversion, wave cloud appearing as washboard-like patterns can be observed in satellite imagery. Even the earliest GOES imagery clearly showed these cloud patterns (Conover 1964). The wavelength of the clouds is proportional to the wind speed and instability (Fritz 1965). Wavelengths have been observed to increase during the day as diurnal heating proceeds. The longer wavelengths (observable in GOES IR imagery) are more likely to be associated with significant low-level turbulence than the shorter wavelengths (Ellrod 1985). Figure 7 shows mountain waves in the Japanese Himiwari AHI and Multi-Transport Satellite (MTSAT) WV images on 7 July 2015. Their detection was made possible by the higher resolution (2 km) of the AHI WV band.

Evidence of mountain waves at high altitudes is indicated by standing lee wave clouds (sometimes referred to as "banner clouds") composed of cirrus that extend as far as hundreds of kilometers downwind from the mountain ridges. The cirrus plumes

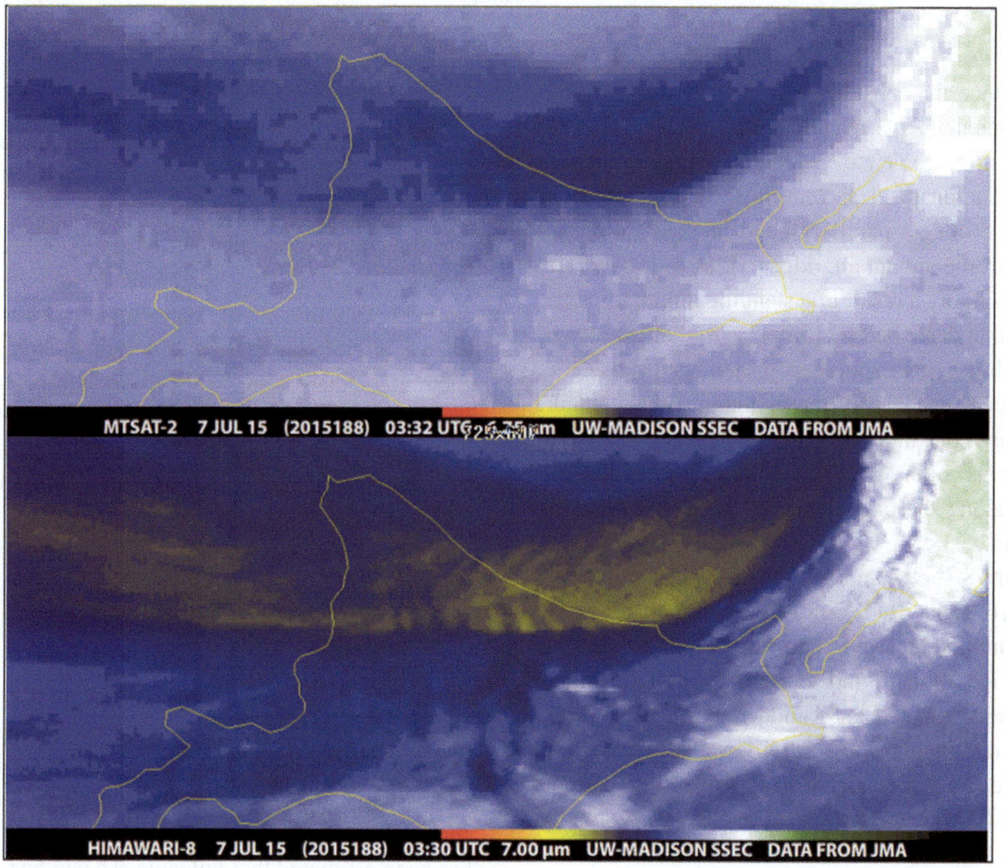

Figure 7
Mountain wave clouds observed in 4-km resolution 7.0-μm WV imagery from Japan's MTSAT (top) and Himiwari AHI 2-km 6.9-μm WV (bottom) on 7 July 2015 at around 0330 UTC (UW-CIMSS Satellite Blog)

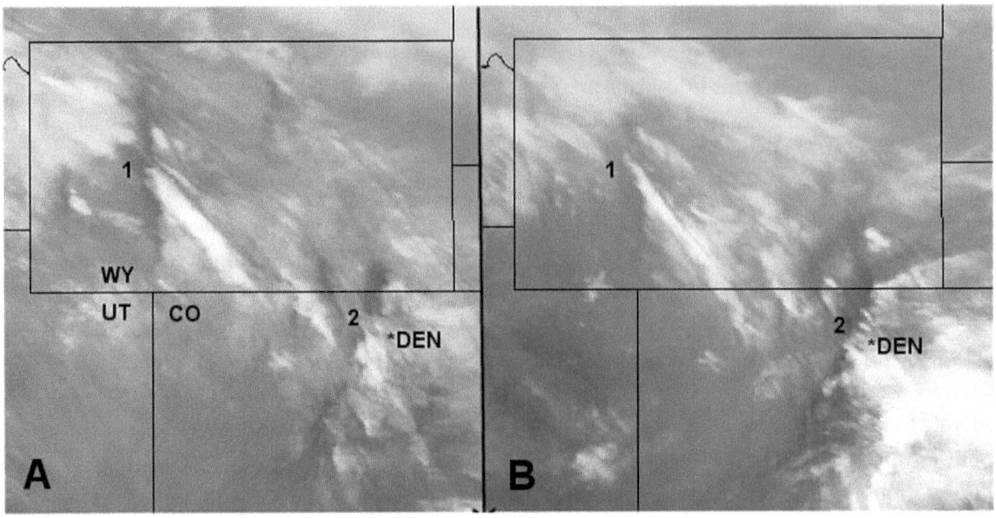

Figure 8
GOES-16 Band 10 (7.3 μm) water vapor imagery at **a** 1157 UTC and **b** 1802 UTC on 29 December 2017. Warm bands associated with katabatic downslope flow can be seen at (1) near the Wind River and Teton Ranges in Wyoming and at (2) near the Front Range in Colorado or the Medicine Bow range in Wyoming. Cold cirrus plumes extending downwind indicate possible mid-high-altitude mountain waves and turbulence

persist since the subsidence of the standing wave is not sufficient to dissipate the ice crystals. There is usually no turbulence in these plumes unless a narrow, stationary clear zone in the cirrus occurs in IR or WV images parallel to and just east of the lee slopes of the mountain ridges (Ellrod 1987). This feature is likely a result of a standing mountain wave, such as that described by Lilly (1978) that extends up into the lower stratosphere. A pronounced subsidence zone occurs just east of the mountain ridges leading to the dark zone in the WV images. Strong downslope windstorms often occur simultaneously with this feature along the east slopes of the Front Range in Colorado.

Figure 8 shows an example of such features in a GOES-16 Band 10 (7.2 μm) image associated with a strong mountain wave and downslope high-wind episode in Colorado and Wyoming on 29 December 2017. Frequent PIREPs of moderate to severe turbulence and mountain wave activity were obtained from aircraft departing or arriving at Denver International Airport (DEN) throughout the day. Additionally, surface winds gusted to more than 50 mph in Northern Colorado and southeast Wyoming.

4.5. Volcanic Hazards Monitoring

Volcanic ash is extremely hazardous to jet aircraft that inadvertently fly through it. The ash is composed primarily of silicate particles that melt when ingested into the combustion chamber of a jet engine, causing severe loss of engine performance and perhaps, a complete shutdown of the engine. Although no known aircraft have been lost due to volcanic ash encounters, there were two well-known incidents that resulted in near-catastrophes to large jet airliners in the late 1980s and early 1990s (e.g. Casadevall 1994). These incidents lead to the formation of the Volcanic Ash Advisory Centers (VAACs) under the International Civil Aeronautical Organization (ICAO) of the World Meteorological Organization (WMO). Since visual sightings of volcanic ash clouds from the ground or from aircraft are limited, remote sensing techniques based on satellite and radar observations are critical to the VAACs and Meteorological Watch Offices (MWOs) in monitoring hazardous ash clouds and issuing advisories such as Volcanic Ash Advisories (VAAs) and SIGMETs to aircraft in flight. It has been estimated that volcanic ash can be present in air routes at altitudes greater than 9 km (30,000 ft) on

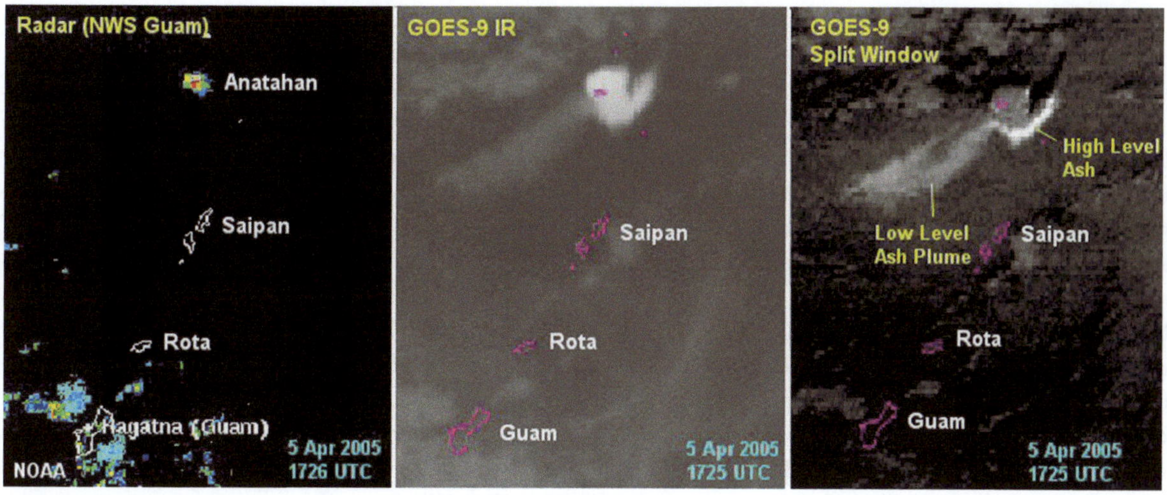

Figure 9
Comparison of radar reflectivity (0.5° scan) from NWS Guam WSR-88D radar (left), GOES-9 band 4 10.7-μm IR (center), and band 5 minus band 4 IR split window (right) at around 1725 UTC, 5 April 2005, following an eruption of Anatahan volcano (Ellrod 2005)

approximately 20 days per year worldwide (Miller and Casadevall 2000).

Volcanic eruptions have been monitored with GOES satellite images since the late 1970's and 1980's using long-wave window IR and visible images. Powerful eruptions such as Mt. St. Helens (May 1980) and El Chichon, Mexico (March 1982) were observed in this way. Although single-band images are effective for a few hours after an eruption, as the ash clouds became thinner, it becomes difficult to distinguish them from thin cirrus clouds or the underlying surface. By the late 1980's, it was learned that the temperature difference in two long-wave IR bands could be used effectively in observing thin ash clouds, initially with AVHRR (Prata 1989), and later with GOES (Rose and Mayberry 2000). Known as the "two-band split-window" (TBSW) or "reverse absorption" technique, it has become a benchmark for observing ash globally, and for measuring the effectiveness of improved multi-spectral techniques.

While the TBSW method is often effective, there are a number of deficiencies that have been identified (Hufford et al. 2000) such as (1) the inability to detect ash when the eruption cloud is opaque (usually within several hours of the eruption), (2) obscuration by ice embedded in the eruption cloud, and (3) the masking of low-level ash clouds by moist tropical atmospheres.

Ash clouds have also been observed by weather radar in a number of instances (e.g. Rose et al. 1995). However, the number of radars near active volcanoes is quite small, and their range is normally limited to about 300 km. There is also the risk of mistaking a thunderstorm for a volcanic eruption, especially in tropical regions. The example in Fig. 9 shows how satellite and radar can complement each other in observing and verifying volcanic eruptions. The Anatahan Volcano in the Marianas Islands in the Western Pacific Ocean erupted strongly on 5 April 2005, producing a dense ash cloud that rose to 15.5 km (51,000 ft). Even at more than 300 km from the location of the radar on Guam, strong reflectivity (50 dBZ) was observed. Due to relatively stable atmospheric conditions, there were no deep convective clouds in the region on this day. The TBSW satellite product from GOES-9 (right panel) showed a strong ash signal (white) around the edge of the high-altitude eruption cloud as well as a lower level ash plume caused by earlier activity. However, the central eruption cloud does not show the classic ash signature in the TBSW image product due to the opaque cloud column consisting of dense water and ice, in addition to ash.

Later improvements in ash detection were achieved by supplementing the split window product with the GOES 3.9-μm (SWIR) IR band (Ellrod et al.

2003). The 3.9-µm band is sensitive during the daytime to small cloud particles, resulting in increased reflectance from ash plumes hours after an eruption. Even at night, some improvements in detection were possible. By incorporating both visible and SWIR imagery with the TBSW (Pavolonis et al. 2005), it was shown that daytime volcanic ash detection could be significantly improved, with nearly one hundred times fewer false alarms globally than with just the use of the TBSW. Developed using multi-spectral data from Aqua MODIS, an algorithm to automatically detect volcanic eruptions using the four-band approach has been developed and has been implemented on GOES-16 ABI and possibly other platforms. Products available with this suite include: ash cloud height, ash loading (integrated mass), ash/dust effective radius, and ash probability, and true color volcanic plume composition (ash vs SO_2).

In an effort to better utilize satellite data for volcanic cloud applications, NOAA, in collaboration with the Cooperative Institute for Meteorological Satellite Studies (CIMSS), has developed the VOLcanic Cloud Analysis Toolkit (VOLCAT). VOLCAT ingests large volumes of data, from many different satellites, to automatically detect volcanic eruptions and send SMS text messages and/or email alerts to relevant users when an eruption is detected. VOLCAT also automatically tracks and characterizes volcanic ash clouds in an effort to improve volcanic ash cloud forecasts (e.g. Chai et al. 2017). The VOLCAT algorithms utilize unique spectral, spatial, and temporal metrics within a supervised learning framework (Pavolonis 2010; Pavolonis et al. 2013, 2015a, b) to automatically detect volcanic ash clouds with an accuracy that is comparable to a skilled human analyst. Figure 10 shows an example of a VOLCAT RGB image derived from Himiwari-8 AHI depicting the ash plume from Kambalny Volcano, Russia, drifting south–southwestward over the northwest Pacific Ocean at 1030 UTC on 25 March 2017. The densest portion of the ash plume is outlined by the brown contour. The yellow color of the plume indicates the presence of both volcanic ash and SO_2 (a combination of input from the red and green color guns representing contributions from the 12–11-µm TBSW and 11–8.5-µm band differences, respectively). The green lines are contrails from commercial jet aircraft that passed through the region earlier, as detected by the 11–8.5-µm IR difference which shows cloud phase.

The use of hyperspectral instruments such as the advanced infrared radiometric sounder (AIRS) or infrared atmospheric sounding interferometer (IASI) provides considerable benefits in both detection of volcanic ash and reduction of false alarms due to airborne dust and surface noise. One such example is a correlation method demonstrated with IASI data (Clarisse et al. 2010) that was significantly better than using benchmark BTD thresholds alone. Current and future deployment of such instruments in GEO orbits (such as the Geostationary Interferometric Infrared Sounder (GIIRS) on FY-4) provides a great promise for improved global ash detection.

For operational, real-time detection of erupting volcanoes and timely notification for aircraft in adjacent airspace, a response time of 5 min or less is desirable (Hufford et al. 2000). Even with GEO satellites, this has not been possible in the past. However, the improved rapid scan capabilities of the GOES-R ABI and Himiwari AHI imagers makes this requirement potentially achievable with 1-min mesoscale imaging frequency for GOES-16 ABI (2.5-min scanning for AHI), provided that a volcano suspected of imminent eruption is being properly monitored. It has been shown that rapid cooling of volcanic clouds during an eruption as observed with AHI 11-µm IR window band is significantly larger (by > 10 standard deviations) than the growth of convective clouds determined from a large database, providing confidence in their detection. An example occurred with the eruption of Kambalny Volcano on the Kamchatka Peninsula of eastern Russia on 25 March 2017. Based on Himiwari AHI 2.5 min scans, the growth rate observed was more than 13 standard deviations above the mean, signaling a likely eruption (Pavolonis 2018).

In addition to the detection of ash, remote sensing of sulfur dioxide (SO_2) emitted from volcanoes can be important to help estimate climatic effects, as well as to warn aircraft in flight about volcanic activity. However, since the SO_2 cloud is sometimes at a higher altitude than the ash, its movement can be in a completely different direction, depending on existing wind shears. Ultraviolet (UV) detectors originally

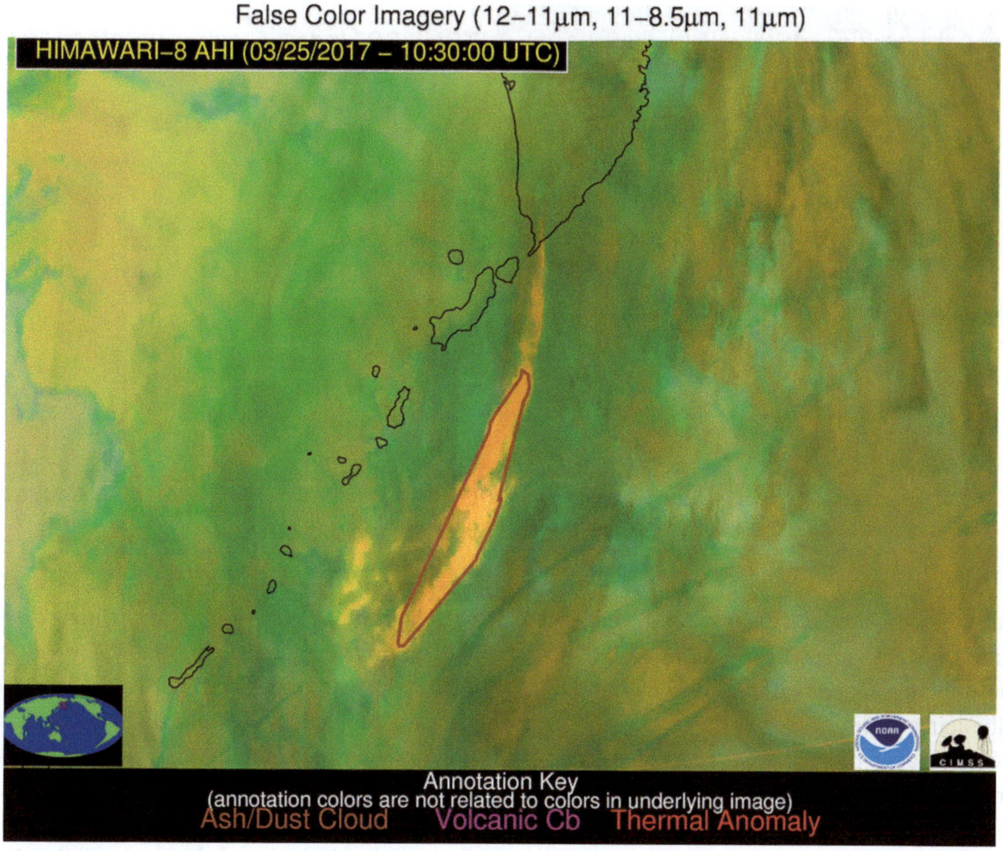

Figure 10
An RGB multi-spectral image from the VOLcanic Cloud Analysis Toolkit (VOLCAT) system on 1030 UTC, 25 March 2017 showing the ash plume from Kambalny Volcano, Kamchatka Peninsula, Russia. See text for more explanation (Bachmeier 2017b)

designed to observe total ozone concentrations (such as NASA's Total Ozone Mapping Spectrometer or TOMS) have been adapted to observe volcanic total column SO_2 (Seftor et al. 1997). They rely on backscattering from SO_2 in a UV band which differs enough from the ozone signature to allow discrimination between the two. Certain IR bands previously available on the GOES Sounder and MODIS (especially a water vapor absorption band centered at 7.3 µm) and now the GOES-R ABI and the Himiwari AHI are sensitive to SO_2 absorption (Prata et al. 2004; Ackerman et al. 2008). Figure 11 shows the similarity between a two-band difference image (based on 7.3–9.7 µm bands) from the GOES Sounder with a TOMS SO_2 Index product showing coverage of high-altitude SO_2 from Soufriere Hills Volcano, Montserrat, on 13 July 2003 (Ellrod and Schreiner 2004).

5. *Aircraft Icing*

Aircraft icing is a significant hazard to aircraft leading to loss of performance, especially among smaller general aviation and commuter class aircraft. Icing intensity is based on cloud parameters such as the liquid water concentration (LWC), temperatures, and drop size distribution, as well as aircraft type, speed, and duration within the clouds.

Observations of the tops of layered stratiform clouds can provide clues for the presence of possible icing conditions. Aside from simply observing sub-freezing CTTs, the presence of super-cooled liquid water (SLW) can be observed in multi-spectral satellite imagery due to its accumulation near the cloud tops (Rauber and Tokay 1991). Early attempts to identify SLW regions showed promise (Ellrod 1996; Thompson et al. 1997). Some shortcomings of

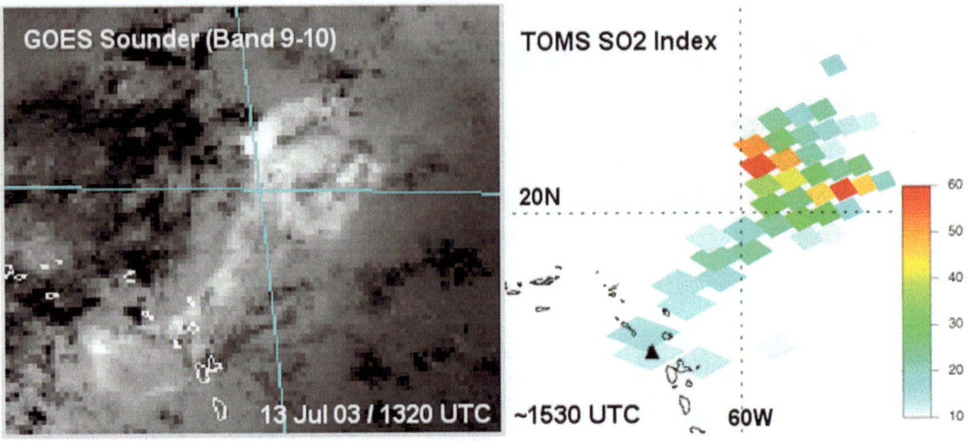

Figure 11
Image of GOES-8 Sounder band 9 (9.7 μm) minus band 10 (7.4 μm) at 1320 UTC, 13 July 2003 (left), versus NASA TOMS SO_2 Index at approximately 1530 UTC (From Ellrod and Schreiner 2004)

these early icing products, however, were the obscuration of SLW clouds by cirrus, as well as the lack of information regarding icing intensity and the base of the icing layer.

The GOES icing product was later enhanced by an improved four-band icing algorithm and the addition of the Cloud Top Pressure (CTP) derived from the GOES sounder to show the maximum altitude ranges of SLW clouds (Ellrod and Bailey 2007). The improved four-band icing algorithm (referred to as "ICECAP") corrected the visible imagery brightness for viewing angle and filtered much of the false icing due to cirrus day and night using empirical thresholds based on the visible and 3.9-, 10.7- and 12.0-μm IR data. There was still some false detection present, such as when there was thin cirrus over a snow cover. Verification for a 2-year period ending April 2005 showed excellent POD, especially at altitudes < 12 kft, but PODn (the ability to correctly observe non-icing regions) was unacceptably low.

A system that derived pixel level cloud properties based on four spectral GOES bands at 0.63, 3.9, 10.8, and 12.0 μm was used to successfully identify SLW clouds (Smith et al. 2000). If a cloud was determined to be liquid with temperatures below 0 °C, it was denoted as SLW with the potential to produce aircraft icing. The algorithm also worked well in the presence of thin cirrus (coverage < 5%), detecting 95% of SLW cases based on PIREPs. Later improvements to the algorithm (Smith et al. 2003; Minnis 2004) led to the prototype candidate for the GOES-R program, described by Smith et al. (2012). The system produces a Flight Icing Threat (FIT) index based on a combination of a satellite-derived icing mask, icing probability and intensity. An example of the GOES FIT product is shown by Fig. 12. Probabilities (low, medium, and high) are shown for light (LGT) and moderate or greater (MOG) icing.

6. Concluding Remarks

Since the 1970s, data from geostationary meteorological satellites have proven their worth as valuable supplements to land-based and aircraft observations in the detection and short-range prediction of aviation hazards such as fog and low stratus, turbulence, thunderstorms, volcanic ash, and aircraft icing. Recent improvements in spectral and spatial resolution, image frequency, and navigational accuracy and stability have increased their value even more. For example, the new imagers can provide full disk coverage every 10–15 min. Not to be overlooked is the recent ability to relay weather products directly to aircraft cockpits for display using electronic flight bags (EFBs). Initially, only text products could be transmitted but rapid advances in ground-to-air satellite communications now allow raw imagery and

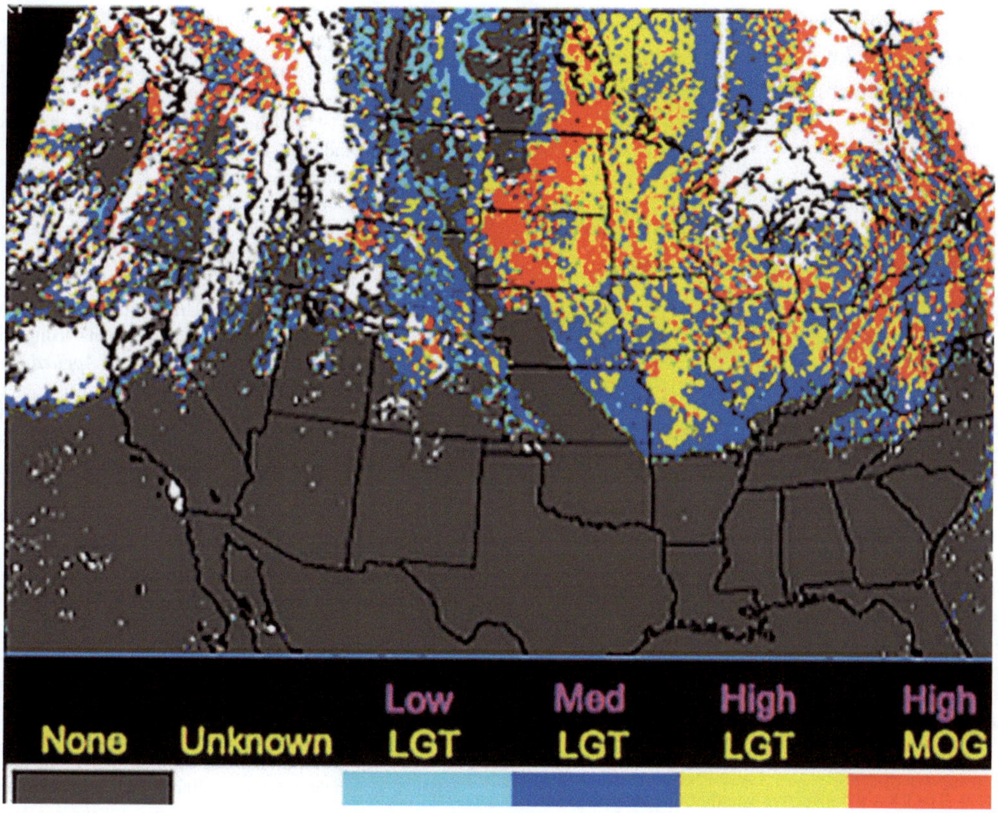

Figure 12
Flight icing threat (FIT) derived from GOES at 1745 UTC, 8 November 2008 (From Smith et al. 2012)

derived products to be displayed in real time, resulting in increased situational awareness by aircrews. One example is the CDO product described in Sect. 4.2.1, which is now commercially available for satellite uplink to aircraft (Kessinger et al. 2017).

In addition to multi-spectral imagers, geostationary sounders such as GOES-VAS have shown the potential to monitor destabilization to predict possible thunderstorm development or microburst activity and to provide gradient and atmospheric motion vectors (AMVs) in remote areas for use in NWP. With the exception of China's Geostationary Interferometric Infrared Sounder (GIIRS) on the FY-4 series (Yang et al. 2017), there are currently no sounders in geostationary orbit, but hyperspectral sounders (with many hundreds of spectral bands) are planned for future spacecraft such as Europe's MTG, and possibly future GOES. In conclusion, the future looks bright for monitoring inflight and surface aviation weather conditions from space.

NOTE: Details concerning the suite of derived products from GOES-16 ABI and GLM are included in the Algorithm Theoretical Bases Documents (ATBD) available from NOAA/NESDIS at: (1) https://www.goes-r.gov/products/baseline.html (for Baseline products such as cloud top height, and volcanic ash detection), and: (2) https://www.goes-r.gov/products/option2.html (for future products such as aircraft icing, fog and low clouds, SO_2 detection, convective initiation, enhanced-V/overshooting tops, and tropopause folding turbulence prediction).

Acknowledgements

The authors would like to thank Dr. Michael Pavolonis (NOAA/CIMSS) for his contributions to the section on volcanic ash detection. Many of the images in this paper were obtained from the University of Wisconsin CIMSS blog pages on the use of improved satellite image data from GOES-R (16) and Himiwari. We also acknowledge the comments of two anonymous reviewers that greatly improved the quality of the paper.

References

Ackerman, S., Schreiner, A., Schmit, T., Woolf, H., Li, J., & Pavolonis, M. (2008). Using the GOES Sounder to monitor upper level SO_2 from volcanic eruptions. *Journal of Geophysical Research: Atmospheres, 113*. https://doi.org/10.1029/2007jd009622.

Adler, R. F., & Fenn, D. D. (1979). Thunderstorm intensity as determined from satellite data. *Journal of Applied Meteorology, 18*, 502–517.

Atkins, N. T., & Wakimoto, R. M. (1991). Wet microburst activity over the southeastern United States: Implications for forecasting. *Weather and Forecasting, 6*, 470–482.

Bachmeier, S. (2017a). Tornadoes and Large Hail in Minnesota and Wisconsin. CIMSS Satellite Blog, Cooperative Institute for Meteorological Satellite Studies, 16 May 2017. http://cimss.ssec.wisc.edu/goes/blog/archives/category/severe-convection/page/6.

Bachmeier, S. (2017b). Eruption of Kambalny Volcano in Kamchatka, Russia. CIMSS Satellite Blog, Cooperative Institute for Meteorological Satellite Studies, 25 March 2017. http://cimss.ssec.wisc.edu/goes/blog/archives/category/volcanic-activity/page/2.

Bass, R. (2017). The FAA's convective weather research program. In *AMS 18th Conf. on Aviation, Range and Aerospace Meteorology (ARAM)*, 23 January 2017, Seattle, Washington. https://ams.confex.com/ams/97Annual/webprogram/Paper313429.html.

Bedka, K., Brunner, J., Dworak, R., Feltz, W., Otkin, J., & Greenwals, T. (2010). Objective satellite-based detection of overshooting tops using infrared window channel brightness temperature gradients. *Journal of Applied Meteorology and Climatology, 49*, 181–202.

Bessho, K., et al. (2016). An introduction to Himiwari-8/9—Japan's new-generation geostationary meteorological satellites. *Journal of the Meteorological Society of Japan, 94*, 151–183.

Bézy, J.-L., Aminou, D., Bensi, P., Stuhlman, R., Tjemkes, S., & Rodriguez, A. (2005). Meteosat Third Generation. *European Space Agency Bulletin, 123*, 28–32.

Byers, H. R., & Braham, R. R. (1949). *The thunderstorm* (p. 247). Washington, DC: U. S. Government Printing Office.

Calvert, C., & Pavolonis, M. (2011). GOES-R Advanced Baseline Imager (ABI) Theoretical Basis Document for Low Cloud and Fog, Version 2.0. NOAA/NESDIS Center for Satellite Applications and Research, p. 72.

Caracena, F., & Flueck, J. A. (1988). Classifying and forecasting microburst activity in the Denver area. *Journal of Aircraft, 25*, 525–530.

Casadevall, T. (1994). The 1989–1990 eruption of Redoubt Volcano, Alaska: Impacts on aircraft operations. *Journal of Volcanology and Geothermal Research, 62*, 301–316.

Chai, T., Crawford, A., Stunder, B., Pavolonis, M., Draxler, R., & Stein, A. (2017). Improving volcanic ash predictions with the HYSPLIT dispersions model by assimilating MODIS satellite retrievals. *Atmospheric Chemistry and Physics, 17*, 2865–2879.

Clarisse, L., Prata, F., Lacour, J.-L., Hurtmans, D., Clerbaux, C., & Coheur, P.-F. (2010). A correlation method for volcanic ash detection using hyperspectral infrared measurements. *Geophysical Research Letters, 37*, L198065. **(p. 5)**.

Conover, J. H. (1964). The identification and significance of orographically induced clouds observed by TIROS satellites. *Journal of Applied Meteorology, 3*, 226–234.

DeMaria, M., DeMaria, R. T., Knaff, J., & Molenar, D. (2012). Tropical cyclone lightning and rapid intensity change. *Monthly Weather Review, 140*, 1828–1842.

Donovan, M., Williams, E., Kessinger, C., Blackburn, G., Herzegh, P., Bankert, R., et al. (2008). The identification and verification of hazardous convective cells over oceans using visible and infrared satellite observations. *Journal of Applied Meteorology and Climatology, 47*, 164–184.

Dworak, R., Bedka, K., Brunner, J., & Feltz, W. (2012). Comparison between GOES-12 overshooting-top detections, WSR-88D radar reflectivity and severe storm reports. *Weather and Forecasting, 27*, 684–699.

Eckert, M. (2017). GOES-16 enhancements to IDSS for San Francisco International Airport (SFO) on 03/03/17. *NWA Specialized Operations Committee—IDSS Success Stories*. http://nwas.org/wp-content/uploads/2016/08/DSS-Success-Story-SFO_GOES-16-1.pdf.

Ellrod, G. P. (1985). Indicators of high altitude, non-convective turbulence observed in satellite images. In *Proc. of the 2nd Intl. Conf. on the Aviation Weather System* (pp. 277–284). Boston, MA: Amer. Meteor. Soc.

Ellrod, G. P. (1987). Identifying high altitude mountain wave turbulence and strong chinook wind events with satellite imagery. In *Proc. Of the AIAA 25th Aerospace Sciences Meeting*, Jan. 12–15, Reno (p. 7). Washington, DC: Amer. Inst. For Aeronautics and Astronautics.

Ellrod, G. P. (1989). Environmental conditions associated with the Dallas microburst storm determined from satellite soundings. *Weather and Forecasting, 4*, 469–484.

Ellrod, G. P. (1990). A water vapor feature related to severe thunderstorms. *National Weather Digest, 15*, 19–29.

Ellrod, G. P. (1995). Advances in the detection of fog at night using GOES multispectral infrared imagery. *Weather and Forecasting, 10*, 606–619.

Ellrod, G. P. (1996). The use of GOES-8 multi-spectral imagery for the detection of aircraft icing regions. In *8th Conf. on Satellite Meteorology and Oceanography*, Atlanta, GA, Amer. Meteor. Soc., (pp. 168–171).

Ellrod, G. P. (2002). Estimation of low cloud base heights at night from satellite infrared and surface temperature data. *National Weather Digest, 26*, 39–44.

Ellrod, G. P. (2005). Remote sensing of volcanic ash. Tutorial developed for *National Weather Association Remote Sensing Committee*. http://www.ellrodweather.com/volcano/ash.htm.

Ellrod, G. P., & Bailey, A. (2007). Assessment of aircraft icing potential and maximum icing altitude from geostationary meteorological satellite data. *Weather and Forecasting, 22*, 160–174.

Ellrod, G. P., Connell, B., & Hillger, D. (2003). Improved detection of airborne volcanic ash using multi-spectral infrared satellite data. *Journal of Geophysical Research, 108*(D12), 4356.

Ellrod, G. P., & Gultepe, I. (2007). Inferring low cloud base heights at night for aviation using satellite infrared and surface temperature data. *Pure and Applied Geophysics, 164*, 1193–1205.

Ellrod, G. P., Maturi, E., & Steger, J. (1989). Detection of fog at night using dual channel GOES-VAS imagery. In *Proc. 12th Conf. on Wea. Analysis and Forecasting,* Monterey, California, Amer. Meteor. Soc. (pp. 515–520).

Ellrod, G. P., & Schreiner, A. (2004). A first look at volcanic ash detection in the GOES-12 era. In *11th Conf. on Aviation, Range, and Aerospace Meteorology*, Hyannis, MA, 4–8 October 2004, Paper #8.13. Boston, MA: Amer. Meteor. Soc.

Eyre, J. R., Brownscombe, J. L., & Allam, R. J. (1984). Detection of fog at night using Advanced Very High Resolution Radiometer (AVHRR) imagery. *Meteorological Magazine, 114,* 187–201.

Federal Aviation Administration (FAA). (2017). NextGen Weather FAQ: Weather Delays. https://www.faa.gov/nextgen/programs/weather/faq/.

Fritz, S. (1965). The significance of mountain lee waves as seen from satellite pictures. *Journal of Applied Meteorology, 4,* 31–37.

Goodman, S. J., Blakeslee, R. J., Koshak, W. J., Mach, D., Bailey, J., Buechler, D., et al. (2013). The GOES-R Geostationary Lightning Mapper (GLM). *Atmospheric Research, 125–126,* 34–49.

Gravelle, C., Mecikalski, J., Line, W., Bedka, K., Petersen, R., Sieglaff, J., et al. (2016). Satellite convective toolkit to "bridge the gap" between severe weather watches and warnings: An example from the 20 May 2013 Moore, Oklahome tornado outbreak. *Bulletin of the American Meteorological Society, 97*(1), 69–84.

Gultepe, I., Pagowski, M., & Reid, J. (2007). Using surface data to validate a satellite based fog detection scheme. *Weather and Forecasting, 22,* 444–456.

Gurka, J. J. (1978). The use of enhanced visible imagery for predicting the time of fog dissipation. In *Proc. Conf. on Weather Forecasting and Analysis and Aviation Meteorology*, Silver Spring, MD, Amer. Meteor. Soc., (pp. 343–346).

Han, H., Lee, S., Im, J., Kim, M., Lee, M.-I., Ahn, M., et al. (2015). Detection of convective initiation using Meteorological Imager onboard Communications, Ocean, and Meteorological Satellite based on machine learning approaches. *Remote Sensing, 7,* 9184–9204.

Hartung, D. C., Sieglaff, J. M., Cronce, L. M., & Feltz, W. F. (2013). An intercomparison of UW cloud-top cooling rates with WSR-88D radar data. *Weather and Forecasting, 28,* 463–480.

Heymsfield, G. M., & Blackmer, R. H. (1988). Satellite-observed characteristics of Midwest severe thunderstorm anvils. *Monthly Weather Review, 116,* 2200–2224.

Hillger, D., Kopp, T., Lee, T., Lindsey, D., Seaman, C., Miller, S., et al. (2013). First-light imagery from Suomi NPP VIIRS. *Bulletin of the American Meteorological Society, 94,* 1019–1029.

Hufford, G., Salinas, L., Simpson, J., Barske, E., & Pieri, D. (2000). Operational implications of airborne volcanic ash. *Bulletin of the American Meteorological Society, 81*(4), 745–755.

Kessinger, C., Megenhardt, D., Blackburn, G., Olivo, J., Lin, L., Hoang, V., Nayote, M., Sievers, K., Ritter, A., Wolf, D., Matz, O., Scheinhartz, R., & Cahall, J. (2017). Displaying convective weather products on an electronic flight bag. *Journal of Air Traffic Control*, Fall Issue, 52–61.

Kim, D., & Ahn, M. (2014). Introduction of the in-orbit test and its performance for the first meteorological imager of the Communications, Ocean and Meteorological Satellite. *Atmospheric Measurement Techniques, 7,* 2471–2485.

Knox, J., Bachmeier, S., Carter, W., Tarantino, J., Paulik, L., Wilson, E., Bechdol, G., & Mays, M. (2010). Transverse cirrus bands in weather systems: a grand tour of an enduring enigma. *Weather Magazine, 65*(2), 35–41. https://doi.org/10.1002/wea.417.

Lay, E. H., Holzworth, R. H., Rodger, C. J., Thomas, J. N., Pinto, O., & Dowden, R. L. (2004). WWLL global lightning detection system: Regional validation study in Brazil. *Geophysical Research Letters, 31,* L03102. https://doi.org/10.1029/2003GL018882.

Lee, Y.-K., Li, Z., Li, J., & Schmit, T. (2014). Evaluation of the GOES-R ABI LAP retrieval algorithm using the GOES-13 sounder. *Journal of Atmospheric and Oceanic Technology, 31*(1), 3–19.

Lee, T. F., Turk, F. J., & Richardson, K. (1997). Stratus and fog products using GOES-8-9 3.9 μm data. *Weather and Forecasting, 12,* 664–677.

Li, Z., Li, J., Menzel, W. P., Schmit, T. J., Nelson, J. P., III, Daniels, J., et al. (2008). GOES sounding improvement and applications to severe storm nowcasting. *Geophysical Research Letters, 35,* L03806. https://doi.org/10.1029/2007gl032797.

Lilly, D. K. (1978). A severe downslope windstorm and aircraft turbulence event induced by a mountain wave. *Journal of the Atmospheric Sciences, 34,* 59–77.

Lindstrom, S. (2017). Harvey and GLM Lightning. *CIMSS Satellite Blog*, Cooperative Institute for Meteorological Satellite Studies, 25 August 2017. http://cimss.ssec.wisc.edu/goes/blog/archives/date/2017/08/page/2.

Liu, C., & Heckman, S. (2011). The application of total lightning detection and cell tracking for severe weather prediction. In *91st American Meteorological Society Annual Meeting*, Seattle, 1–10.

Martin, D. W., Kohrs, R. A., Mosher, F. R., Medaglia, C. M., & Adamo, C. (2008). Over-ocean validation of the Global Convective Diagnostic. *Journal of Applied Meteorology and Climatology, 47,* 525–543.

McCann, D. W. (1983). The enhanced-V, a satellite observable severe storm signature. *Monthly Weather Review, 111,* 887–894.

Mecikalski, J., & Bedka, K. (2006). Forecasting convective initiation by monitoring the evolution of moving cumulus in daytime GOES imagery. *Monthly Weather Review, 134,* 49–77.

Mecikalski, J., Williams, J. K., Jewett, C. P., Ahijevych, D., LeRoy, A., & Walker, J. R. (2015). Probabilistic 0–1-h convective initiation nowcasts that combine geostationary satellite observations and numerical weather prediction model data. *Journal of Applied Meteorology and Climatology, 54,* 1039–1059. https://doi.org/10.1175/JAMC-D-14-0129.1.

Menzel, W. P., Holt, F. C., Schmit, T. J., Aune, R., Schreiner, A. J., Wade, G. S., et al. (1998). Application of GOES-8/9 soundings to weather forecasting and nowcasting. *Bulletin of the American Meteorological Society, 10,* 2059–2077.

Menzel, W. P., & Purdom, J. F. W. (1994). Introducing GOES-I: The first of a new generation of geostationary operational environmental satellites. *Bulletin of the American Meteorological Society, 75,* 757–780.

Merk, D., & Zinner, T. (2013). Detection of convective initiation using Meteosat SEVIRI implementation in and verification with the tracking and nowcasting algorithm Cb-TRAM. *Atmospheric Measurement Techniques, 6,* 1903–1918.

Miller, T., & Casadevall, T. (2000). Volcanic ash hazards to aviation: Encyclopedia of volcanoes. In H. Sigurdsson (Ed.), (pp. 915–930) San Diego, California: Academic.

Miller, S. D., Schmit, T., Seaman, C., Lindsey, D., Gunshor, M., Kohrs, R., et al. (2016). A sight for sore eyes: The return of true color to geostationary satellites. *Bulletin of the American Meteorological Society, 97,* 1803–1816.

Minnis, P., et al. (2004). Real-time cloud, radiation, and aircraft icing parameters from GOES over the USA. In *13th Conf. on Satellite Oceanography and Meteorolology*, Norfolk, Virginia, Amer. Meteor. Soc., P7.1. http://ams.confex.com/ams/pdfpapers/79179.pdf.

Smith, W. L. Jr, Minnis, P., Bernstein, B. C., McDonough, F., & Khaiyer, M. M. (2003). Comparison of super-cooled liquid water cloud properties derived from satellite and aircraft measurements. In *Proc. in-flight icing/de-icing int. conf.*, Chicago, IL, Federal Aviation Administration 2003-01-2156.

Smith, W. L. Jr, Minnis, P., & Young, D. F. (2000). An icing product derived from operational satellite data. In *Proc. AMS 9th conference on aviation, range and aerospace meteorology*, 11–15 September 2000 (pp. 256–259). Orlando, FL: Amer. Meteor. Soc.

Mohr, T. (2014). Preparing for the use of new generation geostationary meteorological satellites. *WMO Bulletin, 63,* 42–44.

Morel, P., Desbois, M., & Szewach, G. (1978). A new insight into the troposphere with the water vapor channel of Meteosat. *Bulletin of the American Meteorological Society, 59,* 711–714.

Mosher, F. R. (2001). A satellite diagnostic of global convection. In *11th Conf. on Satellite Meteorology and Oceanography* (pp. 416–419). Madison, Wisconsin: Amer. Meteor. Soc.

Mosher, F. R. (2002). Detection of deep convection around the globe. In *10th Conf. on Aviation, Range, and Aerospace Meteorology* (pp. 289–292). Portland, Oregon: Amer. Meteor. Soc.

Negri, A. J., & Adler, R. F. (1981). Relation of satellite-based thunderstorm intensity to radar-estimated rainfall. *Journal of Applied Meteorology, 20,* 288–300.

Pavolonis, M. J. (2010). Advances in extracting cloud composition information from spaceborne infrared radiances—A robust alternative to brightness temperatures. Part I: Theory. *Journal of Applied Meteorology and Climatology, 49*(9), 1992–2012.

Pavolonis, M. J. (2018). Personal communication.

Pavolonis, M. J., Feltz, W. F., Heidinger, A. K., & Gallina, G. M. (2005). A daytime complement to the reverse absorption technique for improved automated detection of volcanic ash. *Journal of Atmospheric and Oceanic Technology, 23,* 1422–1444.

Pavolonis, M. J., Heidinger, A., & Sieglaff, J. (2013). Automated retrievals of volcanic ash and dust cloud properties from upwelling infrared measurements. *Journal of Geophysical Research-Atmospheres, 118*(3), 1436–1458.

Pavolonis, M. J., Sieglaff, J. M., & Cintineo, J. L. (2015a). Spectrally enhanced cloud objects (SECO): A generalized framework for automated detection of volcanic ash and dust clouds using passive satellite measurements, Part I: Multispectral analysis. *Journal of Geophysical Research: Atmospheres, 120,* 7813–7841.

Pavolonis, M. J., Sieglaff, J. M., & Cintineo, J. L. (2015b). Spectrally enhanced cloud objects (SECO): A generalized framework for automated detection of volcanic ash and dust clouds using passive satellite measurements, part II: Cloud object analysis and global application. *Journal of Geophysical Research: Atmospheres, 120,* 7842–7870.

Prata, F. (1989). Observations of volcanic ash clouds in the 10–12 μm window using AVHRR/2 data. *International Journal of Remote Sensing, 10,* 751–761.

Prata, F., Schreiner, A., Schmit, T. J., & Ellrod, G. P. (2004). First measurements of volcanic sulfur dioxide from the GOES Sounder: Implications for improved aviation safety. In *Proceedings, 2nd Intl. Conf. on Volcanic Ash and Aviation Safety*, 21–24 June 2004, Alexandria, Virginia, Paper number 3.7.

Pryor K. (2010). Recent developments in microburst nowcasting using GOES. In *17th conf. on satellite meteorology and oceanography* (p. 9.7). Annapolis, Maryland: Amer. Meteor. Soc.

Pryor K. (2012). Microburst nowcasting applications of GOES. In *18th conf. on satellite meteorology, oceanography and climatology* (p. 471). New Orleans, Louisiana: Amer. Meteor. Soc.

Pryor, K. (2014). Downburst prediction applications of meteorological geostationary satellites. In *Proc. SPIE conf. on remote sensing of the atmosphere, clouds, and precipitation V*, Beijing, China. https://doi.org/10.1117/12.2069283.

Pryor, K. (2015). Progress and developments of downburst prediction applications of GOES. *Weather and Forecasting, 30,* 1182–1200.

Pryor, K. (2017). Advances in downburst monitoring and prediction with GOES-16. In *17th conf. on mesoscale processes,* San Diego, CA, Amer. Meteor. Soc., Paper No. 10.6.

Pryor, K., & Ellrod, G. P. (2004a). Recent improvements to the GOES microburst products. *Weather and Forecasting, 19,* 582–594.

Pryor, K., & Ellrod, G. P. (2004b). WMSI—A new index for forecasting wet microburst severity. *Electronic Journal Of Operational Meteorology, 5*(3), 1–25.

Purdom, J. F. W. (1973). Meso-highs and satellite imagery. *Monthly Weather Review, 101,* 180–181.

Purdom, J. F. W. (1976). Some uses of high-resolution GOES imagery in the mesoscale forecasting of convection and its behavior. *Monthly Weather Review, 104,* 1474–1483.

Rauber, R., & Tokay, A. (1991). An explanation for the existence of supercooled water at the top of cold clouds. *Journal of Atmospheric Science, 48,* 1005–1023.

Rose, W. I., Kostinski, A. B., & Kelley, L. (1995). Real time C band radar observations of 1992 eruption clouds from Crater Peak/Spurr Volcano, Alaska. *U. S. Geological Survey Bulletin 2139,* p. 19.

Rose, W. I., & Mayberry, G. C. (2000). Use of GOES thermal infrared imagery for eruption scale measurements, Soufriere Hills, Montserrat. *Geophysical Research Letters, 27,* 3097–3100.

Schmetz, J., Pili, P., Tjemkes, S., Just, D., Kerkmann, J., Rota, S., et al. (2002). An introduction to Meteosat Second Generation (MSG). *Bulletin of the American Meteorological Society, 83,* 977–992.

Schmetz, J., Tjemkes, S. A., Gube, M., & van de Berg, L. (1997). Monitoring deep convection and convective overshooting with METEOSAT. *Advances in Space Research, 19,* 433–441.

Schmit, T., Goodman, S., Gunshor, M., Sieglaff, J., Heidinger, A., Bachmeier, S., et al. (2015). Rapid refresh information of significant events: Preparing users for the next generation of geostationary operational satellites. *Bulletin of the American Meteorological Society, 96,* 561–575.

Schmit, T., Griffith, P., Gunshor, M., Daniels, J., Goodman, S., & Lebair, W. (2017). A closer look at the ABI on the GOES-R series. *Bulletin of the American Meteorological Society, 98*(4), 681–698.

Schmit, T., Gunshor, M., Menzel, W. P., Gurka, J., Li, J., & Bachmeier, S. (2005). Introducing the next-generation Advanced

Baseline Imager on GOES-R. *Bulletin of the American Meteorological Society, 86,* 1079–1096.

Schmit, T., Li, J., Gurka, J. J., Goldberg, M. D., Schrab, K. J., Li, J., et al. (2008). The GOES-R advanced baseline imager and the continuation of current sounder products. *Journal of Applied Meteorology and Climatology, 47,* 2696–2711.

Schultz, C. J., Petersen, W. A., & Carey, L. D. (2009). Preliminary development and evaluation of lightning jump algorithms for the real-time detection of severe weather. *Journal of Applied Meteorology and Climatology, 48,* 2543–2563.

Seftor, C. J., Hsu, N., Herman, J., Bhartia, P., Torres, O., Rose, W., et al. (1997). Detection of volcanic ash clouds from Nimbus 7/Total Ozone Mapping Spectrometer. *Journal of Geophysical Research, 102*(D14), 16749–16760.

Sikdar, D. N., Suomi, V. E., & Anderson, C. E. (1970). Convective transport of mass and energy in severe storms over the United States—An estimate from a geostationary altitude. *Tellus, 22,* 521–532.

Smith, W. L., Jr., Minnis, P., Fleeger, C., Spangenberg, D., Palikonda, R., & Nguyen, L. (2012). Determining the Flight Icing Threat to aircraft with single-layer cloud parameters derived from operational satellite data. *Journal of Applied Meteorology and Climatology, 51,* 1794–1810.

Smith, W. L., Jr., Suomi, V., Menzel, W., Woolf, H., Sromovsky, L., Revercomb, H., et al. (1981). First sounding results from VAS-D. *Bulletin of the American Meteorological Society, 62,* 232–236.

Srivastava, R. C. (1987). A model of intense downdrafts driven by the melting and evaporation of precipitation. *Journal of Atmospheric Science, 44,* 1752–1773.

Suomi, V. E., & Krauss, R. J. (1978). The spin scan camera system: Geostationary meteorological satellite workhorse for a decade. *Optical Engineering, 17,* 6–13.

Tag, P., Bankert, R., & Brody, L. (2000). An AVHRR multiple cloud-type classification package. *Journal of Applied Meteorology, 39,* 125–134.

Thompson, G., Bullock, R., & Lee, T. F. (1997). Using satellite data to reduce spatial extent of diagnosed icing. *Weather and Forecasting, 12,* 185–190.

University of Wisconsin CIMSS. (2017). GOES-R Fog Product Examples, 17 October 2017. http://fusedfog.ssec.wisc.edu/?cat=18.

Wakimoto, R. M. (1985). Forecasting dry microburst activity over the High Plains. *Monthly Weather Review, 113,* 1131–1143.

Walker, J. R., MacKenzie, W. M., Jr., Mecikalski, J. R., & Jewett, C. P. (2012). An enhanced geostationary satellite-based convective initiation algorithm for 0–2-h nowcasting with object tracking. *Journal of Applied Meteorology and Climatology, 51,* 1931–1949. https://doi.org/10.1175/JAMC-D-11-0246.1.

Wimmers, A. J., & Feltz, W. (2010). Tropopause folding turbulence product. GOES-R Algorithm Theoretical Basis Document (ATBD). NOAA Center for Satellite Applications and Research. https://www.goes-r.gov/products/ATBDs/option2/Aviation_Turbulence_v1.0_no_color.pdf.

Wimmers, A. J., Griffin, S., Gerth, J., Bachmeier, S., & Lindstrom, S. (2018). Observations of gravity waves with high-pass filtering in the new generation of geostationary imagers and their relation to aircraft turbulence. *Weather and Forecasting, 33,* 139–144.

Wimmers, A. J., & Moody, J. L. (2001). A fixed-layer estimation of upper tropospheric specific humidity from the GOES water vapor channel: Parameterization and validation of the altered brightness temperature product. *Journal of Geophysical Research, 106*(D15), 17115–17132.

Wimmers, A. J., & Moody, J. L. (2004a). Tropopause folding at satellite-observed spatial gradients: 1. Verification of an empirical relationship. *Journal of Geophysical Research, 109,* D19306. https://doi.org/10.1029/2003JD004145.

Wimmers, A. J., & Moody, J. L. (2004b). Tropopause folding at satellite-observed spatial gradients: 2. Development of an empirical model. *Journal of Geophysical Research, 109,* D19307. https://doi.org/10.1029/2003JD004146.

Yang, J., Zhang, Z., Wei, C., Lu, F., & Guo, Q. (2017). Introducing the new generation of Chinese geostationary weather satellites, Fengyun-4. *Bulletin of the American Meteorological Society, 98,* 1637–1659.

(Received October 30, 2017, revised February 16, 2018, accepted February 24, 2018, Published online March 9, 2018)

Pure Appl. Geophys. 176 (2019), 2045–2055

https://doi.org/10.1007/s00024-018-1772-6

Pure and Applied Geophysics

Hazard Avoidance Products for Convectively-Induced Turbulence in Support of High-Altitude Global Hawk Aircraft Missions

Sarah M. Griffin[1] and Christopher S. Velden[1]

Abstract—A combination of satellite-based and ground-based information is used to identify regions of intense convection that may act as a hazard to high-altitude aircraft. Motivated by concerns that Global Hawk pilotless aircraft, flying near 60,000 feet, might encounter significant convectively-induced turbulence during research overflights of tropical cyclones, strict rules were put in place to avoid such hazards. However, these rules put constraints on science missions focused on sampling convection with onboard sensors. To address these concerns, three hazard avoidance tools to aid in real-time mission decision support are used to more precisely identify areas of potential turbulence: Satellite-derived Cloud-top height and tropical overshooting tops, and ground-based global network lightning flashes. These tools are used to compare an ER-2 aircraft overflight of tropical cyclone Emily in 2005, which experienced severe turbulence, to Global Hawk overflights of tropical cyclones Karl and Matthew in 2010 that experienced no turbulence. It is found that the ER-2 overflew the lowest cloud tops and had the largest vertical separation from them compared to the Global Hawk flights. Therefore, cold cloud tops alone cannot predict turbulence. Unlike the overflights of Matthew and Karl, Emily exhibited multiple lightning flashes and a distinct overshooting top coincident with the observed turbulence. Therefore, these tools in tandem can better assist in identifying likely regions/periods of intense active convection. The primary outcome of this study is an altering of the Global Hawk overflight rules to be more flexible based on the analyzed conditions.

Key words: Aircraft hazard avoidance, high-altitude observations, satellite observations.

1. Introduction

Given its flight range and cruising altitude of approximately 60 kft, the unmanned Global Hawk (GH) drone aircraft can provide an abundance of high-altitude observations. These observations can be useful to atmospheric research and forecasting, especially when studying tropical cyclones (TCs) which can extend high into the upper troposphere and even lower stratosphere, and often exist over the open ocean hundreds of miles from the nearest conventional observations. TC forecasters and numerical model guidance can also benefit from real-time GH-provided information on storm structure, intensity and motion. For example, the data has been used to confirm the strengthening of TC Edouard in 2014 over the open Atlantic ocean (National Hurricane Center 2014) as well as the motion of TC Hermine in 2016 (National Hurricane Center 2016) near the Florida coast. Finally, high-altitude observations also provide researchers with information on TC behavior and dynamics, such as the data gathered during 27 flights over TCs as part of four different field experiments between 1998 and 2010 (Cecil et al. 2014). Two aircraft were utilized during these field experiments: the National Aeronautics and Space Administration (NASA) ER-2, a piloted aircraft that flies at approximately 65,000 ft (FL650), and the NASA unmanned GH.

Prior to the start of another research project in 2012 that utilized the GH, the NASA Hurricane and Severe Storm Sentinel (HS3) field experiment (Braun et al. 2016), new weather hazard avoidance rules were implemented for GH overflights of intense tropical convection (e.g., TCs). These rules included:

1. When flying below Flight Level (FL) 500 [50Kft in pressure altitude (PA) coordinates, 15.24 km], do not approach thunderstorms within 25 nautical miles (nm).
2. When flying above FL500, do not approach reported lightning within 25 nm if cloud tops are at FL500 or higher. Aircraft should maintain at

[1] Cooperative Institute for Meteorological Satellite Studies, University of Wisconsin-Madison, Madison, WI, USA. E-mail: sarah.griffin@ssec.wisc.edu

least 10,000 ft vertical separation from reported lightning if cloud tops are below FL500.
3. No overflight of cloud tops higher than FL500.

It is believed these rules were implemented as a result of an ER-2 flight over the center of TC Emily on 17 July 2005. The pilot encountered significant turbulence during an overpass of the TC eyewall, a ring of intense convection surrounding the TC center (American Meteorological Society 2015a). The turbulence encounter corresponded with an ER-2 Doppler radar (EDOP; Heymsfield et al. 2001) estimated updraft speed greater than 20 m s^{-1} (Cecil et al. 2010). Figure 1 shows the GOES satellite infrared (IR) imagery from TC Emily at 0750 UTC on 17 July 2005 and 1 h of the ER-2 flight track 1 h before and approximately during the time of the observed turbulence. Based on the IR brightness temperatures (BTs) alone, it is not immediately evident that the ER-2 would encounter turbulence in this case.

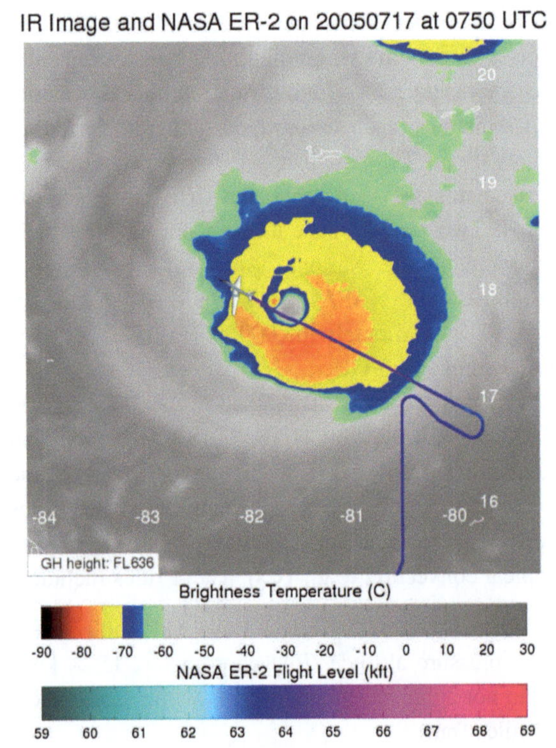

Figure 1
Satellite infrared (IR) imagery from TC Emily and previous hour of the ER-2 flight track at approximately the time of the observed turbulence, 0750 UTC on 17 July 2005

While the above rules were put into place to protect the GH from weather hazards such as turbulence, as the GH is not designed to fly in moderate or severe turbulence (Phil Hall, GH pilot, personal communication), high cloud tops are common in TCs (Houze 2010) and not always associated with turbulence. Therefore, these rules could put a constraint on obtaining valuable observations over a TC core. The purpose of this paper is to describe new hazard avoidance products that can be used toward mission decision support to distinguish areas of possible high-altitude turbulence from areas that are safe for the GH to overfly. The outline is as follows: Sect. 2 describes the three elements of weather hazard avoidance developed to identify areas of high-altitude turbulence which are used in Sect. 3 to compare the ER-2 overflight of TC Emily on 17 July 2005 with non-turbulent GH overflights of TC Karl and TC Matthew on 17 September 2010 and 24 September 2010, respectively. Section 4 outlines the modified GH flight rules resulting from the analysis in Sect. 3, and Sect. 5 provides some concluding remarks.

2. *High-Altitude Weather Hazard Avoidance Products*

Three specific products are described that, when utilized together, can be used to identify weather hazards for high-altitude aircraft missions in tropical regions. The products focus on deep vigorous convection that is not uncommon in the tropics, especially in TCs, and can be associated with aircraft hazards such as turbulence and lightning.

2.1. *Very High Cloud-Top Heights*

Satellite-derived BTs based on IR imagery are often used as a proxy to identify areas of deep and active convection. However, cold IR BTs do not necessarily infer active convection. Other cloud properties need to be taken into account. For example, the cloud emissivity, optical depth, and type can be used to help discriminate convectively-active cumulonimbus from passive thick cirrus clouds. These properties can also be used to estimate the cloud-top heights (CTH), which are more useful than the IR BTs for aviation purposes. CTH can be analyzed using most geostationary or polar-orbiting satellites.

This study utilizes IR imagery from the Geostationary Operational Environmental Satellite (GOES) satellite series (GOES 11-13) to calculate CTH. The first step involves the GOES-R Advanced Baseline Imager (ABI) Cloud Height Algorithm (ACHA; Heidinger 2011) that utilizes the IR BT data and a radiative transfer model to derive cloud-top temperatures (CTT). The CTTs are then converted to cloud-top pressures (CTP) using an atmospheric temperature profile provided by numerical weather prediction model data. The CTPs are then converted to CTH in pressure altitude (PA) coordinates using the method from Griffin et al. (2016). PA, the desired metric in the aviation community, is defined as the height of a given atmospheric pressure in the 1976 International Civil Aviation Organization standard atmosphere (American Meteorological Society 2015b).

As will be described in Sect. 3, the application of these CTH alone is not sufficient to detect hazardous regions to aviation. Very high-altitude CTHs are one indicator of possible concerns, but additional information can augment the probabilities of encountering hazardous conditions.

2.2. *Tropical Overshooting Tops*

Detection of overshooting tops (OTs) in satellite imagery has been linked to vigorous updrafts and shown to correlate with severe weather and tropospheric turbulence (Bedka et al. 2010). The presence of these features above a high cirrus cloud canopy can be a distinguishing feature for determining areas of active convection vs. quiescent conditions at high-altitude flight levels. Tropical overshooting tops (TOTs) can be identified using an objective satellite-based detection algorithm (Monette et al. 2012), which is a derivative of an OT detection algorithm developed by Bedka et al. (2010). TOTs are identified as pixels colder than 215 K in the GOES 10.7-μm IR BTs. The mean BT of the anvil cloud surrounding each TOT candidate pixel is computed using pixels at an 8 km radius in 16 radial directions. At least 9 of 16 anvil pixels must have a BT colder than 225 K, and only these pixels are used to calculate the mean anvil BT. TOT strength can be defined using the BT difference (BTD) between the coldest TOT pixel and the mean anvil BT. Candidate TOTs are divided into two categories based on their BTDs: those with a BTD less than or equal to − 9 K and those with a BTD greater than − 9 K, based on Monette et al. (2012). For application to TCs, TOTs with a BTD greater than − 9 K are only identified within 100 km of the TC center, where the coldest cloud tops associated with the central dense overcast are more prevalent (Velden et al. 1998). Further information about the TOT algorithm can be found in Griffin (2017).

2.3. *Lightning*

Electrical activity in clouds is another sign of active convection. Lightning can be detected using ground-based networks such as the Vaisala Long-range Lightning Detection Network (LLDN). The LLDN uses the large electromagnetic signals associated with cloud-to-ground lightning return strokes, as well as large pulses in cloud discharges, to identify the timing and location of lightning discharges using as few as two sensors. The LLDN employs a subset of the approximately 200 sensors from the United States (US) National Lightning Detection Network and the Canadian Lightning Detection Network (Cummins et al. 2008).

The accuracy of LLDN lightning data can be described using the detection efficiency and location accuracy. The detection efficiency over the Atlantic Ocean is shown in Fig. 2, adapted from Pessi et al. (2009). Detection efficiency is highest near the US coastline and decreases with increasing distance from land. For a given location, the detection efficiency during the day, which is defined as between 12 UTC and 22 UTC, is lower than during the night, defined as 00 UTC to 10 UTC. Location accuracy is about 1 km near the US coast, falling off uniformly to about 30 km near the 5% detection efficiency boundary. More information about the Vaisala LLDN can be found in Pessi et al. (2009). Improved coverage over oceanic regions is becoming available from new satellite-based instruments, such as the GOES-R Lightning Mapper.

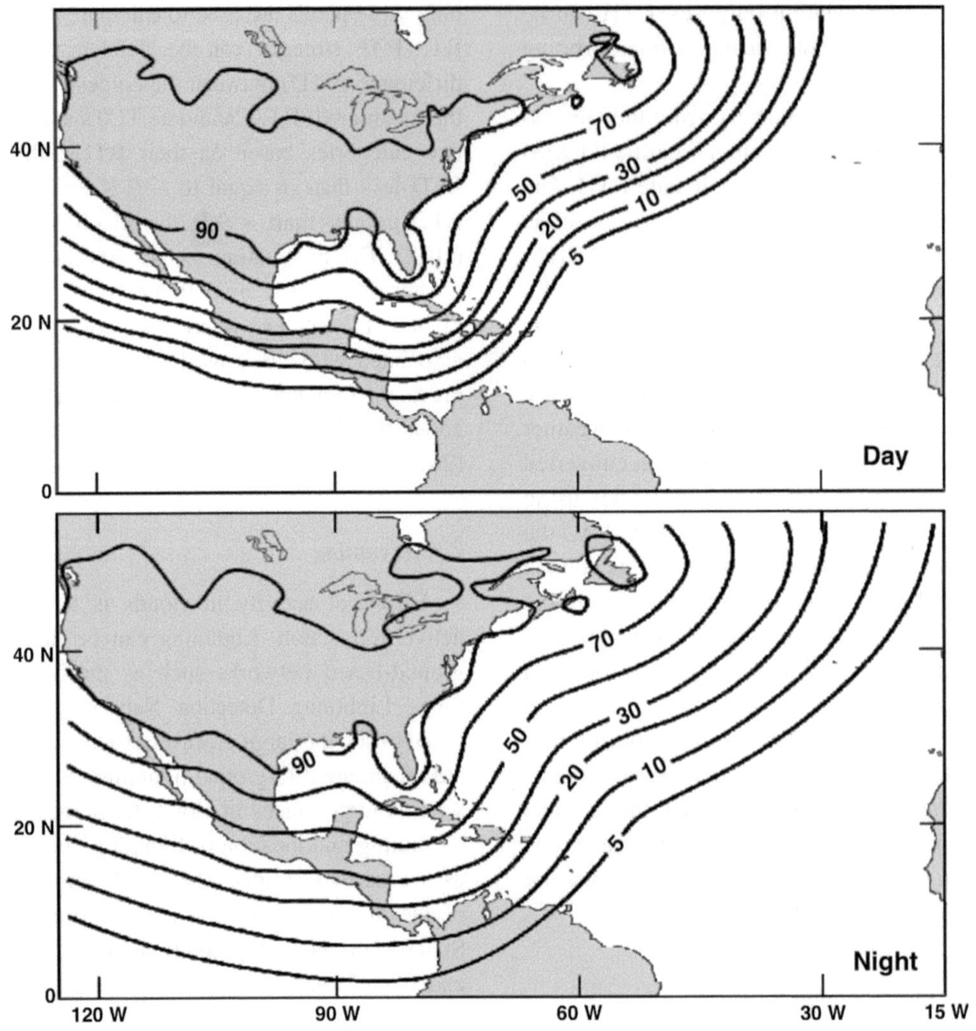

Figure 2
Vaisala Long-range Lightning Detection Network detection efficiency for both day (12 UTC to 22 UTC) and night (00 UTC to 10 UTC). Adapted from Pessi et al. (2009)

3. Comparison of TC Overflight Cases

In this section, we utilize the products described above to characterize and compare selected overflights of TCs by high-altitude aircraft. During TC Emily in 2005, a NASA ER-2 overflew the storm as part of a research mission, and encountered severe turbulence noted by the pilot. In 2010, two TCs (Karl and Matthew) were overflown by the GH as part of a research campaign, and in neither case was any turbulence detected by the pilotless aircraft.

3.1. Cloud-Top Heights

The CTH during the three flights are analyzed using GOES-11 imagery for Emily and GOES-13 imagery for Karl and Matthew. The ACHA CTH products for these three overflights were not available in real time, and have been post-processed for this study. Figure 3 illustrates the vertical cross-sections of each flight. The solid black lines represent the aircraft height during flight, with the gray curtain underneath them representing the 10,000 ft 'safety' barrier outlined in the second GH flight rule noted previously. The dashed black lines represent FL500,

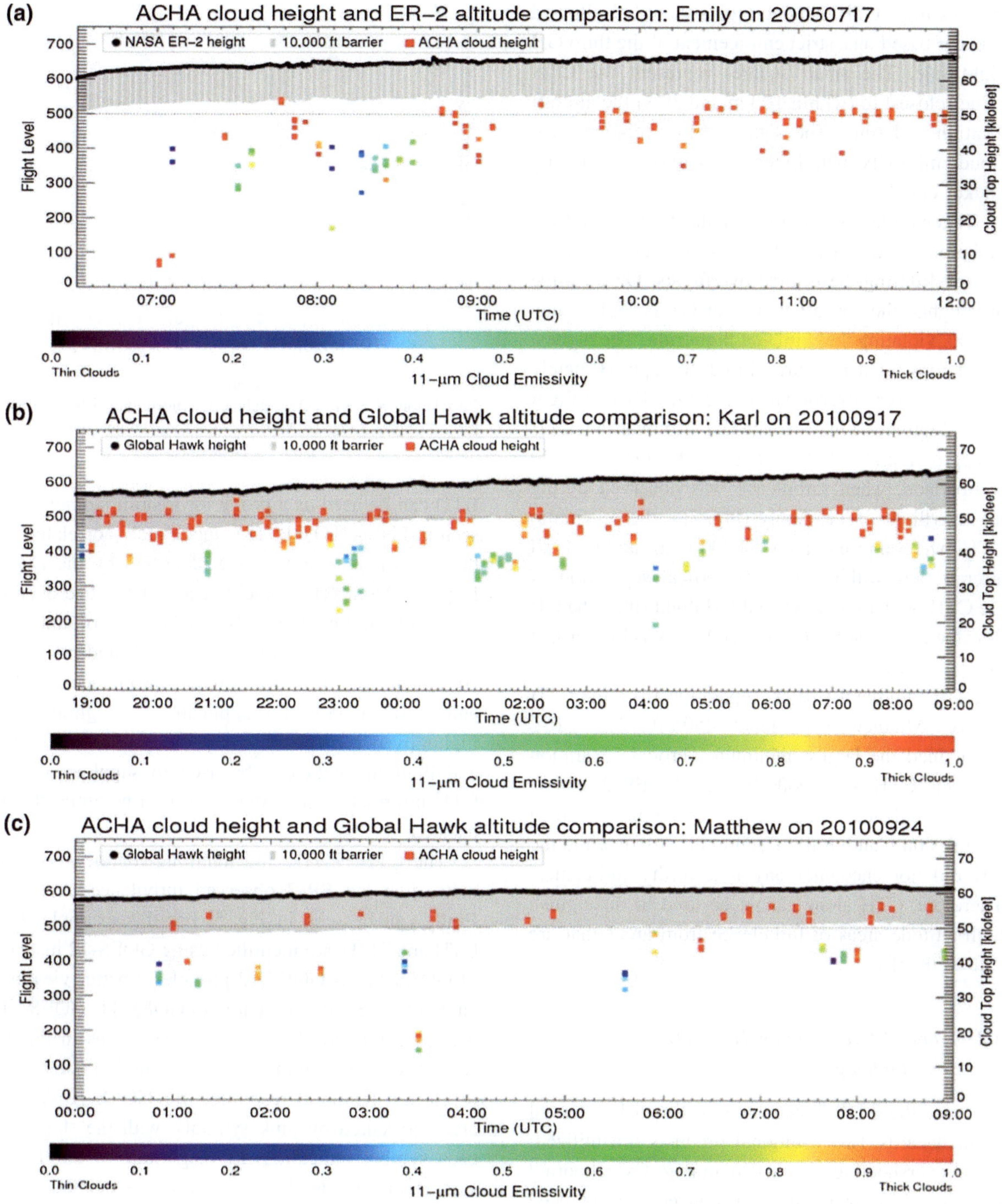

Figure 3
Vertical cross-section of CTH and aircraft height for the overflights of **a** TC Emily in 2005, **b** TC Karl in 2010, and **c** TC Matthew in 2010. The solid black line and gray curtain underneath represents the aircraft height and 10,000 ft vertical separation, respectively. The small squares represent the CTH derived from GOES coincident with the aircraft location and time, colored by cloud emissivity (thickness)

the maximum CTH that overflights should be allowed to occur based on a strict enforcement of the third GH flight rule. The small squares represent the CTH values closest to (within 10 km and 30 s) the aircraft location and time. The square colors represent the cloud emissivity, with higher emissivity representing thicker clouds.

Taken on face value, Fig. 3 indicates that multiple violations of the GH flight rules outlined in Sect. 1 occurred during these three overflights. On all three overflights, the aircraft flew over CTHs higher than FL500, violating the third GH flight rule. The highest cloud tops, at a pressure altitude of approximately 56,800 ft, were overflown during the GH overflight of TC Matthew (Fig. 3c). The GH also overflew CTH of approximately 54,800 ft over TC Karl (Fig. 3b). In comparison, when turbulence was observed during the overflight of TC Emily (Fig. 3a), the maximum CTHs were approximately 54,200 ft. In addition, all aircraft flew within 10,000 ft vertical separation of the CTH, violating the second GH flight rule. The GH flew the closest to the CTH over TC Karl (Fig. 3b), at one point with only a $\sim$ 2350 ft vertical separation. The GH minimum vertical separation from the CTH of TC Matthew was about 4500 ft. The ER-2 maintained the largest minimum vertical separation from the CTH of $\sim$ 9600 ft. Yet, the ER-2 experienced turbulence strong enough for the pilot to deem the mission plan unsafe (Cecil et al. 2010), while the GH did not encounter any noteworthy turbulence. Therefore, CTH alone cannot be used to distinguish high-altitude areas of turbulence from areas that are safe to overfly.

3.2. *Tropical Overshooting Tops (TOTs) and Lightning*

Since the CTH alone cannot distinguish potential flight hazards, two additional products are utilized: cloud overshooting tops and lightning. Overshooting tops are a known cause of convectively-induced turbulence (Bedka et al. 2010), with the probability of experiencing moderate or greater turbulence increasing as distance from the overshooting top decreases. This turbulence is potentially caused by thunderstorm updrafts generating vertically propagating gravity waves (Fovell et al. 1992; Lane et al. 2012). TOT

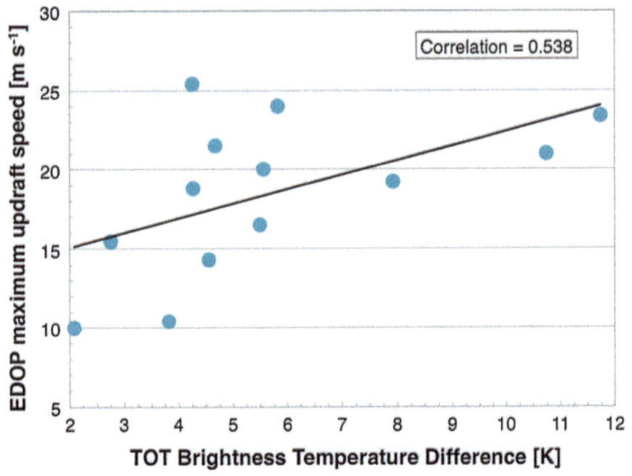

Figure 4
Correlation between TOT brightness temperature difference and ER-2 Doppler Radar (EDOP) measured maximum updraft speed from 13 TC updrafts

BTD can be used as a proxy for updraft strength. As seen in Fig. 4, TOT BTD is moderately correlated to the maximum updraft speed identified by the ER-2 Doppler Radar (Heymsfield et al. 2010). The occurrence of lightning is also indicative of intense convection (Carey and Rutledge 2000) and has been correlated with reported aircraft turbulence (Wolff and Sharman 2008). For applications to high-altitude flights over tropical areas, especially oceanic, TOTs and lightning data can be used to supplement the CTH information to better inform on potential flight hazards.

The CTH, TOTs, and lightning for the approximate time the ER-2 observed turbulence over TC Emily can be seen in Fig. 5. For this example, the CTH and TOTs are identified using GOES-12 instead of GOES-11, as GOES-12 provides a better viewing angle of TC Emily compared to GOES-11. GOES-11 was used for the plots in Fig. 3, as it was in rapid-scan mode with 5-min temporal resolution compared to 15-min temporal resolution for GOES-12. TOTs are represented by pink symbols, with the shape of each symbol indicating the magnitude of the TOT. Larger magnitude TOTSs can be associated with stronger updrafts (Monette et al. 2014), potentially increasing the possibility of experiencing turbulence. Lightning flashes are represented by cyan crosses, with the most recent flashes indicated by larger crosses. Lightning flashes are defined as the average location of all lightning strokes within 75 km and 1 s

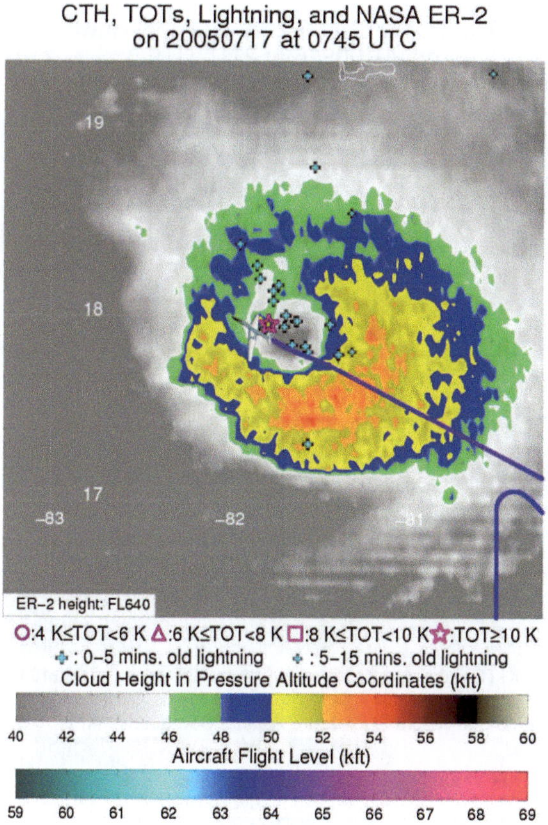

Figure 5
ER-2 flight track height and CTH, TOTs, and lightning from 17 July 2005 at 0745 UTC for the overflight of TC Emily, with the plane symbol located at the approximate location turbulence was experienced

of each other. At the time of observed turbulence, the ER-2 is in the vicinity of a small area of high CTH in the inner northwest eyewall and an associated TOT with a large magnitude (BTD of 11.74 K) commensurate with an aircraft-measured updraft of 23.5 m s^{-1} (Heymsfield et al. 2010). In addition, multiple lightning flashes are observed near the TOT, further indicating active convection in the vicinity of the ER-2 when turbulence was experienced.

As a comparison with Fig. 5, Fig. 6 depicts the CTH, TOTs, and lightning from the GH overflights of TC Matthew and TC Karl when no turbulence was experienced. Figure 6a is coincident with the GH overflying some of the highest CTHs observed in TC Matthew. Unlike the ER-2 over TC Emily in Fig. 5, no TOTs or lightning is observed in the vicinity of the GH track. So, while the CTHs are very high, they could be the result of dense cirrus outflow from previous convection in this region that was indicated around 05 UTC (not shown) (McFarquhar and Heymsfield 1997). Figure 6b depicts TC Karl during the GH's smallest vertical separation from the CTH. Once again, high CTH but no TOTs are observed in the vicinity during this GH overflight. A few lightning flashes are observed; however, the flashes are fewer in number and older than the lightning flashes observed in TC Emily (the lightning network detection efficiency for both TC Emily and TC Karl was approximately 30%). Therefore, these cases indicate that all three products, best used in tandem, are essential for identifying areas of vigorous convection which could potentially cause hazardous conditions for high-altitude aircraft.

4. *Update to GH flight rules*

As a result of the above analysis, the existing GH flight rules were amended to somewhat relax the restrictions on overflying TCs with very high CTH. The updated flight rules allow overflights if no significant convective overshoots or frequent lightning is observed, essentially removing the rule advising against the overflight of CTH higher than FL500. A 25 nm spatial separation between the GH and TOTs/ frequent lightning is still advised, or at least 10,000 ft vertical separation is to be maintained, if the CTH is below FL500 and the GH is above FL500. However, if the CTH is above FL500 and TOTs but no lightning is observed, only a 5,000 ft vertical separation is necessary. The GH is clear to overfly CTHs below FL500 with TOTs, but no frequent lightning.

These modified rules were in place during more recent TC field programs that employed the GH for high-altitude data collection. The remote pilots had access to the three aforementioned products in real time, with updates every 1–30 min depending on the product and available scanning frequency. Some real-time examples from the NASA HS3 and the National Oceanic and Atmospheric Administration (NOAA) Sensing Hazards with Operational Unmanned Technology (SHOUT) GH overflights of TCs are shown in Figs. 6 and 7. A good example of GH hazard avoidance occurred during the overflight of TC Dolly as part of an HS3 mission. Figure 7 shows that the

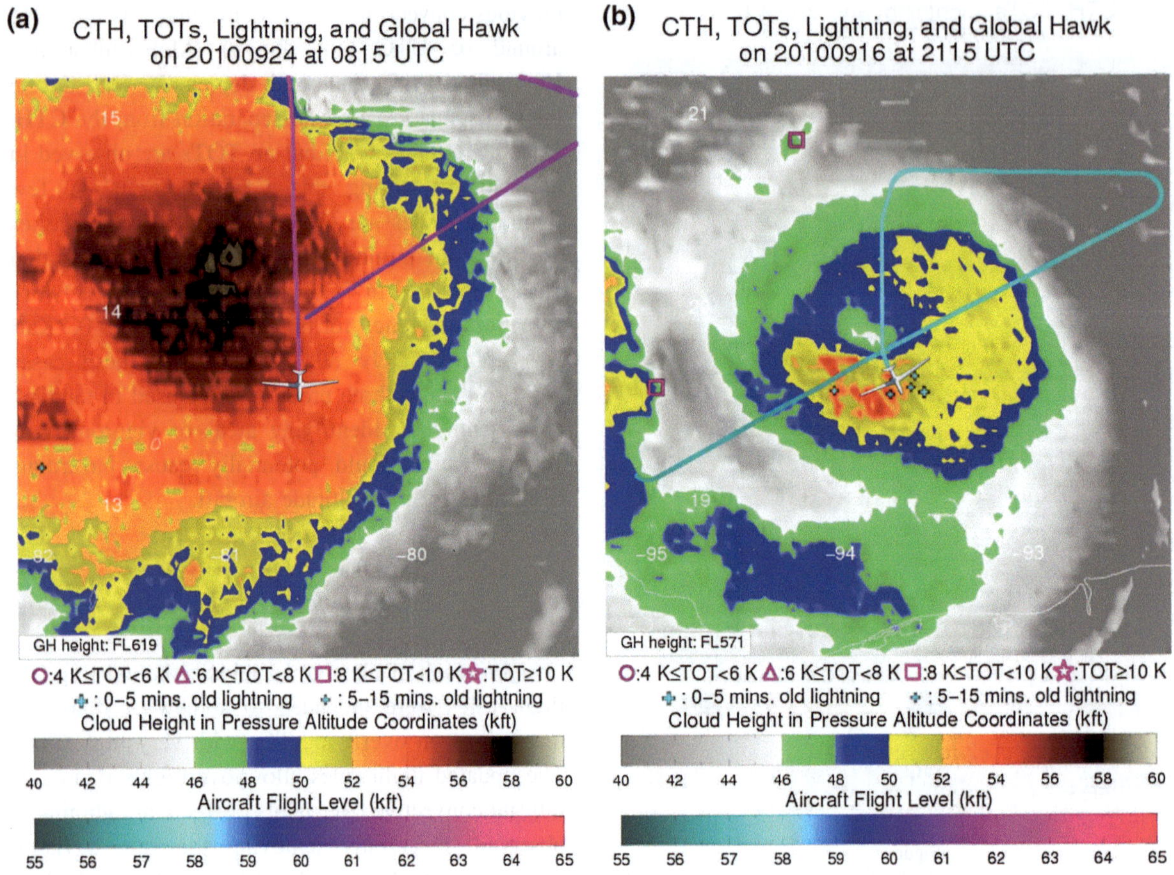

Figure 6

GH flight track height and CTH, TOTs, and lightning from **a** 24 September 2010 at 0815 UTC during the overflight of TC Matthew and **b** 16 September 2010 at 2115 UTC during the overflight of TC Karl. **a** Depicts the time of highest CTH overflown by the GH, and **b** depicts the time of minimum vertical separation between the GH and CTH

GH overflies CTH above FL520 where no TOTs are observed; however, it turns away from the region associated with very high CTH and frequent lightning. Figure 8 depicts the GH overflight of TC Matthew in 2016 during SHOUT. While the CTH associated with TC Matthew was very high (around FL580), neither TOTs nor lightning was observed, and no turbulence was reported.

5. Discussion and Conclusions

This study explores the use of satellite and lightning products to improve the analysis of conditions for high-altitude ($\sim$ 60kft) aircraft during overflights of tropical cyclones (TCs). Three primary tools are identified: Cloud-Top Heights (CTH), Tropical Overshooting Tops (TOTs), and lightning data. CTH and TOTs are identified using geostationary satellite algorithms, while lightning is detected using the Vaisala Long-range Lightning Detection Network. In the case studies examined (and other experiences based on real-time applications), it is found that no product by itself is an adequate predictor of high-altitude flight hazards. However, a combination of all three products when viewed together can convincingly identify likely regions/periods of intense convection in TCs with possible associated above-cloud turbulence vs. quasi-undisturbed conditions.

The motivation for this study was to re-examine existing GH flight rules that were thought to be too strict during the 2012 field phase of the NASA HS3

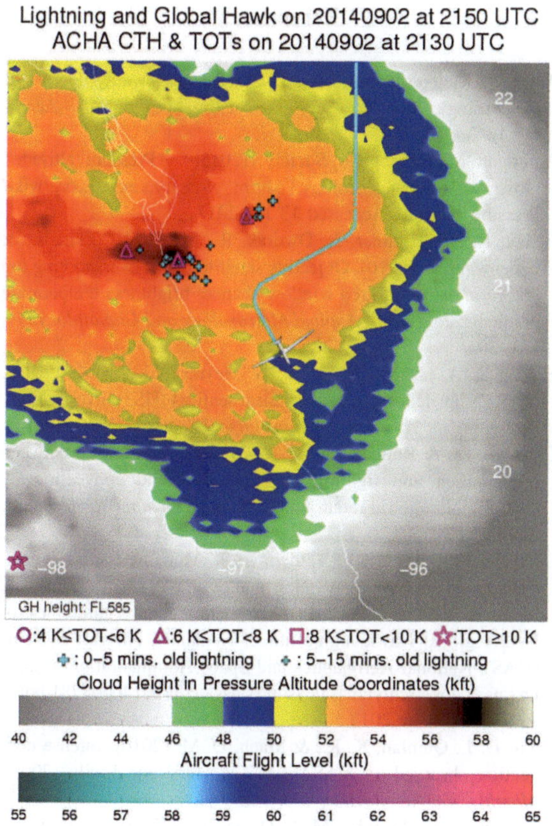

Figure 7
GH flight track height, TOTs and lightning from 2 September 2014 at 2150 UTC for the overflight of TC Dolly. CTH and TOTs are from 2 September 2014 at 2130 UTC

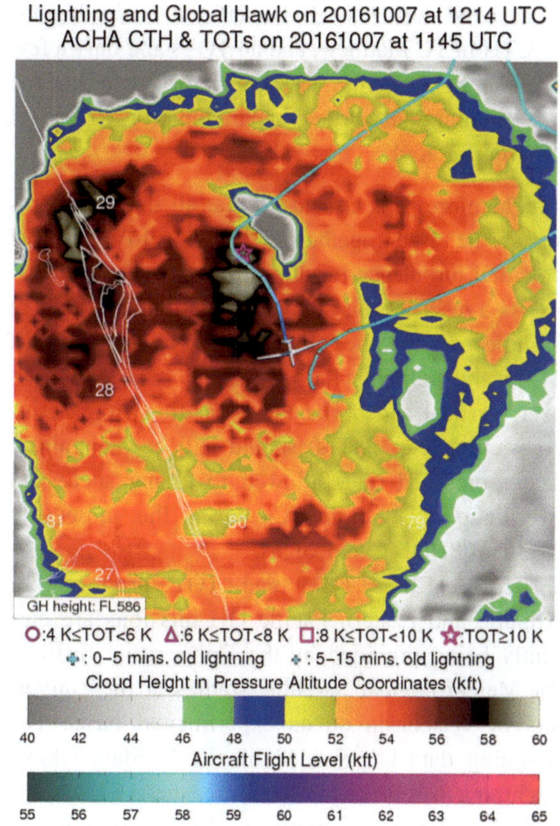

Figure 8
GH flight track height, TOTs and lightning from 7 October 2016 at 1204 UTC for the overflight of TC Matthew. CTH and TOTs are from 7 October 2016 at 1145 UTC

experiment. Based on these rules, GH overflights of TCs were somewhat restricted, potentially limiting the usefulness of the experiment. However, later in HS3, a few flight legs were allowed over very high CTH if no concurrent TOTs or lightning was observed. Detectors onboard the GH indicated no significant turbulence during these overflights.

The origin of the flight rules is believed to have come from a single ER-2 overflight of TC Emily (2005) that experienced severe turbulence when overflying a convective updraft of at least 20 m s^{-1}. The conditions associated with the turbulent ER-2 overflight of TC Emily are compared to the GH overflights of TCs Matthew and Karl in 2010 that did not experience noteworthy turbulence. Significant findings include:

1. The CTH of the convective cell in TC Emily's eyewall overflown at the time of the observed turbulence was lower than most of the high CTH overflown by the GH during TCs Matthew and Karl. Therefore, cold IR brightness temperatures or high CTH are a not sufficient condition for restricted flight paths.
2. The vertical separation between the ER-2 and TC Emily's CTH at the time of the observed turbulence was larger than the minimum vertical separation between the GH and the highest CTH from TCs Karl and Matthew. If the cloud tops appear non-convective and no lightning is observed, then a smaller "safe zone" above the clouds can be allowed.
3. At the approximate time and location of the ER-2 turbulence report during TC Emily, a strong TOT was detected along with multiple lightning flashes that were coincident with a strong convective updraft. No TOTs or lightning flashes were

detected during the GH overflight of the highest CTH in TC Matthew, and no TOTs and only a few older lightning flashes were observed during the GH minimum vertical separation from the CTH over TC Karl.

Therefore, it is concluded that all three products should be used together to best identify areas of potential turbulence resulting from TC convection.

As a result of this study, the GH flight rules were modified to lift the maximum CTH constraint if TOTs and lightning flashes are not detected. However, vertical and horizontal separation from active convection (e.g., TOTs) should still be maintained. These modified rules allowed for many successful overflights of multiple TCs during the subsequent HS3 and SHOUT campaigns.

The real-time CTH and TOT products for Atlantic and Eastern Pacific GH missions are currently being provided by the Cooperative Institute for Meteorological Satellite Studies at the University of Wisconsin on an experimental basis. The lightning data is provided by the Unidata Global Lightning Network. In the future, the ACHA CTH will become an operational GOES product at NOAA/NESDIS. At this time, the TOT product is not on an operational track at NOAA/NESDIS, but will be continued to be supported by CIMSS for specific applications. The coverage of lightning data over the tropical Atlantic and eastern Pacific will expand with the deployment of the Lightning Mapper on the GOES-R series of geostationary satellites. Therefore, it will be possible to expand the research and findings of this study to more general transoceanic aviation applications.

Acknowledgements

The authors wish to acknowledge the encouragement of several individuals to the advancement of this work: Ed Zipser, Scott Braun, Dan Cecil, Gerald Heymsfield, Jason Dunion, Gary Wick, Peter Black, Kris Bedka, Andy Heidinger, Wayne Feltz, Amber Emory. Support for this research was provided by the following programs: NASA HS3, NOAA GOES-R Risk Reduction, and NOAA SHOUT.

References

American Meteorological Society. (2015a). "Eyewall" glossary of meteorology. http://glossary.ametsoc.org/wiki/Eyewall. Accessed 13 Jun 2017.

American Meteorological Society. (2015b). "Pressure altitude" glossary of meteorology. http://glossary.ametsoc.org/wiki/Pressure_altitude. Accessed 13 Jun 2017.

Bedka, K. M., Brunner, J., Dworak, R., Feltz, W., Otkin, J., & Greenwald, T. (2010). Objective satellite-based overshooting top detection using infrared window channel brightness temperature gradients. *Journal of Applied Meteorology and Climatology, 49,* 181–202.

Braun, S. A., Newman, P. A., & Heymsfield, G. M. (2016). NASA's hurricane and severe storm sentinel (HS3) investigation. *Bulletin of the American Meteorological Society, 97,* 2085–2102.

Carey, L. D., & Rutledge, S. A. (2000). The relationship between precipitation and lightning in tropical island convection: A C-band polarimetric radar study. *Monthly Weather Review, 128,* 2687–2710.

Cecil, D. J., Zipser, R. J., Velden, C. S., Monette, S. A., Heymsfield, G. M., Braun, S. A., Newman, P. A., Black, P. G., Black, M. L., & Dunion, J. P. (2014). Weather avoidance guidelines for NASA global Hawk high-altitude Unmanned Aircraft Systems (UAS). National Aeronautics and Space Administration. https://ntrs.nasa.gov/archive/nasa/casi.ntrs.nasa.gov/20140008781.pdf. Accessed 13 Jun 2017.

Cecil, D. J., Quinlan, K. R., & Mach, D. M. (2010). Intense convection observed by NASA ER-2 in Hurricane Emily (2005). *Monthly Weather Review, 138,* 765–780.

Cummins, K. L., Murphy, M. J., Demetriades, N. W. S., Pifer, B., Pessi, A., & Businger, S. (2008). *Modeling and calibration of Vaisala's operational long range lightning detection network.* 20th International Lightning Detection Conference, USA, Tucson AZ, 2008. https://pdfs.semanticscholar.org/79a5/9df76ae1f95caf631470a52d49553490e4c3.pdf. Accessed 13 Jun 2017.

Fovell, R., Durran, D., & Holton, J. R. (1992). Numerical simulations of convectively generated stratospheric gravity waves. *Journal of Atmospheric Science, 49,* 1427–1442.

Griffin, S. M. (2017). Climatology of tropical overshooting tops in North Atlantic tropical cyclones. *Journal of Applied Meteorology and Climatology, 56,* 1783–1796.

Griffin, S. M., Bedka, K. M., & Velden, C. S. (2016). A method for calculating the height of overshooting convective cloud tops using satellite-based IR imager and CloudSat cloud profiling radar observations. *Journal of Applied Meteorology and Climatology, 55,* 479–491.

Heidinger, A. K. (2011). ABI cloud height. NOAA NESDIS Center for satellite applications and research algorithm theoretical basis doc. http://www.goes-r.gov/products/ATBDs/baseline/Cloud_CldHeight_v2.0_no_color.pdf. Accessed 13 Jul 2017.

Heymsfield, G. M., Halverson, J. B., Simpson, J., Tian, L., & Bui, T. P. (2001). ER-2 Doppler radar investigations of the eyewall of Hurricane Bonnie during the Convection and Moisture Experiment-3. *Journal of Applied Meteorology, 40,* 1310–1330.

Heymsfield, G. M., Tian, L., Heymsfield, A. J., Li, L., & Guimond, S. (2010). Characteristics of deep tropical and subtropical convection from nadir-viewing high-altitude airborne Doppler radar. *Journal of Atmospheric Science, 67,* 285–308.

Houze, R. A., Jr. (2010). Clouds in tropical cyclones. *Monthly Weather Review, 138,* 293–344.

Lane, T. P., Sharman, R. D., Trier, St. B, Fovell, R. G., & Williams, J. K. (2012). Recent advances in the understanding of near-cloud turbulence. *Bulletin of the American Meteorological Society, 93,* 499–515.

McFarquhar, G. M., & Heymsfield, A. J. (1997). Parameterization of tropical cirrus ice crystal size distributions and implications for radiative transfer: Results from CEPEX. *Journal of Atmospheric Science, 54,* 2187–2200.

Monette, S. A., Velden, C. S., Griffin, K. S., & Rozoff, C. M. (2012). Examining trends in satellite-detected tropical overshooting tops as a potential predictor of tropical cyclone rapid intensification. *Journal of Applied Meteorology and Climatology, 51,* 1917–1930.

Monette, S. A., Velden, C. S., Zipser, E. J., Cecil, D. J., Black, P. G., Braun, S. A. & Heymsfield, G. M. (2014). Differentiating tropical cyclones exhibiting strong updrafts using combined satellite, lightning, and global Hawk observations. *31st Conference on Hurricanes and Tropical Meteorology*, USA, San Diego, CA. https://ams.confex.com/ams/31Hurr/videogateway.cgi/id/26889?recordingid=26889

National Hurricane Center (2014). Hurricane Edouard discussion 16. http://www.nhc.noaa.gov/archive/2014/al06/al062014.discus.016.shtml. Accessed 28 Jul 2017.

National Hurricane Center (2016). Tropical Storm Hermine Discussion 16. http://www.nhc.noaa.gov/archive/2016/al09/al092016.discus.016.shtml. Accessed 28 Jul 2017.

Pessi, A. T., Businger, S., Cummins, K. L., Demetriades, N. W. S., Murphy, M., & Pifer, B. (2009). Development of a long-range lightning detection network for the Pacific: Construction, calibration, and performance. *Journal of Atmospheric and Oceanic Technology, 26,* 145–166.

Velden, C. S., Olander, T. L., & Zehr, R. M. (1998). Development of an objective scheme to estimate tropical cyclone intensity from digital geostationary satellite infrared imagery. *Weather and Forecasting, 13,* 172–186.

Wolff, J. K., & Sharman, R. (2008). Climatology of upper-level turbulence over the contiguous United States. *Journal of Applied Meteorology and Climatology, 47,* 2198–2214.

(Received October 12, 2017, revised November 30, 2017, accepted January 4, 2018, Published online January 10, 2018)

Pure Appl. Geophys. 176 (2019), 2057–2079

https://doi.org/10.1007/s00024-018-1975-x

Pure and Applied Geophysics

R2O Transition of NCAR's Icing and Turbulence Algorithms into NCEP's Operations

HUI-YA CHUANG,[1] YALI MAO,[1,2] and BINBIN ZHOU[1,2]

Abstract—National Center for Environmental Prediction (NCEP) started distributing global operational gridded in flight icing, turbulence and convective cloud products as part of World Area Forecast System (WAFS) products in 2007. Simple algorithms were used to derive these products during early stage based on NCEP Global Forecast System (GFS) forecast. These products quickly became essential flight planning tool for international aviation community and are especially important to developing countries that do not have resource to run numerical models themselves. To further improve these products, Environmental Modeling Center (EMC) started collaborating with National Center for Atmospheric Research (NCAR) to transition their aviation research algorithms into NCEP's operations (R2O), particularly Forecast Icing Potential (FIP) and Graphical Turbulence Guidance (GTG) algorithms. The initial attempt is to apply FIP to GFS forecast to potentially replace WAFS icing product. Extensive evaluation demonstrated FIP outperformed original WAFS icing product and, with support from Aviation Weather Center (AWC) and Federal Aviation Administration (FAA), EMC replaced US WAFS icing product with FIP in 2015. EMC recently also implemented GTG with 2017 GFS upgrade but GTG will not replace WAFS turbulence until 2019. This paper will describe the methodology which EMC used to transition NCAR's aviation research algorithms into NCEP's operations. It will also describe how EMC generates icing analysis data to be used as truth for performing objective verification. Several case studies will be presented and the methodology and results for objective validation will be discussed. Finally, future collaboration plan with NCAR and implementation plans to continue to improve WAFS products will be stated.

Key words: Aviation R2O, aviation, R2O, aviation product improvement, world area forecast system, WAFS, global gridded icing forecast, global gridded turbulence forecast, global gridded icing analysis, verification of in flight icing forecast, verification of turbulence forecast, improvement of global icing forecast, improvement of global turbulence forecast, collaboration between EMC and NCAR, case studies for in flight icing forecast, case studies for turbulence forecast, FIP, GTG, FIS, CIP, G-FIP, GGTG, G-FIS, G-CIP, speed up of GTG algorithm, speed up R2O process, Collaborative partnership, Efficient R2O Transition.

[1] NOAA/NCEP, Environmental Modeling Center, 5830 University Research Court, College Park, MD 20740, USA. E-mail: hui-Ya.Chuang@noaa.gov

[2] IMSG, Rockville, MD, USA.

1. Introduction

There are two World Area Forecast Centers (WAFC), Washington and London, each providing real-time meteorological information broadcasts for aviation purpose to back up each other. These broadcasts are supervised by International Civil Aviation Organization (ICAO) to fulfill requirements of the ICAO Annex 3 covering meteorological information which is necessary for flight (International Civil Aviation Organization 2016). The WAFC Washington provides ICAO with global forecast of temperature, wind, height, relative humidity, and precipitation to using forecast from its Global Forecast System (GFS http://www.emc.ncep.noaa.gov/GFS/doc.php). In response to ICAO's initiatives to improve global flight safety, both WAFCs started to generate global gridded forecasts of flight hazard products, including in flight icing, turbulence and cumulonimbus clouds out to 36 h forecast, as part of WAFS products in 2007 (Trojan 2007, Turp 2006).

In-flight icing (Gultepe 2018) occurs when aircrafts fly through supercooled clouds and cloud droplets freeze and build up on leading edges of wings. The ice alters airflow over the wings and tail, reduces the lift force that keeps the plane in the air, and can potentially cause aerodynamic stall, a condition that can lead to temporary loss of control. Fortunately, most modern large commercial aircrafts are equipped with deicing equipment which requires burning of fuel. Therefore, most commercial aircrafts can fly through predicted in flight icing environment as long as they know to carry enough fuel ahead of time. It is noted that aircraft icing issues due to high-altitude ice water content can be problematic and is subject of many studies, although it is not the scope of this study. This means prediction of in-flight icing is important for flight planning. Another issue important for flight planning is being able to avoid flying through severe turbulence and hence importance of having capability to predict such an event.

Although international airlines quickly adapted these new gridded WAFS forecast products in their daily flight planning, the occasional large differences between these products generated by two WAFCs confused ICAO users. These differences are attributed not only to the differences in different global forecast models but also to different aviation algorithms used to derive these products. The short term solution was for both centers to start issuing the same single authoritative blended WAFS products in 2011. For icing and turbulence forecast, the blending method used by both centers is simply taking the average of products from two centers for average products and taking the maximum for maximum products. More details are described in methodology Sect. 2.5. It is noted here that the blending is only done on WAFS icing, turbulence and cumulonimbus clouds products. Two WAFCs have been working on comparing and possibly unifying icing and turbulence algorithms. More details will be provided later.

First implementation of WAFC Washington aviation hazard products in 2007 were generated by developing a post processing package that used low resolution GFS output with 50 km horizontal and 50 hPa vertical resolutions. These resolutions were highest resolution available to public at the time although the model native resolution was at approximately 35 km. The algorithm used to derive icing product was a simplified version of NCAR's Forecast Icing Potential (FIP) (Bernstein et al. 2005) algorithm (Wolff 2009) while Ellrod Index (Ellrod 1992) was used to derive turbulence product. Therefore, WAFC Washington turbulence product is not able to explicitly predict mountain wave component and has been relying on WAFC London to predict mountain wave turbulence in blended products. In addition to this shortcoming, the use of low resolution data in deriving aviation products appeared to lose skills quickly when going up to cruise attitude. More details on comparisons of old WAFS icing and turbulence algorithms with new ones will be described in Sect. 3 on verification results. The cumulonimbus cloud product for the WAFC Washington is derived in GFS model using Slingo algorithm (Slingo 1987). Aviation Weather Center (AWC) was recently tasked to start working on improving WAFS cumulonimbus cloud product. Therefore, improvement of this product will not be discussed in this paper.

To further improve WAFC Washington aviation hazard products, the Environmental Modeling Center (EMC) of National Center for Environmental Prediction (NCEP) NOAA started to collaborate with NCAR to transition their aviation algorithms into operational GFS production system in 2009. NCAR's aviation algorithms (Bernstein et al. 2005; Gultepe 2018) include in-flight icing, turbulence, and convection analysis and forecast. The icing and turbulence algorithms have long been coupled with North America Rapid Refresh Model (RAP) (Benjamin et al. 2016) to provide AWC with valuable aviation weather forecast guidance over Continental United States. EMC incorporated a global version of FIP (G-FIP) developed by NCAR into its Unified Post Processor (UPP) (Chuang 2010) and started generating experimental G-FIP products in late 2011. With support from AWC, EMC implemented G-FIP to replace then operational WAFC Washington icing product in 2015. EMC also implemented Global Forecast Icing Severity (G-FIS) (McDonough 2010) to complement G-FIP during 2016 GFS upgrade. A global version of Graphical Turbulence Guidance (GTG) (Sharman et al. 2006) was also incorporated into UPP recently. The new global GTG (G-GTG) (Sharman and Pearson 2017) product was implemented with 2017 GFS upgrade but the product is not yet distributed as it is undergoing objective evaluation from GSD and possibly further tuning by NCAR turbulence group. To fulfill ICAO's goal of moving toward probabilistic aviation forecast in the near future, EMC also started generating G-FIS operationally in all its twenty-one Global Ensemble Forecast System (GEFS) (Zhou et al. 2017) members in 2018.

The objective of this paper is to describe the methodology used by EMC to transition NCAR's aviation algorithms into NCEP's operations. It will also describe how EMC generated icing analysis data will be used as truth for performing objective verification. The results for objective verification as well as subjective evaluation based on a few case studies will be discussed. Finally, future collaboration plan with NCAR for planned implementations to improve WAFS products and to meet ICAO requirement will be presented.

2. Methodology

2.1. Description of First Version of WAFC Washington Icing and Turbulence Algorithms Implemented in 2007

In 2007, EMC implemented first version of WAFS aviation hazardous products, including icing potential forecast and turbulence (Trojan 2007), representing WAFC Washington. Both products used GFS output generated from UPP with horizontal resolution of 0.5° and vertical resolution of 50 hPa. The final products were then interpolated to 1.25° WAFS grid on thinned vertical levels prior to dissemination.

2.1.1 WAFC Washington Icing Potential Algorithm 2007–2014

The icing potential algorithm was based on FIP algorithm developed by NCAR funded by AWRP program (Kulesa 2003). Strictly it was one version of FIP but with different strategy of cloud layers and different fuzzy functions of each member (Fig. 1), tuned by EMC staff.

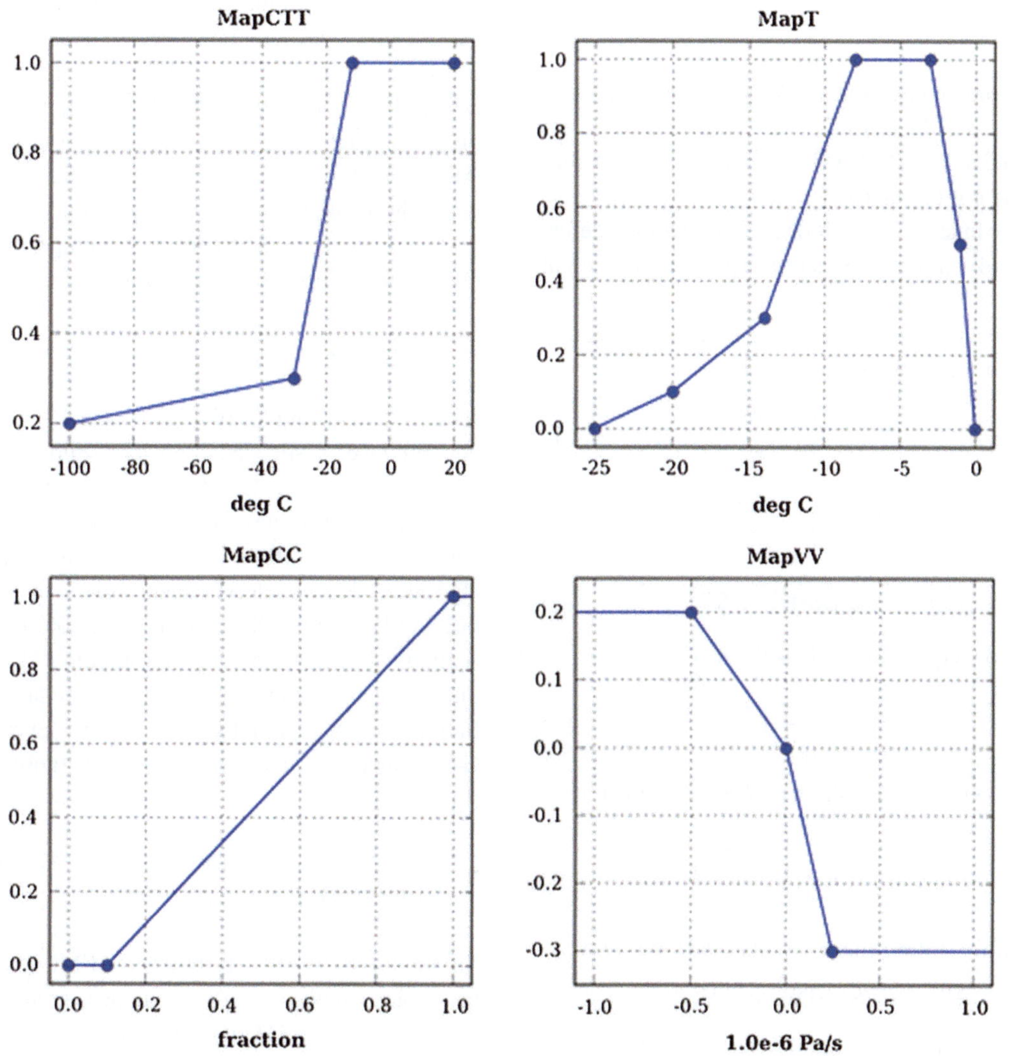

Figure 1
Fuzzy logics interest maps used in G-FIP implemented as operational US WAFS icing product in 2007, where *CTT* represents cloud top temperature, *T* represents temperature, *CC* represents cloud cover, and *VV* represents vertical velocity

Fuzzy logic functions are applied to cloud-related fields/members of model input data and map the members to 0–1 scales, representing the likelihood of icing, respectively. The members are chosen based on cloud physics, research and forecasting experience. To estimate the initial icing potential, temperature, cloud top temperature and cloud coverage are chosen for Icing 2007, while temperature, cloud top temperature and relative humidity are members for FIP 2015. In both versions of icing algorithms, fuzzy function of vertical velocity is applied to adjust the initial icing potential.

2.1.2 WAFC Washington Turbulence Algorithm 2007-Present

To be compatible with WAFC London product, Ellrod Index (Ellrod 1992) was calculated for Clear Air Turbulence (CAT), while mountain wave turbulence was not calculated for GFS.

Ellrod Index is the product of horizontal deformation and vertical wind shear. The total horizontal deformation (DEF) is composed of two parts: shearing deformation (DSH) and stretching deformation (DST) and they are defined as follows.

$$\mathrm{DSH} = \frac{\partial v}{\partial x} + \frac{\partial u}{\partial y} \tag{1}$$

$$\mathrm{DST} = \frac{\partial u}{\partial x} - \frac{\partial v}{\partial y} \tag{2}$$

$$\mathrm{DEF} = \sqrt{\mathrm{DSH}^2 + \mathrm{DST}^2} \tag{3}$$

The convergence (CVG) is defined as:

$$\mathrm{CVG} = -\left(\frac{\partial u}{\partial x} + \frac{\partial v}{\partial y}\right) \tag{4}$$

, and the vertical wind shear (VWS) is defined as:

$$\mathrm{VWS} = \sqrt{\left(\frac{\partial u}{\partial z}\right)^2 + \left(\frac{\partial v}{\partial z}\right)^2} \tag{5}$$

The resulting Ellrod Index (EI) is given as:

$$\mathrm{EI} = \mathrm{VWS} * (\mathrm{DEF} + \mathrm{CVG}) \tag{6}$$

The Table below matches Ellrod index thresholds with levels of severity (https://aviationweather.gov/exp/ellrod/info.php):

Pilot reports	Ellrod index threshold
Light–moderate	4
Moderate	8
Moderate–severe	12

2.2. Description of New WAFC Washington Icing and Turbulence Algorithms Incorporated into Unified Post Processor

2.2.1 New Icing (FIP) Algorithms Implemented in 2015

Through collaboration between NOAA and NCAR starting 2010, NCAR developed a global version of FIP (G-FIP) to work with GFS output (McDonough 2010) based on original FIP. The G-FIP was integrated into NCEP's Unified Post Processor (UPP) which derives diagnostic variables on highest model horizontal and vertical resolutions.

G-FIP uses GFS forecast of temperature (T), relative humidity (RH), cloud water mixing ratio (CWM), and vertical velocity (VV). Different from the Thompson microphysics scheme (Thompson and Eidhammer 2014) used in RAP, the Zhao scheme (Zhao and Carr 1997) used by GFS does not produce suspended rain and snow. Another adjustment made in G-FIP was to take into account the higher RH values in GFS when temperature is below − 20 °C. Based on GFS features, G-FIP algorithm defines cloud layers based on different thresholds for three different regions: 80% for tropics (latitude $\leq$ 23.5°), 75% for mid-region (23.5° < latitude < 66°), and 70% for polar region (latitude $\geq$ 66°). Finally, the fuzzy functions of G-FIP members are different from those of FIP (Fig. 2).

2.2.2 New Turbulence Algorithm, G-GTG to Replace WAFS Turbulence in 2019

To improve WAFS turbulence, EMC began to collaborate with NCAR turbulence group in late 2015 to expand GTG globally by coupling it with GFS output. The strategy is to integrate GTG algorithm into UPP to take advantage of high resolution GFS native model data as G-FIP does,

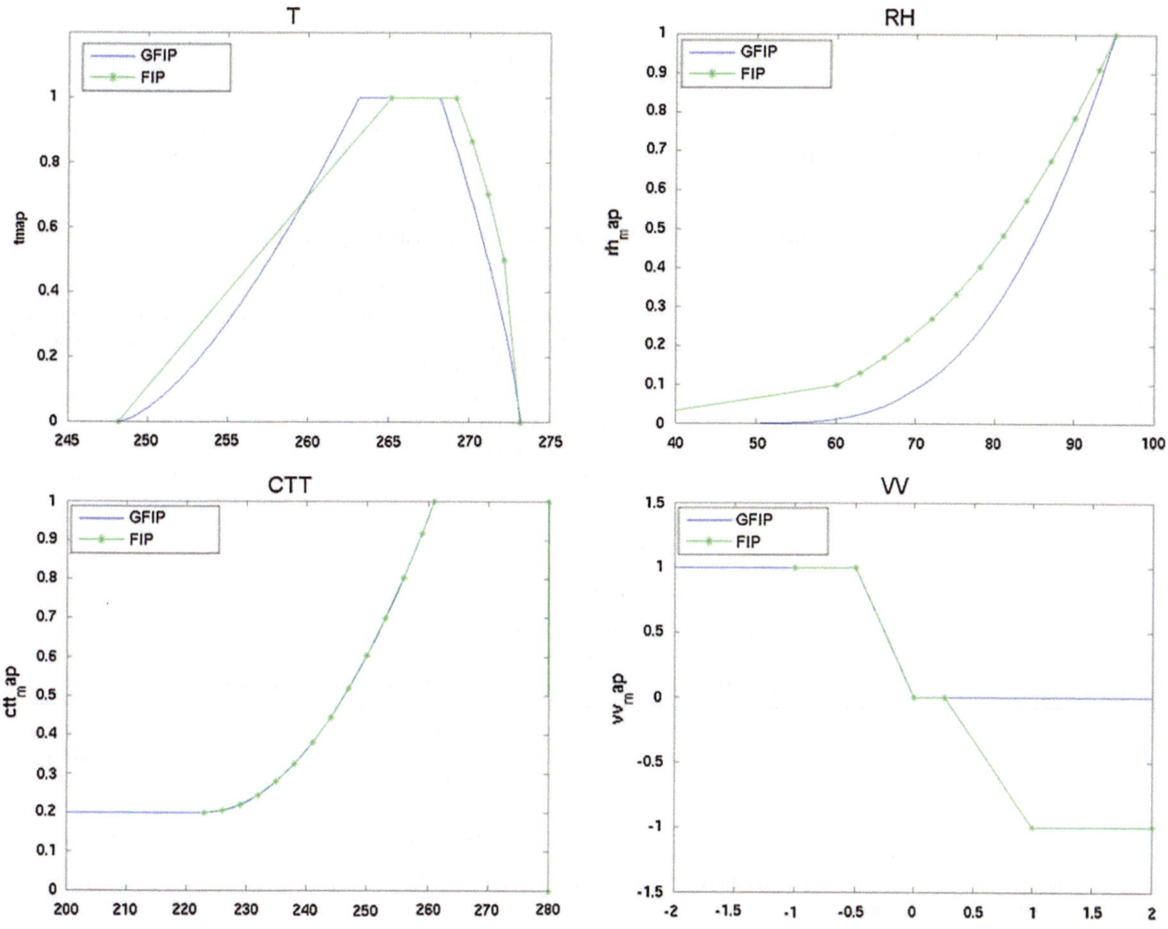

Figure 2

Fuzzy logics interest maps used in G-FIP implemented as operational US WAFS icing product in 2015 versus those used in Conus FIP, where *CTT* represents cloud top temperature, *T* represents temperature, *RH* represents relative humidity, and w represents vertical velocity

and to speed up the product delivery of G-GTG using UPP's MPI framework. With the same index configurations for GFS, it only takes 2.5 min per forecast for UPP to generate G-GTG on GFS native 13 km resolution using fourty-eight processors, while it takes 20 min per forecast if running serially.

The transitioning started from a latest version, GTG3 (GTG version 3, Sharman and Pearson 2017), which generates aircraft independent EDR ($=\varepsilon^{1/3}$, where ε is energy dissipation rate in m^2/s^3). All GTG science parts remain the same when transitioning GTG3 into UPP. For GFS, G-GTG picks different indices for three different vertical regions (low/mid/upper levels) (as shown in Tables 1, 2), computes turbulence diagnostics for each index, and then each diagnostic is mapped to a common turbulence EDR 0–1. Towards deriving the final ensemble of turbulence, the indices are divided into two categories, Clear-Air Turbulence (CAT) and Mountain-Wave Turbulence (MWT). In each category of CAT and MWT, the indices weigh equally. Research shows very similar results when applying static climatological weights and dynamic weights (Sharman et al. 2006). The maximum value of CAT and MWT is used to specify final GTG value.

2.3. Advantages of Using UPP as an Aviation R2O Tool

EMC's strategy to incorporate NCAR's aviation algorithms is to integrate these algorithms into its Unified Post processor (UPP). EMC started to

Table 1

Choice of indices for low/mid/high levels for CAT

		High	Ellrod3 by addition of a divergence trend
Low	Mid	High	NCEP's Nested Grid Model (NGM) Predictor: (Def*Speed)
		High	Shear vector magnitude
		High	Square of horizontal deformation
		High	Absolute value "horizontal divergence"/Ri
Low			Inverse of Richardson number, saturated: 1/Ris
	Mid	High	Structure function-derived eddy dissipation rate
	Mid		Inverse of Ri based on wind shear derived from the thermal wind relation
Low			Lighthill-Ford-Knox spontaneous imbalance of inertia–gravity waves/Ri
	Mid		Inertial-advective wind/Ri
Low			Square of Vertical velocity
		High	Square of Vertical velocity/Ri
Low			Structure function-derived sigma vertical velocity/Ri
		High	Structure function-derived sigma vertical velocity in x direction/Ri
	Mid	High	Frontogenesis function on a constant z surface/Ri

Table 2

Choice of indices on low/mid/high levels for MWT, each diagnostic is multiplied by MWT multiplier

		High	Frontogenesis function on a const z surface
		High	Absolute value "horizontal divergence"
	Mid		NCEP's Nested Grid Model (NGM) Predictor: (Def*Speed)
Low			Inertial-advective wind
	Mid		Square of structure function-derived EDR
Low			EDR derived from Schumann's gravity wave formulation

There are some changes to GTG3 when integrated into NCEP UPP, for two main reasons. One is to speed up GTG product delivery; the other is to be compatible with UPP features

develop UPP in 2006 to be used as the common processor for all NOAA's operational models. The request to have a common post processor within NOAA came from NOAA forecasters and customers as they were often confused by different algorithms used to derive the same fields. They felt a fair comparison cannot be made between model output from different models until the same post processing algorithms were applied across all models. EMC decided to use its Mesoscale post processor as the foundation for UPP and expanded it to interface with other NOAA operational models. The first implementation of UPP took place in 2007 when GFS implemented UPP operationally as its post processor. The UPP currently supports the following eight operational models: GFS, the Global Ensemble Forecast System (GEFS), the North America Mesoscale Forecast System (NAM Rogers 2017), the Short Range Ensemble Forecast System (SREF, Du et al. 2015), the Hurricane Weather Research and Forecasting Model (HWRF), the Rapid Refresh Model (RAP, Benjamin et al. 2016), the High Resolution Rapid Refresh Model (HRRR, https://rapidrefresh.noaa.gov/hrrr), the NOAA GFS Aerosol Component (NGAC, Lu et al. 2016).

There are several advantages to this new strategy of utilizing UPP as the aviation Research to Operations (R2O) tool. First, because UPP carries and derives variables on highest resolution horizontal and vertical grids, one can expect more accurate aviation forecast products when these products are derived within UPP. This is especially true at cruising altitude and near tropopause where most models place several vertical layers to resolve tropopause. However, because most users can only assess GFS output with 50 hPa vertical resolution, aviation products derived using this low resolution GFS output could easily miss important forecast signals at cruising altitude.

Second, algorithms added to UPP can often be applied to most of its supported eight models. In the application for aviation algorithms, minor tuning may be necessary. For example, the GTG algorithm is sensitive to model resolution so different configuration or weighting function may be used for different models. Nevertheless, UPP provides a centralized unified framework to maintain and exercise aviation algorithms for all operational models. The bug fixes and improvement in performance or efficiency of FIP and GTG packages will only need to be applied once within UPP, instead of to many different packages used for different models. In addition, since UPP also supports NCEP's ensemble modeling system, NCEP is able to output probabilistic aviation products by generating aviation products in all NCEP's GEFS members. Finally, UPP uses message passing interface (MPI) to speed up operational post processing by taking advantage of available multiple processors on today's High Performance Computing (HPC) system. The number of processors is used to decompose Y-axis direction into multiple subdomains and computation is done independently over smaller subdomain to save time. This feature is especially important for adapting GTG algorithm into NCEP's operational production suite. The estimated run time for generating GTG on current GFS 13 km native horizontal resolution is about 20 min per forecast, without use of parallel framework. Unfortunately, this long run time will not adhere to operational product delivery time line. With use of MPI, UPP is able to generate GTG within 2.5 min per forecast. The use of UPP to generate GTG and FIP has enabled us to deliver operational aviation products much faster while still deriving these products on highest model resolution.

2.4. *Adaptation of FIP and GTG Algorithm into UPP and NCEP's Production Suite*

EMC obtained two versions of FIP algorithms from NCAR in 2010. The first version was then operational version of FIP for Rapid Update Cycle (RUC) model (Benjamin et al. 2004). The second version was developed by McDonough (2010) to work for simple Zhao microphysics (1997) used in GFS. He also divided the earth into three climate zone based on latitude bands and each zone used different interest maps for temperature, relative humidity, cloud top temperature, and vertical velocity. More details can be found in McDonough (2010). The two versions were merged, recoded to use MPI and Fortran 90, and integrated to UPP. The adaptation of FIP into UPP is relatively simple because FIP employed column model concept.

EMC obtained GTG3 from NCAR in 2015 and began integration of GTG algorithm into UPP. Because GTG algorithm used differentiation, averaging, and filtering throughout its package, proper halo exchanges must be used to ensure these operations were performed correctly across boundaries of each MPI task when integrating GTG into UPP MPI framework. In addition, these operations needed to be adapted to work for different staggered grid types supported by UPP, including Arakawa A, B, and E grids (Arakawa and Lamb 1977).

Once FIP and GTG are integrated into UPP, EMC set up real time parallel to generate these two products from operational GFS output four cycles each day. EMC then worked with NCAR developers, AWC, and FAA to distribute and evaluate performance of these experimental products. The verification results will be discussed in Sect. 3. After a period of evaluation, EMC will coordinate with AWC and FAA to schedule operational implementation of these products from UPP. Current operational implementation paradigm at NCEP usually requires UPP to be implemented together with each major model upgrade which it supports. For WAFS products, a separate package to perform blending of UK's and US's hazard products will need to modified to ingest new UPP-generated FIP and GTG as the new official US's hazard products. As mentioned in introduction, UPP-generated Global FIP replaced WAFS icing forecast product after 2015 GFS upgrade. While UPP-generated G-GTG was implemented (but not disseminated) with 2017 GFS upgrade, the replacement of WAFS turbulence product with G-GTG is scheduled to take place in late 2018 due to the following two reasons. First, WAFS users need a period of time to become accustomed to new products. Second, because GFS has been going through annual upgrade in recent years, this delay in implementation of post processing products is necessary to allow for tuning of these products.

2.5. *Methodology to Blend WAFS Products from WAFC Washington and WAFC London*

Both WAFCs have agreed to apply the same methodology to blend maximum and mean WAFS hazard products. The approach is very simple. The mean icing and turbulence forecast products are obtained by taking the average of those from two centers while maximum products are obtained by taking the maximum.

3. *Validation Data and Evaluation Method*

3.1. *Validation Data: Development of Global Current Icing Potential Product (G-CIP) to be used as Verifying Truth for Icing Forecast Product*

Although a handful of observation data, such as Cloud Sat, can be inferred to indicate current icing environment, there are no direct measurement of in-flight icing observations with adequate global coverage at a given time. The PIREPs data is often used to perform icing verification over CONUS but its coverage is sparse over much of the globe. Therefore, global verification of in-flight icing has been difficult at best. Due to lack of global coverage of many observation data, it is a common practice for Global model developers to perform objective verification against its own analysis (White 2017) to obtain global verification statistics.

NCAR developed CONUS version of Current Icing Potential (CIP) to provide real time current icing environment (Bernstein et al. 2005 and Gultepe 2018) diagnose for users to make route specific decisions to avoid flying through hazardous icing condition. It combines RAP model output with observation data from satellite, pilot report, radar, surface, and lightning with final three sets of data as optional data. Similar to FIP algorithm, the CIP algorithm uses fuzzy logic and decision-tree logic that are derived from cloud physics and forecaster and onboard flight experiences from field program. However, a major difference between FIP and CIP algorithms is observation data is used in determining the locations of clouds and precipitation along with model output.

Because it is a common practice for developers of global models to verify their forecast against their own analysis, EMC developed Global version of CIP to be used as verifying analysis for all global icing forecast. Furthermore, the major purpose of product verification at NCEP is to prove skill is not degraded after a product upgrade. Therefore, using Global version of CIP as the truth for both old and new icing products is a good way to evaluate whether icing forecast skill improves. Note that verification scores computed using analysis will be slightly more favorable than those computed using observation data. EMC's strategy to expand CONUS CIP to Global CIP is as follows. First, GFS is used as background model data instead of RAP. Secondly, EMC requested NESDIS to make global composite of geostationary satellite image data which was implemented operationally in 2014. Data from five geostationary satellites are used to make this global composite satellite data and they are GOES-East, GOES-West, Meteosat at 0, Meteosat at 63, and MTSAT. The first two satellites are operated by NOAA while the two Meteosats are operated by European Organization for the Exploitation of Meteorological Satellites. The MTSAT is operated by Japan. Third, existing NCEP operational Global METAR data is used. METAR data is routine weather report of sensible weather elements measures at airports or other stations. In addition, optional NCEP in house PIREPS, radar, and lightning data are used wherever they are available. Prior to operational implementation of global composite of geostationary data in 2014, EMC was able to start generating experimental G-CIP data in 2013 and started to perform global verification of G-FIP and other icing forecast products using parallel global composite of geostationary data.

The verification is based on grid to grid. Since WAFS icing is at 1.25° while G-CIP is at 0.25°, G-CIP was up-scaled to match WAFS icing using bilinear interpolation option. WAFS icing potential has two sets of values, mean/max, when it is converted from high-resolution forecast to 1.25° per ICAO standards. Therefore, two curves will be provided, representing mean and max, respectively.

3.2. *Objective Evaluation Method*

Three category verification scores are derived for icing potential forecast. Category verification uses contingency table, converting continuous forecast values to dichotomous (yes/no) icing events. However, as the validation data, G-CIP itself is icing potential with continuous values from 0 to 1, not a yes or no event. To work it out, G-CIP is pre-defined with four thresholds, 0.2 0.4 0.6 0.8, presenting categories of low, medium, high and extreme high potential, respectively. Then there are two steps to derive the statistical scores. First G-FIP is verified on yes/no events of the four categories of G-CIP, respectively, using a set of increasing probability thresholds of G-FIP 0.1 0.2 0.3 0.4 0.5 0.6 0.7 0.8 0.9. Then for each threshold of G-FIP, all yes/no events of the four categories of G-CIP are added up.

First of all, Equitable Threat Score (ETS, or Gilbert Skill score GSS, see Table 3) is used for verifying icing potential forecast in comparison to other events, icing event is lower frequent event. In addition, for large scale grid-to-grid verification of icing over global score, the number of random hits could be very large. The equitable threat score not only excludes the random hits but also treats the correct rejection (or no event forecasts) and correct forecasts equally. This implies that using ETS, correct forecasts of less frequent events are weighted more strongly than high those of frequent events (Wilks 2006).

Table 3

2 × 2 verification contingency table

		Observed	
		Yes	No
Forecast	Yes	a	b
	No	c	d

PODy (Hit rate) = a/(a + c), POFD (False alarm rate) = b/(b + d)

ETS = (a − r)/(a − r + b + c), where, $r = (a + b)(a + c)/n$, where n is sample size

Second, frequency bias is derived for icing verification, defined by the ratio of total predicted icing events divided by total observed icing events. The best score is 1. If the bias is larger than 1.0, icing events are over-predicted to imply higher false alarm rate, otherwise are under-predicted to imply higher missing rate. An over-prediction system means higher false alarms but not necessarily higher hits.

Third, many aviation weather forecasters are more interested in evaluating Receiver Operating Characteristic (ROC), another skill measure in icing potential forecast. A ROC curve is constructed by plotting the hit rate (PODy see Table 3) as y-axis against false alarm rate (POFD see Table 3) as X-axis in a 1.0 × 1.0 coordinate system for all events. The ROC score is the area between the ROC curve and the diagonal line, with the perfect score of 0.5. A positive ROC area means that the forecast hit rates is higher than that of false alarm, a 0 ROC area means no score or similar to climatology forecast, and a negative area means that it is worse than climatology forecast. ROC measures resolution, measuring forecast discrimination from two alternative outcomes. It helps developers to improve the forecast through calibration and helps aviation community to decide thresholds of forecast icing representing low/medium/high potential.

4. *Verification Results and Discussions*

Extensive daily Global objective verifications of new G-FIP generated by UPP against G-CIP along with verification of then operational US WAFS Icing and blended WAFS icing products were performed from 2013 to the implementation of new G-FIP in 2015. Prior to 2013, EMC performed objective verification comparison of above mentioned products using AWC's CONUS CIP as truth. Several case studies were also performed as a sanity check and to support qualitative validation. In Sect. 4.1, authors will first show a few case studies to demonstrate good performance of new G-CIP products and that of to be implemented G-GTG products. In Sect. 4.2, objective verification results of all WAFS old and new icing forecast prior and since 2017 will be presented. GSD recently performed objective verification of G-GTG and their results will be shown in summary section.

4.1. *Qualitative Validation via Case Studies*

4.1.1 *Case Studies for G-FIP and G-FIS*

When EMC started to generate experimental G-FIP from UPP in 2011, numerous comparisons were made between G-FIP and CONUS FIP and CIP generated from RUC forecast for sanity check. It was reassuring that the two products had similar synoptic patterns although they use forecast from different models as well as slightly different versions of FIP algorithms. Figure 3 shows such a comparison from a more recent case valid at 13Z on Jan 16 2018 for both G-FIS and G-FIP as compared with CONUS current icing severity and current icing potential analysis products, respectively. The current icing product plots from AWC ADDS web site also display PIREP report of icing with various degree of severity as symbolized under color bars. Note that this is an early morning case and hence one does not expect many PIREP reports. In this particular case, both G-FIP and G-FIS 7-hour forecasts and CONUS CIP indicated high potential of 3 moderate icing along the front from eastern Texas to Kentucky as well as from British Columbia to eastern Oregon. This agreement in forecast and analysis was further supported by PIREP of moderate icing in southeastern Texas and light icing in central Tennessee. Although both global and CONUS products were in agreement that there was severe icing in eastern central Texas, G-FIP combined G-FIS indicated sever icing in central Texas while CONUS analysis product indicated light to moderate icing over the same area. Unfortunately there was no PIREP there to indicate the icing condition at that time.

The second case involved an icing incident that took place at 07Z on November 14 2016 near Bergen Norway reported by a Jettime Avions de Transport Regional ATR-72-212A on behalf of Scandinavian Airlines (2014). The flight was initially cleared to fly to FL170 out of Bergen Norway but encountered icing condition while climbing through FL150. The crew received a "Degraded Perf" caution before the stick shaker was activated. The aircraft stopped climbing at FL160 at 07Z and began to descend as both wings dropped. The crew regained control manually, turned flight toward northwest while descending to FL100, and eventually landed safely at Aalesund. As shown in Fig. 4a, 24 h blended WAFS icing forecast predicted approximately 70% icing potential near Bergen at 06 Z. Note that WAFS forecasts are only generated three hourly. Therefore, we have chosen to show available WAFS forecast closest to the time of this incident. In addition, authors were able to re-run UPP to generate GFS icing severity from the same forecast cycle and lead time as shown in Fig. 4b. This figure shows severe icing was predicted near Bergen in addition to large icing potential as indicated in blended WAFS icing forecast.

4.1.2 *Case Studies for G-GTG*

Aviation Weather Center NOAA alerted us to a severe turbulence flight incident that took place near Pueblo Colorado at 00Z on February 28 2017. An American Airline Boeing 737 from San Diego to Chicago had to make an emergency landing at the Denver Airport after hitting extreme turbulence at FL370 that injured five passengers. The pilots were also concerned the plane may have been damaged by the turbulence. Figure 5 showed this turbulence event was predicted by EMC's then experimental G-GTG products consistently with both 12 h as well as 36 h lead time. Unfortunately, EMC's G-GTG products were experimental at the time while undergoing evaluation and possible further tuning. Therefore, they were only distributed to our collaborators instead of to general public to provide guidance for this event.

An Eva Airways Boeing 777–300 from Taipei Taiwan to Chicago O'Hare Airport was enroute at FL310 about 180 nautical miles east southeast of Fukuoka Japan when the aircraft encountered turbulence causing the aircraft to drop and climb 550 feet and 375 feet, respectively. A cabin crew was seriously injured while eight other crews and three passengers received minor injuries. The flight went on to land at Chicago O'Hare Airport safely. The incident took place at 14Z on November 22 2017. However, because EMC only archived model output

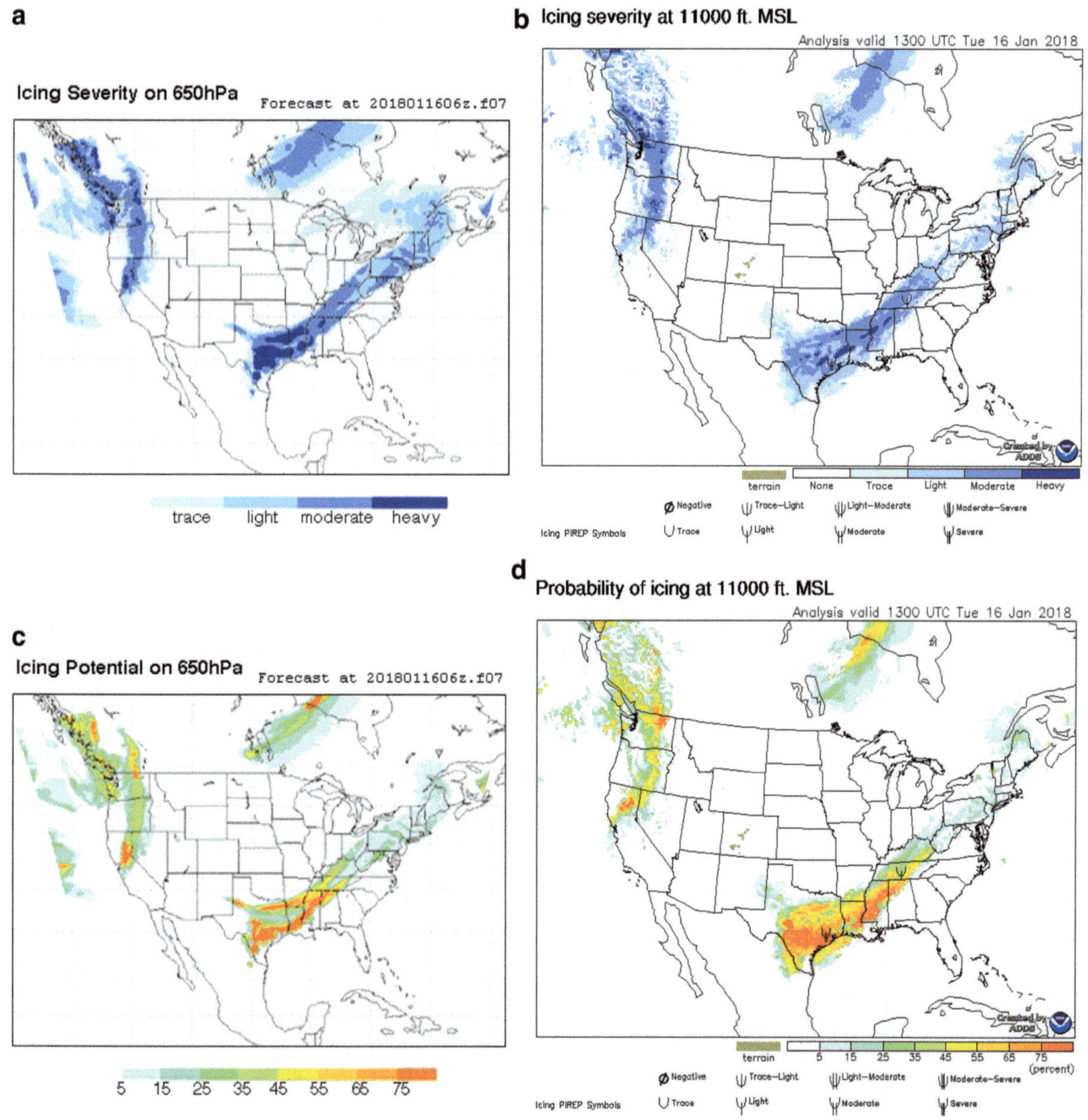

Figure 3
Comparisons of G-FIS (**a**) and G-FIP (**c**) with Conus current icing severity (**b**) and Conus current icing potential (**d**), all valid at 13Z January 16 2018. Plots on the right-hand side panels also showed PIREPs of icing in various degrees of severity levels as indicated under color bars

six hourly, we can only re-run UPP to generate G-GTG valid at 12Z and then 18Z on November 22 2017, respectively. Figure 6 shows UPP generated G-GTG predicted moderate turbulence southeast of incident location 2 h prior to the event happening and then northwest of incident 4 h after event happening. Through interpolation, it is reasonable to infer G-GTG would have predicted this turbulence event at the location close to incident.

4.2. Objective Verification Results and Discussion of Icing Forecast

NCEP's global icing products include icing potential and icing severity predictions from both

a

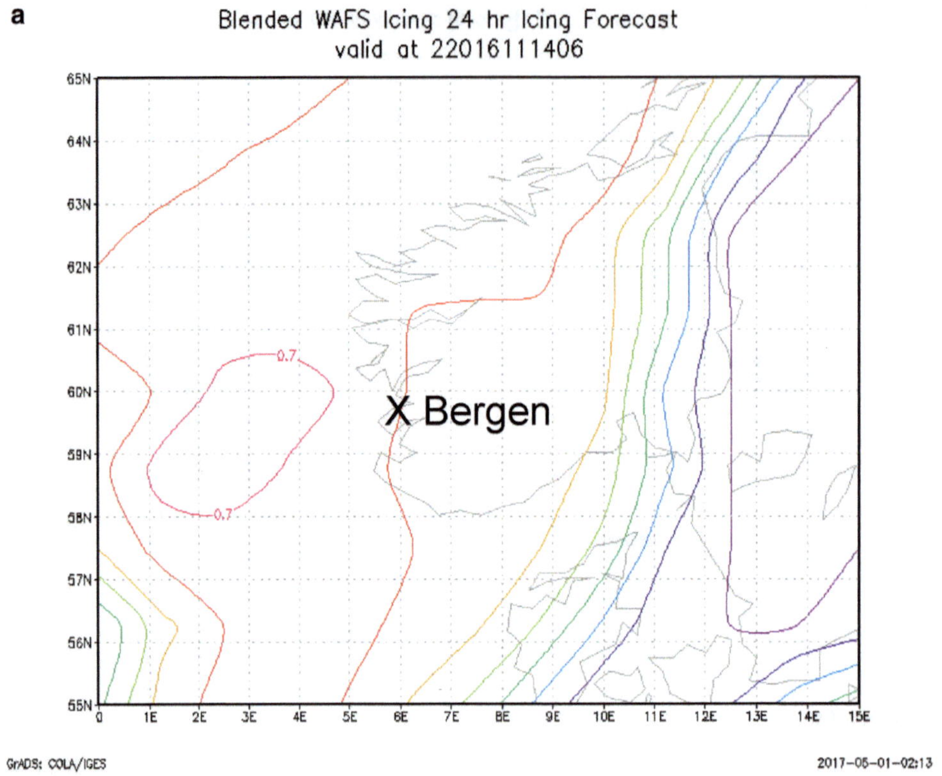
Blended WAFS Icing 24 hr Icing Forecast
valid at 22016111406
65N
64N
63N
62N
61N
60N
59N
58N
57N
56N
55N
0
1E
2E
3E
4E
5E
6E
7E
8E
9E
10E
11E
12E
13E
14E
15E
0.7
0.7
X Bergen
GrADS: COLA/IGES
2017-05-01-02:13

b

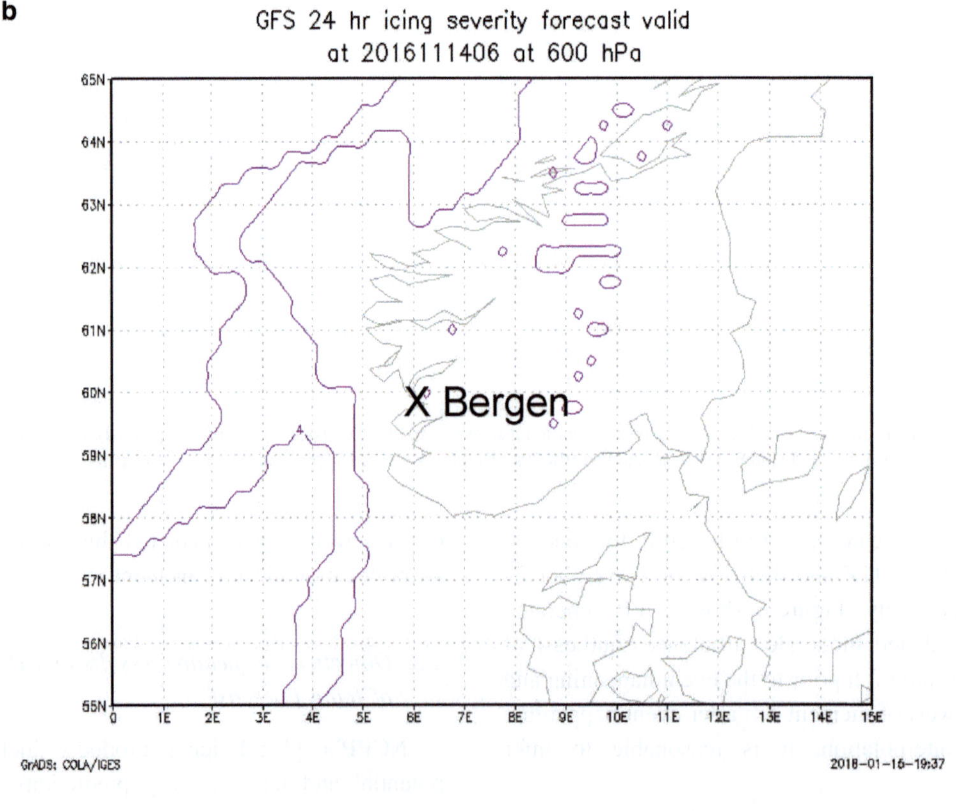
GFS 24 hr icing severity forecast valid
at 2016111406 at 600 hPa
65N
64N
63N
62N
61N
60N
59N
58N
57N
56N
55N
0
1E
2E
3E
4E
5E
6E
7E
8E
9E
10E
11E
12E
13E
14E
15E
4
X Bergen
GrADS: COLA/IGES
2018-01-16-19:37

◄Figure 4
Twenty-four hour forecast of Blended WAFS icing potential product (**a**) and GFS icing severity (**b**) predicted 70% potential of severe in-flight icing (in purple lines) occurring near Bergen, Norway

GFS deterministic model and GFS ensemble forecast (GEFS). In this section, only icing potential forecasts from deterministic model and ensemble forecasts are objectively evaluated due to a lack of truth data in icing severity data.

The objective evaluation of icing potential forecast is based on grid-to-grid comparison in NCEP's unified verification system which was established for verifying both operational forecast and parallel forecast (test) for all models at NCEP. In other words, the gridded icing potential forecast at each grid is verified against gridded icing analysis at the same grid. In the case of development and implementation of WAFS icing products, their performance and skills were evaluated and compared to make sure the new products are superior to the old

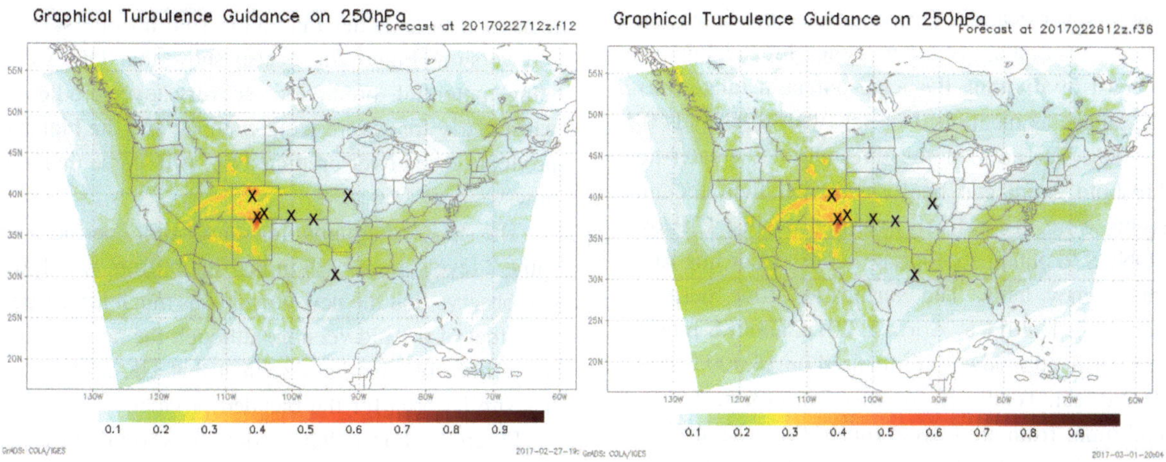

Figure 5
12 h (L) and 36 h of G-GTG forecast over CONUS, both valid at 00Z February 28 2017, predicted severe turbulence event that caused American airline 1296 to make an emergency landing at the Denver Airport after hitting extreme turbulence near Pueblo Colorado at FL370. PIREPs of turbulence locations were marked with x on both plots

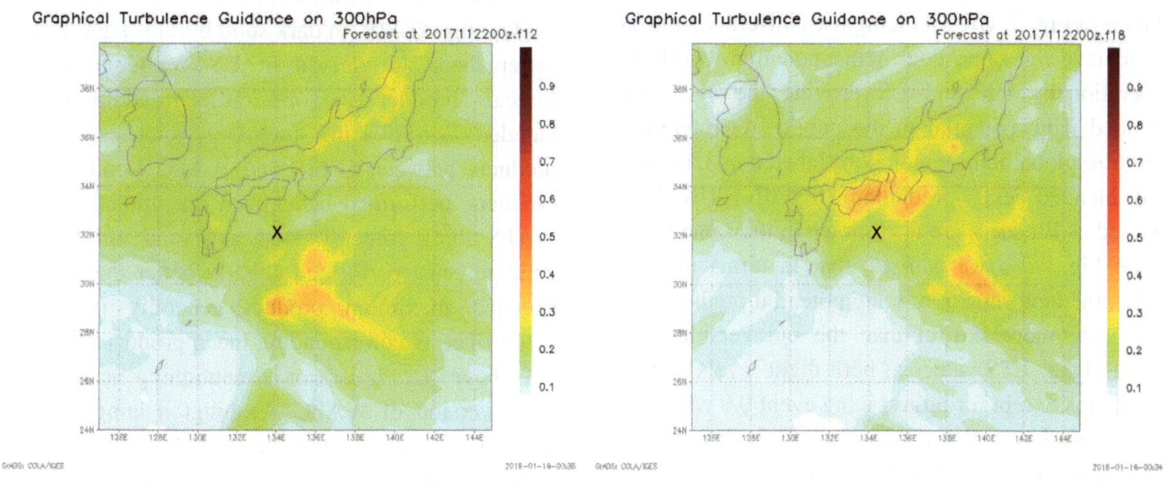

Figure 6
12 h G-GTG forecast (L) valid at 12Z November 22 2017 and 18 h G-GTG forecast (R) valid at 18Z November 22 2017. The severe turbulence incident took place at 14Z November 22 2017 at the location marked with X

version. In this section, the results of three types of evaluations will be presented.

4.2.1 *Evaluation and Comparison of Old and New Global Icing Potential Predictions*

Because global current icing potential analysis (G-CIP) data were under development until 2013, AWC's CIP analysis data over CONUS was first used as truth in the grid to grid verification during the early development stage of global icing forecast. This means Global icing prediction was only evaluated over CONUS at EMC until 2013. To apply the category verification, the continuous predicted icing potential is first converted to dichotomous icing event forecast by dividing the icing potential range into 9 sub-ranges as mentioned in 3.2 (with thresholds 0.1, …. 0.9). For each of the thresholds, a greater-than threshold event can be formed, and then the nine events were, respectively, verified against CIP over CONUS to obtain appropriate scores according to the contingency table. The verification time period is from Aug 1, 2011 to Apr. 15, 2012, about 8 months. The verification levels are focused on 14,000 feet and 10,000 feet. Totally 12 forecast times (3, 6, 9,…36 h lead time) from all 4 cycle runs (00, 06,12 and 18Z) were verified at 8 validation times (00, 03, 06, 09, 12, 15, 18 and 12Z) each day. The resultant performances of old (labeled as by USFIP) and new version (labeled as GFIP) expressed by Frequency Bias and Equitable Threat Score (ETS) for all sub-ranges are shown in Fig. 7. UK Met icing uses different strategy to achieve the same goal, icing potential, as GFIP. Operationally UK and US provide the averaged/blended icing potential to the public. As reference, UK Met global icing prediction (labeled as UK Met) and blended icing prediction (USFIP + UK Met, labeled as Blended) are also shown in the same plots. From Fig. 7a, b we can see that in terms of ETS value, the new version of icing potential prediction is obviously more skillful than the old version with much higher ETS values at both flight levels and all thresholds except for largest icing event 0.9 where all predictions have no skill. It is also shown that the old version is worse than UK Met's prediction while the new version is better than UK Met's and is close to the blended product at both flight levels for all thresholds.

In terms of bias, it can be seen from Fig. 7c, d that the old version is over-all under-predicted at both flight levels for all thresholds, particularly when threshold is larger than 0.5, the ratio bias is close to 0, implying that the old version has very low hit rate and almost misses all severe icing events. The new version on the other hand, is generally over predicted but not too much. Over entire potential range, its ratio bias is no larger than 2.0. This means that the new version has pretty good skill in hit rate and lower missing rate with a little bit higher in false alarms. This is seen as better than almost no hit rate skill in old version. Figure 7a, b also show that UK Met's icing potential prediction is very over-predicted, particularly for severe icing events while the blended products eased over-predicted but still worse than the new version.

The ROC diagrams for same time period and same 2 levels are shown in Fig. 8a, b, from which we can observe that the new version has much larger area than that of the old version, although the new version has a little bit smaller size than that UK Met which has similar area for the blended product.

Figure 9 shows ROC scores of several different icing forecast products when verifying against G-CIP for the winter 2014–2015 prior to 2015 implementation of G-FIP to replace first version WAFC Washington icing forecast product at 600 hPa and 400 hPa. In term of mean product at 600 hPa, it was evident that G-FIP (in dark solid blue) is significantly better than old version of US Washington icing forecast product (in solid red). In addition, new blended WAFS icing forecast, which is blended products between G-FIP and WAFC London icing products, performed better then operational blended WAFS icing forecast. Both new and old blended icing forecast performed slightly better than G-FIP. Blended maximum product generally performed better than its corresponding mean product at threshold above 0.2. It is worth mentioning at 400 hPa, older version of WAFC Washington icing forecast showed no skill at all during this period while both mean and maximum G-FIP still retained good skills which was especially true for maximum G-FIP product.

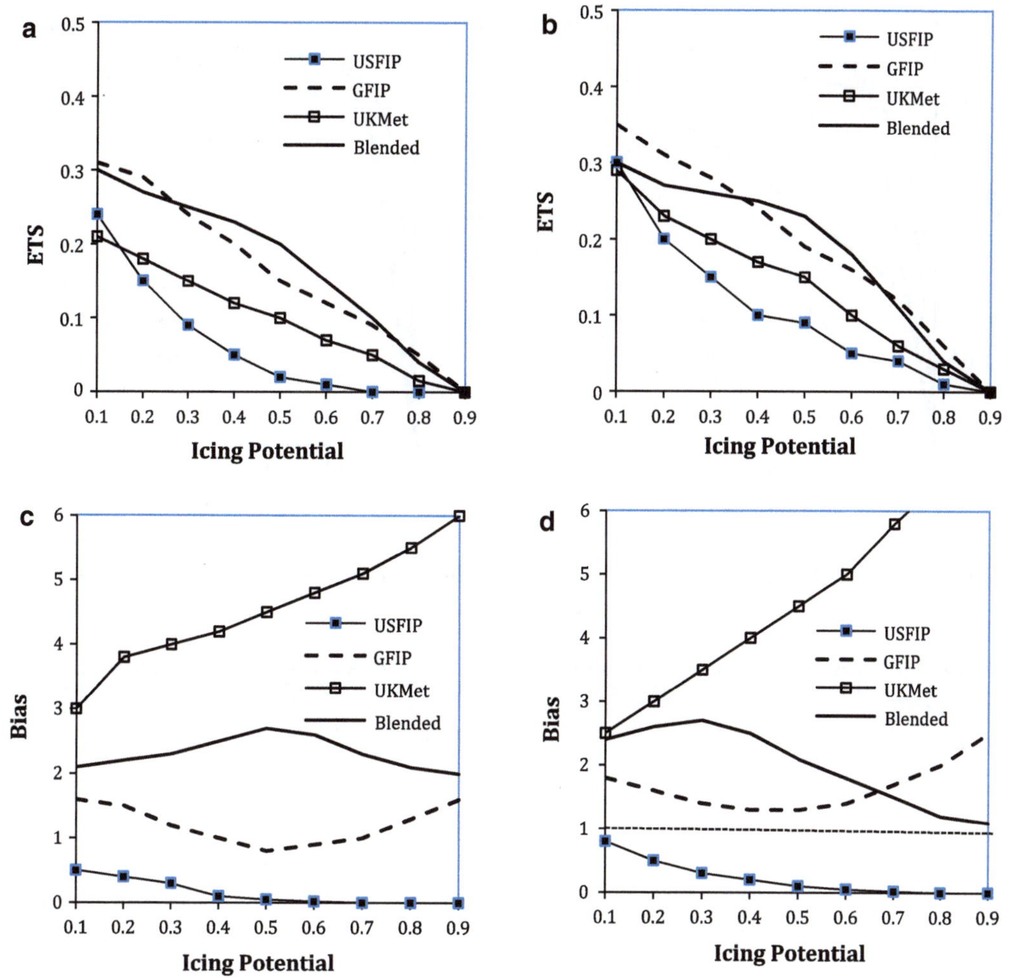

Figure 7
Equitable Threat Scores (**a**, **b**) and Bias (**c**, **d**) evaluated against CIP over CONUS for icing potential forecasts from four different products at FL140 (7**a**, **c**) and FL100 (7**b**, **d**)

4.2.2 Evaluation of Blended WAFS Product Against G-CIP

After G-CIP and G-FIP were implemented in 2015, the objective verification was focused on blended WAFS mean and blended WAFS max products based on blending of new G-FIP and UK Met icing potential product. The results of the evaluation are continuously updated and displayed on NCEP's WAFC verification web site (http://www.emc.ncep.noaa.gov/gmb/icao/). Here just ROC diagrams at 3 flight levels (FL100, FL140, FL180) and 3 different forecast hours (12 h, 24 h and 36 h) from May 2015 to Feb. 2017 are shown in Fig. 10. It can be seen that both blended mean and blended max are skillful in that both have very high ROC area at all forecast hours and all three flight levels. In addition, the blended mean and max ROC curves at all 3 forecast hours and 3 flight levels are very close to each other although the ROC areas at FL140 are a little larger than Fl100 and FL180, and ROC areas decrease with increased forecast time.

4.2.3 Ensemble Evaluation of GEFS Icing

The purpose of ensemble verification is to see whether the ensemble forecast can provide enough uncertainty information and is better than one single

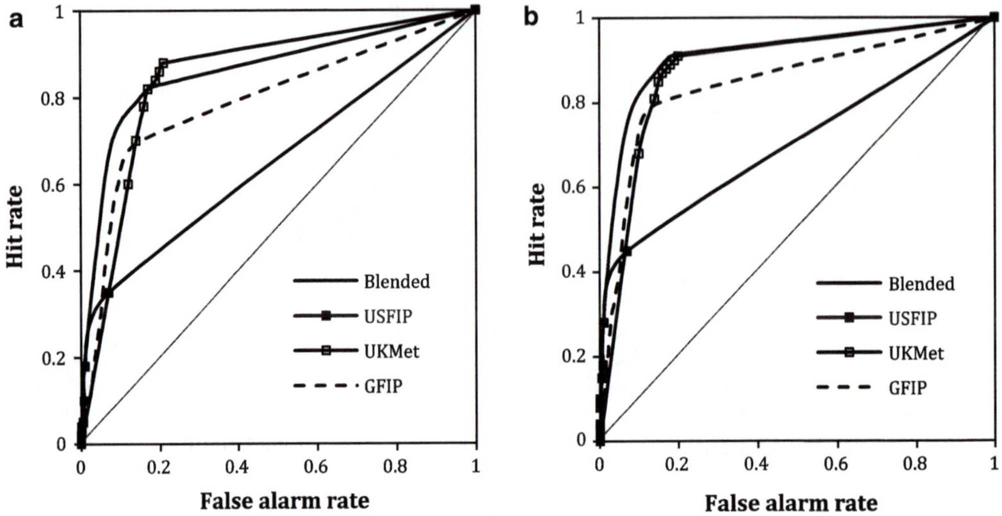

Figure 8
ROC curves evaluated against CIP over CONUS for icing potential forecasts from four different products at FL140 (8**a**) and FL100 (8**b**)

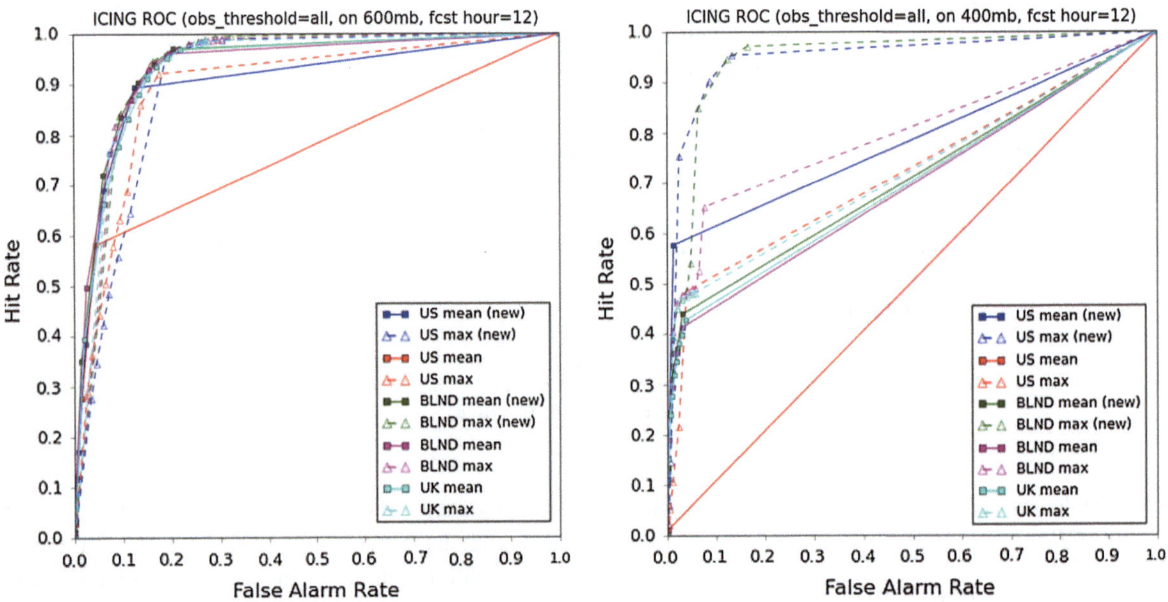

Figure 9
ROC curves evaluated against G-CIP globally for all WAFS icing potential forecasts at 600 hPa (L) and 400 hPa (R) during winter 2014–2015

model forecast. During and after the new G-FIP was implemented in all members of NCEP's global ensemble forecast system, the performance of ensemble prediction of icing potential was evaluated against G-CIP and compared to the deterministic GFS model icing potential prediction. The detailed probabilistic verification method can be seen in Zhou and Du (2010) where the ensemble verification has two aspects. One is ensemble system evaluation by estimating the over-all ensemble system performance, particularly the member diversity, RMSE and mean error distribution, and bias. The second aspect is event-based forecast skills, particularly in comparison to the single model forecast. Following

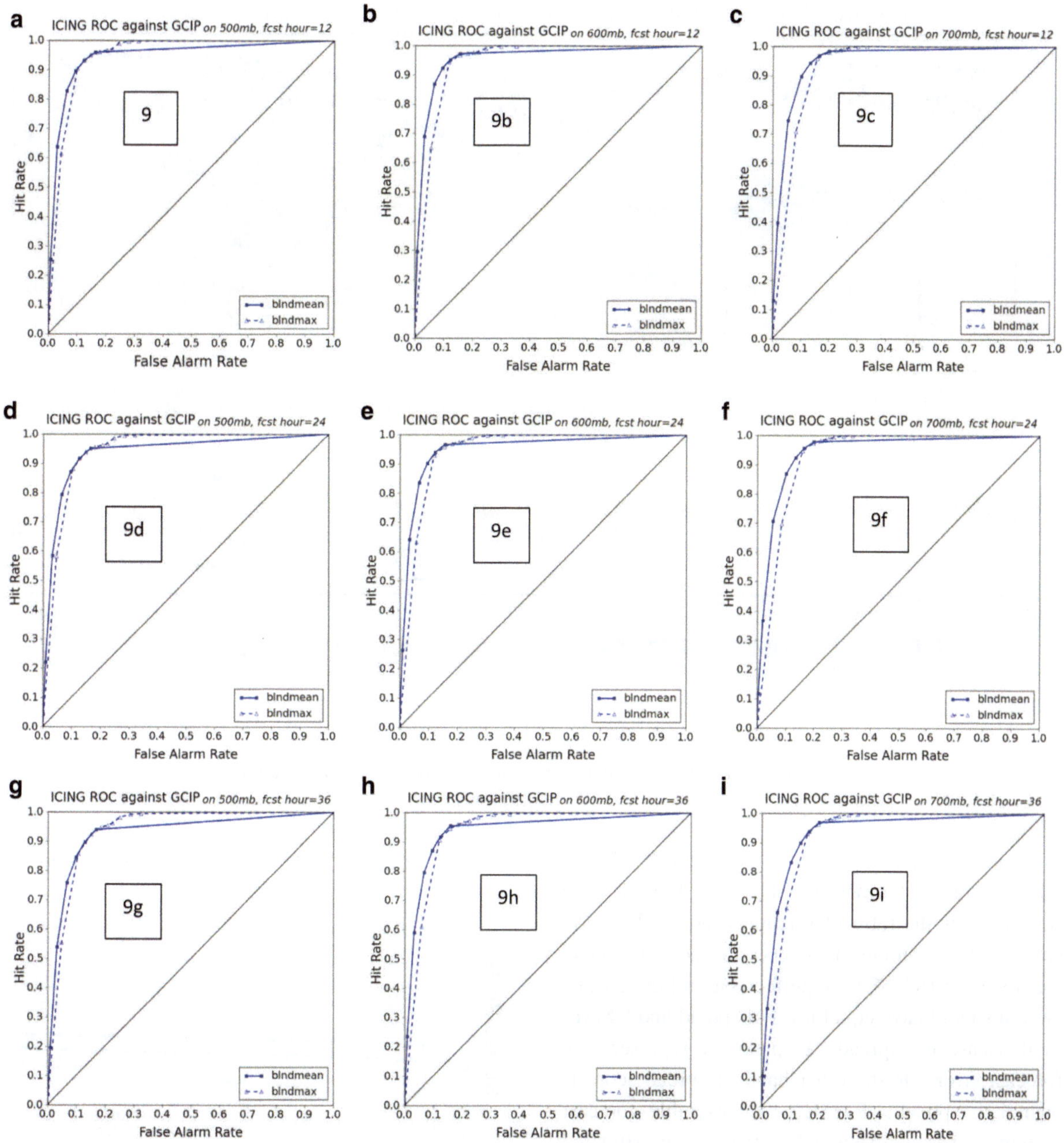

Figure 10

ROC curves for blended mean (solid line) and blended max (dash line) at three flight levels (left for FL180, middle for FL140, right for FL100) and three forecast times (upper for 12 h, middle for 24 h and lower for 36 h)

plots show some of the ensemble verification results in 5 months from June to October 2016, for all forecast running times and all validation times.

Figure 11 is mean error global distribution at FL180 (18,000 feet) for 24 forecast hours. The positive mean error indicate over prediction while negative value under-prediction. It can be seen that overall the ensemble icing potential mean at FL180 is a little bit over-predicted over lower lands/oceans and higher over-predicted in tropical regions but under-predicted in mountain regions, particularly in Himalaya and Rocky mountains. The member diversity

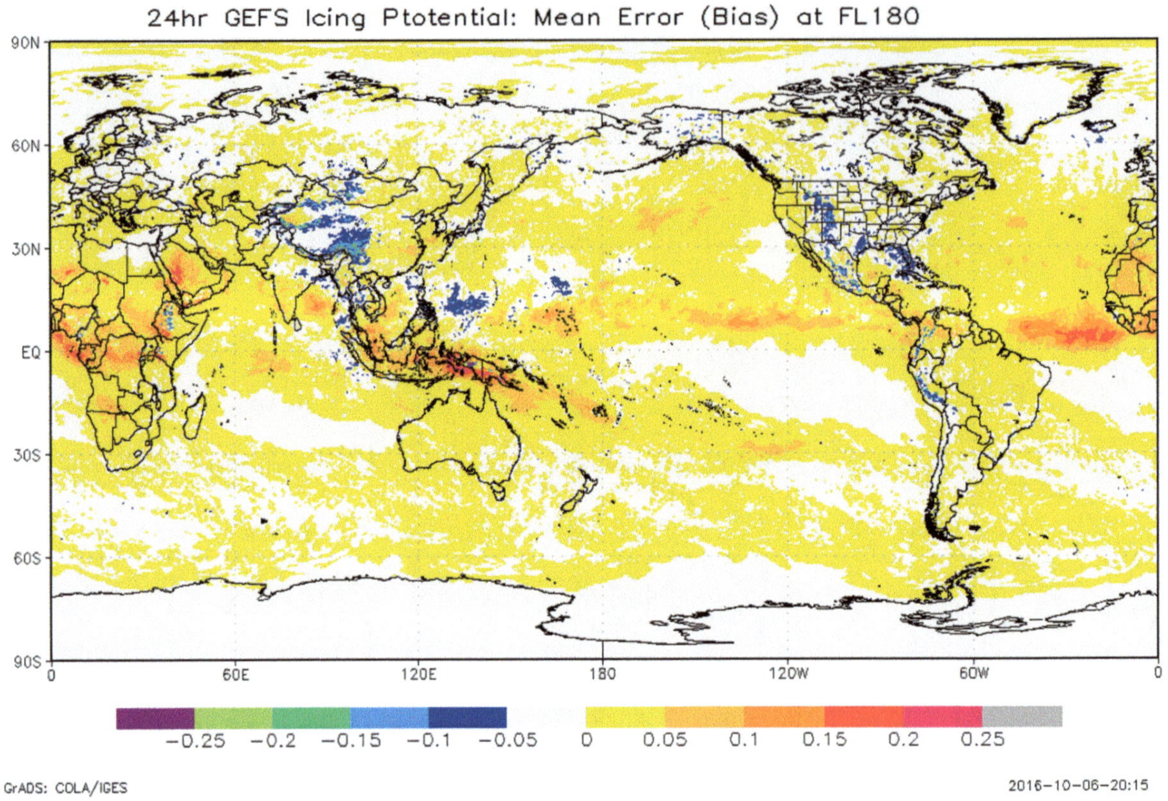

Figure 11
Mean Error (mean value—G-CIP) over flight level 18,000 feet for 24 forecast hours

expressed as skill score at three flight levels for different forecast times are shown in Fig. 12. The skill score is the ratio of spread (standard deviation among all ensemble members) to the root mean square error (RMSE). A perfect ensemble forecast system should have equal level of spread and RMSE. In this case the spread can precisely represent the forecast error. On the other hand, if spread score is over 1.0 (spread > RMSE), the ensemble forecast system is over diverse and the error is over estimated by the spread, while spread score is under 1.0 (spread < RMSE), the ensemble forecast system is under-diverse and the error is under estimated by the spread. We can see that current GEFS icing potential prediction is under-diverse, and the spread may under-estimate the actual forecast error for all forecast hours and this situation becomes worse for lower flight levels although it is improved a little for the lower level.

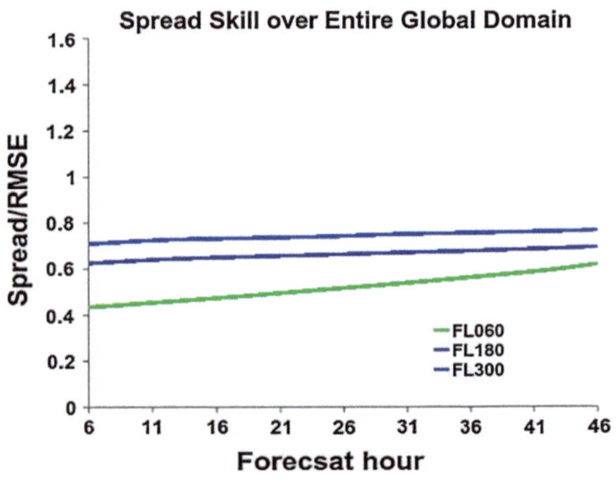

Figure 12
Spread/RMS ratio for three flight levels and different forecast times over entire GEFS Domain

The under dispersion may not be due to ice potential computation method. It could largely be due to the problem in the global ensemble forecast

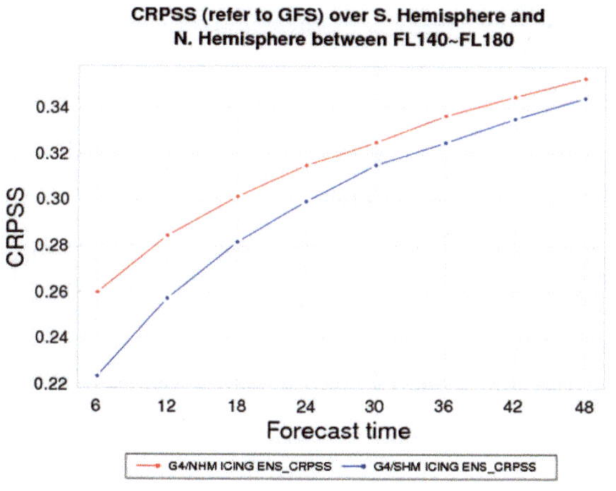

Figure 13
FL140-FL180 averaged CRPSS referenced by G-FIP for various forecast times over south hemisphere (blue) and north hemisphere (red)

system. Currently, almost all ensemble forecasts at NCEP suffer from under dispersion issue. Improvement of the member diversity has been one of goals for ensemble research and development at NCEP.

To compare ensemble forecast to single model forecast (reference), continuous ranked probabilistic skill score (CRPSS) is shown in Fig. 13 where the reference is G-FIP and averaged CRPSS between FL140 and FL180 over north hemisphere and south hemisphere are provided. A positive CRPSS indicates the ensemble forecast has smaller error than that of referenced forecast and negative CRPSS is otherwise. The CRPSS curves shows all positive CRPSS values for entire forecast time range for both south and north hemisphere which is a little better than the south. Moreover, as forecast hour increases, the ensemble forecast becomes better.

Reliability (also called statistical consistency or probability bias, Zhou and Du 2010) is an important measure to indicate whether the forecast probability is statistically consistent with observed frequency of the occurrence of events over the verification time. It also implies confidence of the probability forecast. The reliability curves for 9 icing potential thresholds ($>$ 0.01, 0.2, ...,0.9) at 3 different flight levels over entire domain and all forecast hours are displayed in Fig. 14, showing that the higher ensemble probability forecast is over confident, or too aggressive (forecast probability $>$ observed frequency) while lower ensemble probability is under confident, or too conservative (forecast probability $<$ observed frequency). Such a reliability information is useful to forecasters when they use ensemble probability in icing forecast. As is in single model forecast, ROC area can also be used to show forecast skills in terms of hit rate and false alarm rates. See Fig. 14 (R), where ROC curves for 9 icing potential thresholds at 3 flight levels over entire domain and all forecast hours are shown. We can see that the ensemble icing potential forecasts at all three levels are skillful. Comparing Fig. 14 to Fig. 10 one can conclude that the ensemble icing prediction is more skillful than single model forecast in terms of hit rates and false alarm rates.

Finally, the event-driven icing potential forecast was evaluated. Similar as icing potential is converted to nine events, the ensemble probability forecast for each icing event (range from 0 to 100%) also can be converted to 9 deterministic and category forecasts by probability $>$ 10, 20, ... 90% as thresholds. And then the contingency table can be applied to verify the category forecasts. To do this, therefore, totally 9×9 (81) events can be verified in category way (i.e., according to the contingency table). The ETS at FL180 over entire domain and all forecast hours are summarized in Fig. 15 where only icing potential events with thresholds 0.1, 0.3, 0.5, 0.7 and 0.9 are displayed. It can be seen that using ensemble probability, the performance of weak icing event prediction are generally better than that of severe icing event. We also see that for weak icing events, the best probability thresholds are around 30 $\sim$ 50%, while for severe icing events, the best probability thresholds near 10%. This implies that severe icing events are difficult to predict and prone to miss if using higher ensemble probability. In other words, asking too many members to capture a severe icing event is difficult. This evaluation also provides valuable information about the forecast uncertainty in icing potential ensemble forecast.

5. *Summary and Future Work*

To further improve WAFC Washington aviation hazard products, EMC began to collaborate with NCAR icing and turbulence groups in 2009 to

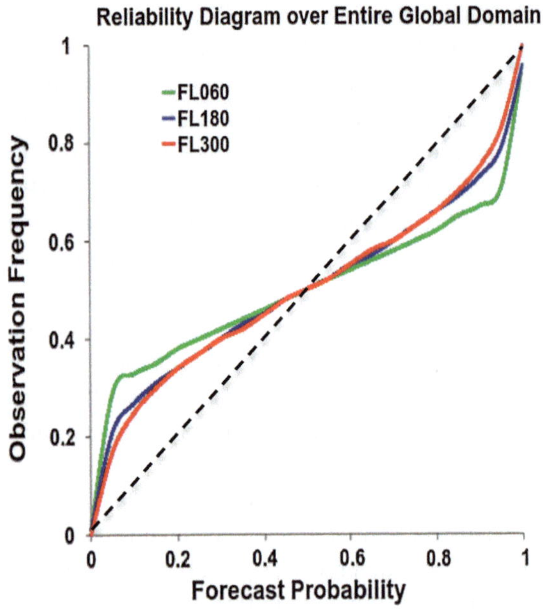

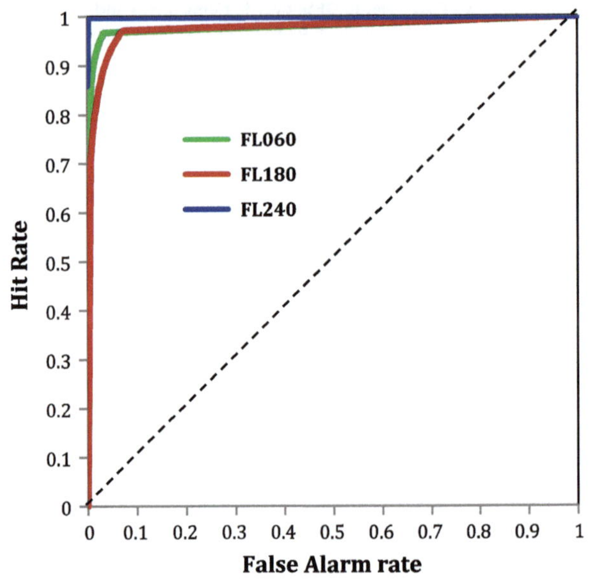

Figure 14
Reliability diagrams (left) and ROC diagrams (right) for three flight levels over entire global Domain and all forecast times (Note here 'observation' refers to analysis)

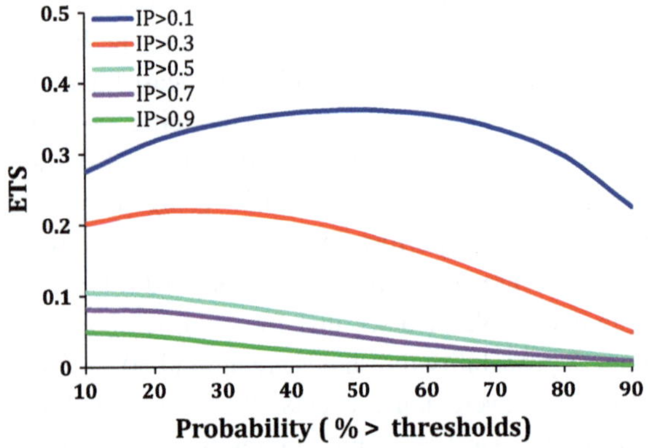

Figure 15
ETS of GEFS icing prediction at FL180 for different ensemble probability thresholds and different icing potential thresholds over entire domain and all forecast hours

transition their icing forecast, icing analysis, and turbulence forecast algorithms into NCEP operations. The new paradigm is to use EMC's Unified Post Processor (UPP) as an aviation research to operation (R2O) tool. NCAR' algorithms are re-worked and incorporated into UPP that supports eight operational models and uses MPI to speed up post-processing when multiple processors are used. This new strategy has been proven successful. First of all, EMC is able to improve global aviation products significantly through the use of high resolution grid available to UPP. Second, because UPP also supports operational Global Ensemble Forecast System (GEFS), EMC is able to start deriving probabilistic icing and turbulence products by scheduling operational implementation for all its twenty-one Global ensemble members to output icing and turbulence products. Finally, much improved efficiency gained by generating aviation product using UPP has not only sped up product delivery time, but also make it feasible to start planning generation of aviation product on very high resolution model such as HRRR. Using UPP to generate Global Graphical Turbulence Guidance (G-GTG) products with fourty-eight processors is eight times faster than generating G-GTG serially.

Adaptation of FIP (Forecast Icing Potential), FIS (Forecast Icing Severity), and GTG algorithms into UPP was performed through multiple year careful collaboration with NCAR icing and turbulence groups. The first effort was to incorporate a global version of FIP algorithm (G-FIP) developed by NCAR (McDonough 2010) into UPP to work with GFS output. In 2011, EMC started generating experimental G-FIP products generated by UPP.

These products were soon distributed to Aviation Weather Center (AWC) and Alaska Aviation Weather Unit for evaluation. Subsequently, Global version of FIS (G-FIS) algorithm was also incorporated into UPP. Because National Weather Service (NWS) also committed itself to performing icing verification for both WAFCs, EMC was tasked to develop Global version of CIP (G-CIP) to be used as verifying truth for WAFS icing forecast.

The global version of GTG (G-GTG) developed by Sharman and Pearson (2017) was the latest to be incorporated into UPP. Integration of G-GTG into UPP was more labor intensive not only due to its much more extensive algorithms but also because it computes many derivatives and spatial averaging which requires extra handling in UPP's MPI environment. More details on algorithms can be found in Sect. 2.2 while information on how algorithms were adapted in UPP can be found in Sect. 2.4.

Qualitative validations via case studies were performed to evaluate the performance of G-FIP, G-FIS, and G-GTG along with objective verification. In Sect. 4.1, two in flight icing cases were shown first to demonstrate good performance of G-FIP and G-FIS. The first case showed how G-FIP and G-FIS 7-hour forecast was able to predict moderate icing event in southeastern Texas and light icing event in central Tennessee that occurred at 13Z on Jan. 16 2018 (Fig. 3). The second case involved a pilot experienced aerodynamic stall while climbing though icing condition at flight level 150 at 07Z on Nov. 14 2016 near Bergen Norway. This incident was predicted by twenty-four hours WAFS icing forecast. As indicated on Fig. 4, 24 h WAFS icing forecast indicated there would be 70% chance of severe icing near Bergen Norway.

Two high impact turbulence cases were also examined to demonstrate great performance of recently implemented G-GTG products. The first case involved a highly publicized case of America Airline hitting extreme turbulence near Pueblo Colorado at FL370 at 00z on Feb 27 2017 that resulted in five injuries. Figure 5 shows this incident was predicted by then experimental G-GTG products consistently starting thirty-six hours ahead of time to 12 h ahead of time. In case 2, EVA air reported hitting severe turbulence near Fukuoka Japan at FL310 at 14Z on November 22 2017 that caused its Boeing 777 to first drop 550 feet and then climb 375 feet. As results, a cabin crew was seriously injured and eight other crews and three passengers received minor injuries. Although EMC only stores GFS output at 12Z and 18Z (Fig. 5) on the event day, it was evident from these forecasts that a severe turbulence could have happened at 14Z near Fukuoka Japan.

Objective verification of G-FIP against CONUS CIP at early development stage and then against G-CIP after 2014 indicated it out-performed older version of WAFC Washington icing forecast in terms of equitable threat score, bias score, and ROC score. Therefore, EMC implemented G-FIP to become new WAFC Washington icing forecast product in 2015 with support from AWC and FAA. After this implementation, EMC continuously verifies blended WAFS mean and max icing forecast against G-CIP and publishes updated statistics on a web site every few month. In this paper, authors showed averaged ROC scores from May 2015 to Feb 2017 (Fig. 10) for various forecast times and flight levels and concluded both blended WAFS mean and max forecast have been very skillful at all examined levels and forecast times since 2015 G-FIP implementation. The skills of ensemble icing forecast were also examined. Although probabilistic icing forecast showed problems of under dispersion, which is most likely due to inherent under dispersion problem with GEFS, it does show that ensemble icing forecast is more skillful than deterministic icing forecast in terms of hit rates and false alarm rates.

GSD verification group recently finished their objective verification of recently implemented G-GTG against both EDR and PIREP data from July to October of 2017. Note that although G-GTG was implemented operationally, it has not replaced WAFC Washington turbulence product. The verification results indicated G-GTG performed better than blended WAFS turbulence product in all thresholds above 0.1. When verified against EDR data, G-GTG is significantly better than blended WAFS turbulence product in terms of ROC scores (Fig. 16). Furthermore, G-GTG on higher resolution of 0.25 degree grid has slightly better skills than G-GTG mapped to operational WAFS resolution of 1.25 degree grid. Authors wish to credit Forecast Impact and Quality

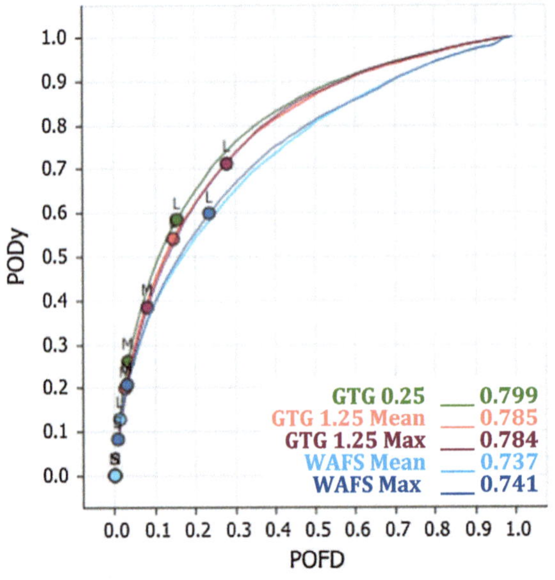

Figure 16
Global verification of G-GTG and WAFS turbulence products against EDR data expressed in ROC curves for all cycles, forecast hours, and all levels 150–400 hPas from July to October of 2017. The Y axis PODY indicates probability of detection and X-axis indicates probability of false detection. The letters L, M, and S mark the forecast thresholds for light, moderate, and severe turbulences, respectively This plot is courtesy of Forecast Impact and Quality Assessment Section (FIQAS) within NOAA/OAR/ESRL/GSD

Assessment Section (FIQAS) within NOAA/OAR/ESRL/GSD for providing these verification results. Furthermore, this research was performed in response to requirements and funding by the FAA and views expressed in this publication are those of the authors and do not necessarily represent the official policy or position of the FAA.

EMC will continue to work with NCAR and AWC to further improve WAFS products and to meet ICAO milestones. With NOAA's plan to replace Global forecast system with Finite Volume Model Version 3 (FV3 https://www.gfdl.noaa.gov/fv3/fv3-documentation-and-references/) in 2019, EMC has begun to modify G-FIP and G-FIS algorithms within UPP to work with more sophisticated GFDL and Thompson microphysics schemes. EMC is also working with NCAR to ensure G-GTG algorithm works as well for FV3 output and perform minor tuning when necessary. The plan is to replace US WAFS turbulence with G-GTG in 2019. It is worth mentioning that UK is also implementing G-GTG to replace UK WAFS turbulence. As a result, WAFS users will see improvement in blended WAFS turbulence forecast product. EMC is collaborating with AWC to develop tools to compute global probabilistic icing severity product to complete ICAO's milestone of start distributing probabilistic aviation product.

Acknowledgements

NCEP WAFS work was previously funded by former NOAA Aviation Service Branch but is currently funded by NOAA Office of Science and Technology Integration (STI). Authors would like to thank NCAR icing and turbulence groups for providing EMC with their latest version of G-FIP and G-GTG algorithms. Authors also wish to thank NOAA Aviation Weather Center and EMC management for their support in providing feedback and for coordinating implementations to meet ICAO milestones. Finally, FAA's funding support to NCAR to collaborate with EMC on this R2O effort and to GSD to validate G-GTG products is much appreciated.

References

Arakawa, A., & Lamb, V. R. (1977). *Computational design of the basic dynamical processes of the UCLA general circulation model. Methods of computational physics* (Vol. 17, pp. 173–265). New York: Academic Press.

The aviation herald. Available online. http://www.avherald.com

Benjamin, S., et al. (2004). An hourly assimilation and model-forecast cycle: The RUC. *Monthly Weather Review, 132,* 495–518.

Benjamin, S., et al. (2016). A North American hourly assimilation and model forecast cycle: The rapid refresh. *Monthly Weather Review, 144,* 1669–1694.

Bernstein, B., et al. (2005). Current icing potential: Algorithm description and comparison with aircraft observations. *Journal of Applied Meteorology Science, 44,* 969–986.

Chuang, H.-Y. (2010). *2010: Development and application of the unified post processor for WRF NMM, WRF ARW, and GFS.* Boulder: WRF Workshop.

Du J., DiMego, G., Zhou, B., Jovic, D., Ferrier, B., Yang, B. (2015). Regional ensemble forecast systems at NCEP, 23rd Conf. on numerical weather prediction and 27th Conf. on weather analysis and forecasting, Chicago, IL, Amer. Meteor. Soc., June 29–July 3

Ellrod, G. P., & Knapp, D. I. (1992). An objective clear-air turbulence forecasting technique: verification and operational use. *Weather and Forecasting, 7,* 150–165.

The global forecast system. Available online. http://www.emc.ncep.noaa.gov/GFS/doc.php.

Gultepe, I., et al. (2018). A meteorological supersite for aviation and cold weather applications. Pure and Applied Geophysics. https://doi.org/10.1007/s00024-018-1880-3 (**this issue**)

International Civil Aviation Organization (2016). "REVIEW AMENDMENT 77 TO ANNEX 3", MET SG/20 – IP/05, Bangkok

Kulesa, G. J., Pace, D.J., Fellner, W.L., Sheets, J.E., Travers, V.S., Kirchoffer, P.J. (2003). The FAA aviation weather research program's contribution to air transportation safety and efficiency". P9.1 ARMS, AMS. (https://ams.confex.com/ams/pdfpapers/54862.pdf).

Lu, C. H., et al. (2016). The implementation of NEMS GFS aerosol component (NGAC) Version 1.0 for global dust forecasting at NOAA/NCEP. *Geoscientific Model Development, 9,* 1905–1919.

McDonough, F., Politovich, M., Wolff, C. (2010). The Global forecast icing product, Paper AIAA-2010-8111, AIAA Atmospheric and Space Environments Conference, Toronto

Rogers E., et al. (2017). Mesoscale modeling development at the national centers for environmental prediction: Version 4 of the NAM forecast system and scenarios for the evolution to a high-resolution ensemble forecast system, 28th Conf. on weather analysis and forecasting, Seattle

Sharman, R. D., & Pearson, J. M. (2017). Prediction of energy dissipation rates for aviation turbulence. Part I: Forecasting nonconvective turbulence. *Journal of Applied Meteorology and Climatology, 56,* 317–337.

Sharman, R., Tebaldi, C., Wiener, G., & Wolff, J. (2006). An integrated approach to mid- and upper-level turbulence forecasting. *Weather and Forecasting, 21,* 268–287.

Slingo, J. M. (1987). The development and verification of a cloud prediction scheme for the ECMWF model. *Quarterly Journal of the Royal Meteorological Society, 113,* 899–927.

Thompson, G., & Eidhammer, T. (2014). A study of aerosol impacts on clouds and precipitation development in a large winter cyclone. *Journal of Atmospheric Science, 71,* 3636–3658.

Trojan, G. (2007). Grib aviation products: WAFC Washington Progress Report

Turp, D.J., Macadam, I., Bysouth, C., Jerrett,D., (2006). Development of grib icing and turbulence products for WAFC London, WAFC London Progress report

White, G., et al. (2017). Evaluation of May 2016 GFS upgrade, 28th Conf. on weather analysis and forecasting. Seattle

Wilks, D. S. (2006). *Statistical methods in atmospheric sciences.* International Geophysics Series (Vol. 59, p. 627). New York: Academic Press.

Wolff, C., McDonough, F., Politovich, M., Gary, C. (2009). Forecast icing product: Recent upgrades and improvements, 1st AIAA Atmospheric and Space Environments Conference. San Antonio

Zhao, Q. Y., & Carr, F. H. (1997). A prognostic cloud scheme for operational NWP models. *Monthly Weather Review, 125,* 1931–1953.

Zhou, B., & Du, J. (2010). Fog prediction from a multi-model mesoscale ensemble prediction system. *Weather and Forecasting, 25,* 303–322.

Zhou, X., Zhu, Y., Hou, D., Luo, Y., Peng, J., & Wobus, D. (2017). The NCEP global ensemble forecast system with the EnKF initialization. *Weather and Forecasting, 32,* 1989–2004.

(Received March 6, 2018, revised August 1, 2018, accepted August 9, 2018, Published online September 25, 2018)

Pure Appl. Geophys. 176 (2019), 2081–2095

https://doi.org/10.1007/s00024-018-1822-0

Pure and Applied Geophysics

Aviation Turbulence: Dynamics, Forecasting, and Response to Climate Change

LUKE N. STORER,[1] PAUL D. WILLIAMS,[1] and PHILIP G. GILL[2]

Abstract—Atmospheric turbulence is a major hazard in the aviation industry and can cause injuries to passengers and crew. Understanding the physical and dynamical generation mechanisms of turbulence aids with the development of new forecasting algorithms and, therefore, reduces the impact that it has on the aviation industry. The scope of this paper is to review the dynamics of aviation turbulence, its response to climate change, and current forecasting methods at the cruising altitude of aircraft. Aviation-affecting turbulence comes from three main sources: vertical wind shear instabilities, convection, and mountain waves. Understanding these features helps researchers to develop better turbulence diagnostics. Recent research suggests that turbulence will increase in frequency and strength with climate change, and therefore, turbulence forecasting may become more important in the future. The current methods of forecasting are unable to predict every turbulence event, and research is ongoing to find the best solution to this problem by combining turbulence predictors and using ensemble forecasts to increase skill. The skill of operational turbulence forecasts has increased steadily over recent decades, mirroring improvements in our understanding. However, more work is needed—ideally in collaboration with the aviation industry—to improve observations and increase forecast skill, to help maintain and enhance aviation safety standards in the future.

1. *Introduction*

Atmospheric turbulence is a major aviation hazard, causing damage to aircraft and injury to passengers and crew. Turbulence is part of the chaotic atmosphere, and the chaotic nature poses a great challenge in understanding and forecasting turbulence. Aviation turbulence is incompletely understood and difficult to forecast, making it a significant hazard. The USA National Transportation Safety Board (NTSB) records the average number of air-carrier-related injuries as 58 per year (FAA 2017b). However, Sharman et al. (2006) suggest that this number is an underestimate, as not all injuries are reported. They state that in the period 1980–2008, there were 234 turbulence accidents, resulting in 298 serious injuries and three fatalities on United States operated air carriers. 184 of the serious injuries involved flight attendants and 114 involved passengers (FAA 2017b). These turbulence injuries will come at a cost to the airlines in two ways: (1) through compensation being paid, which amounts to over $10 million per year, and (2) through lost working days by injured cabin crew, which is over 7000 days per year (Sharman and Lane 2016). Although most of the flight is spent at the cruise phase (around 33,000–39,000 ft), this is also the most vulnerable part of the flight as passengers and crew are unbuckled, so encountering any turbulence would be more likely to result in injury. As a result, most incidents occur above 10,000 ft (Sharman et al. 2006).

Kauffmann (2002) analysed, in detail, the impact that turbulence has on the aviation industry. He presents figures from the USA Federal Aviation Administration (FAA), who state that there were 342 reports of turbulence affecting flights on major air carriers over the period 1981–1997, with three fatalities, 80 serious injuries, and 769 minor injuries. These figures are similar to those of Sharman et al. (2006) and, again, may underestimate the real statistics of turbulence injuries. These values could also be higher in the future, as climate change is likely to increase the frequency of turbulence around the world, particularly in the mid-latitudes (Williams and Joshi 2013; Williams 2017; Storer et al. 2017). The cost of turbulence to the aviation industry is significant and comes from many sources, one of which is preventing aircraft from flying on the

[1] Department of Meteorology, University of Reading, Earley Gate, Reading RG6 6BB, UK. E-mail: luke.storer@pgr.reading.ac.uk

[2] Met Office, FitzRoy Road, Exeter EX1 3PB, UK.

optimum route. This cost is as much as $16 million a year, as it is estimated that 5% of flights are forced to fly non-optimal routes (Search Technology 2000). More importantly, it is also suggested that 15% of the diversions could have been avoided with improved turbulence detection. One form of turbulence detection could be using LIght Detection And Ranging (LIDAR) technology on the front of an aircraft. The current RADAR technology used by aircraft is unable to detect turbulence unless hydrometeors are present, whereas for clear-air turbulence, they are absent. However, LIDAR technology is able to sense clear-air turbulence using non-hydrometeor particles. This technology could warn pilots of turbulence ahead, enabling them to divert or put on the seatbelt sign, so passengers and crew are aware of the upcoming danger. At the moment, however, Kauffmann (2002) calculates that LIDAR technology costs more to install than the saving would be, and therefore, rolling it out is not worth the investment. Satellites can be used to avoid some kinds of turbulence, particularly turbulence associated with convection (Mecikalski et al. 2007; Francis and Batstone 2013). There is also the potential for the new generation of satellites (e.g., GOES-16) with higher spatial and temporal resolutions to improve the avoidance of turbulence. This improvement could come from better identification of deep convection and the ability to resolve gravity waves that would otherwise be invisible to the older generation GOES satellites (Wimmers et al. 2018; Nunez 2018).

Forecasting turbulence is another possibility to mitigate injuries and damage. Turbulent eddies in the atmosphere occur on scales ranging from the planetary scale down to millimeters, but only eddies of approximately 100 m in size impact aviation (Sharman et al. 2006). This is a scale that is not explicitly simulated in forecasts, because the numerical models have resolutions too coarse to resolve the individual eddies. Therefore, turbulence diagnostics are used. These diagnostics generally work on the principle that the energy associated with turbulence on aviation-affecting scales cascades down from the larger scales that can be explicitly resolved by numerical models.

The different kinds of turbulence that impact aviation are outlined in Fig. 1. Vertical wind shear instabilities, mountain waves, and convection are the three main sources that we consider in this study. Turbulence in and near clouds can also cause injuries to passengers and crew, but can easily be detected visually by pilots and using on-board RADAR. Boundary-layer turbulence is not considered here, because it influences only a small portion of the flight after take-off and before landing.

This review paper examines the current understanding of the different sources of aviation turbulence (Sect. 2), how turbulence might change in the future because of climate change (Sect. 3), and the current methods of turbulence forecasting (Sect. 4). The paper concludes with a summary and discussion (Sect. 5).

2. *Turbulence Sources*

2.1. *Clear-Air Turbulence*

Clear-air turbulence (CAT) is defined as high-altitude aircraft bumpiness in regions devoid of significant cloudiness and away from thunderstorm activity (Chambers 1955). Far from mountains, CAT is generally accepted to result from shear instabilities. Wind shear is, therefore, a major source of CAT and is one of the best understood sources. Figure 1 indicates this type of turbulence and its association with the jet stream. To understand why shear causes turbulence, we must define the Richardson number:

$$Ri = \frac{N^2}{(\partial U/\partial z)^2} = \frac{(g/\theta)(\partial\theta/\partial z)}{(\partial U/\partial z)^2}, \qquad (1)$$

where N^2 is the Brunt–Väisälä frequency squared, U is horizontal wind speed, z is altitude, g is gravitational acceleration, and θ is potential temperature. The Richardson number is a nondimensional number with the numerator representing the stratification and the denominator representing the vertical wind shear. It follows from theoretical considerations that instability occurs when $Ri < 0.25$, so instability is favoured by large vertical wind shear (denominator) and weak stratification (numerator). In computational calculations using gridded data, numerical models rarely reach $Ri = 0.25$ due to the coarse resolutions, and therefore, thresholds of turbulence are model specific. To overcome this, Williams (2017) and

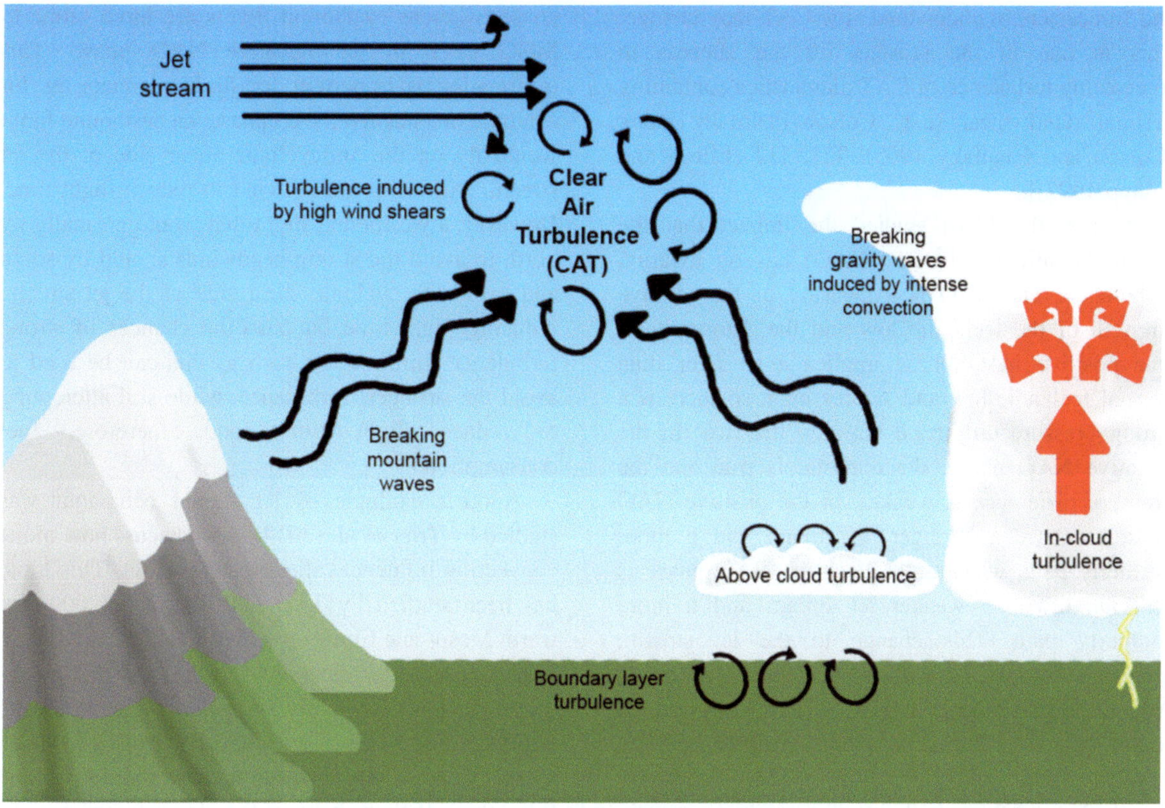

Figure 1
Plot of the main sources of turbulence that impact aviation: shear turbulence caused by high wind shear, breaking gravity waves induced by intense convection, and breaking mountain waves. From Marlton (2016)

Storer et al. (2017) chose thresholds based on the distribution of turbulence in the atmosphere. For example, they assume that severe turbulence is found in 0.01% of the atmosphere, and therefore, they take the top 0.01% (99.9–100%) of the probability distribution to be severe turbulence. Therefore, each threshold is specific to each model and resolution.

It is possible for turbulence to be produced when the environmental Richardson number is much larger than the theoretical critical value, if a local effect reduces the Richardson number locally. For example, gravity waves can cause CAT by reducing the Richardson number (*Ri*) locally in an environment that would not normally produce turbulence, initiating the Kelvin–Helmholtz shear instability and leading ultimately to overturning and breaking billows. The various sources of gravity waves are discussed by Williams et al. (2003, 2005, 2008). In particular, gravity waves can be produced by convection (as discussed in Sect. 2.2) and spontaneous loss of geostrophic balance as the flow evolves, as described by the Lighthill–Ford theory (Lighthill 1952; Ford 1994; Knox et al. 2008; McCann et al. 2012). There is a direct connection in certain circumstances that links deformation to the generation of inertia–gravity waves (Knox et al. 2008). In some cases, this could explain the success of deformation-based CAT diagnostics such as Ellrod and Knapp (1992) TI1 and TI2 (Knox 1997).

It is also possible for the environmental Richardson number to be reduced on a much larger scale, in regions of the atmosphere with particularly strong wind shear such as the jet streams. Strong vertical wind shears around the jet stream increase the denominator in Eq. (1) which, therefore, decreases the Richardson number until it reaches a critical value and turbulence is produced. Therefore, understanding the behaviour of the jet stream will help researchers and forecasters to understand how CAT may change. This is one of the reasons for the success in forecasting turbulence of CAT diagnostics containing vertical wind shear [e.g., Colson–Panofsky index (Colson and Panofsky 1965), TI1, TI2 (Ellrod and Knapp 1992)].

Kim et al. (2016) studied the impact that the North Atlantic Oscillation (NAO) has on aviation turbulence. The NAO is a measure of the relative strength of the Icelandic low and the Azores high. The positive NAO phase implies a stronger than normal Icelandic low and Azores high, so there is a strong pressure difference between the two. In the negative NAO phase, the opposite is true and the pressure difference is weaker. In the positive NAO phase, we see a stronger jet stream and a more northerly path, whereas the negative NAO phase is associated with a weaker jet stream and a more southerly path. This change to the jet stream, therefore, has an impact on turbulence for trans-Atlantic flights. Kim et al. (2016) used wind-optimal routes to find the fastest possible flight path between London (LHR) and New York (JFK). The study found that eastbound flights fly more frequently through regions of CAT than westbound flights, and, therefore, experience more turbulence in both the positive NAO phase and the negative NAO phase. The reason for this is that eastbound flights utilise the jet stream more to benefit from strong tailwinds, so the wind-optimal routes would fly in stronger vertical shear regions more for eastbound flights than westbound flights. Westbound wind-optimal routes avoid the strong headwinds of the jet stream, and would also avoid the stronger vertical shear associated with the jet stream and, therefore, encounter less turbulence.

According to Kim et al. (2016), in the positive NAO phase, westbound flights experience more moderate-or-greater (MOG) CAT than in the negative NAO phase, because some of the westbound flights detour northward to be on the cyclonic shear side of the northerly shifted jet stream, which is more susceptible to MOG turbulence. In contrast, eastbound flights in the negative NOA phase fly through the cyclonic shear side of the southerly shifted jet. Therefore, the study suggests that westbound flights are more prone to MOG CAT in the positive NAO phase, whereas eastbound flights are more prone to MOG CAT in the negative NAO phase. This information is important for flight planners, as for example in a positive NAO phase, an eastbound flight would fly on the anticyclonic shear side of the jet stream, still using the tailwind to reduce flight time. However, a westbound flight that would normally fly north to avoid the strong headwinds should fly south and avoid the cyclonic shear side of the jet stream, reducing flight time but also the chances of strong turbulence. Information such as this can be used to avoid the strongest turbulence, while still attempting to reduce flight times and, therefore, fuel consumption.

Another example of jet stream behaviour was studied by Trier et al. (2012), who discuss how moist convection influences the upper level jet. This topic has been studied by Trier and Sharman (2009) for warm Mesoscale Convective Systems (MCSs). Latent heat release perturbs an anticyclone, and this mechanism accounts for nearly all of the magnitude of the upper level jet. This is a similar mechanism found in a cold weather outbreak, so a mid-latitude cyclone (like the MCS) enhances the downstream anticyclone. The study suggests that, although not as dominant as MCSs, mid-latitude cyclones account for 30–50% of the strength of the southerly jet stream. Without the moist convection and the perturbation of the anticyclone, the wind shear, and therefore, resultant CAT would not be as strong. This is different to CIT, because it is the indirect effect of convection on the jet stream and the increased shear that causes turbulence. The other mechanism is associated with strong convection in the mid-latitude cyclone. With cloud tops below the flight level, the cloud tops generate gravity waves induced by the convection. This mechanism is discussed further in Sect. 2.2.

2.2. *Convective Turbulence*

Understanding the relationship between buoyancy and shear is very important in understanding where and why turbulence forms. Lane et al. (2012) explore our current understanding of near-cloud turbulence or Convectively Induced Turbulence (CIT). They explain that the FAA regulations at the time were not sufficient for avoiding severe turbulence. For example, guideline 5 states: "Do avoid by at least 20 miles (laterally) any thunderstorm identified as severe or giving an intense radar echo, especially under the anvil of a large cumulonimbus". Guideline 6 states: "Do clear the top of a known or suspected severe thunderstorm by at least 1000 ft altitude for each 10 knots of wind speed at the cloud top".

A possible source of CIT is unstable upper tropospheric thunderstorm outflow, similar to that described in Trier and Sharman (2009). Those authors proposed a mechanism for turbulence formation after completing simulations using a convection-permitting model. Their proposed mechanism is the formation of strong vertical wind shear in the outflow regions of Mesoscale Convective Systems (MCS), which we know reduces *Ri* and leads to Kelvin–Helmholtz instability, which is a well-known source of turbulence. In the simulations, they also found that strong vertical wind shear created regions of strong static instability. This was caused by differentially advecting equivalent potential temperature gradients, which were influenced by the adiabatic cooling in the convective updrafts.

The above static instabilities are different from the traditional CIT mechanisms that generally result from high-frequency gravity wave breaking (Lane et al. 2003; Lane and Sharman 2008) or from reductions in *Ri* as a result of propagating gravity waves (Sharman et al. 2012). Propagating gravity waves induced by convection are indicated as a source of turbulence in Fig. 1. Lane et al. (2003) studied the traditional mechanisms forming turbulence above deep convection, and found that gravity waves formed when the overshooting top returns back down. As these waves propagate up, the changes in vertical wind shear and buoyancy with height can change the wavelength of gravity waves and the tilt. The phase speed remains the same above the jet as within the jet based on linear theory, and the decreasing speed above the jet could reach the critical level (Lane et al. 2012). This is found when $U(z) - c = 0$, where $U(z)$ is vertical wind shear and $c = \omega/k$, with ω being the wave frequency and k being the horizontal wave number. Therefore, the gravity wave can break if wave frequency is low, or wind shear is high.

Since wind shear is an important part of the gravity wave breaking, the jet stream plays an important role. The smaller the jet width, the larger the wind shear, and therefore, the closer to the cloud top the gravity wave breaking would be. Lane et al. (2012) also showed turbulence can be found in cirrus bands (banding caused by thermal instabilities), and also ducted gravity waves that propagate far from the updraft, reducing the Richardson number and causing turbulence much farther away than the 20 km outlined in the guidelines. The new guidelines from FAA (2017a) no longer reference the acceptable height above the thunderstorm pilots can fly. Guideline 2 now states: "Don't attempt to fly under a thunderstorm even if you can see through to the other side. Turbulence and wind shear under the storm could be hazardous". Guideline 3 states "Don't attempt to fly under the anvil of a thunderstorm. There is a potential for severe and extreme clear-air turbulence". Guideline 14 states "Do avoid by at least 20 miles any thunderstorm identified as severe or giving an intense radar echo. This is especially true under the anvil of a large cumulonimbus".

2.3. Mountain-Wave Turbulence

Mountain-wave turbulence is similar to Convectively Induced Turbulence, as gravity waves produced by the terrain (instead of convection) propagate and then break when a critical vertical wind shear value is reached (Nastrom and Fritts 1992; Wurtele et al. 1996). This is also indicated in Fig. 1 and shows its similarity to CIT. A study by Epifanio and Qian (2008) used a high-resolution ensemble to study turbulence produced by breaking mountain waves and found that their results were consistent with the previous work by Fritts et al. (1994, 1996), suggesting that shear is an important factor in the breaking of gravity waves. Fritts et al. (1996) ran simulations to understand the turbulence mechanism in mountain waves, and found that the dominant source of instability, and therefore turbulence, was the wind shear. This came from the mean wind field, and the differential vertical advection of the mean shear by the wave field. Similar to convective turbulence, the vertical wind shear and its interaction with propagating gravity waves is an important mechanism in turbulence production. This understanding will help forecasters in the future.

Wolff and Sharman (2008) show that MOG turbulence is typically found where topographic heights are above 1.5 km. They also analysed the low-level wind direction when mountain-wave turbulence is produced, and concluded that the wind direction must be within 30° of the perpendicular to the mountain range orientation. For north America, this makes the Rocky Mountains a prime location for MOG MWT, as the mountain range is oriented north–south and frequently experiences westerly low-level flow with mountain heights above 1.5 km. Greenland is also an area with high mountain ranges with the potential to cause turbulence for trans-Atlantic flights.

3. *Climatology and Response to Climate Change*

Jaeger and Sprenger (2007) used reanalysis data to understand upper tropospheric clear-air turbulence in the Northern Hemisphere, because of its role in stratosphere–troposphere exchange (Shapiro 1976, 1978, 1980). The authors create a climatology of the Ellrod and Knapp (1992) index (TI1), Brunt–Väisälä frequency squared (N^2), Richardson number (*Ri*), and potential vorticity (PV). Negative potential vorticity is a turbulence predictor as it is related to isentropic inertial instability. The study found that TI1 is largest north of the jet stream, and Kelvin–Helmholtz instability indicated by *Ri* is near the jets, which is what we would expect with larger wind shear in that region. Symmetric instability is most frequent south of the jets, and is particularly associated with anticyclonic jets. Hydrostatic instability is only slightly dependent on the jet position, and is most common over land where convection and gravity wave activity are most prevalent (mountain wave and deep convection gravity waves). The study also showed that winter (DJF) has the highest turbulence frequency, which follows the understanding that the jet stream is stronger in winter. The study also found long-term trends in the frequency of diagnosed turbulence over the reanalysis period. Figure 2 shows that over the reanalysis period, there is an increasing trend of turbulence in the northern Hemisphere. They were also able to relate inter-annual variability to the North Atlantic Oscillation (NAO) and Pacific/North American flow, which we know from Sect. 2.1 influences CAT.

We know that the climate system is changing due to anthropogenic forcing, and these changes may have an impact on turbulence in the future. Collins et al. (2013) showed that the changes in temperature are not uniform around the world. Importantly for turbulence, the upper troposphere and lower stratosphere respond differently to anthropogenic forcing. The tropical upper troposphere is predicted to warm faster than the tropical surface, due to an increase in latent heat release. Latent heat is released during convection, and in a warmer climate the atmosphere can hold more moisture, and as a result more convection and subsequent latent heat release will warm the troposphere. The lower stratosphere, however, will cool with the increasing greenhouse gases (Fels et al. 1980). This cooling is related to changes in infrared radiation.

The upper tropospheric and lower stratospheric changes discussed above lead to an increase in the equator-to-pole temperature gradient at flight cruising levels, as shown in Fig. 3. Particularly in the RCP8.5 simulation (right), where greenhouse gas emissions are highest, we see the warming in the tropics, and the cooling at the poles, increasing the equator-to-pole temperature gradient. This increase in meridional temperature gradient will induce a thermal wind response, resulting in an increase in vertical wind shear and, therefore, turbulence in the mid-latitudes. These changes to the jet stream are shown in Fig. 4, which is taken from Delcambre et al. (2013). The changes to the jet stream are predominant in the mid-latitudes and at airline cruise altitudes, making their impact the largest in the busiest flight regions around the world. In addition to modifying turbulence, the increased wind speeds are also expected to modify flight times (Williams 2016). Another impact of anthropogenic forcing is the 20th-century release of chlorofluorocarbons (CFCs), which destroy atmospheric ozone (O_3). The loss of ozone reduces the lower stratospheric temperature, further increasing the equator-to-pole temperature gradient. This effect enhances the predicted change arising from carbon dioxide.

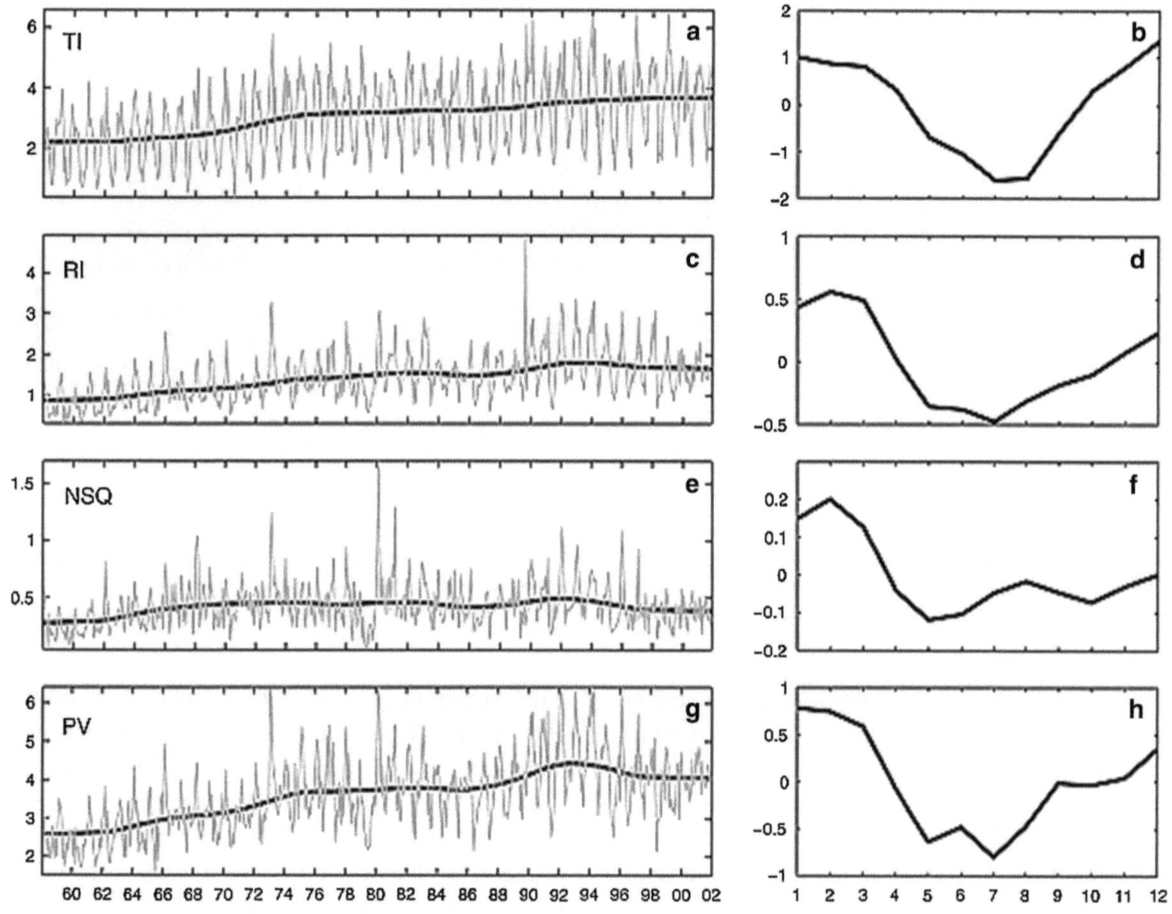

Figure 2
(Left) TI, Ri, N^2, and PV (from top to bottom) frequency time series (grey lines) and nonlinear trend estimates from STL analysis (bold black line) for the North Atlantic sector from 90°W–10°E and 30°–70°N in the tropopause region (%). (Right) Mean seasonal cycle component of the turbulence indicators from STL decomposition (Δ%). All panels are for the time period 1958–2001. Note the different scales. From Jaeger and Sprenger (2007)

Several studies have reported that climate change will act to increase clear-air turbulence in the future, according to climate model simulations (Williams and Joshi 2013; Williams 2017; Storer et al. 2017). The first study to look at this (Williams and Joshi 2013) focused on north Atlantic moderate-or-greater turbulence and showed that it would increase in frequency with climate change by around 40–170%. Williams (2017) then furthered the study to see how climate change might influence turbulence in five strength categories from light to severe, finding that all would increase in frequency with climate change. Storer et al. (2017) then extended the work further to see how climate change will impact CAT globally, through all four seasons, and also at two flight levels in eight geographic regions. They found that CAT will increase globally in all four seasons. This is shown in Fig. 5, with the largest and most statistically significant increases confined to the mid-latitudes around the jet stream. Some regions locally will see increases of several hundred per cent for example at 200 hPa (39,000 ft.) the north Atlantic will see an increase of 143.3% in moderate turbulence and 181.4% in severe turbulence. The study also showed that at 200 hPa (1) in winter, severe CAT by 2050–2080 will be as common as moderate CAT in the control period, and (2) for a range of turbulence strengths from light to moderate-to-severe,

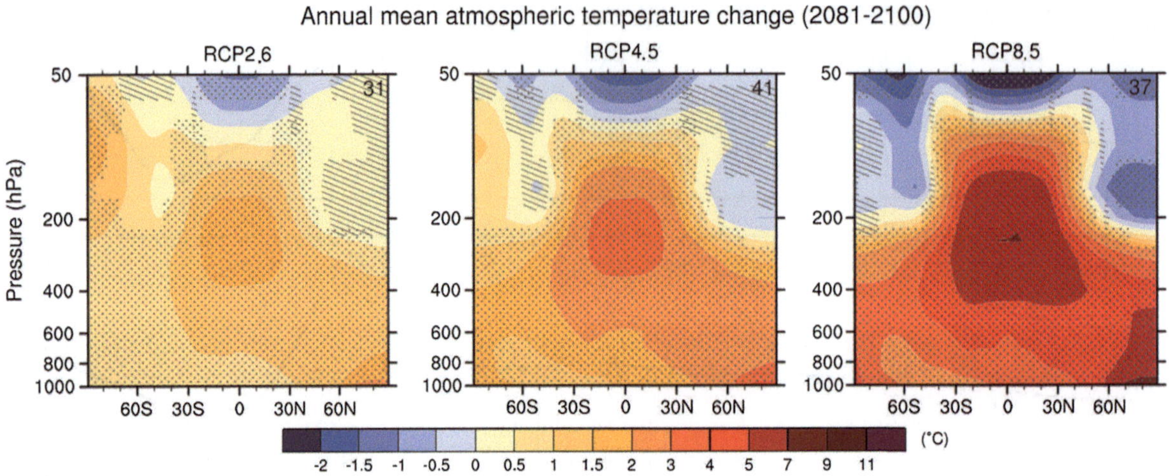

Figure 3
Modelled zonal-mean annual-mean ensemble-mean future temperature changes for climate change scenario RCP2.5 (left), RCP4.5 (middle), and RCP8.5 (right). Hatching indicates the regions where the ensemble-mean change is less than one standard deviation of internal variability. Stippling indicates regions where the ensemble-mean change is greater than two standard deviations of internal variability and where at least 90% of the models agree on the sign of the change. From Collins et al. (2013). Available under https://creativecommons.org/licenses/by/3.0/

summertime CAT by 2050–2080 will be as common as wintertime CAT in the control period. Aircraft manufacturers and forecasters need to be prepared to prevent an increase in damage to aircraft and injuries to passengers and crew.

Currently, there is a gap in the literature looking at the response of Convectively Induced Turbulence and Mountain-Wave Turbulence to climate change. However, we know that deep convection can produce turbulence, and therefore, it is likely that CIT would increase if convection was increased. Price and Rind (1994) showed an increase of 5–6% of global lightning activity with every 1°C of global warming. Lightning activity has been used to characterise turbulence events (Gill and Stirling 2013; Meneguz et al. 2016), and therefore, if we see an increase in lightning, we would expect to see an increase in convection and possibly an increase in CIT as a result. Price and Rind (1994) show that lightning activity and convection will increase particularly around the tropics, which is the area where CAT is not expected to increase (Fig. 5). Therefore, combining CAT and CIT, we would see a global increase in all aviation turbulence. This trend of increasing lightning with climate change is supported by Reeve and Toumi (1999) who show an increase in lightning activity of 40% for every 1 K of average land wet-bulb temperature. It could be possible to study the change in CIT to climate change. Convective precipitation accumulation was used by Gill and Stirling (2013) as a convective indicator for turbulence forecasts, and therefore, assessing the changes of this quantity in climate models could indicate how climate change might impact CIT.

4. Forecasting

Currently, the World Area Forecast Centres (WAFCs) in London (Met Office) and Washington (NOAA) produce operational turbulence forecasts for aviation. This includes a $t + 24$ h significant weather (SIGWX) chart four times a day (0000, 0006, 0012, 0018 UTC) which displays multiple aviation hazards including icing, CAT, as well as the location of convective cloud. The WAFCs also produce a gridded turbulence forecast four times a day at five pressure levels with a lead time $t + 6$ h to $t + 36$ h (Gill 2014).

4.1. Clear-Air Turbulence

Using diagnostics to forecast clear-air turbulence allows flight planners to avoid turbulent regions. Fahey (1993) showed that strategic planning can lead

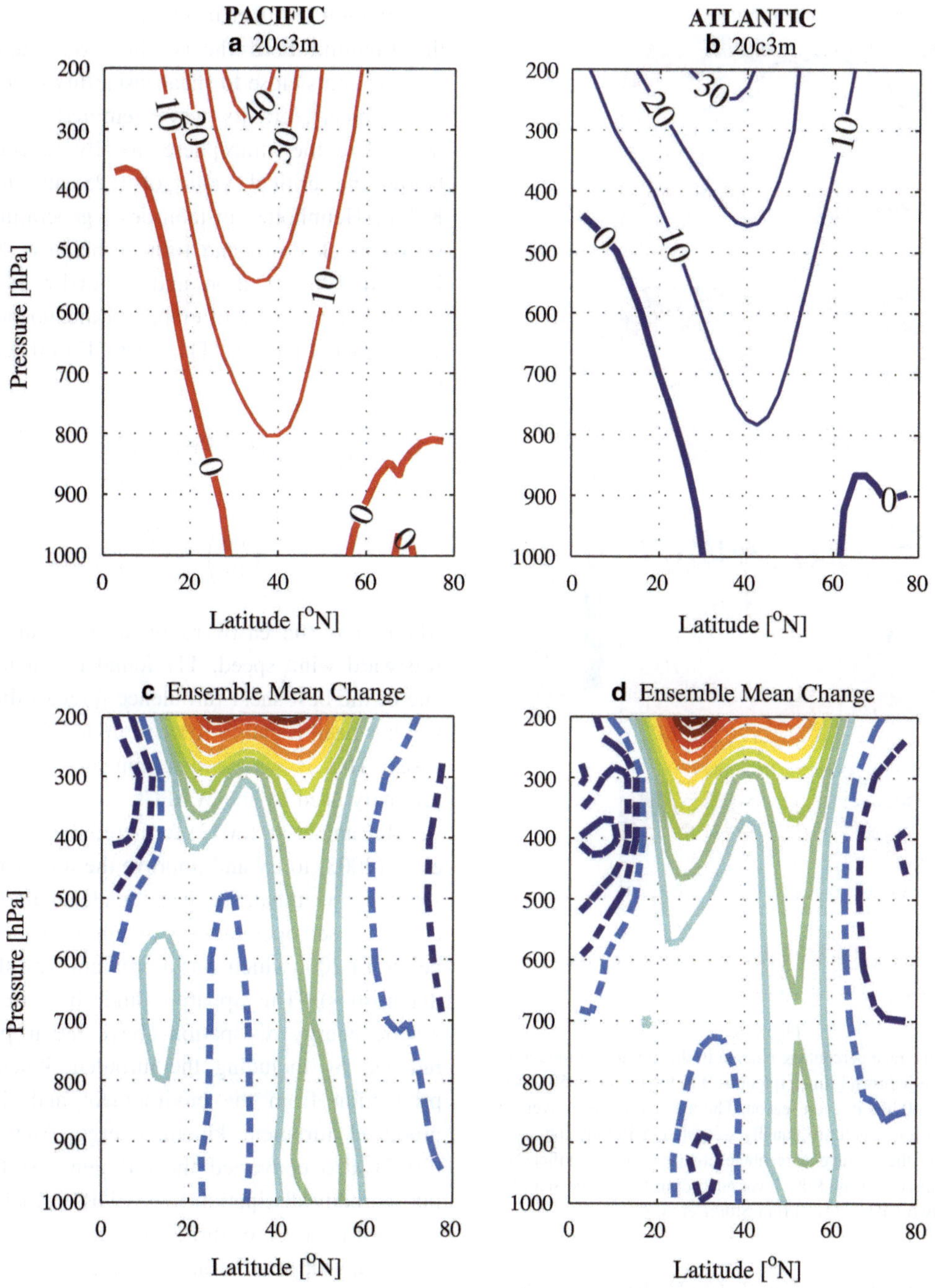

Figure 4
Modelled zonal-mean annual-mean ensemble-mean future zonal wind changes. The zonal mean is taken over the Pacific (left) and Atlantic (right). The upper row shows a control period contoured every 10 m s^{-1} and the lower row shows the future change contoured every 0.25 m s^{-1}. From Delcambre et al. (2013). ©American Meteorological Society. Used with permission

to not only a reduction in injuries, but also costs. The use of diagnostics is also the only way to operationally forecast turbulence, as the turbulent eddies that affect aviation are smaller than the resolution of global atmospheric models. Turbulence predictors that have a deformation term are particularly good at

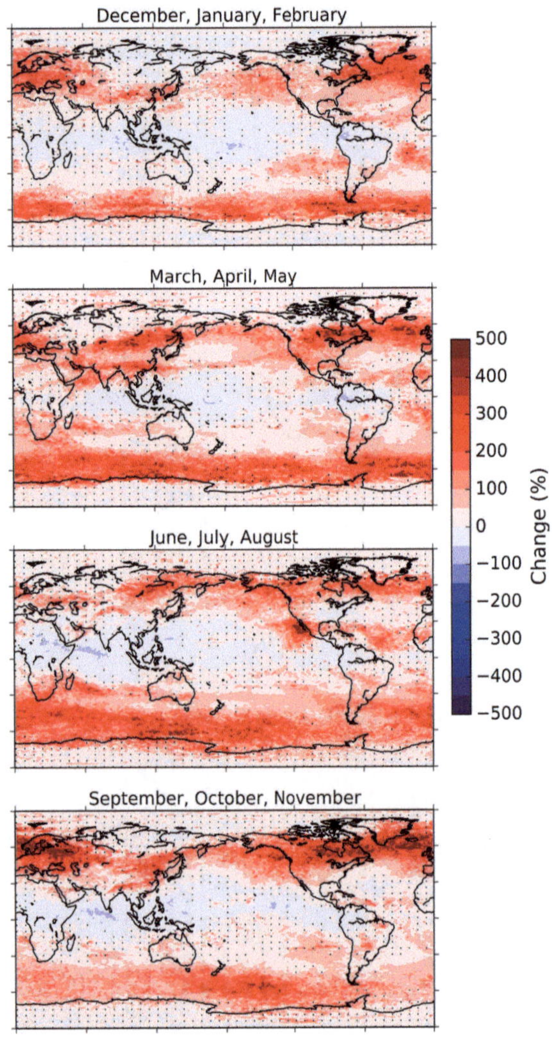

Figure 5
Maps of the average percentage change in the amount of moderate CAT from pre-industrial times (picontrol) to the period 2050–2080 (RCP8.5) at 200 hPa in each season. The average is taken over 20 CAT diagnostics, which are equally weighted. Stippling indicates regions where the average percentage change is not significantly different from zero at the 90% level according to the one-sample, two-tailed t test. From Storer et al. (2017)

forecasting CAT. Knox et al. (2008) tried to understand the previous work that linked deformation to Kelvin–Helmholtz instabilities. This included frontogenesis that could initiate Kelvin–Helmholtz instabilities, or mesoscale waves that would then break down and form turbulence (Mancuso and Endlich 1966). They found that neither could fully explain the relationship between deformation and turbulence, so instead found a link between deformation and inertia–gravity wave generation via the Lighthill–Ford theory. In certain atmospheric environments close to shear instability (low Richardson number), gravity waves emitted by imbalance destabilize the atmosphere, locally reducing Ri to below the critical value of 0.25 and leading to Kelvin–Helmholtz instabilities, generating waves which break down and form turbulence (Miles and Howard 1964; Dutton and Panofsky 1970). This explains the success of empirical diagnostics such as Turbulence Index 1 (TI1) from Ellrod and Knapp (1992):

$$\mathrm{TI1} = \left[\left(\frac{\partial u}{\partial x}-\frac{\partial v}{\partial y}\right)^2+\left(\frac{\partial v}{\partial x}+\frac{\partial u}{\partial y}\right)^2\right]^{1/2} \times \left[\left(\frac{\partial u}{\partial z}\right)^2+\left(\frac{\partial v}{\partial z}\right)^2\right]^{1/2}, \tag{2}$$

where u is the eastward wind speed and v is the westward wind speed. TI1 found in equation (2) is one of the best shear turbulence forecast diagnostics, and has been found to forecast up to 75% of all CAT cases. This is why a deterministic TI1 forecast is routinely used by the WAFCs.

McCann et al. (2012) furthered the work by Knox et al. (2008) to try and improve the forecast based on spontaneous imbalance. They made an algorithm that could be used operationally by improving the forecast below FL200 (which was a problem with the current diagnostics). The apparent high bias in turbulent kinetic energy dissipation above the tropopause is reduced by including the turbulent kinetic energy production from the environment and the locally produced turbulent kinetic energy. McCann et al. (2012) also converted the turbulent kinetic energy into an eddy dissipation rate (EDR), which is now a standard measure of turbulence in the atmosphere as it does not depend on aircraft size.

There are many turbulence diagnostics, each with its own its strengths and weaknesses. All forecast CAT and are unable to forecast convective turbulence or mountain-wave turbulence. Since each individual diagnostic is unable to forecast every CAT event, Sharman et al. (2006) generated the Graphical Turbulence Guidance (GTG) system for forecasting turbulence. They selected multiple turbulence diagnostics and then used PIlot REPorts (PIREPS) to understand the diagnostics' performances. A weighting system can then be added depending on the skill of each diagnostic to produce the best overall forecast possible. The study showed that this was an improvement over the single turbulence diagnostic, and is, therefore, worth doing. Over the years, this forecast has been improved, and more recently, a mountain-wave turbulence predictor has been added to the GTG system. At the moment CIT is not forecast by the GTG system, but it has been shown that it is a useful addition (Gill and Stirling 2013).

4.2. *Convective Turbulence*

Convection is one of the main turbulence sources. It is not yet forecast using numerical models operationally, but, instead, forecasters manually put the location of convective clouds on the significant weather (SIGWX) charts used by pilots and flight planners. Gill and Stirling (2013) showed that using a convective diagnostic from numerical weather prediction output can forecast many convective turbulence events. Therefore, combining the shear turbulence predictors and the convective predictors (similar to GTG) offers the promise to greatly improve the forecast skill. The convective indicators used in the study included convective precipitation rate, convective precipitation accumulation, and Convective Available Potential Energy (CAPE). Gill and Stirling (2013) concluded that it is possible to improve turbulence forecasts by not only combining different predictors for the same turbulence source (e.g., TI1, Brown index, etc.) but also combining predictors for different turbulence sources as well (e.g., clear-air turbulence and convective turbulence). They suggest that more work is needed in this area, but there is the potential to greatly increase the forecast skill by adding in these extra turbulence diagnostics and forecasting a greater proportion of turbulence events.

4.3. *Mountain-Wave Turbulence*

An algorithm for forecasting mountain-wave turbulence was described by Turner (1999). Mountain waves are a special case of gravity waves and they are smaller than the numerical resolution of the models, so the wave stress has to be parameterised. To do this, the surface winds and stability are used to work out a wind stress, and then, the wind stress is used to work out the vertical profile. The wind stress is computed for every grid point, taking into account the abruptness of orography (as that is the location of the mountains). The surface stress is passed upwards from the surface (unless a hydraulic jump, critical level, or saturation stress is diagnosed), and therefore, less stress is transferred to the next level. Above the maximum point that gravity wave stress can be sustained, the waves break and thus turbulence forms, so the models can use the wave stress at each model level to calculate the location and height at which the gravity wave stress is too high. This is, therefore, the turbulence location in the model. For mountain-wave formation, the higher the gravity wave propagates, the more likely it is to break. This is because the maximum gravity wave stress which the atmosphere can hold decreases with increased wind shear and lower air density. Mountain-wave turbulence is found to be associated most strongly with the mid-latitude westerlies and mountains perpendicular to the flow.

The new GTG3 system includes mountain-wave turbulence diagnostics alongside the typical CAT diagnostics (Sharman and Pearson 2017). These authors have also used a different approach to forecasting MWT and use a combination of low-level wind combined with CAT diagnostics. The authors produced 14 new MWT diagnostics, which showed an increased forecast skill in MWT prone areas, supporting their use in the new GTG system.

4.4. *Ensemble Forecasting*

One of the main problems with the turbulence forecasting methods outlined in Sects. 4.1, 4.2, and 4.3 is that they do not convey uncertainty. To resolve this issue, Gill and Buchanan (2014) and Buchanan (2016) trialled the use of ensemble forecasting for aviation turbulence. An ensemble is a collection of forecast runs, each of which is considered to be equally likely. Using an ensemble, the uncertainty in the forecast can be found. In these studies, the Met Office Global and Regional Ensemble Prediction System (MOGREPS) was used, which was made operational in 2008 (Bowler et al. 2008). Multiple turbulence predictors were considered, including the Dutton (1980) index, the Ellrod and Knapp (1992) indices, the Brown (1973) index, the Lunnon index (Roach and Dixon 1970), the Richardson number, and two convective predictors. Using the Derived Equivalent Vertical Gust (DEVG) as a truth in the verification, the studies found that using the ensemble forecast, they increased the forecast skill for most of the turbulence diagnostics.

The studies by Gill and Buchanan (2014) and Buchanan (2016) then combined the predictors in a manner similar to Sharman et al. (2006), using an iterative scheme to maximise the forecast skill. Again, the studies showed that the ensemble forecast was more skillful than a single model deterministic forecast. A probabilistic forecast would also be more useful for pilots and flight planners, as understanding the uncertainty of the forecast can help them to make the best decision possible to maintain safety of passengers and crew, while still flying the optimum routes and reducing flight times and fuel consumption. Further work is needed in ensemble forecasting before it can be used operationally, such as studying whether multi-model ensembles can improve the forecast. In addition, more research is needed to include a mountain-wave turbulence predictor alongside the convective and shear turbulence predictors. However, it is clear that ensembles can be beneficial and should eventually be used operationally in the future.

4.5. Forecast Verification

Turbulence forecast verification is difficult, because objective verification data sets are limited. Some previous studies resolved this issue using PIREPs (Kim and Chun 2011; Schwartz 1996) to identify regions of turbulence based on a semi-quantitative scale from light to extreme, but these can be unreliable (Kane et al. 1998). PIREPs are subjective, in the sense that a more experienced pilot may catagorise an event as moderate, but an inexperienced pilot may record it as severe. In addition, PIREPs are aircraft-dependent, so a smaller aircraft will experience stronger turbulence than a larger aircraft in the same volume of air. PIREPs also have poor spatial reliability as they tend to be located in turbulence, so null turbulence events are rarely recorded (Kane et al. 1998).

To avoid the above problems, there are two main aircraft-independent measures, which are the Eddy Dissipation Rate (EDR) and the Derived Equivalent Vertical Gust (DEVG). The measures are calculated using high-resolution automated aircraft data, available at 4 s intervals. The EDR is calculated following Haverdings and Chan (2010) using:

$$\mathrm{EDR} = \frac{\sigma_w}{\sqrt{1.05 V_a^{\frac{2}{3}} \left(\omega_1^{-\frac{2}{3}} - \omega_2^{-\frac{2}{3}} \right)}}, \tag{3}$$

where σ_w is the running mean of the vertical wind shear over a 10 second time window, V_a is the true airspeed, and ω_1 and ω_2 are cut-off frequencies set at 0.1 and 2 Hz, respectively. The DEVG (Tenenbaum 1991; Gill 2014, 2016; Kim et al. 2017) is calculated following Truscott (2000) using:

$$\mathrm{DEVG} = \frac{Am|\Delta n|}{V}, \tag{4}$$

where $|\Delta n|$ is the peak modulus value of the deviation of the aircraft acceleration from 1g in units of g, m is the total mass of the aircraft (metric tonnes), V is the calibrated airspeed at the time of the observation (knots), and A is an aircraft-specific parameter that varies with flight conditions and can be calculated using the following:

$$A = \overline{A} + c_4(\overline{A} - c_5)\left(\frac{m}{\overline{m}} - 1\right) \tag{5}$$

and:

$$\overline{A} = c_1 + \frac{c_2}{c_3 + H(kft)}, \tag{6}$$

where H is the altitude (thousands of feet), $\overline{m}$ is the reference mass of the aircraft (metric tonnes), and parameters c_1–c_5 depend on the aircraft's flight profile as outlined by Truscott (2000).

Unlike PIREPs, EDR and DEVG are aircraft-independent, and therefore, all observations from multiple different aircraft can be combined to create a consistent observational database. Also unlike PIREPs, EDR and DEVG report all results including null reports. Because it is automated, the entire flight is recorded, and therefore, the only limit in coverage is the location the aircraft fly. Therefore, if every commercial aircraft had this capability, then all turbulence events could be recorded, and forecast verification improved. There are limitations to using the DEVG data set, as aircraft manoeuvres and active control techniques can enhance or dampen vertical accelerations of aircraft, leading to over- or under-representation of the vertical gusts (WMO 2003). However, despite these limitations, using an aircraft-independent measure (such as EDR or DEVG) is preferable to pilot reports (PIREPs).

5. Summary

Atmospheric turbulence is a major aviation hazard and we do not yet fully understand it. As a result, it causes many injuries to passengers and crew all around the world, with millions of dollars lost due to aircraft damage, planes being in the wrong place from a diverted flight, loss of hours from cabin crew, and compensation. This study has reviewed how turbulence that affects aviation can come from three main sources. The first is shear turbulence and is mainly found around the jet stream. The second is Convectively Induced Turbulence (CIT), and this forms around deep convection. However, there are multiple mechanisms that produce turbulence from convection, one of which is similar to the mechanism for the third source of turbulence known as Mountain-Wave Turbulence, which occurs when orography perturbs the atmosphere and produces gravity waves that can lead to turbulence.

This study has also reviewed how climate change is expected to have a large impact on clear-air turbulence in the future, with several studies finding that CAT will increase with climate change, particularly around the mid-latitudes in all seasons.

This makes turbulence forecasting very important in the future, to maintain the safety of passengers and crew. Currently, only one turbulence index known as the Ellrod and Knapp (1992) Turbulence Index 1 (TI1) is used by the two World Area Forecast Centres (WAFCs), but others have been included in the Graphical Turbulence Guidance (GTG) forecast. GTG combines many indices but is not yet used fully operationally in the WAFCs. Ensemble forecasting has also been developed to improve the forecasts, as currently only a deterministic forecast is created and that does not show the uncertainty in the forecast. The use of next-generation satellites (e.g., GOES-16) could help to improve nowcasting of turbulence, because with higher spatial and temporal resolutions, convection and gravity waves are more clearly resolved, aiding avoidance on the short time scales.

We know that the current methods of aviation turbulence forecasting are imperfect, and not all turbulence events are predicted. Therefore, continuing research to improve the forecast systems should be a high priority in the future. In addition, the aviation industry must make sure that the aircraft of the future are up to the challenge. This is because turbulence is not the only aviation hazard, with take-off weight restrictions due to high temperatures or aircraft icing also major hazards that might change in the future. In addition, the airline industry's desire to choose the most time-effective route to reduce fuel consumption and costs may force some pilots to fly into more turbulence regions. By making sure that all these aspects are taken into account, new and better working processes can be developed to make sure that the impact aviation has on the climate, and the impact climate change may have on aviation, are both reduced. With the volume of air travel also expected to increase over time, resulting in more congested skies, the research conducted now is vital to making sure that the industry is prepared for the future.

Acknowledgements

L.N.S. acknowledges support through a Ph.D. studentship from the Natural Environment Research Council SCENARIO Doctoral Training Partnership (reference NE/L002566). P.D.W. acknowledges support through a University Research Fellowship from the Royal Society (reference UF130571). The authors would also like to thank Dr. Graeme Marlton for providing Fig. 1.

References

Bowler, N. E., Arribas, A., Mylne, K. R., Robertson, K. B., & Beare, S. E. (2008). The MOGREPS short-range ensemble prediction system. *Quarterly Journal of the Royal Meteorological Society, 134*(632), 703–722.

Brown, R. (1973). New indices to locate clear-air turbulence. *Meteorological Magazine, 102*, 347–361.

Buchanan, P. (2016). Aviation turbulence ensemble techniques. In: *Aviation turbulence*, pp. 285–296. Springer.

Chambers, E. (1955). Clear air turbulence and civil jet operations. *The Aeronautical Journal, 59*(537), 613–628.

Collins, M., Knutti, R., Arblaster, J., Dufresne, J.-L., Fichefet, T., Friedlingstein, P., et al. (2013). *Long-term climate change: projections, commitments and irreversibility*. Cambridge: Cambridge University Press.

Colson, D., & Panofsky, H. (1965). An index of clear air turbulence. *Quarterly Journal of the Royal Meteorological Society, 91*(390), 507–513.

Delcambre, S. C., Lorenz, D. J., Vimont, D. J., & Martin, J. E. (2013). Diagnosing Northern Hemisphere jet portrayal in 17 CMIP3 global climate models: Twenty-first-century projections. *Journal of Climate, 26*(14), 4930–4946.

Dutton, M. (1980). Probability forecasts of clear-air turbulence based on numerical-model output. *Meteorological Magazine, 109*(1299), 293–306.

Dutton, J. A., & Panofsky, H. A. (1970). Clear air turbulence: A mystery may be unfolding. *Science, 167*(3920), 937–944.

Ellrod, G. P., & Knapp, D. I. (1992). An objective clear-air turbulence forecasting technique: Verification and operational use. *Weather and Forecasting, 7*(1), 150–165.

Epifanio, C. C., & Qian, T. (2008). Wave-turbulence interactions in a breaking mountain wave. *Journal of the Atmospheric Sciences, 65*(10), 3139–3158.

FAA. (2017a). Air traffic plans and publications. https://www.faa.gov/air_traffic/publications. Accessed 09 Feb 2018.

FAA. (2017b). Turbulence: Staying safe. https://www.faa.gov/travelers/fly_safe/turbulence/. Accessed 09 Feb 2018.

Fahey, T. H. (1993). Northwest airlines atmospheric hazards advisory and avoidance system. In: *International conference on aviation weather systems, 5th, Vienna, VA*, pp. 409–413.

Fels, S., Mahlman, J., Schwarzkopf, M., & Sinclair, R. (1980). Stratospheric sensitivity to perturbations in ozone and carbon dioxide: Radiative and dynamical response. *Journal of the Atmospheric Sciences, 37*(10), 2265–2297.

Ford, R. (1994). Gravity wave radiation from vortex trains in rotating shallow water. *Journal of Fluid Mechanics, 281*, 81–118.

Francis, P. N., & Batstone, C. (2013). *Developing a satellite product to identify severe convective storms hazardous to aviation.* Satellite Applications Technical Memo 11.

Fritts, D. C., Garten, J. F., & Andreassen, Ø. (1996). Wave breaking and transition to turbulence in stratified shear flows. *Journal of the Atmospheric Sciences, 53*(8), 1057–1085.

Fritts, D. C., Isler, J. R., & Andreassen, Ø. (1994). Gravity wave breaking in two and three dimensions: 2.Three-dimensional evolution and instability structure. *Journal of Geophysical Research: Atmospheres, 99*(D4), 8109–8123.

Gill, P. G. (2016). Aviation turbulence forecast verification. In: *Aviation turbulence*, pp. 261–283. Springer.

Gill, P. G. (2014). Objective verification of World Area Forecast Centre clear air turbulence forecasts. *Meteorological Applications, 21*(1), 3–11.

Gill, P. G., & Buchanan, P. (2014). An ensemble based turbulence forecasting system. *Meteorological Applications, 21*(1), 12–19.

Gill, P. G., & Stirling, A. J. (2013). Including convection in global turbulence forecasts. *Meteorological Applications, 20*(1), 107–114.

Haverdings, H., & Chan, P. W. (2010). Quick access recorder data analysis software for windshear and turbulence studies. *Journal of Aircraft, 47*(4), 1443–1447.

Jaeger, E., & Sprenger, M. (2007). A Northern Hemispheric climatology of indices for clear air turbulence in the tropopause region derived from ERA40 reanalysis data. *Journal of Geophysical Research Atmospheres, 112*(D20).

Kane, T., Brown, B., & Bruintjes, R. (1998). Characteristics of pilot reports of icing. In: *Preprints: 14th conference on probability and statistics*, pp. 11–16.

Kauffmann, P. (2002). The business case for turbulence sensing systems in the US air transport sector. *Journal of Air Transport Management, 8*(2), 99–107.

Kim, J.-H., Chan, W. N., Sridhar, B., Sharman, R. D., Williams, P. D., & Strahan, M. (2016). Impact of the North Atlantic Oscillation on transatlantic flight routes and clear-air turbulence. *Journal of Applied Meteorology and Climatology, 55*(3), 763–771.

Kim, J.-H., & Chun, H.-Y. (2011). Statistics and possible sources of aviation turbulence over South Korea. *Journal of Applied Meteorology and Climatology, 50*(2), 311–324.

Kim, S.-H., Chun, H.-Y., & Chan, P. W. (2017). Comparison of turbulence indicators obtained from in situ flight data. *Journal of Applied Meteorology and Climatology, 56*(6), 1609–1623.

Knox, J. A. (1997). Possible mechanisms of clear-air turbulence in strongly anticyclonic flows. *Monthly Weather Review, 125*(6), 1251–1259.

Knox, J. A., McCann, D. W., & Williams, P. D. (2008). Application of the Lighthill–Ford theory of spontaneous imbalance to clear-air turbulence forecasting. *Journal of the Atmospheric Sciences, 65*(10), 3292–3304.

Lane, T. P., & Sharman, R. D. (2008). Some influences of background flow conditions on the generation of turbulence due to gravity wave breaking above deep convection. *Journal of Applied Meteorology and Climatology, 47*(11), 2777–2796.

Lane, T. P., Sharman, R. D., Clark, T. L., & Hsu, H.-M. (2003). An investigation of turbulence generation mechanisms above deep convection. *Journal of the Atmospheric Sciences, 60*(10), 1297–1321.

Lane, T. P., Sharman, R. D., Trier, S. B., Fovell, R. G., & Williams, J. K. (2012). Recent advances in the understanding of near-cloud turbulence. *Bulletin of the American Meteorological Society, 93*(4), 499–515.

Lighthill, M. J. (1952). On sound generated aerodynamically. I. General theory. *Proceedings of the Royal Society of London A: Mathematical Physical and Engineering Sciences, 211*, 564–587.

Mancuso, R., & Endlich, R. (1966). Clear air turbulence frequency as a function of wind shear and deformation. *Monthly Weather Review, 94*(9), 581–585.

Marlton, G. J. (2016). *On the development, characterisation and applications of a balloon-borne atmospheric turbulence sensor.* Ph.D. thesis, University of Reading.

McCann, D. W., Knox, J. A., & Williams, P. D. (2012). An improvement in clear-air turbulence forecasting based on spontaneous imbalance theory: The ULTURB algorithm. *Meteorological Applications, 19*(1), 71–78.

Mecikalski, J. R., Berendes, T. A., Feltz, W. F., Bedka, K. M., Bedka, S. T., Murray, J. J., et al. (2007). Aviation applications for satellite-based observations of cloud properties, convection initiation, in-flight icing, turbulence, and volcanic ash. *Bulletin of the American Meteorological Society, 88*(10), 1589–1607.

Meneguz, E., Wells, H., & Turp, D. (2016). An automated system to quantify aircraft encounters with convectively induced turbulence over europe and the northeast atlantic. *Journal of Applied Meteorology and Climatology, 55*(5), 1077–1089.

Miles, J. W., & Howard, L. N. (1964). Note on a heterogeneous shear flow. *Journal of Fluid Mechanics, 20*(2), 331–336.

Nastrom, G. D., & Fritts, D. C. (1992). Sources of mesoscale variability of gravity waves. Part I: Topographic excitation. *Journal of the Atmospheric Sciences, 49*(2), 101–110.

Nunez, R. (2018). Integrating GOES-16 satellite into convective porosity determination at CWSU Houston. In *Sixth aviation, range, and aerospace meteorology special symposium*, Austin, TX: American Meteorological Society.

Price, C., & Rind, D. (1994). Possible implications of global climate change on global lightning distributions and frequencies. *Journal of Geophysical Research Atmospheres, 99*(D5), 10823–10831.

Reeve, N., & Toumi, R. (1999). Lightning activity as an indicator of climate change. *Quarterly Journal of the Royal Meteorological Society, 125*(555), 893–903.

Roach, W., & Dixon, R. (1970). A note on the paper on the influence of synoptic development on the production of high level turbulence. *Quarterly Journal of the Royal Meteorological Society, 96*(410), 758–760.

Schwartz, B. (1996). The quantitative use of pireps in developing aviation weather guidance products. *Weather and Forecasting, 11*(3), 372–384.

Search Technology. (2000). *A pilot-centered turbulence assessment and monitoring system, TAMS, Phase II Final Report*. Norcross, GA.

Shapiro, M. (1976). The role of turbulent heat flux in the generation of potential vorticity in the vicinity of upper-level jet stream systems. *Monthly Weather Review, 104*(7), 892–906.

Shapiro, M. (1978). Further evidence of the mesoscale and turbulent structure of upper level jet stream-frontal zone systems. *Monthly Weather Review, 106*(8), 1100–1111.

Shapiro, M. (1980). Turbulent mixing within tropopause folds as a mechanism for the exchange of chemical constituents between the stratosphere and troposphere. *Journal of the Atmospheric Sciences, 37*(5), 994–1004.

Sharman, R., Trier, S., Lane, T., & Doyle, J. (2012). Sources and dynamics of turbulence in the upper troposphere and lower stratosphere: A review. *Geophysical Research Letters, 39*(12), L12803.

Sharman, R., & Lane, T. (2016). *Aviation turbulence: Processes, detection, prediction*. Berlin: Springer.

Sharman, R., & Pearson, J. (2017). Prediction of energy dissipation rates for aviation turbulence. Part I: Forecasting nonconvective turbulence. *Journal of Applied Meteorology and Climatology, 56*(2), 317–337.

Sharman, R., Tebaldi, C., Wiener, G., & Wolff, J. (2006). An integrated approach to mid-and upper-level turbulence forecasting. *Weather and Forecasting, 21*(3), 268–287.

Storer, L., Williams, P., & Joshi, M. (2017). Global response of clear-air turbulence to climate change. *Geophysical Research Letters, 44*(19), 9976–9984.

Tenenbaum, J. (1991). Jet stream winds: Comparisons of analyses with independent aircraft data over southwest Asia. *Weather and Forecasting, 6*(3), 320–336.

Trier, S. B., & Sharman, R. D. (2009). Convection-permitting simulations of the environment supporting widespread turbulence within the upper-level outflow of a mesoscale convective system. *Monthly Weather Review, 137*(6), 1972–1990.

Trier, S. B., Sharman, R. D., & Lane, T. P. (2012). Influences of moist convection on a cold-season outbreak of clear-air turbulence (CAT). *Monthly Weather Review, 140*(8), 2477–2496.

Truscott, B. (2000). EUMETNET AMDAR AAA AMDAR Software Developments Technical Specification. *Doc. Ref. E_AMDAR/TSC/003. Met Office: Exeter, UK.*

Turner, J. (1999). Development of a mountain wave turbulence prediction scheme for civil aviation. *Met Office Forecasting Research Technical Report*, p. 265.

Williams, P. D., Read, P., & Haine, T. (2003). Spontaneous generation and impact of inertia–gravity waves in a stratified, two-layer shear flow. *Geophysical Research Letters, 30*(24), 2255.

Williams, P. D. (2016). Transatlantic flight times and climate change. *Environmental Research Letters, 11*(2), 024008.

Williams, P. D. (2017). Increased light, moderate, and severe clear-air turbulence in response to climate change. *Advances in Atmospheric Sciences, 34*(5), 576–586.

Williams, P. D., Haine, T. W., & Read, P. L. (2005). On the generation mechanisms of short-scale unbalanced modes in rotating two-layer flows with vertical shear. *Journal of Fluid Mechanics, 528*, 1–22.

Williams, P. D., Haine, T. W., & Read, P. L. (2008). Inertia–gravity waves emitted from balanced flow: Observations, properties, and consequences. *Journal of the Atmospheric Sciences, 65*(11), 3543–3556.

Williams, P. D., & Joshi, M. J. (2013). Intensification of winter transatlantic aviation turbulence in response to climate change. *Nature Climate Change, 3*, 644–648.

Wimmers, A., Griffin, S. M., Bachmeier, A. S., Gerth, J., & Lindstrom, S. (2018). Resolving gravity waves with Himawari-8 and GOES-16 imagery at the new limit of resolution and the application to aircraft-scale turbulence. In: *Sixth aviation, range, and aerospace meteorology special symposium*, Austin, TX: American Meteorological Society.

WMO. (2003). Aircraft meteorological data relay (AMDAR) reference manual.

Wolff, J., & Sharman, R. (2008). Climatology of upper-level turbulence over the contiguous united states. *Journal of Applied Meteorology and Climatology, 47*(8), 2198–2214.

Wurtele, M., Sharman, R., & Datta, A. (1996). Atmospheric lee waves. *Annual Review of Fluid Mechanics, 28*(1), 429–476.

(Received October 30, 2017, revised February 22, 2018, accepted February 24, 2018, Published online March 8, 2018)

Pure Appl. Geophys. 176 (2019), 2097–2120

https://doi.org/10.1007/s00024-018-1787-z

Pure and Applied Geophysics

Sensitivity of Numerical Simulations of a Mesoscale Convective System to Ice Hydrometeors in Bulk Microphysical Parameterization

ZHAOXIA PU,[1] CHAO LIN,[1] XIQUAN DONG,[2] and STEVEN K. KRUEGER[1]

Abstract—Mesoscale convective systems (MCSs) and their associated cloud properties are the important factors that influence the aviation activities, yet they present a forecasting challenge in numerical weather prediction. In this study, the sensitivity of numerical simulations of an MCS over the US Southern Great Plains to ice hydrometeors in bulk microphysics (MP) schemes has been investigated using the Weather Research and Forecasting (WRF) model. It is found that the simulated structure, life cycle, cloud coverage, and precipitation of the convective system as well as its associated cold pools are sensitive to three selected MP schemes, namely, the WRF single-moment 6-class (WSM6), WRF double-moment 6-class (WDM6, with the double-moment treatment of warm-rain only), and Morrison double-moment (MORR, with the double-moment representation of both warm-rain and ice) schemes. Compared with observations, the WRF simulation with WSM6 only produces a less organized convection structure with a short lifetime, while WDM6 can produce the structure and length of the MCS very well. Both simulations heavily underestimate the precipitation amount, the height of the radar echo top, and stratiform cloud fractions. With MORR, the model performs well in predicting the lifetime, cloud coverage, echo top, and precipitation amount of the convection. Overall results demonstrate the importance of including double-moment representation of ice hydrometeors along with warm-rain. Additional experiments are performed to further examine the role of ice hydrometeors in numerical simulations of the MCS. Results indicate that replacing graupel with hail in the MORR scheme improves the prediction of the convective structure, especially in the convective core region.

Key words: Microphysical parameterization, mesoscale convective systems, cloud, precipitation, WRF model.

[1] Department of Atmospheric Sciences, University of Utah, University of Utah, 135 S 1460 E, Rm. 819, Salt Lake City, UT 84112, USA. E-mail: Zhaoxia.Pu@utah.edu

[2] Department of Hydrology and Atmospheric Sciences, University of Arizona, Tucson, AZ, USA.

1. Introduction

Mesoscale convective systems (MCSs) are essential for regional and global precipitation and water cycles. They are also of great interests in aviation meteorology. Long-lasting, slow-moving MCSs are the major causes of heavy rainfall and flooding. Ice clouds can also be very dangerous for the airplane. Therefore, accurate forecasts of convection initiation, evolution, and associated precipitation are of great importance, yet they are often not produced by operational numerical models, partly due to uncertainty in the representation of cloud microphysical processes in numerical weather prediction (NWP) models.

Since uncertainties of microphysical parameterization (MP) represent an important error source in MCS and cloud simulations, many previous studies have been devoted to developing various microphysical parameterization schemes. Basically, these MP schemes differ in their treatments of the number and size distributions of hydrometeor species. Earlier MP schemes, such as the Kessler scheme (Kessler 1969), were simple and included only warm-rain hydrometeors (e.g., rainwater, and cloud water). Complexity was then added to MP schemes by including different ice categories, such as ice, hail, snow, and graupel (e.g., Lin et al. 1983; Rutledge and Hobbs 1983; Zhao and Carr 1997; Thompson et al. 2008; Morrison et al. 2009; Lim and Hong 2010; Mansell et al. 2010), when the importance of ice processes in convective evolution was realized (Fovell and Ogura 1988; McCumber et al. 1991; Liu et al. 1997). Many recent MP schemes include warm-rain and ice hydrometeors. According to their treatment of the hydrometeor size distribution, MP schemes are divided into two types: a bin MP scheme, which uses dozens, or even hundreds, of particle size bins to represent the actual size spectra of different hydrometeor species (Li et al. 2009); and a bulk MP scheme, which assumes that the hydrometeor size distributions follow specific functional forms for both single-moment (1M hereafter) and multiple-moment (e.g., double-moment; 2M hereafter) representations of the hydrometeor size distributions (Milbrandt and Yau 2005a, b; Morrison et al. 2009; Philips et al. 2007). Specifically, a 1M scheme predicts mixing ratios, whereas a 2M scheme forecasts both mixing ratios and number concentrations of the hydrometeor size distributions for various hydrometeor species. Different MP treatments can significantly influence the eventual storm structure and evolution (Weisman 1992, 1993; James et al. 2006; Li and Pu 2008; Adams-Selin et al. 2013).

Owing to their large computational expenses, bin MP schemes have limited weather and climate forecast applications. Bulk MPs, especially single-moment schemes, are more practical for NWP models because they are computationally efficient and are simple in terms of their treatment of particle size distributions. However, the increase in available computing power has also made it possible to implement 2M schemes in NWP models (Morrison 2010). Among the available 2M schemes in the NWP models, such as those included in an advanced research version of the Weather Research and Forecasting (WRF) model (Skamarock 2008), some are partial 2M because they treat part of the hydrometeors as double-moment species while others are still 1M species. For example, the Thompson scheme (Thompson et al. 2008) represents rainwater as a 2M species; the WRF double-moment 6-class scheme (WDM6 hereafter; Lim and Hong 2010) treats warm-rain hydrometeors as 2M species and ice-phase hydrometeors as 1M species. Meanwhile, 2M schemes represent both warm-rain and ice hydrometeors as 2M species. Obviously, the major difference between the partial 2M and 2M schemes here is the representation of the ice species. Therefore, the important question is: compared to the 1M scheme, whether 2M schemes are necessary for better prediction of MCSs? Furthermore, since some 2M schemes, such as WDM6 only includes 1M ice hydrometeors, what is the impact of a 2M representation of ice hydrometeors on the numerical representation of MCSs?

The answer to these questions is controversial, based on previous studies. For instance, Morrison et al. (2009) found that double moments for rain can produce the same results as those with a 2M representation of rain and ice scheme. In contrast, Weverberg and Vogelmann (2012) pointed out that the simulations of surface precipitation and cold pools depend not only on a 2M representation of rain, but also on 2M representations of all precipitating hydrometeors.

In order to further explore the above question and also to evaluate the influence of 2M schemes on accurately simulating MCSs, we perform a series of numerical simulations with 1M, partial 2M, and 2M MP schemes. This study is designed not only to examine the sensitivity of numerical simulations to the various MP schemes, but also to obtain insight into the impact of warm-rain, warm-rain and ice, and ice only hydrometeors on the accurate simulation of MCSs. Specifically, considering different moment representations of warm-rain and ice as well as the popularity in previous evaluation (e.g., Hong and Lim 2006; Morrison et al. 2009), we choose the WRF single-moment 6-class scheme (WSM6 hereafter; Hong and Lim 2006), the WDM6 scheme, and the Morrison double-moment scheme (MORR hereafter; Morrison et al. 2009) for numerical simulations of an MCS observed during the Department of Energy and NASA's Midlatitude Continental Convective Clouds Experiment (MC3E). NEXRAD radar reflectivity, the NCEP Stage IV precipitation dataset (Baldwin and Mitchell 1996; Lin and Mitchell 2005), surface Mesonet observations, and a cloud classification algorithm (Steiner et al. 1995 and Feng et al. 2011) are used to validate simulation results.

The paper is organized as follows. Section 2 briefly describes the MCS case, the data, and a cloud classification method used for verification, and Sect. 3 introduces the experimental design and present the simulated results, respectively. Effects of 2M representations of warm-rain and ice hydrometeors are diagnosed in Sect. 4, while additional numerical experiments are performed to further examine the impact of ice hydrometeors on numerical simulations of the MCS in Sect. 5. Discussion and conclusions are provided in Sect. 6.

2. An MCS Case and Observations

2.1. An MCS Case

An MCS case during 23–24 May 2011 was selected for this study. The convective system was supported by an upper-level trough and low-level moisture supplied from the Gulf of Mexico. It consisted of four main convective episodes that interacted with each other through dynamic and thermodynamic processes. For this study, we chose the second convective episode (system), from 2100 UTC 23 May 2011 to 0600 UTC 24 May 2011, considering the availability of the radar-derived data products.

The convective system originally formed from isolated convective cells in western Oklahoma (Fig. 1a) under the influence of the dry line and outflow boundary produced by the previous convective episode in northeastern Oklahoma. Figure 2 shows the surface relative humidity (shading) and temperature (red contour lines) at 1.5 m height, and 10 m wind vectors at 2100 UTC 23 May 2011. The dry line and three convective cells were clearly observed from NEXRAD reflectivity as shown in Fig. 1a. After the initiation, the convective system strengthened in both size and intensity over the next several hours, until it became a multi-cell mesoscale convective system (MCS) with an NE–SW orientation at 2300 UTC 23 May 2011 (Fig. 1b). This MCS then combined with the first convective episode at 0000 UTC 24 May 2011 and evolved into a mature phase with cells organized in a W–E direction. The mature stage lasted about 3–4 h (Fig. 1c, d). Finally, this MCS began to dissipate at about 0400 UTC 24 May 2011 and slowly moved out of the Oklahoma region at about 0600 UTC 24 May 2011.

According to its structure and evolution with time, this MCS can be categorized into four different stages: initiation (2100 UTC 23 May 2011, Fig. 1a), developing (2200–2300 UTC 23 May 2011, Fig. 1b), mature (0000–0400 UTC 24 May 2011, Fig. 1c, d), and dissipating (0400 UTC 24 May 2011, figure not shown).

2.2. Observations and a Cloud Classification Algorithm

The observational data available for the above-mentioned MCS include NEXRAD 3-dimensional radar reflectivity, the NCEP Stage IV precipitation dataset, and surface Mesonet observations. Specifically, radar reflectivity data are obtained from three NEXRAD radars (KTLX, KINX, and KVNX) that cover most of the Oklahoma region. The radars operate at a wavelength of 10 cm (S band) in a preprogrammed sequence of 360° azimuthal sweeps at various elevation angles (Klazura and Imy 1993). Surface data are obtained from the Oklahoma Mesonet, a network of 127 automated weather stations designed to observe mesoscale meteorological phenomena (Brock et al. 1995) over the entire state of Oklahoma. In this study, temperature, humidity, and wind data are used to identify the dry line (e.g., Fig. 2). The NCEP Stage IV rainfall analysis is a mosaic of the regional multisensor analysis produced by the NWS River Forecast Centers. It combines precipitation estimates from about 150 NEXRAD with about 5500 h rain gauge measurements over the conterminous United States (Baldwin and Mitchell 1996; Lin and Mitchell 2005). It has been used at NCEP for forecast verification and data assimilation purposes.

In order to validate the model-simulated cloud properties, a radar reflectivity-based cloud classification algorithm is developed following the work of Steiner et al. (1995) and Feng et al. (2011). This algorithm can classify the convective system into a convective core (CC), stratiform rain (SR), non-precipitating transitional anvil cloud (ACtrans), and anvil cloud (AC) from NEXRAD radar reflectivity observations. Among the products, the CC is characterized by strong upward motion and intense precipitation. The SR region is known for its relatively homogeneous precipitation and weak horizontal reflectivity gradients. The transitional anvil is a stage between the precipitating SR and ACtrans regions, with an echo base (DBZ = 0) below 3 km and a top above 6 km. The AC region is represented

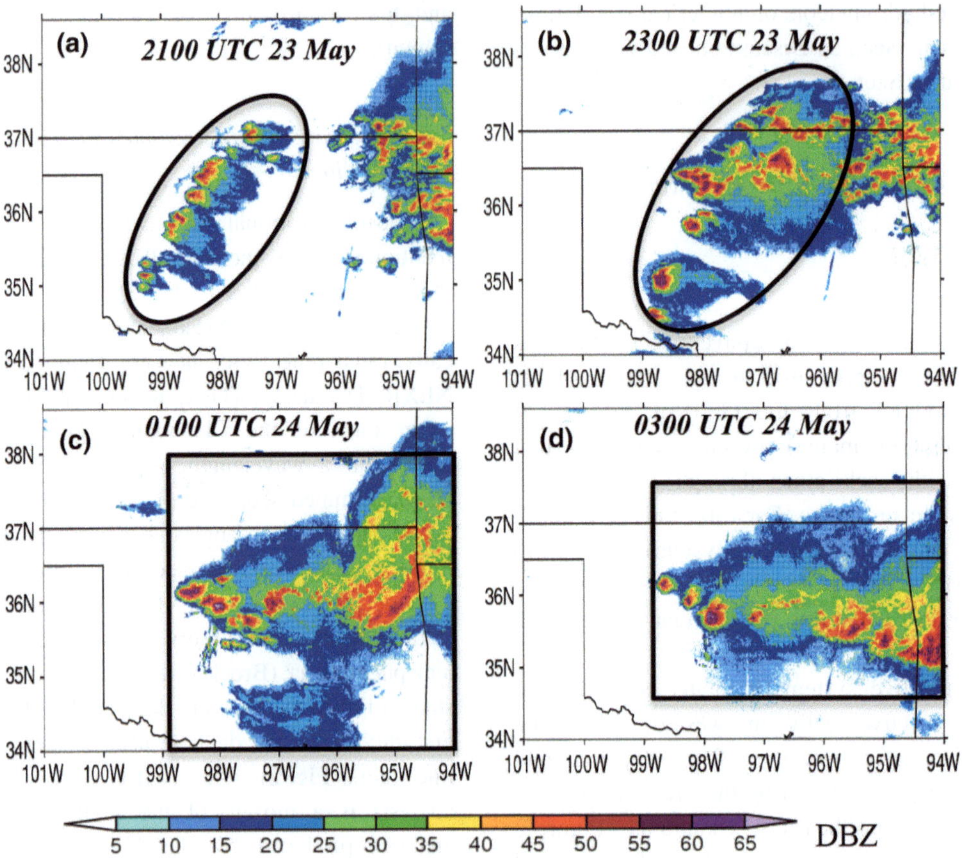

Figure 1

NEXRAD composite radar reflectivity observations (dBZ) at **a** 2100 UTC 23 May 2011, **b** 2300 UTC 23 May 2011, **c** 0100 UTC 24 May 2011, **d** 0300 UTC 24 May 2011. The elliptical or rectangular boxes indicate the mesoscale convective system of concern

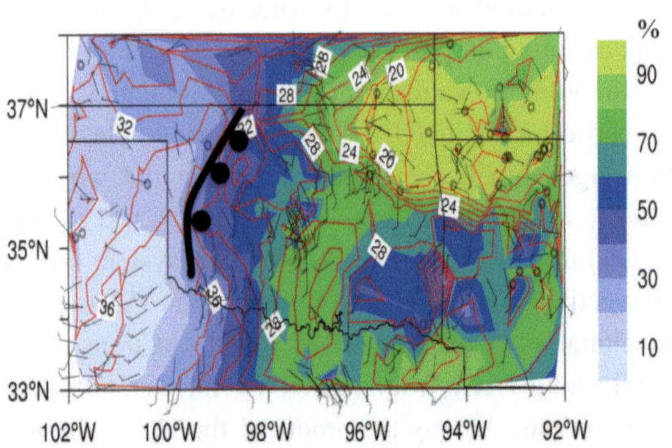

Figure 2

Surface wind, 2M temperature (contours in °C), relative humidity (shaded, unit: %), and wind barbs from surface Mesonet observations at 2100 UTC 23 May 2011. The bold black line shows the dry line, with large black dots corresponding to the locations of three convection initiation cells

by radar reflectivity that has an echo base of more than 3 km (Feng et al. 2011). Both ACtrans and AC are non-precipitating clouds. Wu et al. (2013) confirmed that this cloud classification algorithm is able to classify the dynamic and thermodynamic properties of MCSs.

3. Numerical Simulations and Results

3.1. Simulation Design

The WRF model [see details in Skamarock et al. (2008)] version 3.4.1 is used. Figure 3 displays the location and area coverage of model domains used in all simulations. The blue box inside d03 denotes the verification area. The model configurations and physical parameterization schemes (except for MP

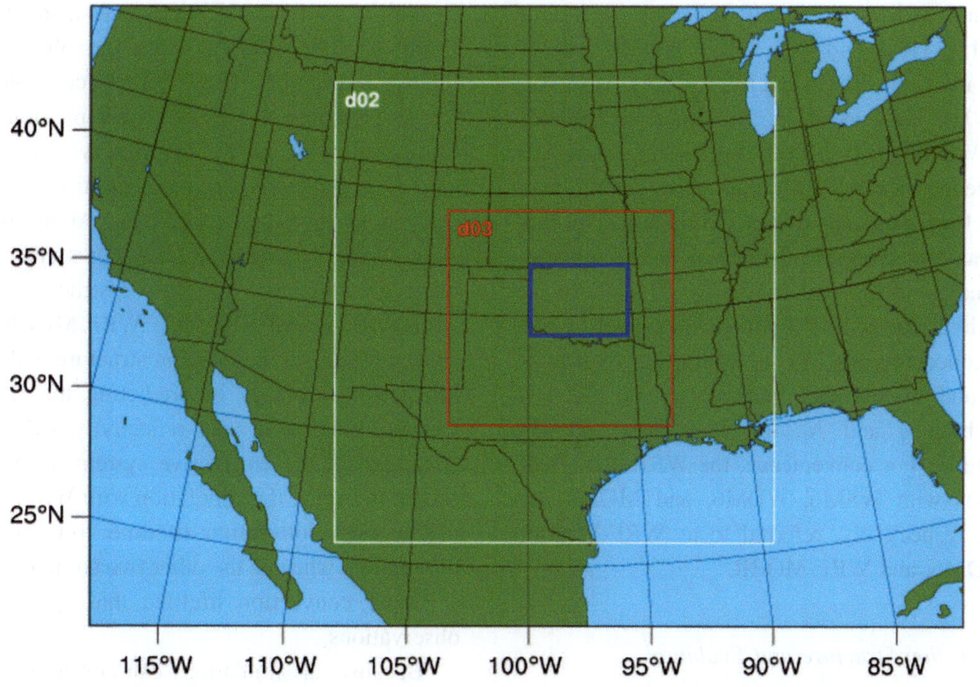

Figure 3

Location of model domains used in the WRF simulations. Horizontal domain sizes for the first (outermost), second (d02), and third (d03) domains are 12, 4, and 1.33 km, respectively. The blue box inside of d03 denotes the verification area

Table 1

WRF model configuration and simulation design

	D01	D02	D03
Horizontal resolution	12 km	4 km	1.33 km
Vertical levels	46 sigma levels	46 sigma levels	46 sigma levels
Cumulus scheme	Kain–Fritsch (Kain 2004)	None	None
Longwave radiation scheme	RRTM (Mlawer et al. 1997)	RRTM (Mlawer et al. 1997)	RRTM (Mlawer et al. 1997)
Shortwave radiation scheme	Dudhia (1989)	Dudhia (1989)	Dudhia (1989)
Land surface scheme	RUC (Smirnova et al. 2000)	RUC (Smirnova et al. 2000)	RUC (Smirnova et al. 2000)
Planetary boundary layer (PBL) scheme	MYJ (Janjic 1990)	MYJ (Janjic 1990)	MYJ (Janjic 1990)

schemes) for numerical simulations are listed in Table 1. All simulations initiate at 0000 UTC 23 May 2011 and end at 1200 UTC 24 May 2011 with the initial and boundary conditions derived from NCEP North American Mesoscale Model (NAM) analysis. The initial time is 20 h before the convection initiation to eliminate any possible strong influence from initial conditions.

In order to examine influences of the different moment representations of warm-rain and ice hydrometeors on simulating MCS properties, three

Table 2

Mass variables and number variables of the three microphysical schemes (WSM6, WDM6, MORR)

Microphysical scheme	Mass variables	Number variables
WSM6	q_c q_r q_i q_s q_g	
WDM6	q_c q_r q_i q_s q_g	N_n N_c N_r
Morrison 2-moment (MORR)	q_c q_r q_i q_s q_g	N_r N_i N_s N_g

c cloud water, *r* rain water, *i* ice, *s* snow, *g* graupel, *q* mixing ratio, *N* number concentration, N_n CCN number experiments with different MP schemes, including WSM6 (1M, Hong and Lim 2006), WDM6 (partial 2M, Lim and Hong 2010), and MORR (2M, Morrison et al. 2005), are conducted. As listed in Table 2, WSM6, WDM6, and MORR utilize the same six hydrometeor species, including three warm-rain species (cloud, rain, and vapor) and three ice species (ice, snow, and graupel). WSM6 is a 1M scheme, which predicts just the mixing ratios (q) of these six kinds of hydrometeors; WDM6 predicts both q and number concentrations (N) for warm-rain hydrometeors (cloud water and rainwater); and MORR predicts both q and N_0 for warm-rain and ice hydrometeors. For convenience, the WRF numerical simulations with WSM6, WDM6, and MORR are hereafter, respectively, referred to as WRF-WSM6, WRF-WDM6, and WRF-MORR.

3.2. Convection Structure and Evolution

Figure 4 shows snapshots of composite radar reflectivity at different MCS stages from various simulations. Compared with the NEXRAD observations (Fig. 1a), all experiments (Fig. 4a–c) predict convection initiation, although there are time and location errors. At the developing stage, WRF-WSM6 (Fig. 4d) produces multiple convective cells. The coverage of the convection system is small, which is a common problem of the WSM6 scheme, as it tends to produce an isolated intense convective core compared with other schemes (Thompson et al. 2006; Weisman et al. 2009). WRF-WDM6 (Fig. 4e) predicts an organized convection system even though its areal coverage is also smaller compared with observations. WRF-MORR (Fig. 4f) produces the proper size area of convection coverage. Both WRF-WDM6 and WRF-MORR produce a stronger convective system in terms of maximum radar reflectivity (dBZ) compared with NEXRAD observations.

The differences become more distinguishable in terms of the coverage and structure of the convective system at the observed mature stage (Fig. 4g, h, i). While observations (Fig. 1c) still show that the convective system is mature, it begins to dissipate in WRF-WSM6 (Fig. 4g). WRF-MORR (Fig. 4i) produces reasonable coverage of the convective system but with a much stronger convective core. In addition, WRF-MORR produces a squall line-like storm, slightly different from the observed multi-supercell structure. These differences contribute to the overestimation of the precipitation amount. Compared with NEXRAD observations (Fig. 1c), all the simulations produce additional convective cells in northwestern Oklahoma. At the dissipation stage, WRF-WDM6 (Fig. 4k) shows clear improvements in simulated convective structure compared with WRF-WSM6 (Fig. 4j). WRF-MORR (Fig. 4l) produces realistic convection structure and coverage, but performs poorly in predicting the strength [in terms of maximum radar reflectivity (dBZ) and area coverage] of the convective system. In addition, as shown in Fig. 4, the simulation with WSM6 produces the shortest convection lifetime from initiation to dissipation, whereas the other two simulations predict a proper convection lifetime that agrees with the observations.

Because all simulations can correctly predict the convection initiation of the MCS but produce large discrepancies in the developing and mature stages, the rest of the evaluation will emphasize the latter two stages.

Figure 5 compares the vertical structures of the MCS observed by NEXRAD radar and simulated by the WRF simulations at 2300 UTC 23 (developing stage) and 0100 UTC 24 May 2011 (mature stage). As illustrated in Fig. 5a, e, the NEXRAD radar echo tops can extend up to 13 km at both developing and mature stages near the edge of the convective core. The NEXRAD reflectivities range from 15 to 35 dBZ above the melting level in both the rearward and forward directions. WRF-MORR (Fig. 5d, h) reproduces similar vertical structures to radar observations at both stages, though it generates much broader moderate radar reflectivity in a range between 35 and 50 dBZ, especially in the middle and upper troposphere. The simulations from both WRF-WSM6 (Fig. 5b, f) and WRF-WDM6 (Fig. 5c, g), however, underestimate the echo tops. All the simulations exhibit high biases against NEXRAD observations in the middle to upper troposphere, which is a common problem for the current model with bulk MP, mainly because of the unrealistically high amount of graupel in the simulations (Lang et al. 2011).

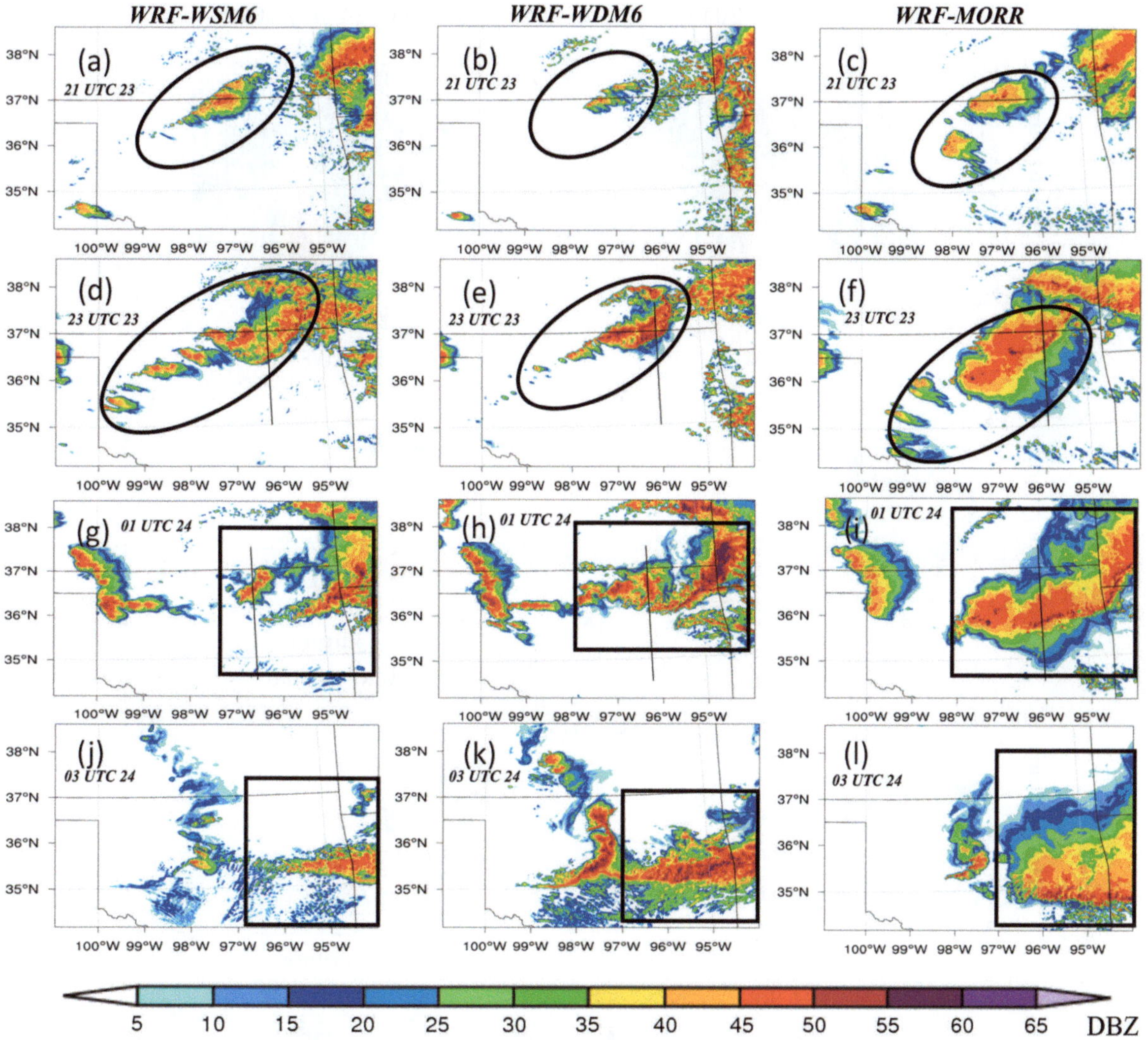

Figure 4
Simulated composite radar reflectivities (dBZ) at **a–c** 2100 UTC 23 May 2011; **d–f** 2300 UTC 24 May 2011; **g–i** 0100 UTC 24 May 2011; and **j–l** 0400 UTC 24 May 2011, from **a**, **d**, **g**, **j** WRF-WSM6; **b**, **e**, **h**, **k** WRF-WDM6; and **c**, **f**, **i**, **l** WRF-MORR

3.3. Cloud Types

The convective cloud distribution reveals the internal 3-D components and structure of the convective systems. Figure 6 shows snapshots of the classified components derived from NEXRAD reflectivity(Fig. 6a, b) and WRF simulations (Fig. 6c–h) at the developing and mature stages. In the NEXRAD-derived products, relatively discrete areas of CC appear in the center of the convective system, with a surrounding large SR. AC covers the outer region of the convective system. Narrow but continuous ACtrans appear between the stratiform and anvil clouds. Compared with NEXRAD-derived products, all simulations produce a much organize and larger CC. WRF-WSM6 (Fig. 6c, d) and WRF-WDM6 (Fig. 6e, f) both produce less SR. There are almost no ACtrans between CC and AC. In contrast, WRF-MORR (Fig. 6g, h) produces a cloud distribution that agrees with NEXRAD-derived products reasonably, with CC in the core region, a large SR surrounding the CC, and clear ACtrans between SR and AC.

In order to examine the internal cloud structure, a time series of the cloud component fractions (CC, SR,

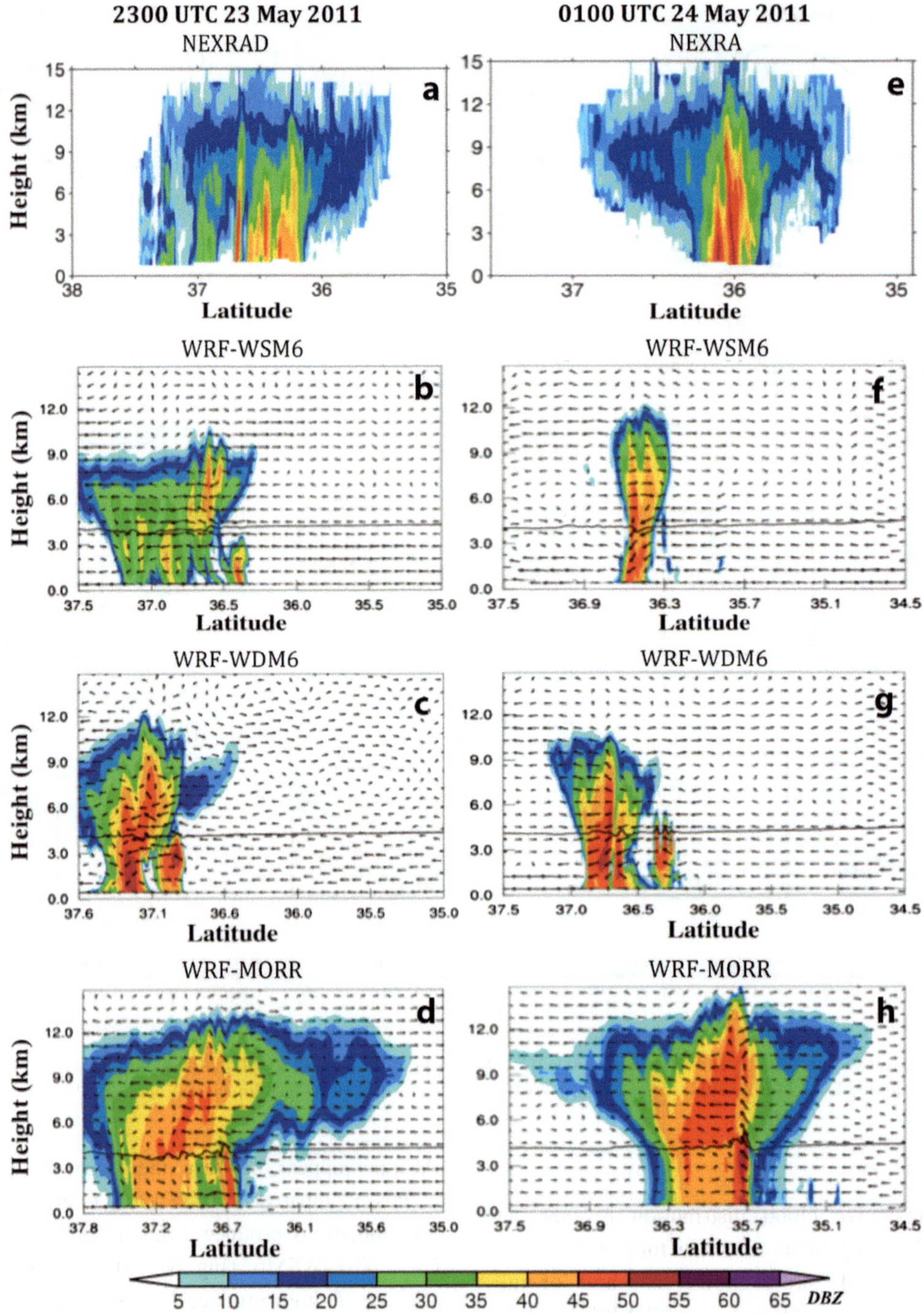

Figure 5

North–south vertical cross section (along the thin lines crossing the convective system in Fig. 4d–f) of radar reflectivity (dBZ; **a–d**) at 2300 UTC 23 May 2011 and **e–h** at 0100 UTC 24 May 2011, from **a**, **e** NEXRAD observations; **b**, **f** WRF-WSM6; **c**, **g** WRF-WDM6; and **d**, **h** WRF-MORR. The thick black lines denote melting levels

ACtrans, and AC against total cloud coverage) over the MCS is shown in Fig. 7. The mean cloud fractions averaged over the whole-time period are summarized in Table 3. Compared with the NEXRAD-derived data, all simulations overestimate the CC coverage (Fig. 7a). Specifically, WRF-WDM6

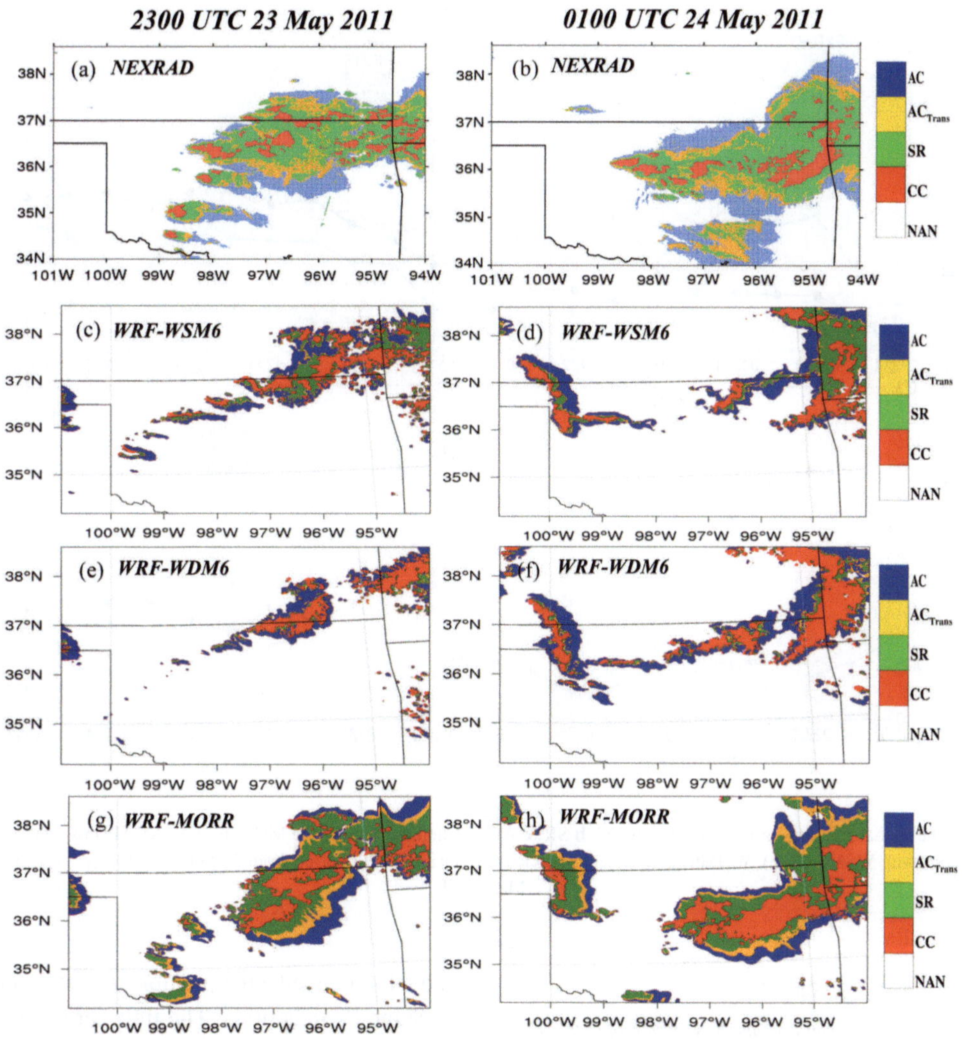

Figure 6

Classified components of the convective system: convective core (CC), stratiform (SR) region, transitional anvil clouds (AC_{trans}), and anvil clouds (AC), following the method of Feng et al. (2011), from **a**, **b** NEXRAD radar observations and the WRF simulations (**c**, **d** WSM6; **e**, **f** WDM6; and **g**, **h** MORR) at (**a**, **c**, **e**, **g**) 2300 UTC 23 May 2011 and (**b**, **d**, **f**, **h**) 0100 UTC 24 May 2011

predicts the largest CC fraction overall (Table 3). In addition, the fraction of CC derived from NEXRAD is relatively small at the initiation and developing stages but increases at the mature stage and then decreases at the dissipation stage. WRF-MORR captures this feature with a larger CC ratio at the mature stage and a smaller fraction at the initiation and dissipation stages, while both WRF-WSM6 and WRF-WDM6 miss it.

The fraction of SR (Fig. 7b) derived from NEXRAD is about 3 times larger than that of CC (Table 3). WRF-MORR (red) matches the NEXRAD-derived SR fraction well, but the SR/CC ratio is unreasonable due to the large CC fractions. Both WSM6 and WDM6, especially WDM6, underestimate the SR fractions. The extremely large CC fraction and small SR fraction produced by WDM6 confirm the unreasonable precipitation distribution it produces. WSM6 and WDM6 also heavily underestimate the ACtrans fraction (Fig. 7c and Table 3), implying that both simulations produce an unreasonable internal cloud structure. In particular, WRF-WDM6 does not reproduce a realistic cloud structure, while WRF-MORR produces a much better cloud structure.

Reprinted from the journal

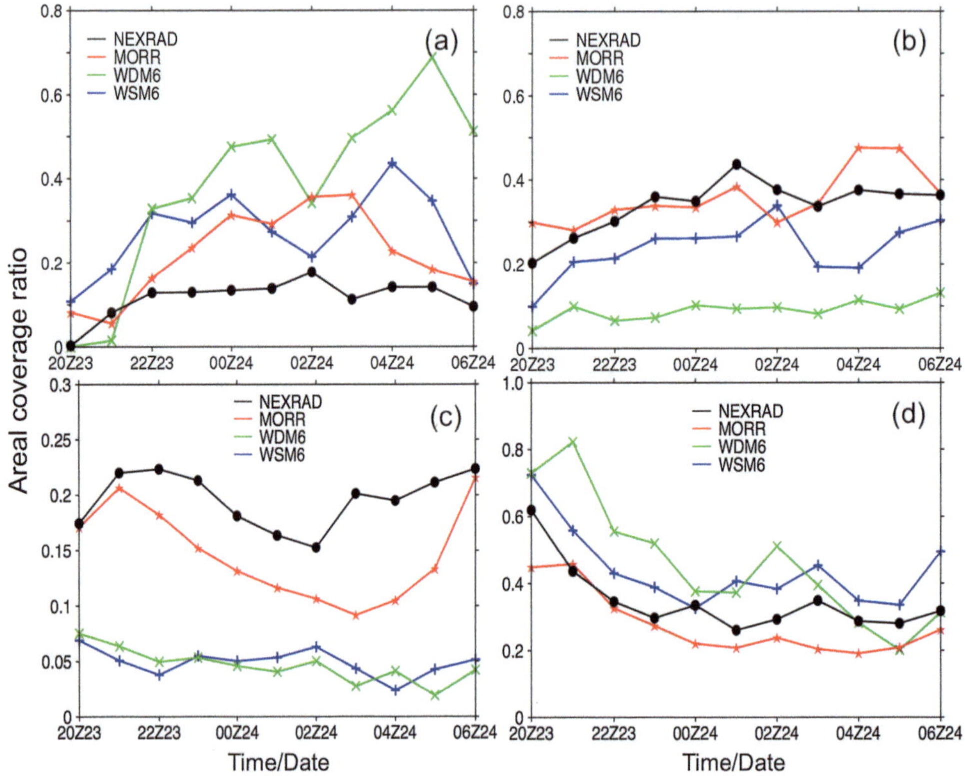

Figure 7
Time series of the spatial fractions of the classified **a** CC, **b** SR, **c** AC_{trans}, and **d** AC over the MCS from NEXRAD observations (black) and WRF simulations with WSM6 (blue), WDM6 (green), and MORR (red) during a 10-h period from 2000 UTC 23 May 2011 to 0600 UTC 24 May 2011

Table 3

Mean cloud fractions averaged over the time period of 2000 UTC 23 May 2011–0600 UTC 24 May 2011, derived from NEXRAD observations and WRF simulations

	CC	SR	ACtrans	AC
NEXRAD	0.116805	0.339358	0.196422	0.347416
WRF-WSM6	0.272633	0.237330	0.0491956	0.440842
WRF-WDM6	0.387878	0.0902116	0.0463075	0.461604
WRF-MORR	0.220521	0.356896	0.146489	0.276095

3.4. Precipitation Rate and Quantitative Precipitation Forecasting

To quantitatively evaluate the rainfall rate produced in the WRF simulations, hourly precipitation rates obtained from the NCEP Stage IV rainfall dataset (http://www.emc.ncep.noaa.gov/mmb/ylin/pcpanl/stage4/) are used for comparison. The WRF model-simulated precipitation output is interpolated at Stage IV data grids. The time series of the rainfall rate is then obtained by averaging the precipitation amount over all data grids in the verification region (as shown in Fig. 3). The area-averaged hourly precipitation amounts from both Stage IV and simulations are compared in Fig. 8. The precipitation peak occurs at the mature stage (at about 0200 UTC 24 May 2011) in Stage IV observations, with a small precipitation amount during the initiation and dissipation stages. The precipitation produced by WRF-MORR agrees well with this pattern, while slightly overestimating the precipitation amount at the same time. Both WRF-WDM6 and WRF-WSM6 perform poorly. WRF-WDM6 produces a large amount of precipitation (overestimation) during the dissipation stage but heavily underestimates the precipitation amount during the mature stages. Except for the first 3 h, WRF-WSM6 heavily underestimates the precipitation amount over the entire MCS period. Specifically, compared with Stage IV data, the total accumulated average precipitation from 2000 UTC 23–0600 UTC 24 May 2011 is heavily underestimated in WRF-WSM6, by 43% of the amount.

The quantitative precipitation forecast (QPF) skill is also evaluated for different simulations. The equitable threat scores (ETS) are computed using the following equation:

$$\mathrm{ETS} = \frac{A}{A+B+C}. \quad (1)$$

For a given threshold, A represents the number of grid points that exceed the threshold in both the model forecast and the observations; B denotes the number of grid points that exceed the threshold in the model forecast but not in the observations; and C is the number of grid points in which the model forecast does not reach the threshold while the observations do (Wilks 1995).

Precipitation verification is performed with different thresholds: 1, 5, and 10 mm h^{-1}. The time series of ETS is shown in Fig. 9. Among all simulations, WRF-MORR shows the best ETS scores most of the time. WDM6 performs better than WSM6, especially for medium-to-heavy rainfall (Fig. 9b, c). Overall, the simulation with WDM6 outperforms that with WSM6. The simulation with MORR performs the best, for this particular MCS case.

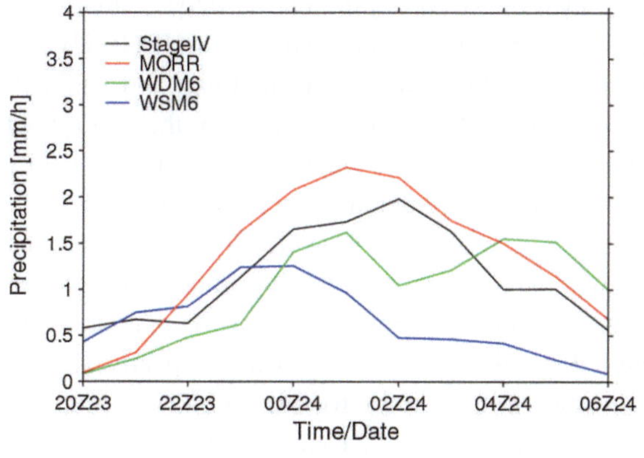

Figure 8
Same as Fig. 7, except for the area-averaged hourly precipitation rates (mm h^{-1}) from the NCEP Stage IV rainfall dataset (black) and WRF simulations

4. *Effects of Double-Moment Prediction of Hydrometeors*

The results so far show that the numerical simulations of the MCS are sensitive to different MP schemes. Compared with NEXRAD observations, the simulation with the 1M scheme produces the most disorganized convection distribution, the shortest convection lifetime, and the lowest echo top. It also heavily underestimates the precipitation. The cloud component distributions and fractions are also unreasonable. By adding 2M representations of warm-rain hydrometeors, WRF-WDM6 shows a clear improvement in predicting the convective system by producing a proper convection lifetime and an organized convection structure. Even though the

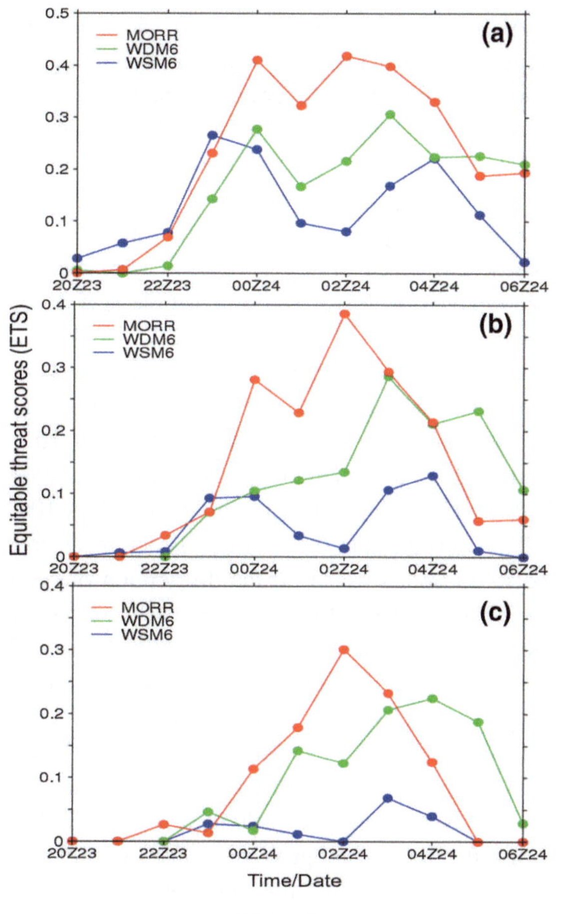

Figure 9
Same as Fig. 7, except for equitable threat scores (ETS) of 1-h accumulated precipitation from three WRF simulations with a threshold of **a** 1 mm, **b** 5 mm, and **c** 10 mm

precipitation amount is larger than that in WRF-WSM6, WRF-WDM6 still underestimates the precipitation amount compared with observations. The cloud distribution and fraction are also unrealistic. The echo top is lower than that in NEXRAD observations. The simulation with the 2M MP schemes (WRF-MORR) is much better, as it predicts the proper convection lifetime, echo tops, and cloud component distributions and fractions, as well as precipitation amount.

To better understand why the simulations produce different convection properties and to associate these properties with different moment representations of warm-rain and ice hydrometeors, further diagnoses are made in this section.

4.1. Cold Pools

As mentioned in the introduction, the shape and strength of the cold pool influence the eventual MCS structure, longevity, and development (e.g., James et al. 2006; Adams-Selin et al. 2013). Previous studies have shown that simulations with 2M microphysical schemes could result in a better cold pool intensity forecast than those with 1M schemes, because 2M schemes help produce a more realistic evaporation rate due to realistic particle size distributions (Morrison et al. 2009). Typically, a cold pool is indicated by temperature decreases and pressure increases, as well as local horizontal divergence. The 2-m air temperature, sea level pressure, and 10-m wind field from simulations at the developing stage (2300 UTC 23 May 2011) are illustrated in Fig. 10a–c. It is apparent that the size and structure of cold pools correspond well with the distribution of composite radar reflectivity of the simulated convective system (Fig. 4d, e, f). For instance, WRF-MORR produces a larger size of cold pool, corresponding to the larger area coverage of the simulated convective system. Figure 10d (WRF-WSM6), e (WRF-WDM6), and f (WRF-MORR) illustrate the temperature cross sections, the melting level, and 0 dBZ radar reflectivity. The figures show that the cold pool edge matches the convection edge (about 36.3 N) in WRF-WSM6 (Fig. 10d), while its intensity (in terms of the temperature differences between the cold pool and its environment) is weaker than that in WRF-WDM6 (Fig. 10e) and WRF-MORR (Fig. 10f). In addition, there are apparent inversion layers at about 0.8 km height in the stratiform cloud region. The inversion layers cause discontinuities of vertical structures of the cold pool thus reduce the strength of the cold pool and the lifting. As the result, they prevent the triggering of new convective cells in WRF-WSM6. This explains at least in part why WRF-WSM6 predicts a shorter lifetime of the convective system compared with the observations.

The cold pool edge simulated by WRF-WDM6 (Fig. 10e) overlaps with the convection system border. The intensity of the cold pool is stronger and its size is larger than that in WRF-WSM6. Lim and Hong (2010) noted that a simulation with WDM6 showed well-organized downdrafts with a strengthened cold pool as compared to the result with WSM6 in an idealized 2D thunderstorm simulation. WRF-MORR (Fig. 10f) produces a cold pool edge that is away from the radar echo edge. This is because of the presence of light precipitation in the direction the convection is moving. The cold pool from WRF-MORR is also larger than that from WRF-WSM6 and is more homogeneous. In addition, both WRF-MORR and WRF-WDM6 produce a deeper cold pool, compared with WRF-WSM6.

4.2. Number Concentration (N), Hydrometeors, and Diabatic Cooling

According to Houze (2004), cold pools result from diabatic cooling by evaporation, sublimation, and melting of precipitation particles. Smull and Houze (1987) and Lafore and Moncrieff (1989) also suggest that the evaporation, melting, and sublimation all contribute substantially to the cold pool evolution. Specifically, diabatic cooling is due mainly to rainwater evaporation and the melting and sublimation of ice-phase hydrometeors (i.e., graupel and snow). Thus, rainwater mixing ratio and number concentration has implications for the diabatic cooling. Figures 11 and 12 show, respectively, the rainwater mixing ratio for all simulations and the intercept parameter N_{0r} (only for WRF-WDM6 and WRF-MORR) in the developing and mature stages. It is obvious that WRF-WSM6 (Fig. 11a) produces a much smaller rainwater mixing ratio (q_r) than the

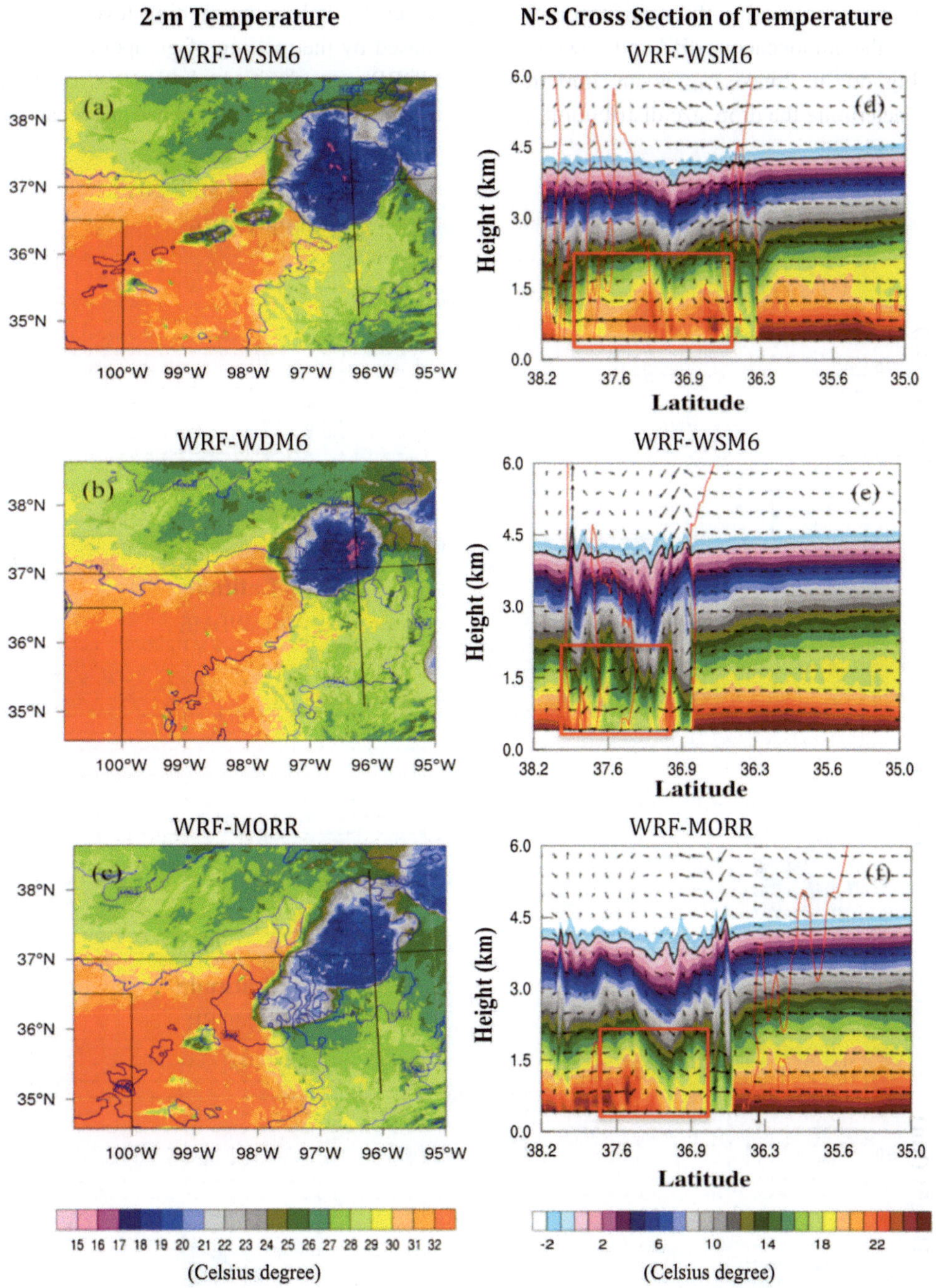

Figure 10
WRF-simulated 2M air temperature (°C) at 2300 UTC 23 May 2011 with **a** WRF-WSM6, **b** WRF-WDM6, and **c** WRF-MORR; **d**, **e**, **f** are the north–south vertical cross sections of temperature (°C) along the black line showing in the left figures. The thick black lines in **d**, **e**, **f** are the melting level. The red line indicates the 0 dBZ radar reflectivity. The thick red boxes indicate the cold pool edges

other two simulations do. Compared with WRF-WSM6, which uses a fixed number concentration for rainwater, WRF-WDM6 and WRF-MORR predict an N_{0r} that decreases from the convective core to the outer region. The larger N_{0r} is located just below the melting level (in the zoomed-in red box in Fig. 12b, d). These regions correspond to the stratiform region as indicated by the homogeneous dBZ and downward motion in Fig. 5d, h. Since N_{0r} is an important parameter to determine the drop size distribution, the larger N_{0r} below the melting level is believed to be caused by the melting of graupel and snow. Because MORR treats the ice hydrometeors as 2M species, the melting process of graupel and snow can contribute to

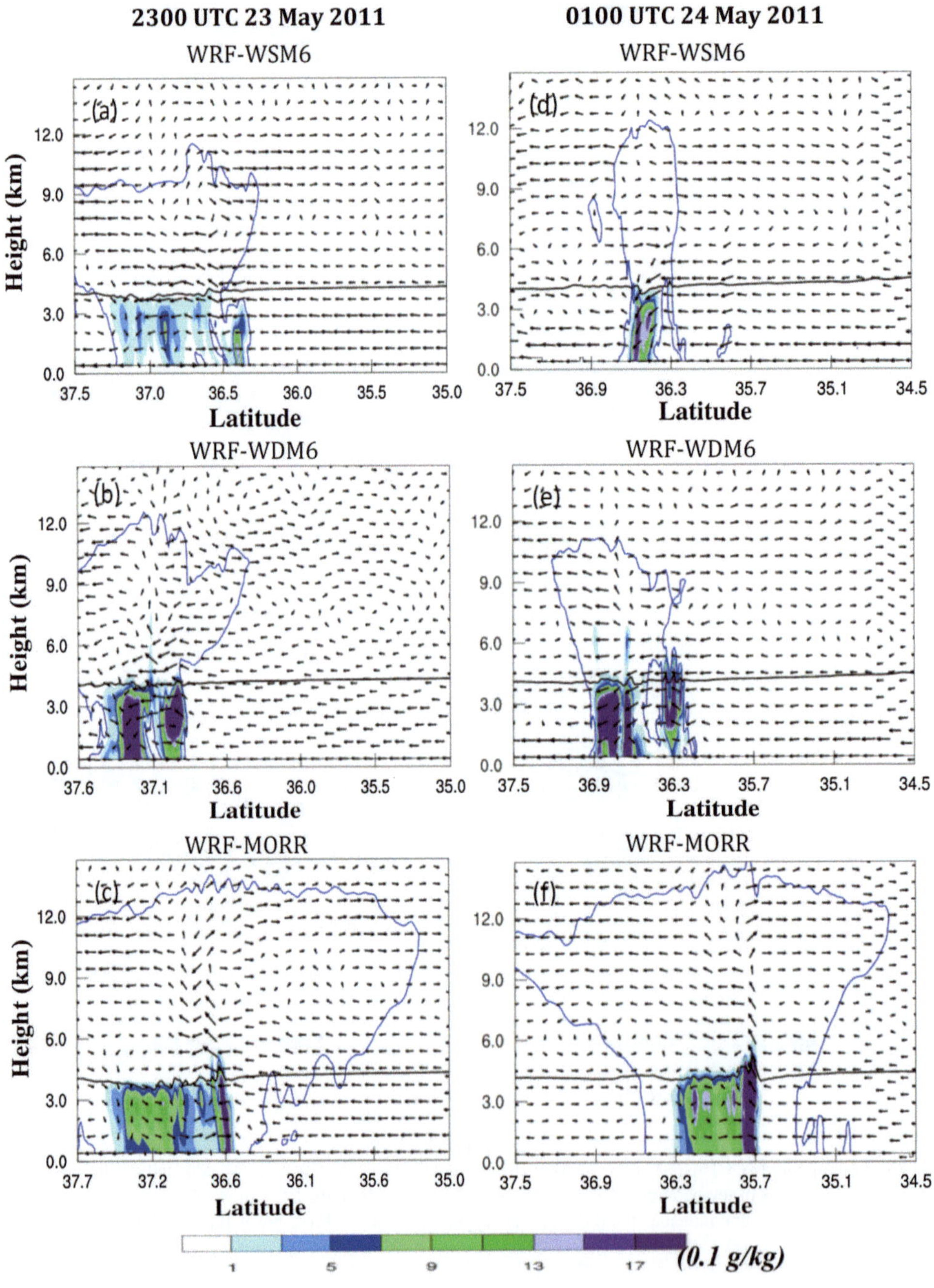

Figure 11
Same as Fig. 10**d–f** except for cross sections of rainwater mixing ratio (0.1 g kg^{-1}) from **a**, **d** WRF-WSM6, **b**, **e** WRF-WDM6; **c**, **f** WRF-MORR, at 2300 UTC 23 May 2011 (**a–c**) and 0100 UTC 24 May 2011 (**d–f**)

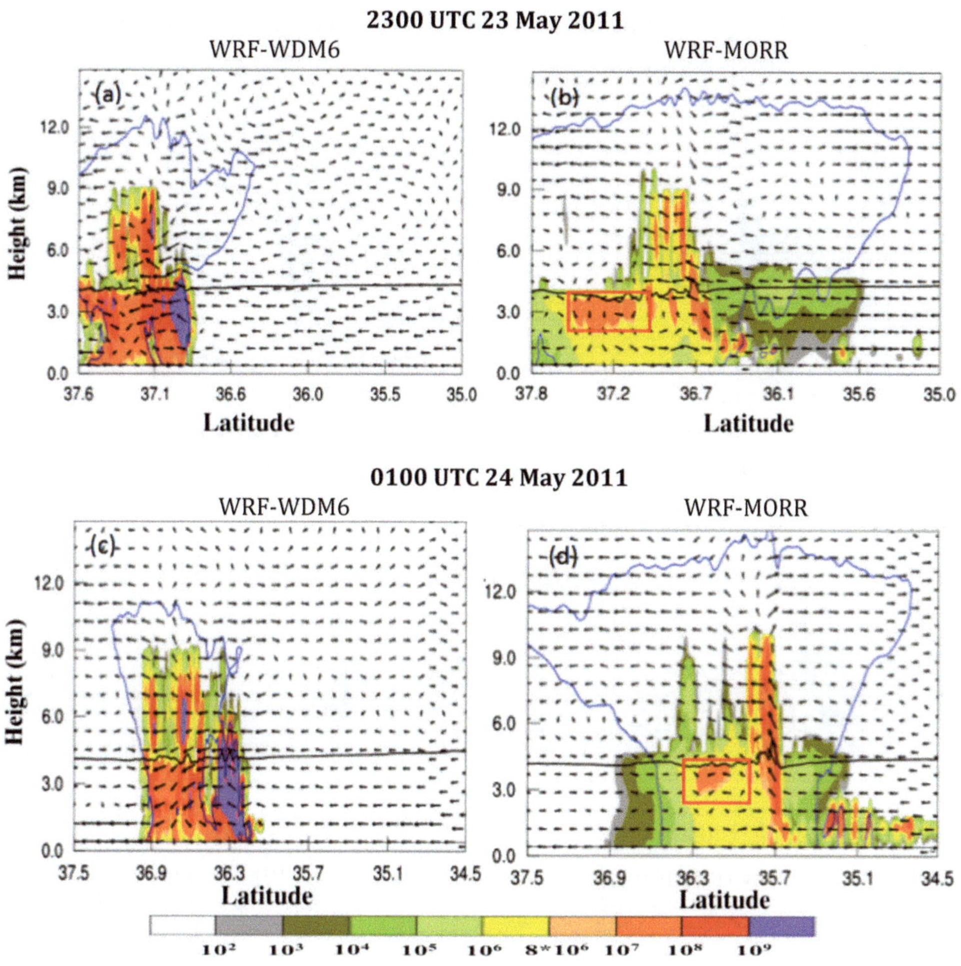

Figure 12
Same as Fig. 11 except for cross sections of rainwater intercept parameter N_{0r}. **a**, **c** WRF-WDM6; **b**, **d** WRF-MORR at **a**, **b** 2300 UTC 23 May 2011 and **c**, **d** 0100 UTC 24 May 2011

the increase of rain number concentrations, thus resulting in the decrease of graupel and snow number concentrations (Morrison et al. 2009). In addition, the size-sorting mechanism (which is not handled in 1M representations) allows the larger hydrometeors to fall faster than the smaller hydrometeors thus they have better chances to reach the melting level and below, resulting in more graupel and snow melting. Due to these effects, larger rainwater intercept parameters (Fig. 12b, d) and larger rainwater mixing ratios (Fig. 11c, f) appear in WRF-MORR. The change in the intercept parameters and rainwater mixing ratios of graupel and snow also contributes to the abrupt change in dBZ around the melting level (as shown in Fig. 5d, h).

The smaller q_r from WRF-WSM6 implies less evaporation and weak cooling. In contrast, larger N_{0r} and q_r in WDM6 result in a high evaporation rate and large amounts of rainwater that could arrive at the ground, contributing to a strong (Fig. 10e) and continuous cold pool that triggers new convective cells and maintains a realistic convection lifetime (Figs. 4b, e, h, k, 6), and a larger precipitation amount (Fig. 8). Similar to WDM6, even stronger effects of N_{0r} and q_r are found in MORR. Meanwhile, the sublimation and melting of ice-phase hydrometeors around the melting level is also important for sustaining cold pools, especially deep cold pools. Taking graupel as an example, because it dominates ice hydrometeors, Fig. 13 shows the graupel mixing

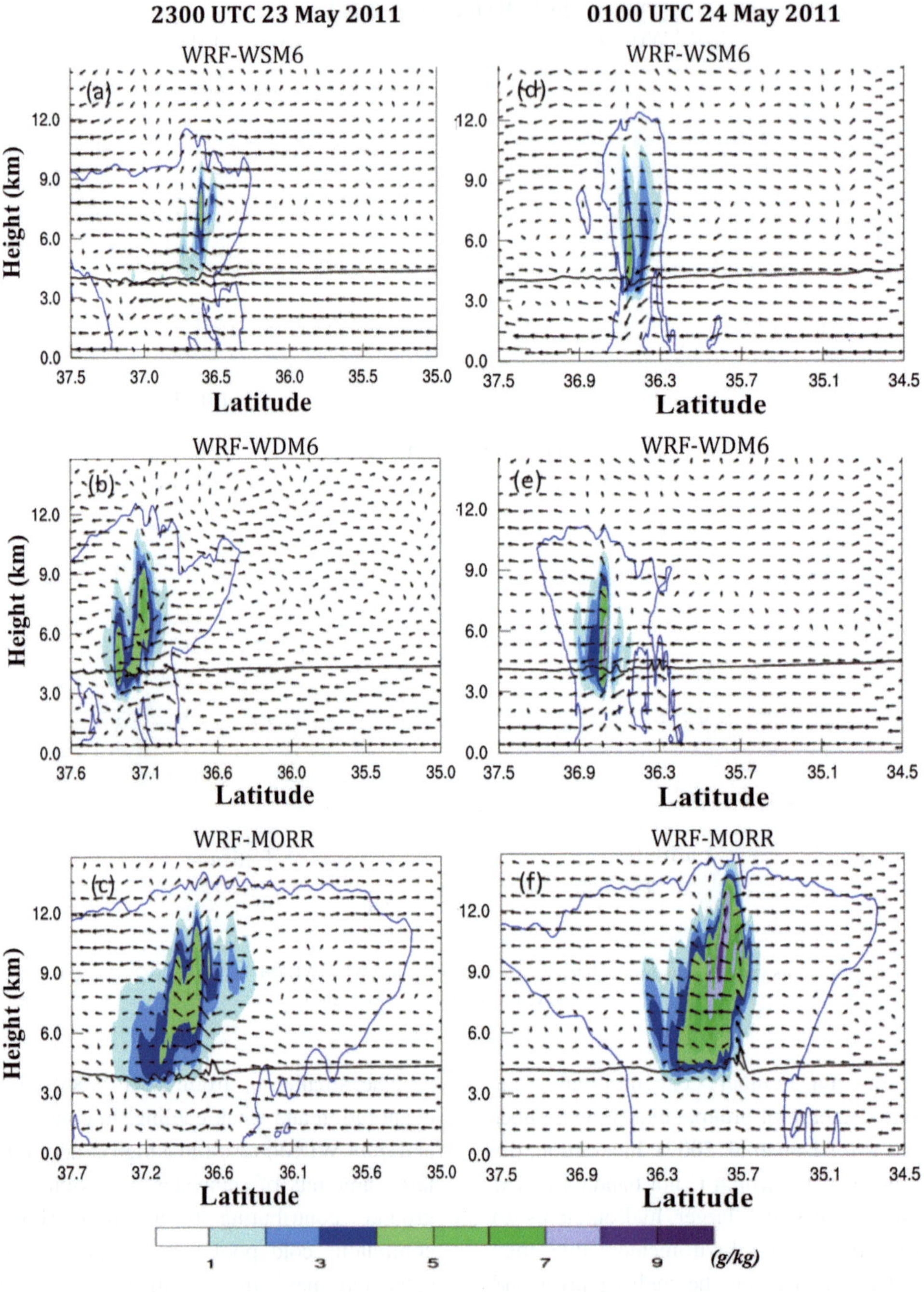

Figure 13
Same as Fig. 11, except for the graupel mixing ratio (g kg^{-1})

ratio (q_g) vertical cross section for the simulations. It is clear that q_g in WRF-MORR (Fig. 13c) is larger than that in WRF-WSM6 (Fig. 13a) and WRF-WDM6 (Fig. 13b). The smaller graupel mixing ratio produced by WRF-WSM6 may be caused by the relatively small amounts of supercooled water (Fig. 11a, d) above the melting level available for the riming processes. Due to the small amounts of ice-phase hydrometer melting and sublimation around the melting level, WRF-WSM6 produces shallow and small cold pools (Fig. 10d). In addition, because of small rainwater amounts, WRF-WSM6 produces a cold pool with inadequate depth and time to trigger new cells. This also seems to be the reason why WRF-WSM6 produces the shortest convection life cycle. In WRF-MORR, graupel melting and sublimation below the melting level due to the size-sorting mechanism in the SR region also help the simulation sustain cooling in the mid-troposphere and contribute to a deep cold pool (as shown in Fig. 10f) that supplies continuous forcing to trigger new convective cells.

The total microphysical latent heating confirms the above analysis of hydrometeor distribution. Figure 14 shows the cross section of the microphysical latent heating for the three simulations. It is clear that diabatic cooling is present in the SR region (in the zoomed-in black box). The cooling is less in WRF-WSM6 (Fig. 14a) than in the other two simulations (Fig. 14b, c). WRF-WDM6 (Fig. 14b) produces very strong cooling in the SR region due to a very large N_{0r}. The elevated cooling region around the melting level in the SR region from WRF-MORR (Fig. 14c) is due to the melting of graupel and snow and the large N_{0r}. Figure 14c also confirms the extension of deep cool pools in WRF-MORR (as shown in Fig. 10). Overall, the simulation with the 2M scheme shows the combined effects of rainwater evaporation in the CC region and the melting/sublimation of ice-phase hydrometeors in the SR region that produce a strong, deep cold pool and supply continuous forcing to trigger new cells.

4.3. Hydrometeor Advection

As stated in Sect. 4, WRF-WSM6 and WRF-WDM6 produce relatively small percentages of stratiform cloud, while WRF-MORR shows better agreement with observations (Figs. 4, 8). Previous observational studies claimed that hydrometeor advection is important for maintaining large SR and AC regions in convective systems, and these two regions are responsible for the large MCS coverage (Biggerstaff and Houze 1991; Rutledge and Houze 1987). In order to understand the reason for the differences in cloud structure produced by various simulations, hydrometeor advection is examined.

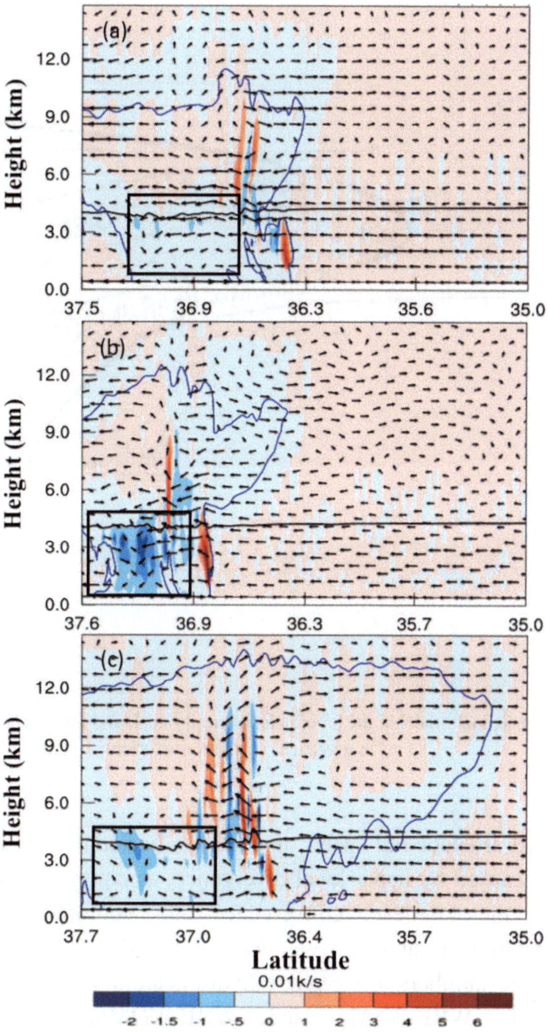

Figure 14
Same as Fig. 11, except for cross sections of the diabatic heating (Ks^{-1}) at 2300 UTC 23 May 2011 for **a** WRF-WSM6, **b** WRF-WDM6, and **c** WRF-MORR

According to Houze et al. (1989), the basic dynamic mechanism of hydrometeor advection in a typical MCS is the relative flow from the convective core to the outer region in the upper levels (Fig. 15). Fortunately, as shown in Fig. 5, all the experiments successfully simulated the upper-level flow from the inner core to the forward and rearward (front to rear flow) regions.

Hydrometeor advection is influenced by two main factors: one is the number of ice-phase hydrometeors available in the core regions, and the other is the relative value of the horizontal wind and

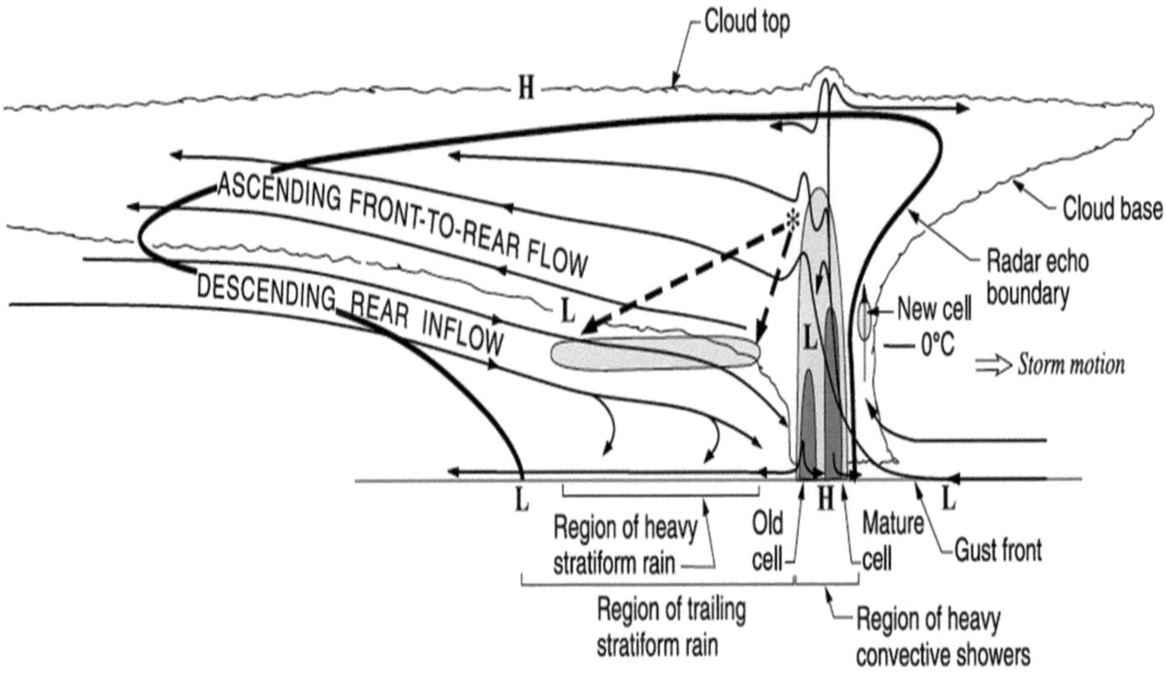

Figure 15
Conceptual model of the kinematic, microphysical, and radar echo structure of a convective line with trailing-stratiform precipitation viewed in a vertical cross section oriented perpendicular to the convective line (from Houze et al. 1989)

hydrometeor fall speed; upper-level horizontal wind is caused by upper-level divergence, as compensates to the low-level convergence and updrafts. As mentioned above, WRF-MORR produces many more numbers of ice-phase hydrometeors, compared with WRF-WSM6 and WRF-WDM6. These hydrometeors act as the source of hydrometeors to be advected to produce larger SR and AC regions. This is one reason the simulation with MORR produces larger and more realistic SR and AC regions. In addition, the upper-level horizontal wind speed is larger in WRF-MORR (Fig. 13c, f) than in WRF-WSM6 (Fig. 13a, d) or WRF-WDM6 (Fig. 13b, e), and the hydrometeor fall speed in WRF-MORR is less than in WRF-WSM6 or WRF-WDM6 due to the different graupel fall velocity–diameter (*V–D*) relationships used in the three microphysical schemes. The relatively high horizontal wind speed and small hydrometeor fall speed enable the hydrometeors to spread into a larger region. This accounts for the larger cloud coverage produced by WRF-MORR.

5. *Sensitivity to Ice Hydrometeors*

So far, we have examined the sensitivity of numerical simulations of the MCS to different microphysical schemes. As mentioned, the major difference between MORR and the other two schemes (WDM6 and WSM6) is the 2M representation of ice hydrometeors. Therefore, the sensitivity of numerical simulations of the MCS to ice hydrometeors is worth further investigation.

The MORR microphysical scheme provides the flexibility of using hail or graupel as the dense riming ice species. The main difference between hail and graupel is that hail has a larger density and a faster fall speed. Bryan and Morrison (2012) indicated that simulations with hail produced more realistic results than those with graupel for a squall line convective system. As shown in Sect. 4, the WRF-MORR simulation produces an unrealistically large CC area with a positive radar reflectivity bias aloft. In order to evaluate the influences of different types of dense ice in the simulations, a new experiment (H-MORR) using the Morrison scheme with hail (instead of

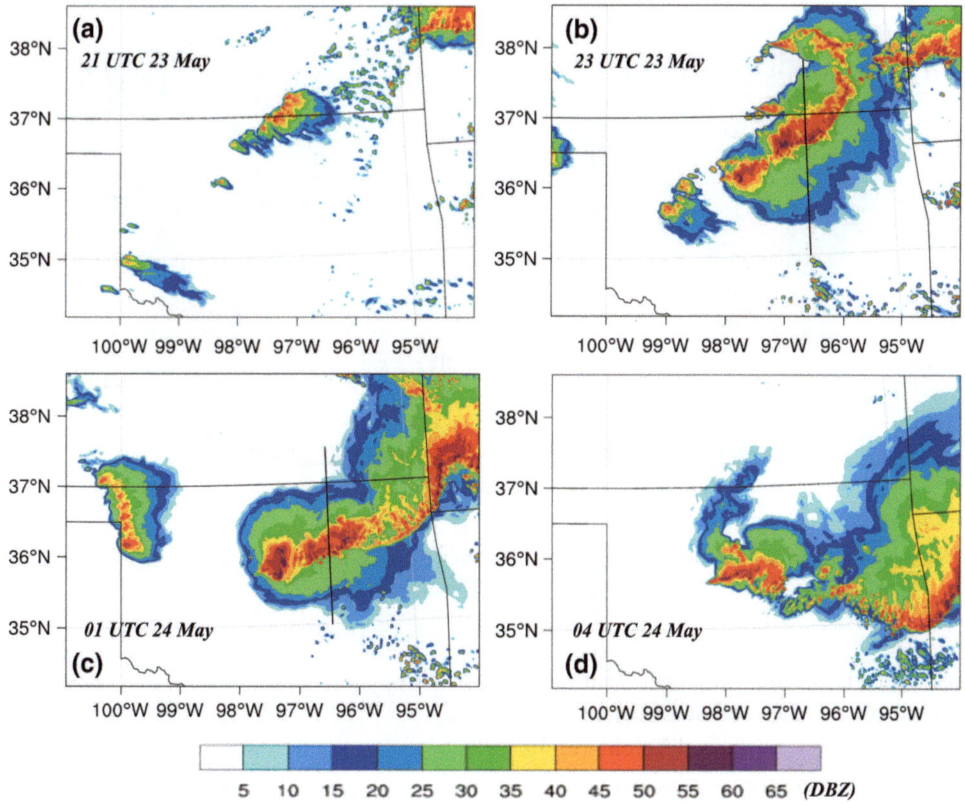

Figure 16
Composite radar reflectivity (dBZ) from the H-MORR simulation at **a** 2100 UTC 23 May 2011, **b** 2300 UTC 23 May 2011, **c** 0100 UTC 24 May 2011, and **d** 0400 UTC 24 May 2011

graupel as in WRF-MORR) as the dense ice species is conducted.

Figure 16 shows snapshots of composite radar reflectivity at the initiation stage (Fig. 16a), developing stage (Fig. 16b), mature stage (Fig. 16c), and dissipating stage (Fig. 16d). Compared with the results from WRF-MORR (Fig. 4c, f, i, l), H-MORR produces the same convection life cycle and similar convection coverage, but a different intensity distribution. The largest difference between H-MORR and WRF-MORR is the size of the region with medium-to-high radar reflectivity values (> 40). The high dBZ zone is much narrower in H-MORR than in WRF-MORR. However, this narrow zone better agrees with the NEXRAD observations (Fig. 1). Overall, H-MORR performs better than WRF-MORR for the convection structure, although it also overestimates the dBZ values compared with the NEXRAD observations.

To illustrate the vertical structure of the MCS, the dBZ cross sections are shown in Fig. 17a, b. Note that there are high dBZ values aloft from the melting level to the middle and upper troposphere in the SR region due to the large amount of graupel accumulation caused by the slower fall speed and graupel melting in WRF-MORR (Fig. 5d, h). This problem, however, disappears in H-MORR. The high dBZ in H-MORR is concentrated into a narrower region but is more realistic compared to NEXRAD observations (Fig. 5a, e). The sizes of the convective system produced by H-MORR and WRF-MORR are similar.

In addition, compared with Stage IV precipitation data, H-MORR produces more intense precipitation than WRF-MORR does. It also concentrates more precipitation in the CC region and less in the SR region (figure not shown).

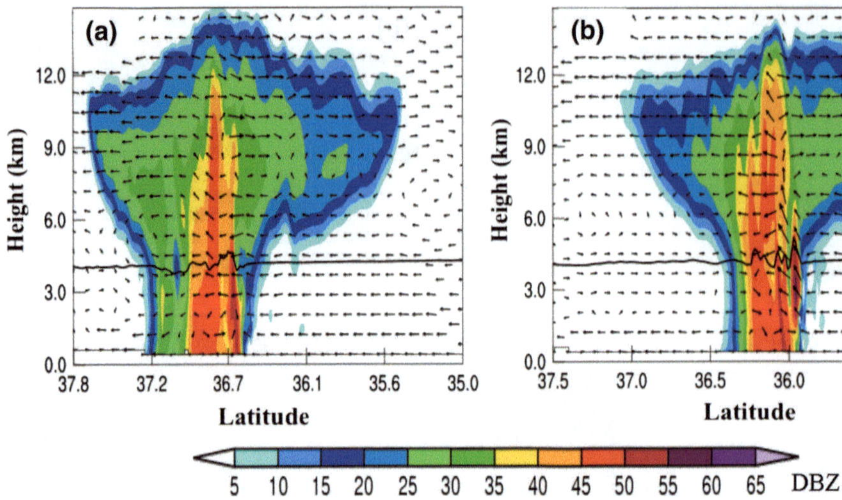

Figure 17
North–south cross sections (along the black lines shown in Fig. 16b, c) of radar reflectivity (dBZ) from the H-MORR simulation at **a** 2300 UTC 23 May 2011 and **b** 0100 UTC 24 May 2011

It seems that replacing graupel with hail improves the simulated convection structure. But why does the use of hail produce different results than the use of graupel? In order to obtain additional insights, the hydrometeor numbers and distributions are examined. Figure 18 shows the time series of domain-averaged hydrometeor paths produced by WRF-MORR and H-MORR. The hydrometeor path is defined as the vertical integral of the hydrometeor mass per unit area.

$$Hp_i = \int_{0}^{Z_{top}} \rho_i q_i \mathrm{d}z = \int_{p_{top}}^{p_{surface}} \frac{p}{RT} q_i \mathrm{d}p = \sum_{l=0}^{l=n} \frac{p(l)q(l)}{RT_l}[(p(l) - p(l+1))], \quad (2)$$

where i represents the specific kind of hydrometeor (rain, graupel, snow, etc.), and l indicates the model level.

Compared with WRF-MORR (Fig. 18a), the snow path in H-MORR (Fig. 18b) is slightly larger, with the maximum value is 0.9 kg m^{-2} in H-MORR versus 0.8 kg m^{-2} in WRF-MORR. Specifically, in H-MORR, the hail path is larger than the snow path before the mature stage (about 0000 UTC 24 May 2011), while the snow path becomes larger at the mature and dissipation stages. These differences may be caused by the different parameters such as the V–D relationship and the density among graupel, hail, and snow. For instance, the density is 100 kg m^{-3} for snow, 400 kg m^{-3} for graupel, and 900 kg m^{-3} for

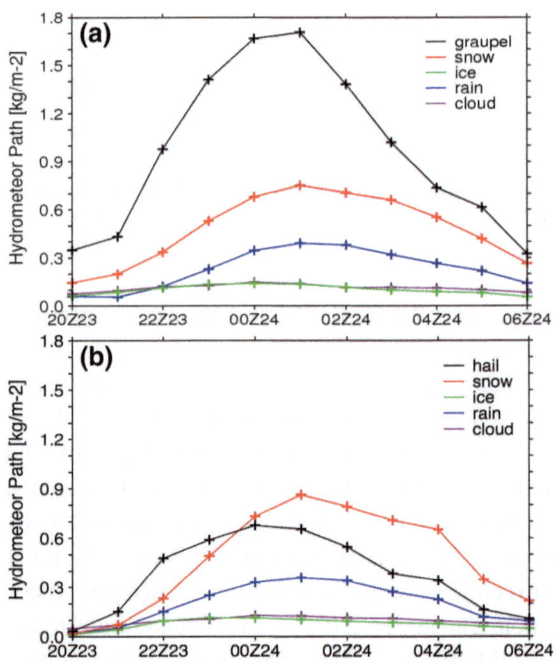

Figure 18
Time series of domain-averaged hydrometeor path (kg m^{-2}) of hail (black line), snow (red line), ice (green line), rain (blue line), and cloud (purple line) from **a** WRF-MORR and **b** H-MORR

hail. The V–D relationship for hail is $V = 114.5D^{0.5}$, $V = 19.3D^{0.37}$ for graupel, and $V = 11.72D^{0.41}$ for snow. Obviously, for the same size of particle, hail has a larger fall speed compared with snow and graupel. The larger fall speed for hail means it has a shorter suspended time. Once hails pass through the melting level, most of them will be melted and converted to rain, and a small amount of hails also arrives at the ground. Because of this, large amounts of hail in the CC region contribute to the intense precipitation there. The domain-averaged hail hydrometeor path is smaller than that of graupel, especially during the mature and dissipating stages of the MCS. The hail mixing ratio and intercept parameter N_{0h} in H-MORR (Fig. 19) support this conclusion. Meanwhile, since the less dense graupel can spread into a larger area, the simulation with graupel overestimates the dBZ values. Furthermore, the snow mixing ratio (Fig. 20a, c) and intercept parameter N_{0s} (Fig. 20b, d) for both WRF-MORR and H-MORR show that the mixing ratio of snow in H-MORR is larger than that in WRF-MORR. Most of the snow is produced in the SR region. In WRF-MORR, a large amount of graupel is available in the SR region, while graupel may collect some snow by aggregation. More snow in H-MORR is transported into the outer region by the middle- and upper-level horizontal wind. A broad convection structure appears in the middle and upper levels as illustrated in radar reflectivity (Fig. 17).

6. Summary and Concluding Remarks

In this study, the sensitivity of numerical simulations of a mesoscale convective system (23–24 May 2011) and its associated clouds and precipitation to various cloud microphysical schemes is examined. Three different microphysical schemes used in the experiments include 1M, partial 2M, and 2M representation of warm-rain and ice hydrometeors.

It is found that all the simulations reproduce the basic properties of this convective event, such as its initiation and development. Compared with NEXRAD observations, the simulation with the 1M microphysical scheme produces the most scattered convection distribution, the shortest convection lifetime, and a lower echo top. It also heavily underestimates the precipitation intensity, and the cloud component distribution and ratio are unrealistic. By adding 2M representations of warm-rain hydrometeors, WRF-WDM6 shows a clear improvement in convection properties by producing a realistic convection lifetime and an organized horizontal convection structure. However, it still underestimates the precipitation amount. The cloud distribution and ratio are also unrealistic, and the echo top is also lower than in NEXRAD observations. With 2M representations of both warm-rain and ice hydrometeors, simulations with MORR schemes produce much better results than the simulations with 1M and partial 2M schemes; they predict a realistic

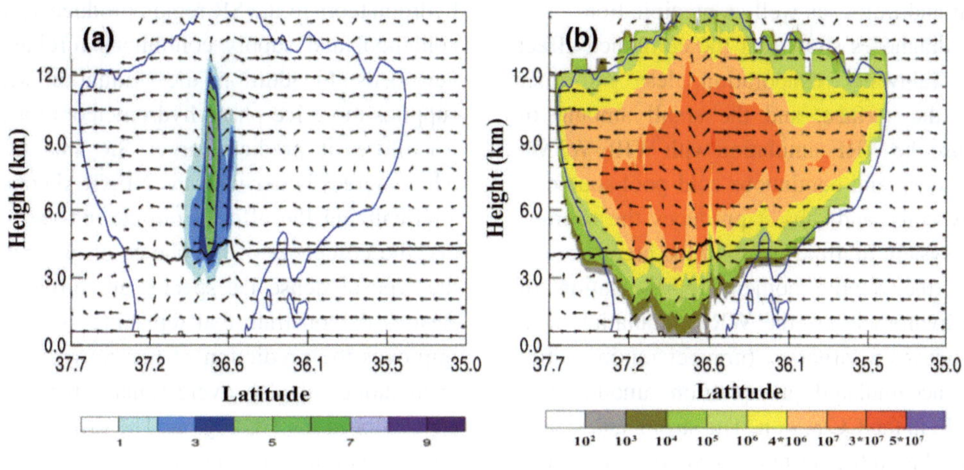

Figure 19
Same as Fig. 17, except for **a** hail mixing ratio (g kg^{-1}) and **b** intercept parameter N_{0h} at 2300 UTC 23 May 2011

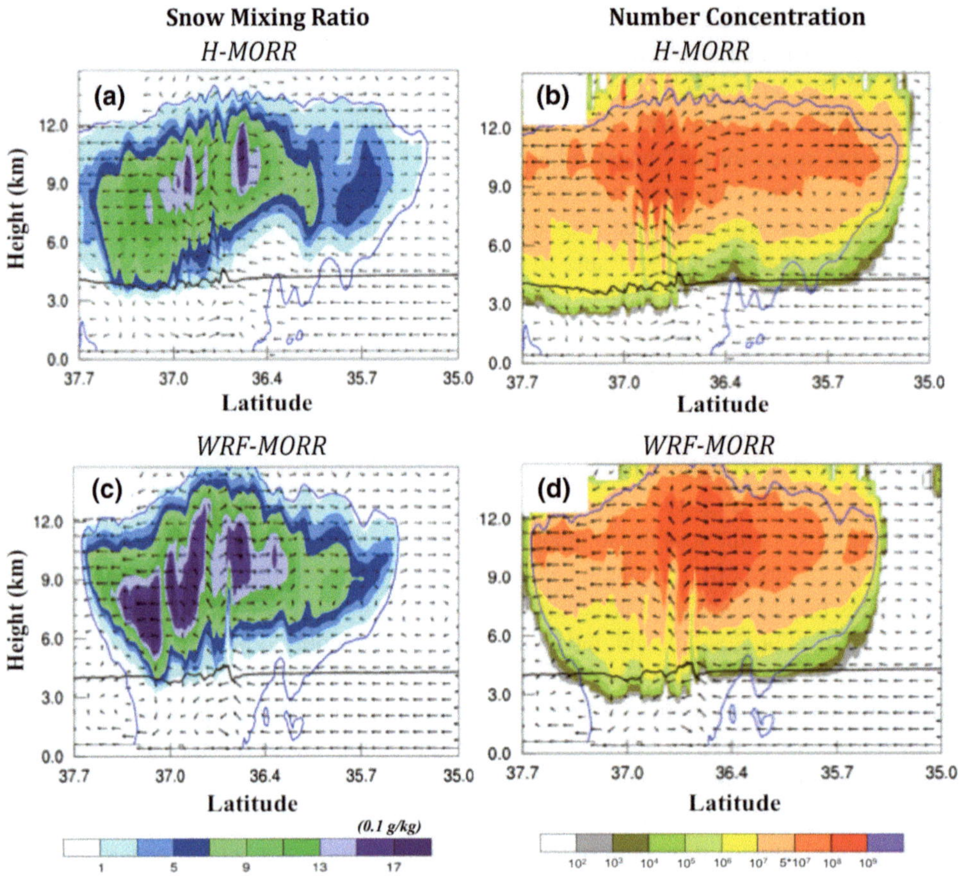

Figure 20
Same as Fig. 17, except for **a**, **c** snow mixing ratio (0.1 g kg^{-1}) and **b**, **d** intercept parameter N_{0s} for **a**, **b** H-MORR and **c**, **d** WRF-MORR at 2300 UTC 23 May 2011

convection lifetime, echo tops, and cloud component distribution and ratios, as well as precipitation.

Further diagnoses show that the combined effect of greater rainwater evaporation, less rainwater arriving at the surface, and the small amount of graupel cause the WRF simulation with WRF-WSM6 that lacks a deep and continuous cold pool source to trigger new cells, and thus it produces the shortest convection system lifetime and the least precipitation. By adding the double-moment representation of warm-rain hydrometeors, the WRF simulation with WDM6 produces a larger q_g (graupel mixing ratio) and larger accumulated precipitation amount than WRF-WSM6 in the CC region. The strong and steady cold pool is able to trigger new convective cells, thus producing a realistic convection lifetime. The combined effects of rainwater evaporation in the CC region and the melting/sublimation of ice-phase hydrometeors in the SR region produce a strong, deep cold pool and supply continuous forcing to trigger new cells. In addition, the simulation with MORR supplies more ice-phase hydrometeors for advection. Therefore, it predicts larger SR and AC regions, which agrees better with the NEXRAD observations.

Realizing the affect from adding ice hydrometeors, the influences of the choice of graupel or hail as the type of dense ice species are further examined. Results demonstrate that replacing graupel with hail improves the prediction of the convection structure, even though it also overestimates the precipitation, especially in the CC region. The differences between the two simulations appear to be caused mainly by the density and V–D relationship differences between hail and graupel. The faster fall speed of hail concentrates the precipitation into a narrower CC region. Since the less dense graupel can spread into a larger area, the simulation with graupel overestimates the dBZ values. Graupel/hail processes appear to play important roles in determining the storm structure.

In addition to the great sensitivity of numerical simulations of MCS to different microphysical schemes, overall results from this study indicate the importance of accurate representation of ice hydrometeors in microphysical parameterizations and its great influence on predicting MCSs and their associated clouds and precipitation. Future work will further confirm these conclusions by studying more cases with more microphysical schemes. In addition, observations of hydrometeor distribution are needed for evaluation purposes (e.g., Pu and Lin 2015). More comprehensive overview and future research regarding the ice-phase precipitation are also addressed in Gultepe et al. (2017)

Acknowledgements

This study was supported by the Office of Science (BER), US Department of Energy Grant no. DE-SC0007060 (Z. Pu, C. Lin, and S. Krueger) and DE-SC0008468 (X. Dong). The first author (Pu) was also supported by NASA NEWS program Award NNX15AD14G. Oklahoma Mesonet data were obtained from the Atmospheric Radiation Measurement (ARM) Program sponsored by the US Department of Energy, Office of Science, Office of Biological and Environmental Research, Climate and Environmental Sciences Division. The authors are grateful to the NCAR WRF model development group. The computing support from the Center for High Performance Computing (CHPC) at the University of Utah is also gratefully acknowledged. Funding was provided by Aeronautics and Space Administration (Grant no. NNX15AD14G)

References

Adams-Selin, R. D., ven den Heever, S. C., & Johnson, R. H. (2013). Impact of graupel parameterization schemes on idealized bow echo simulations. *Monthly Weather Review, 141,* 1241–1262.

Baldwin, M. E., & Mitchell, K. E. (1996). The NCEP hourly multi-sensor U.S. precipitation analysis. Preprints, *11th Conf. on Numerical Weather Prediction* (pp. J95–J96) Norfolk, VA, Amer. Meteor. Soc.

Biggerstaff, M. I., & Houze, R. A., Jr. (1991). Kinematic and precipitation structure of the 10–11 June 1985 squall line. *Monthly Weather Review, 119,* 3034–3064.

Brock, F. V., Crawford, K. C., Elliott, R. L., Cuperus, G. W., Stadler, S. J., Johnson, H. L., et al. (1995). The Oklahoma Mesonet: A technical overview. *Journal of Atmospheric Science, 12,* 5–19.

Bryan, G. H., & Morrison, H. (2012). Sensitivity of a simulated squall line to horizontal resolution and parameterization of microphysics. *Monthly Weather Review, 140,* 202–225.

Dudhia, J. (1989). Numerical study of convection observed during the winter monsoon experiment using a mesoscale two-dimensional model. *Journal of Atmospheric Science, 46,* 3077–3107.

Feng, Z., Dong, X., Xi, B., Schumacher, C., Minnis, P., & Khaiyer, M. (2011). Top-of-atmosphere radiation budget of convective core/stratiform rain and anvil clouds from deep convective systems. *Journal of Geophysical Research.* https://doi.org/10.1029/2011JD016451.

Fovell, R. G., & Ogura, Y. (1988). Numerical simulation of a midlatitude squall line in two dimensions. *Journal of Atmospheric Science, 45,* 3846–3879.

Gultepe, I., Heymsfield, A. J., Field, P. R., & Axisa, D. (2017). Ice-phase precipitation. *Meteorological Monographs, 58,* 6.1–6.36.

Hong, S. Y., & Lim, J.-O. (2006). The WRF single-moment 6-class microphysics scheme (WSM6). *Journal of the Korean Meteorological Society, 42,* 129–151.

Houze, R. A. (2004). Mesoscale convective systems. *Reviews of Geophysics, 42,* RG4003. https://doi.org/10.1029/2004rg000150.

Houze, R. A., Biggerstaff, M. I., Rutledge, S. A., & Smull, B. F. (1989). Interpretation of Doppler weather radar displays of midlatitude mesoscale convective systems. *Bulletin of the American Meteorological Society, 70,* 608–619.

James, R. P., Markowski, P. M., & Fritsch, J. M. (2006). Bow echo sensitivity to ambient moisture and cold pool strength. *Monthly Weather Review, 134,* 950–964.

Janjic, Z. I. (1990). The step-mountain coordinate: Physical package. *Monthly Weather Review, 118,* 1429–1443.

Kain, J. S. (2004). The Kain–Fritsch convective parameterization: An update. *Journal of Applied Meteorology, 43,* 170–181.

Kessler, E., (1969). On the distribution and continuity of water substance in atmospheric circulations. *Meteor. Monogr.*, vol. 32. American Meteorological Society.

Klazura, G. E., & Imy, D. A. (1993). A description of the initial set of analysis products available from the NEXRAD WSR-88D system. *Bulletin of the American Meteorological Society, 74,* 1293–1311.

Lafore, J. P., & Moncrieff, M. W. (1989). A numerical investigation of the organization and interaction of the convective and stratiform regions of tropical squall lines. *Journal of Atmospheric Science, 46,* 521–544.

Lang, S. E., Tao, W.-K., Zeng, X., & Li, Y. (2011). Reducing the biases in simulated radar reflectivities from a bulk microphysics scheme: Tropical convective systems. *Journal of Atmospheric Science, 68,* 2306–2320.

Li, X., & Pu, Z. (2008). Sensitivity of numerical simulation of early rapid intensification of Hurricane Emily (2005) to cloud microphysical and planetary boundary layer parameterizations. *Monthly Weather Review, 136,* 4819–4838.

Li, X., Tao, W.-K., Khain, A. P., Simpson, J., & Johnson, D. E. (2009). Sensitivity of a cloud-resolving model to bulk and explicit bin microphysical schemes. Part I: Comparisons. *Journal of the Atmospheric Sciences, 66,* 3–21.

Lim, K. S. S., & Hong, S. Y. (2010). Development of an effective double-moment cloud microphysics scheme with prognostic cloud condensation nuclei (CCN) for weather and climate models. *Monthly Weather Review, 138,* 1587–1612.

Lin, Y. L., Farley, R. D., & Orville, H. D. (1983). Bulk parameterization of the snow field in a cloud model. *Journal of Climate and Applied Meteorology, 22,* 1065–1092.

Lin, Y., & Mitchell, K. E. (2005). The NCEP stage II/IV hourly precipitation analyses: Development and applications. Preprints, *19th Conf. on Hydrology,* San Diego, CA, American Meteor Society, 1.2. http://ams.confex.com/ams/Annual2005/techprogram/paper_83847.htm.

Liu, C., Moncrieff, M. W., & Zipser, E. J. (1997). Dynamical influence of microphysics in tropical squall lines: A numerical study. *Monthly Weather Review, 125,* 2193–2210.

Mansell, E. R., Ziegler, C. L., & Bruning, E. C. (2010). Simulated electrification of a small thunderstorm with two-moment bulk microphysics. *Journal of Atmospheric Science, 67,* 171–194.

McCumber, M., Tao, W.-K., Simpson, J., Penc, R., & Soong, S.-T. (1991). Comparison of ice-phase microphysical parameterization schemes using numerical simulations of tropical convection. *Journal of Climate and Applied Meteorology, 30,* 985–1004.

Milbrandt, J. A., & Yau, M. K. (2005a). A multimoment bulk microphysics parameterization. Part I: Analysis of the role of the spectral shape parameter. *Journal of Atmospheric Science, 62,* 3051–3064.

Milbrandt, J. A., & Yau, M. K. (2005b). A multimoment bulk microphysics parameterization. Part II: A proposed three-moment closure and scheme description. *Journal of Atmospheric Science, 62,* 3065–3081.

Mlawer, E. J., Taubman, S. J., Brown, P. D., Iacono, M. J., & Clough, S. A. (1997). Radiative transfer for inhomogeneous atmospheres: RRTM, a validated correlated-k model for the longwave. *Journal of Geophysical Research: Atmospheres (1984–2012), 102,* 16663–16682.

Morrison, H. (2010) An overview of cloud and precipitation microphysics and its parameterization in models, *WRF Workshop,* June 21–25, 2010. Boulder, CO. http://www2.mmm.ucar.edu/wrf/users/workshops/WS2010/WorkshopPapers.php. Accessed 26 Dec 2017.

Morrison, H., Curry, J. A., & Khvorostyanov, V. I. (2005). A new double-moment microphysics parameterization for application in cloud and climate models. Part I: Description. *Journal of Atmospheric Science, 62,* 1665–1677.

Morrison, H., Thompson, G., & Tatarskii, V. (2009). Impact of cloud microphysics on the development of trailing stratiform precipitation in a simulated squall line: Comparison of one- and two-moment schemes. *Monthly Weather Review, 137,* 991–1007.

Philips, V. T., Donner, L. J., & Garner, S. T. (2007). Nucleation processes in deep convection simulated by a cloud-system-resolving model with double-moment bulk microphysics. *Journal of Atmospheric Science, 64,* 738–768.

Pu, Z., & Lin, C. (2015). 2015: Evaluation of double-moment representation of ice hydrometeors in bulk microphysical parameterization: Comparison between WRF numerical simulations and UND-Citation data during MC3E. *Geoscience Letters, 2,* 11. https://doi.org/10.1186/s40562-015-0028-x.

Rutledge, S. A., & Hobbs, P. (1983). The mesoscale and microscale structure and organization of clouds and precipitation in midlatitude cyclones. VIII: A model for the "seeder-feeder" process in warm-frontal rainbands. *Journal of Atmospheric Science, 40,* 1185–1206.

Rutledge, S. A., & Houze, R. A., Jr. (1987). A diagnostic modelling study of the trailing stratiform region of a midlatitude squall line. *Journal of Atmospheric Science, 44,* 2640–2656.

Skamarock, W. C., Klemp, J. B., Dudhia, J., Gill, D. O., Barker, D. M, Duda, M. G., et al. (2008). A description of the Advanced research WRF version 3. NCAR Technical Note.

Smirnova, T. G., Brown, J. M., Benjamin, S. G., & Kim, D. (2000). Parameterization of cold-season processes in the MAPS land-surface scheme. *Journal of Geophysical Research, 105,* 4077–4086.

Smull, B. F., & Houze, R. A., Jr. (1987). Rear inflow in squall lines with trailing stratiform precipitation. *Monthly Weather Review, 115,* 2869–2889.

Steiner, M., Houze, R. A., Jr., & Yuter, S. E. (1995). Climatological characterization of three-dimensional storm structure from operational radar and rain gauge data. *Journal of Climate and Applied Meteorology, 34,* 1978–2007.

Thompson, G., Field, P. R., Hall, W. D., & Rasmussen, R. M. (2006). A new bulk microphysical parameterization for WRF (&MM5). In *Proc. Seventh Weather Research and Forecasting Model Workshop* (pp. 1–11).

Thompson, G., Field, P. R., Rasmussen, R. M., & Hall, W. D. (2008). Explicit forecasts of winter precipitation using an improved bulk microphysics scheme. Part II: Implementation of a new snow parameterization. *Monthly Weather Review, 136,* 5095–5115.

Weisman, M. L. (1992). The role of convectively generated rear-inflow jets in the evolution of long-lived mesoconvective systems. *Journal of Atmospheric Science, 49,* 1826–1847.

Weisman, M. L. (1993). The genesis of severe, long-lived bow echoes. *Journal of Atmospheric Science, 50,* 645–670.

Weisman, M. L., Wang, W., & Manning, K. (2009). The use of the RUC DFI initialization for the 2009 WRF-ARW 3 km explicit convective forecasts. In *Proceedings of the 10th Weather Research and Forecasting Model Workshop* (pp. 1–18). NCAR Mesoscale and Microscale Meteorology Division, Boulder, Colo, USA.

Weverberg, K. V., & Vogelmann, A. M. (2012). Sensitivity of idealized squall-line simulation to the level of complexity used in two-moment bulk microphysics schemes. *Monthly Weather Review, 140,* 1883–1907.

Wilks, D. S. (1995). *Statistical methods in atmospheric sciences.* San Diego: Academic Press.

Wu, D., Dong, X., Xi, B., Feng, Z., Kennedy, A., Mullendore, G., et al. (2013). Impacts of microphysical scheme on convective and stratiform characteristics in two high precipitation squall line events. *Journal of Geophysical Research, 118,* 11119–11135.

Zhao, Q., & Carr, F. H. (1997). A prognostic cloud scheme for operational NWP models. *Monthly Weather Review, 125,* 1931–1953.

(Received November 26, 2017, revised January 18, 2018, accepted January 22, 2018, Published online January 31, 2018)

Pure Appl. Geophys. 176 (2019), 2121–2137

https://doi.org/10.1007/s00024-018-1901-2

Pure and Applied Geophysics

Assessment of the Sensitivity to the Thermal Roughness Length in Noah and Noah-MP Land Surface Model Using WRF in an Arid Region

Michael Weston,[1] Naira Chaouch,[1] Vineeth Valappil,[1] Marouane Temimi,[1] Michael Ek,[2] and Weizhong Zheng[3]

Abstract—Atmospheric models are known to underestimate land surface temperature and, by association, 2 m air temperature over dry arid regions during the day due to the treatment of the thermal roughness length also known as roughness length of heat. The thermal roughness length can be controlled by the Zilitinkevich parameter, known as Czil, which is a tunable parameter within the models. Three different scenarios with the WRF model are run to test the impact of the Czil parameter on the simulations using two land surface models: the Noah and Noah-MP models. In this study, a modified version of the Noah-MP model is tested, in which the Czil parameter, and, therefore, the thermal roughness length varies depending on the land cover and vegetation height. The model domain is over the United Arab Emirates (UAE) where the major land cover type is desert. The following configurations are tested: the Noah model with Czil = 0.1, Noah model with Czil = 0.5 and the Noah-MP model with Czil = 0.5 over desert. Results of 2 m air temperature are verified against three stations in the UAE. Mean gross error of the diurnal 2 m temperature was reduced by up to 1.48 and 1.54 °C in the 24 and 48 h forecasts, respectively. This reduced the cold bias in the model. This improvement in air temperature showed to improve the diurnal cycle of relative humidity at the three monitoring stations as well as the duration of the sea breeze in some cases.

Key words: Czil, roughness length, arid, 2 m temperature, WRF, Noah-MP.

1. *Introduction*

Land surface processes play an important role in both regional and mesoscale atmospheric models and climate models as they control the heat, momentum, and water exchanges between the surface and the atmosphere (Pitman 2003). An accurate land surface model that simulates the initial and the progress of the ground state is necessary to properly simulate the diurnal and vertical structure of the planetary boundary layer (PBL). Modeling the interaction between the surface and the atmosphere is critical to reproduce the physical processes between the PBL and the surface-land to simulate the turbulent heat and water exchanges. Several physical representations of land–atmosphere interactions at the surface with different degrees of complexity in terms of number of parameters, number of soil layers, and vegetation characteristics have been proposed in the literature. Different performances among the schemes in portioning the water and energy fluxes depending on the time, spatial scales, and applications were found for both and coupled with atmospheric models runs (Liang et al. 1998; Pitman et al. 1999; Timbal and Henderson-Sellers 1998; Wood et al. 1998). Wood et al. (1998) compared 16 land surface models and showed the importance of the calibration of the models for the prediction of the energy and water fluxes. In addition, different models showed different sensitivity to the same parameter depending on the model structure and locations (Bastidas et al. 2006). Depending on the application, there are trade-offs between models complexity and performances (Bastidas et al. 2006; Hogue et al. 2006).

For regional weather forecasting, the Weather Research and Forecasting (WRF) model with the Advanced Research WRF (ARW) core (Wang et al. 2009; Huang et al. 2008; Skamarock et al. 2008) is widely used for both research and operational applications. It offers a wide range of physical

[1] Masdar Institute, Khalifa University of Science and Technology, P.O. Box 54224, Abu Dhabi, United Arab Emirates. E-mail: mjweston@masdar.ac.ae
[2] NOAA, College Park, MA, USA.
[3] IMSG at NOAA, NCEP/Environmental Modeling Center, College Park, MD 20740, USA.

parameterizations representing the interaction between the atmosphere, the soil, and the surface. The WRF model makes use of a coupled land surface model for lower boundary simulation of the heat and moisture fluxes.

The Noah Land Surface Model (LSM) is widely coupled with general circulation models and mesoscale atmospheric models because of its relative simplicity for real time global and regional weather and hydrology applications (Chen and Dudhia 2001; Ek et al. 2003; Tewari et al. 2004). The Noah Land Surface Model (LSM) has been implemented in the National Centers for Environment Prediction (NCEP) operational Eta Model for global and regional forecasts (Ek et al. 2003). It has also been coupled with the Penn State-NCAR fifth generation Mesoscale Model (Chen and Dudhia 2001). Finally, the Noah Land Surface Model is being used for both research and operational purposes. An updated version of this model, namely Noah-MP, which includes multi-parameterization options, has also been implemented in different mesoscale atmospheric models (Niu et al. 2011).

For arid and semi-arid regions, the surface soil heat flux and the turbulent sensible heat fluxes dominate the surface energy budget because of the lack of precipitation. The thermal and aerodynamic roughness lengths constitute the main parameters for the bulk transfer equations to calculate the turbulent fluxes in the LSMs (Chen et al. 2010, 2011; Zheng et al. 2012). Wei et al. (2013) tested the Noah LSM and found a large bias of the latent heat, total runoff and land surface temperature during the warm, spring, and summer seasons over vegetated areas. Zheng et al. (2012) reported that the Global Forecast System (GFS) model underestimates land surface skin temperatures during daytime over arid regions of the US when the surface roughness for heat was considered equal to the momentum surface length. The decrease of the thermal surface roughness through the use of vegetation-dependent formulation for the calculation of the momentum and thermal roughness lengths reduces the daytime cold bias over arid and semi-arid regions during warm season. Chen et al. (2010) tested different thermal roughness length formulations with Noah LSM model and found an underestimation of the land surface temperature and overestimation of the sensible heat flux. Similar results were also found by other works (Hogue et al. 2005; LeMone et al. 2008). Better performance is obtained when thermal roughness parameter values higher than the default are considered (LeMone et al. 2008). Chen and Zhang (2009) showed the importance of considering variable roughness length for heat and moisture depending on the land cover and vegetation height to better simulate the surface exchange coefficient. Improved results were also obtained when the diurnal variation of the thermal roughness length is considered (Chen et al. 2010, 2011; Yang et al. 2008).

This paper aims to study the sensitivity of WRF to land surface models, namely Noah and Noah-MP, for the simulation of meteorological parameters in the United Arab Emirates (UAE). For this purpose, a sensitivity analysis of WRF model to the thermal roughness parameter and, therefore, to the surface exchange coefficient for heat was conducted for eight events with fog conditions recorded at Abu Dhabi Airport and at other locations in the UAE. Fog is one of the most hazardous weather conditions that affects road and air transportation in the UAE. Fog forms as a result of the strong temperature inversion overnight and the large difference between inland and the sea temperatures, especially during winter (De Villiers and Van Heerden 2007). In the region, limited works addressed this particular issue. Chaouch et al. (2017) compared different PBL schemes in WRF with different lead times under fog conditions in the UAE. Bartoková et al. (2015) used the combination of WRF and a decision tree-based model for fog nowcasting in Dubai. Ajjaji et al. (2008) compared different PBL, radiation, and microphysics schemes during fog conditions in the UAE. This study builds upon the previous investigations and places a particular interest on the effect of the land surface model scheme in WRF simulating different thermodynamic variables during fog and no-fog conditions. The model is verified using in situ observations of air temperature, relative humidity, and wind speed at three airports in the UAE: Abu Dhabi, Dubai, and Al Ain.

2. *Study Area*

The United Arab Emirates (UAE) is located in the northeastern side of the Arabian Peninsula (Fig. 1). It is surrounded by the Arabian Gulf from the north and the west, by the Gulf of Oman from the east, and by Saudi Arabia from the south and west. The UAE is an arid region with an average annual precipitation of around 100 mm (Harris et al. 2014). The eastern and northern parts of the UAE usually record the highest precipitation amount because of the presence of the Al Hajar Mountains in the east, which foster local convective processes and rainfall in the summer season. Vegetation is mostly present in the eastern region due to the summer rainfall and available ground water. Desert dominates the rest of the country except around the Country's large cities.

Fog events are usually reported in the region, especially during fall and winter seasons. They usually lead to a significant degradation in visibility below 100 m. This has an adverse impact on all transportation systems and air traffic (Bartok et al. 2012). Most fog events are observed during the night, and they are characterized by calm wind and humidity that can reach 80%. The fog formation in the region is mainly due to the combination of the advection of warm and moist air from the sea with the radiation cooling during the night (De Villiers and Van Heerden 2007). The sea and land breezes in the area prompt the formation of fog. Two types of sea breeze can occur in the region. The first is the bay breeze, which is due to the temperature gradient between the coastline and the gulf; this fosters the condensation and fog formation inland during the night. The second is the ocean breeze which is controlled by the large-scale processes (Abbs and Physick 1992). During the day, the water vapour from the Gulf and the ocean is transported and then trapped inland during the night because of the low wind speed.

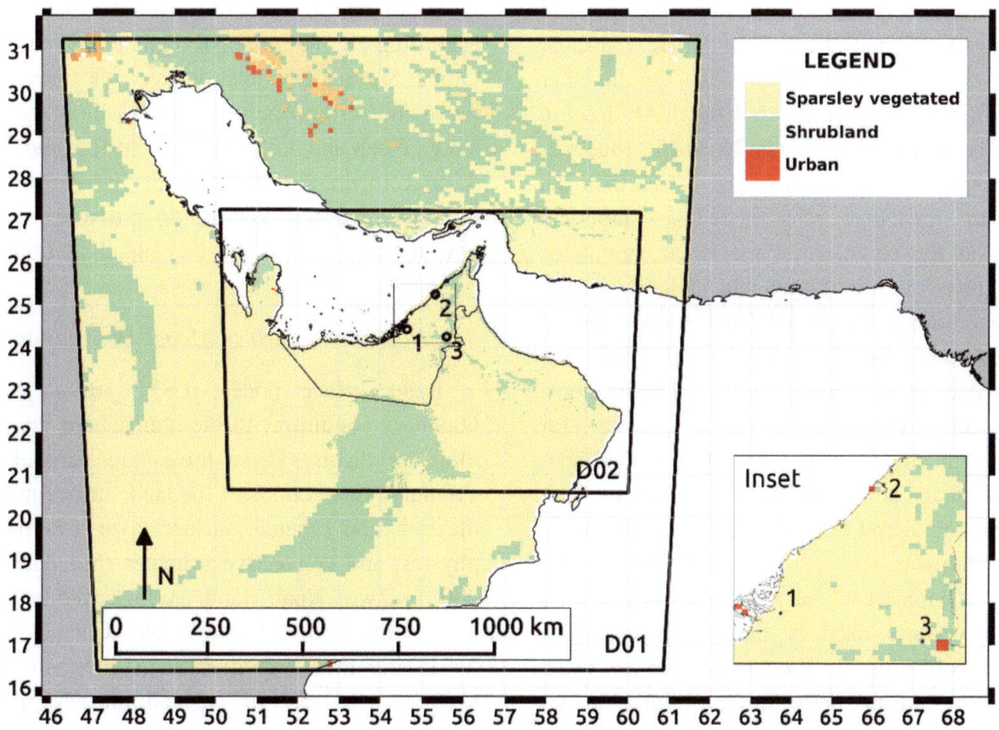

Figure 1
WRF model domains D01 (12 km) and D02 (4 km) with the USGS land cover. White is open water. The three airports used for verification are 1. Abu Dhabi, 2. Dubai, and 3. Al Ain

3. Methodology

We run the WRF model with two main configurations for eight case studies between January and March 2016. The configurations, described in detail below, test the effect of the Zilitinkevich parameter (Czil) in two land surface models in WRF, namely Noah and Noah-MP. The model results are verified for 2 m air temperature at three sites in the United Arab Emirates.

3.1. WRF Model

The Advanced Research WRF (WRF-ARW) mesoscale model V3.7.1 developed by the National Center for Atmospheric Research with one-way nesting has been used. The WRF model is a compressible and non-hydrostatic atmospheric model that uses terrain-following hydrostatic pressure vertical coordinate. Two domains, with 12 and 4 km horizontal spatial resolutions, were considered for the parent and nested domains, respectively. The parent domain extended over the Arabian Peninsula, parts of the Arabian Gulf, and the Sea of Oman (latitude: 16.3°N–31.5°N, longitude: 47°E–61°E). On the other hand, the nested domain covered the UAE Region (Fig. 1). The nest domain for the Noah-MP runs was slightly larger (extended east of the UAE) than the nest domain for Noah runs to reduce numerical instability in the model over the Hajar mountains which occurred during testing when increasing the vertical resolution in the model. This change was not expected to alter the dynamics in the region of interest where the model was verified. The meteorological data from the National Center for Environmental Prediction (NCEP-GFS) at 0.5° spatial resolution and at 06 UTC initialization time were used for the initial and lateral boundary conditions in the parent domain.

Table 1 summarizes all the parameterization schemes that were used in the simulations. The default USGS land-use–land-cover data set was used. For all simulations, the WRF Single-Moment 3-class (WSM3) scheme was used to represent the grid-scale water vapour, cloud, and precipitation processes (Hong et al. 2004). The sub-grid-scale clouds were represented with the Kain–Fritsch scheme (Kain 2004; Kain and Fritsch 1990). The atmospheric heating was calculated using the Rapid Radiative Transfer Model (RRTM) longwave radiation (Mlawer et al. 1997) and RRTMG shortwave radiation (Iacono et al. 2008) schemes. The exchanges of mass, energy, and moisture between land and atmosphere were determined using the Quasi-Normal Scale Elimination (QNSE) scheme (Sukoriansky et al. 2005). The model simulated hourly variables like potential temperature, latent heat fluxes, sensible heat fluxes, three-dimensional wind, pressure, and water vapour at the surface and at different pressure levels. The land cover was basically composed of four classes, namely water, urban and built-up land, shrubland, and barren or sparsely vegetated (Fig. 1).

Eight case studies, January 19th and 22nd, February 15th, 16th, 18th, 19th, and 20th, and March 10th, 2016 were considered for this work The above dates, except the 19th of January, corresponded to recorded fog events at Abu Dhabi Airport. For Dubai and Al Ain, these dates correspond to both fog conditions and no-fog conditions. Fog corresponded to conditions when visibility falls below 1000 m according to the METAR data collected from the Global Surface Observation Network (http://weather.uwyo.edu/surface/meteorogram/). For each case study, Noah and Noah-MP land surface models using 45 and 46 pressure levels with 50 hPa top pressure value were tested. For the 46 pressure level run, a new sigma level at 0.996 was added near the surface.

3.2. Noah Land Surface Model Description

Land surface models (LSM) provide the lower boundary conditions to the atmosphere via heat and moisture fluxes. These fluxes are derived through simulated interactions of the land surface model with the radiation scheme, surface layer scheme, microphysics, and convective schemes (Skamarock et al. 2008). Two Noah land surface model versions namely the Unified Noah (Noah) (Chen and Dudhia, 2001) and the Noah multi-parameterization (Noah-MP) (Niu et al. 2011) land surface model are available in the WRF model. The Noah model is a simple scheme that represents the soil temperature and moisture in four layers up to 1 m below the surface. It calculates the surface energy fluxes over a

Table 1

WRF configuration for control and experiment

WRF parameters	Noah0.1 (control)	NoahmpCzil (experiment)
Data	NCEP-GFS 0.5°	
Initial time	06 UTC	
Domain size (km)		
12	125 × 140	125 × 140
4	187 × 184	253 × 184
Time interval	6 h	
Model time step	60 s	
PBL/surface layer	QNSE/QNSE	
Microphysics	WSM3	
Cumulus	Kain–Fritsch (new Eta)	
LWR	RRTM	
SWR	RRTMG	
LSM	Noah	Noah-MP
Surface layer drag coefficient		Original Noah
Czil	0.1 constant for all LULC	0.5 for desert (see Table 2)
LULC	USGS	
Vertical levels (lowest 3 and 4 sigma levels)	45 (1, 0.993, 0.983)	46 (1, 0.996, 0.993, 0.983)

combined surface layer of both vegetation and soil (Chen and Dudhia, 2001). The Noah-MP scheme is an updated version of the Noah with multi-parameterization options of a number of processes. Updates present in Noah-MP that are relevant to this study are the inclusion of a vegetation canopy, which requires vegetation height, and bare ground fraction to calculate the fluxes to the atmosphere.

In atmospheric modeling, air temperature at 2 m above ground level is derived from land surface temperature, heat flux, and the coefficient of heat exchange as follows (Niu et al. 2011):

$$\mathrm{T2} = \mathrm{LST} - \frac{\mathrm{HFX}}{\mathrm{RHO} \times \mathrm{CP} \times C_H}, \quad (1)$$

where T2 is air temperature at 2 m above the ground (K), LST is land surface temperature (K), HFX is net upward heat flux at the surface (Wm^{-2}), RHO is air density ($\mathrm{kg\ m}^{-3}$), CP is heat capacity of dry air ($\mathrm{J\ kg}^{-1}\ \mathrm{K}^{-1}$), and CH is the exchange coefficient for heat at 2 m above ground (ms^{-1}). RHO and CP are two constants. Changes to the land surface temperature, heat flux, or the coefficient of heat exchange alter the air temperature at 2 m above the ground.

The Zilitinkevich coefficient parameter (Czil) in the WRF model affects land surface temperature and the coefficient of heat exchange. Therefore, the Czil plays a role in diagnosing 2 m air temperature (Zilitinkevich 1995). Czil is a tunable parameter that affects the aerodynamic resistance of the surface layer (Mitchell et al. 2005). It controls the relationship between Z_0m/Z_0t and the roughness Reynolds number (Re^*) (Chen et al. 1997). In other words, it links the type of air flow, whether laminar or turbulent, to the roughness length of heat. As it will be described below, the effect of increasing Czil results in a decrease in the roughness length of heat, Z_0t, which will decrease the coefficient of heat exchange, C–H, and, therefore, in an increase of land surface temperature. This is achieved by controlling the ratio of the roughness length of heat with the roughness length of momentum according to (Eq. 2) (Zilitinkevich 1995; Chen et al. 1997):

$$z_0m/z_0t = \exp(k\mathrm{Czil}\sqrt{(Re^*)}), \quad (2)$$

where Z_0m is momentum roughness length, Z_0t is thermal roughness length, k is the von Karman constant of 0.4, Czil is the Czil parameter, and Re^* is the Reynolds number. The Reynolds number is dependent on frictional velocity. Thus, when Czil is set to a constant, as it is currently in WRF, the ratio of Z_0t to Z_0m varies with frictional velocity [Eq. (3)]:

$$Re^* = (u_0^* z_0 m)/v, \quad (3)$$

where u_0^* is the frictional velocity and v is kinematic viscosity ($\sim 1.5 \times 10^{-5}$). The effect of increasing Czil results into a decrease in the heat roughness length, Z_0t, which will decrease the coefficient of heat exchange, CH according to (Eq. 4) (Niu et al. 2011; Chen et al. 1997). CH is calculated as follows:

$$C_H = k^2/[\ln(z/z_0m) - \psi_m(z/L) + \psi_m(z_0m/L)][\ln(z/z_0t) - \psi_h(z/L) + \psi_h(z_0t/L)], \quad (4)$$

where k is the Von Karman constant, z is the reference height, L is the Monin–Obukhov length, and Ψ is a stability correction function.

The value for Czil has evolved over time. Before the introduction of Czil, assumptions were made on the relationship between Z_0m and Z_0t. As outlined by (Chen et al. 1997), early efforts assumed $Z_0m = Z_0t$, However, this created problems when using land surface temperature to solve the Monin–Obukhov similarity theory, as the boundary conditions were incorrect. To overcome this issue, Z_0t was decreased so that $Z_0m/Z_0t = 10$ (Braud et al. 1993). However, this was a fixed value and was not applicable for all land-cover types. Zilitinkevich (1995) then proposed the tunable value of 0.1, which effectively allowed the ratio of Z_0m/Z_0t to vary with frictional velocity. Chen et al. (1997) verified this value of 0.1 by comparing modeled and measured values of CH, and found it to be the most favorable value for most land-cover types. Other values of 0.01 and 1 were also tested but did not perform as well. The deficiency of this approach (Czil = 0.1) is that land surface temperature can be overestimated during the daytime and underestimated at night time. Finally, Chen and Zhang (2009) proposed that Czil varies with vegetation height as follows:

$$\text{Czil} = 10^{(-kh)}, \quad (5)$$

where k is the Von Karman constant of 0.4 and h is the vegetation height. We use this approach for our vegetated land classes.

3.3. WRF Configuration

For both WRF domains, different simulations were conducted with both versions of Noah LSM model using different heat roughness parameter values and pressure levels. Three different scenarios were tested.

Scenario 1 the control run uses the Unified Noah LSM with 45 pressure levels and Czil equals 0.1 (default in WRF V3.7), referred to hereafter as Noah0.1

Scenario 2 A constant Czil equals 0.5 with the Unified Noah LSM version and referred to hereafter as Noah0.5 was tested

Scenario 3 a variable Czil in Noah-MP LSM depending on the vegetation height as proposed by Chen and Zhang (2009) according to Eq. (5) was tested and referred to NoahmpCzil

For the study domain, the vegetation height formula is only relevant for shrubland and sparsely vegetated areas which are dominant in arid and semi-arid environments. For the sparsely vegetated areas (i.e., desert), a value of 0.5 is applied after a process of trial-and-error (Table 2). A different set of runs using Czil value of 0.8 as suggested by (Zheng et al. 2012) for desert were tested but showed an excess increase of the air temperature at 2 m during daytime. For urban and water areas, the default value of 0.1 is applied, as in the default configuration of WRF (Table 2). Moreover, under this scenario, simulations were conducted with 46 pressure levels unlike the first two scenarios where 45 pressure levels were used. The extra vertical level near the surface affects the local closure of flux calculations and should, therefore, affect the calculation of air temperature.

The Noah-MP LSM code was modified and adapted in WRF to read the land-cover-specific Czil

Table 2

Czil values applied to each LULC

Land use/land cover	Czil
Open water	0.1
Barren or sparsely vegetated (desert)	0.5
Shrubland	0.36
Urban	0.1

values (Table 2) from a lookup table (MPTABLE) when calculating the surface layer drag. This approach is novel in the context of arid regions as the option of using a dynamic Czil value that varies with vegetation height (Eq. 5) is active in WRF version 3.7.1 only for few PBL schemes such as, the YSU and MYNN PBL and surface layer option of MYNN and MM5 Monin–Obukhov schemes. In this work, we used QNSE PBL scheme and QNSE surface layer option which was found to outperform the other schemes in the simulation of the air temperature, humidity, and wind speed during fog events (Chaouch et al. 2017). Hence, the new combination QNSE PBL scheme with a variable Czil value in the Noah-MP was tested in this study.

3.4. Model Verification

Three airport stations, Abu Dhabi Airport (24.43°N, 54.65°E), Dubai airport (25.25°N, 55.33°E), and Al Ain airport (24.25°N, 55.60°E), are used for the validation (Fig. 1). Both Abu Dhabi and Dubai airports are characterized as barren and sparsely vegetation. Both these airports are coastal airports. On the other hand, Al Ain airport is an inland airport located in shrubland vegetation class.

Model results were presented in UTC time and divided into the 24 and 48 h forecasts. The mean gross error and bias in 2 m air temperature and relative humidity were calculated for each case study and averaged per hour for each forecast. For the first 24 h forecast, the comparison between the simulated and the observation data were done for the last 18 h from 12 to 06 UTC of the following day of each case study. The first 6 h were excluded and considered as model spin up time. The 48 h forecast evaluation considered the second 24 h from 07 UTC of day two to 06 UTC of day 3 of each case study. An improvement in model performance should show a decrease in the mean gross error. Each model scenario was verified against in situ observations data at the three sites. Model post processing was conducted using the tools from the NCAR command language [The NCAR Command Language (Version 6.4.0) (Software). (2017)]. Boulder, Colorado: UCAR/NCAR/CISL/TDD. http://dx.doi.org/10.5065/D6WD3XH5) and R statistical software (http://www.r-project.org). Graphs were made using the openair package in R (Carslaw and Ropkins, 2012).

4. Results and Discussion

The analysis of the mean values of simulated air temperature using the control run configuration (Noah with Czil equal to 0.1 and 45 pressure levels, Noah0.1) presented a warm bias just after the sunrise and early morning when compared with in situ observations data taken between October 18th 2016 and February 27th, 2017, at Abu Dhabi Airport (Fig. 2). The simulated diurnal air temperature also showed an early heating during the morning and a quick cooling during the afternoon and early evening. The resulting daytime cold bias is consistent with all the simulations using the Unified Noah LSM version with the control run conditions. The same behavior was also reported by Zheng et al. (2012) using the GFS model and default Noah LSM for the simulated land surface skin temperatures during daytime over arid regions of the United States. They showed that this behavior is mostly due to the surface turbulent heat transfer simulation within Noah LSM which is controlled mainly by the roughness length for heat, and thus, by the Czil value (Eqs. 2–5). The comparison between the WRF simulated air temperatures using the Noah LSM with Czil of 0.1 (Noah0.1, control run) and using Czil of 0.5 (Noah0.5) showed a similar trend and almost no improvement in simulating the air temperature (Fig. 3). The improvement in mean gross error was less than 1 °C when

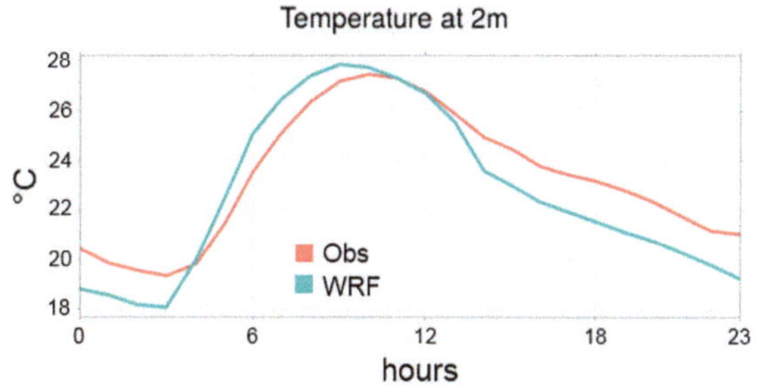

Figure 2
The problem. Average temperature at 2 m at Abu Dhabi airport for winter 2016/2017 (2016-10-18–2017-02-27). WRF shows a warm bias in the morning after sunrise and a cold bias in later afternoon/early evening. Time is in UTC. Site is UTC + 4

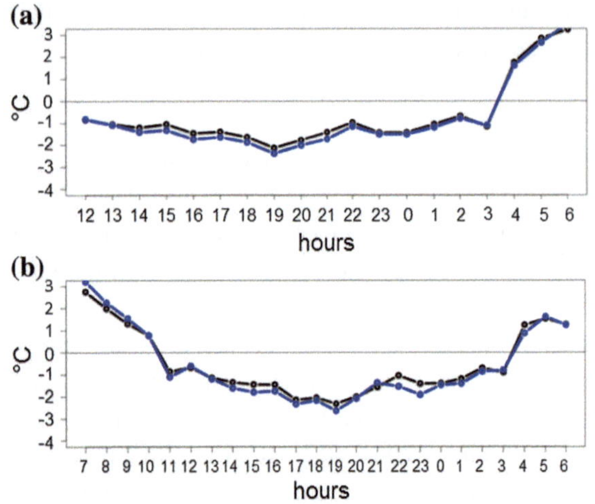

Figure 3
Temperature mean gross error with bias as the sign for Noah with Czil values of 0.1 and 0.5. This is the mean gross error for all case studies at Abu Dhabi Airport. Time is in UTC. Site is UTC + 4

changing Czil from 0.1 to 0.5 using Noah LSM. In WRF V.3.7.1, the Czil parameter is no longer active in the Noah LSM and is instead used within the surface layer schemes. This explains in part the small difference obtained in the simulated 2 m temperature when changing the Czil parameter (Fig. 3), as it has no direct effect on land surface temperature.

The diurnal variability of the simulated air temperature using Noah0.1, Noahmp0.1 (Noah-MP with constant Czil of 0.1) and NoahmpCzil for the case study of February 19th, 2016 showed an improvement of the simulation of the air temperature when Czil value is changed to 0.5 for the desert area and using Noah-MP (Fig. 4). Minor difference between Noah0.1 and Noahmp0.1 simulated air temperature are observed. Using a constant Czil value of 0.1 for the whole domain, presented the same simulated diurnal air temperature trend (i.e., cold bias in later afternoon/evening) with both versions of Noah LSM. The effect of increasing Czil to 0.5 is a decrease in the thermal roughness length, Z_0t, which will decrease the coefficient of heat exchange, CH, which will increase the land surface temperature. Even though the value of 0.1 of Czil was previously obtained by the comparison of modeled and measured values of CH for most land-cover types (Chen et al. 1997).

Results in this study showed the importance of increasing the Czil value to improve the forecasts and decrease the cold bias during daytime within WRF. Similarly, Zheng et al. (2012) adjusted the Czil value to 0.8 in the GFS model to correct for the model bias in land surface temperature for arid region in the US. LeMone et al. (2008) found a better simulation of the latent and sensible heat fluxes and land surface temperature when a value of Czil of 0.5 is considered, especially in the Great Plains in the Oklahoma Panhandle during summer time.

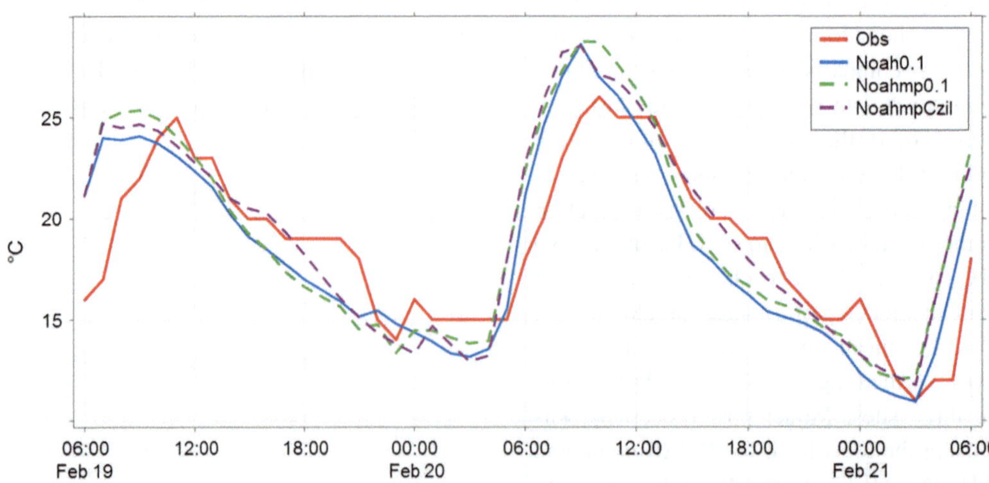

Figure 4
Air temperature at 2 m for three WRF configurations, Noah0.1, Noahmp0.1, and NoahmpCzil at Abu Dhabi Airport. Improvements in cold bias are achieved by changing the Czil and not by changing the LSM. Time is in UTC. Site is UTC + 4

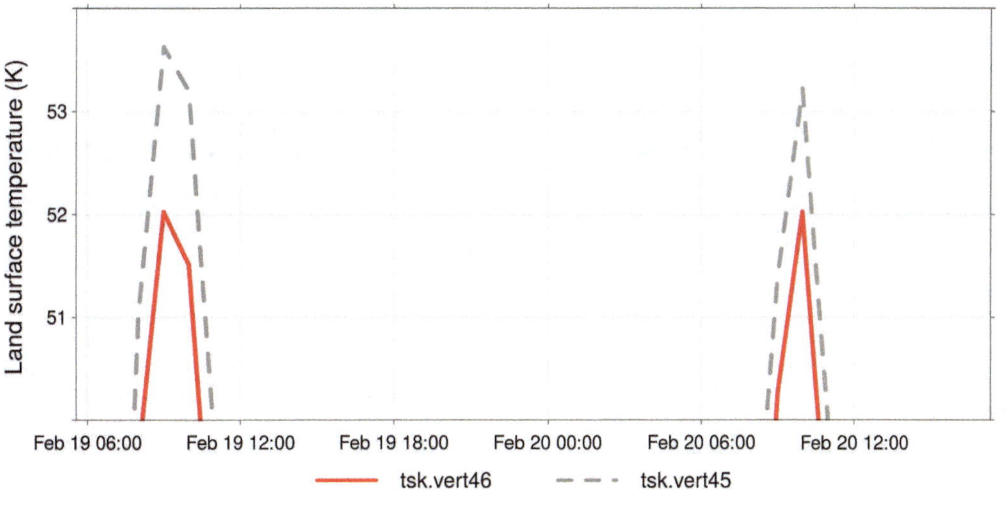

Figure 5
Difference in maximum land surface temperature for Noahmp (with variable Czil) runs with different vertical levels. This is for case study 2016-01-19. The extra vertical level closer to the ground decreases the land surface temperature and dampens the daily maximum air temperature at 2 m. Time is in UTC. Site is UTC + 4

The addition of the extra vertical level near the surface helped to dampen the effect of increasing the daily maximum temperature (Fig. 5). The example of the case study of February 19th, 2016 using both Noah-MP configurations, one with 46 vertical levels where the extra level is closer to the ground and the second with 45 vertical levels as the control run showed lower maximum LST of 1.2–1.5 Kelvin using 46 pressure levels (Fig. 5). Using Eq. (1), this leads to a lower air temperature at 2 m above ground. Chen et al. (1997) also noted that the increase in Czil value induced an increase in the maximum LST and, therefore, increase of the daily maximum temperature (Eq. 1). The extra vertical level near the surface affects the local closure of flux calculations, causing the fluxes to increase and the LST to decrease, and consequently, a decrease in the daily maximum temperature.

The comparison between Noah0.1 (control run) and the NoahmpCzil in simulating the 2 m air temperature showed better results when a variable Czil value is assigned. For Noah0.1, the mean gross error reached 2 °C for the control run during the late afternoon to evening at Abu Dhabi and Dubai airports for both forecast windows (Fig. 6). For NoahmpCzil at Abu Dhabi airport, mean gross error was decreased down to 1.48 °C for the first 24 h forecast and up to 1.54 °C for the 48 h forecast (Fig. 6a, b). However, after sunrise, NoahmpCzil heats up faster than Noah0.1 and reaches higher maximum air temperature values. The same behavior was also noted and described by Chen and Zhang (2009). This causes an increase in the mean gross error of up to 1.43 and 1.56 °C in the 24 and 48 h forecast, respectively. At Dubai airport, the mean gross error improved by up to 0.94 and 1.4 °C for the 24 and 48 h forecasts (Fig. 6c, d), respectively. On the other hand, for Al Ain airport, the mean gross error improvements reached up to 0.61 and 0.49 °C for the 24 and 48 h forecast (Fig. 6e, f), respectively.

The largest improvement in air temperature was achieved between 5 pm to 1 am local time at Abu Dhabi (Fig. 6a). The same performance of both model configurations was observed just before sunrise (at 2 am UTC) for the first 24 h forecast. For the 48 h forecast, NoahmpCzil extends the improvement to 6 am local time of the second day, generating a more reliable forecast. This is also true for the Dubai forecast. For both forecast windows, NoahmpCzil slightly improves the minimum temperature forecast during the night. The improvement in mean gross error starts later in the day than Abu Dhabi, from 8 pm local time instead of 5 pm. The gains at Al Ain are minor which can be due first to the fact that the

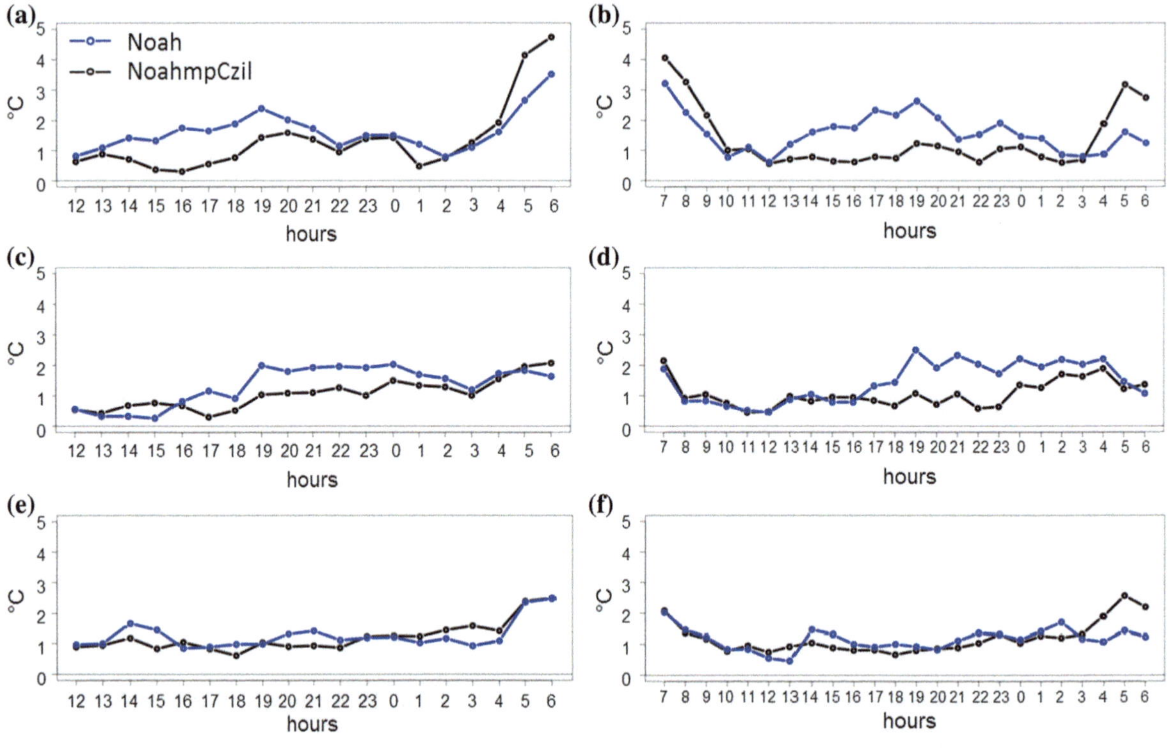

Figure 6
Temperature mean gross error (with bias as the sign) for Noah0.1 (blue line) and NoahmpCzil (black line) at **a** Abu Dhabi 24 h forecast and **b** 48 h forecast, **c** Dubai 24 h forecast and **d** 48 h forecast, and **e** Al Ain 24 h forecast and **f** 48 h forecast. Time is in UTC. Site is UTC + 4

cold bias at Al Ain was not as large as it is in the other land-use–land-cover type. Second, the Czil value of 0.36 is more conservative than 0.5. In addition, the inland dynamics are much different than the two coastal sites.

The mean diurnal air temperature results for all cases and for both forecasts at the three sites are presented in Fig. 7. These results show that NoahmpCzil makes an improvement in the cold bias that occurs in the late afternoon to the evening at Abu Dhabi and Dubai airports. However, it does not make any improvement to the early heating observed just after sunrise, which suggests that the early heating is caused by different physical mechanism in the model. At this point, it is worth noting that Abu Dhabi observations heat up after Dubai, and this may be due to the fog bank over Abu Dhabi. In addition, the model heating rate is faster than the observations at the three sites. This may be due to aerosols that are not adequately accounted for in the radiation scheme in the model. Thus, the real benefit of the NoahmpCzil is the decrease in the cold bias occurring just after the daily maximum temperature.

The analysis of the mean diurnal forecast of the relative humidity at 2 m above the ground also showed an improvement using the NoahmpCzil configuration. The control run with Noah0.1 causes relative humidity to increase too early in the evening when compared to the observations (Fig. 8). This is due to the cold bias in the model as the relative humidity is a diagnostic variable that is calculated from the temperature and water vapour mixing ratio. The NoahmpCzil runs improve the timing of the peak of the relative humidity that occurs later due to the improvement in temperature. The mean gross error of the simulated maximum relative humidity is less than 5% for the 24 h forecast, while the 48 h forecast overestimates relative humidity by over 5% using the NoahmpCzil configuration. This is partially due to the remaining cold bias in the NoahmpCzil 48 h forecast (although decreased), and also to the excess in water vapour simulation (not shown). The excess

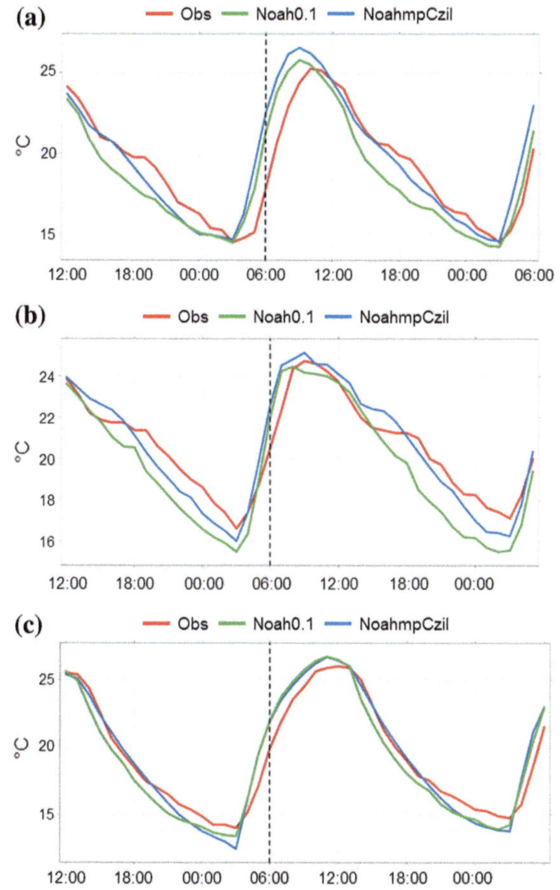

Figure 7
Mean diurnal temperature for all case studies at **a** Abu Dhabi, **b** Dubai, and **c** Al Ain. The black line represents the end of the 24 h forecast and start of the 48 h forecast. Time is in UTC. Site is UTC + 4

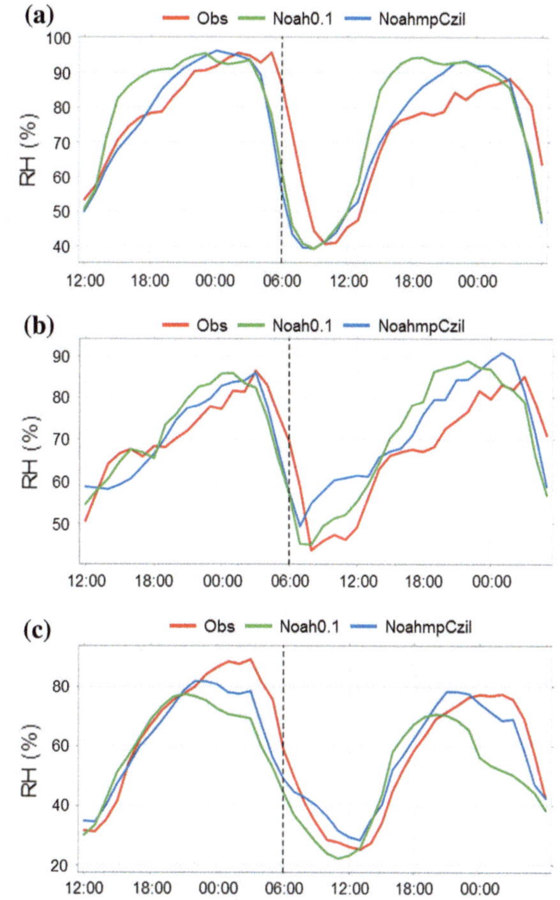

Figure 8
Mean diurnal relative humidity for all case studies at **a** Abu Dhabi, **b** Dubai, and **c** Al Ain. The black line represents the end of the 24 h forecast and start of the 48 h forecast. Time is in UTC. Site is UTC + 4

water vapour is linked to different model parameterisations that fall outside the scope of the Czil tests.

The comparison of the sensible heat flux and land surface temperature between model simulations (due to lack of observations) showed an excess of the surface heat flux between the land and the atmosphere using the Noah0.1 configuration. As expected, by increasing Czil value, a decrease of the mean sensible heat flux by 75–100 Wm^{-2} at Abu Dhabi and Dubai, and about 120 Wm^{-2} in Al Ain are achieved (Fig. 9). This increased the peak LST by about 15 K at Abu Dhabi and Dubai and by 13 K at Al Ain (Fig. 10). As per Eq. (1), this increase in LST and decrease in the surface turbulent exchange coefficient resulted in the decreased bias in the 2 m temperature simulation.

In addition, the increase of the Czil value resulted in an improvement in the diurnal wind pattern by increasing the duration of the sea breeze (wind direction higher than 300°) (Fig. 11). The sea breeze is driven by the temperature gradient between the air over the land and the adjacent Gulf waters. The sea breeze blows when the air over the land is warmer. During nighttime, the wind shifts to a land breeze when the land cools down. In the control simulations, the land cools down too quickly, which appears to result in an early shift of the sea breeze (Fig. 11b). With the NoahmpCzil configuration, the land remains warmer for longer resulting in longer duration of the sea breeze. While the land temperature only represents half the problem in this situation (where sea

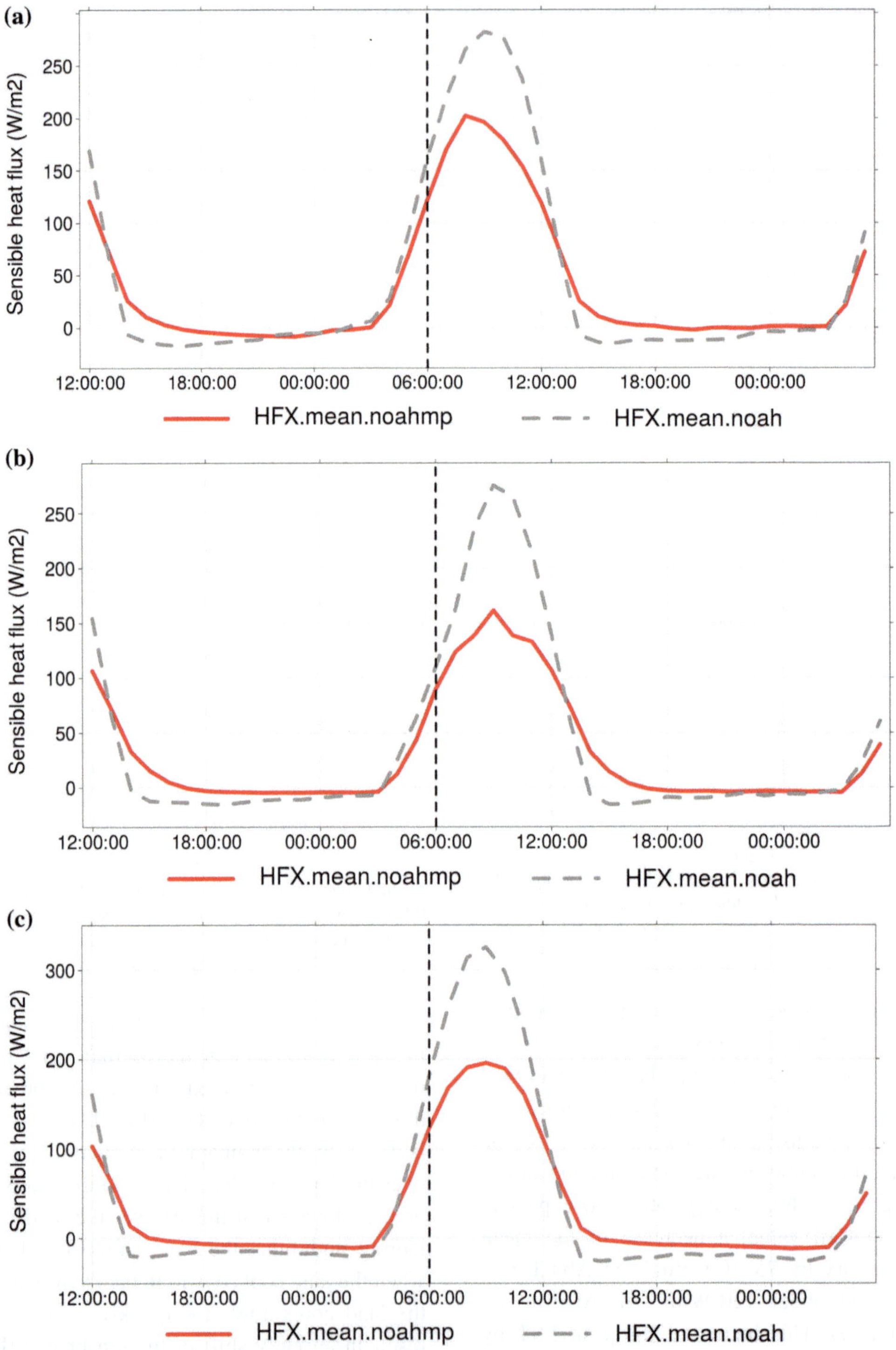

Figure 9
Mean diurnal sensible heat flux for all case studies at **a** Abu Dhabi, **b** Dubai, and **c** Al Ain. The black line represents the end of the 24 h forecast and start of the 48 h forecast. Time is in UTC. Site is UTC + 4

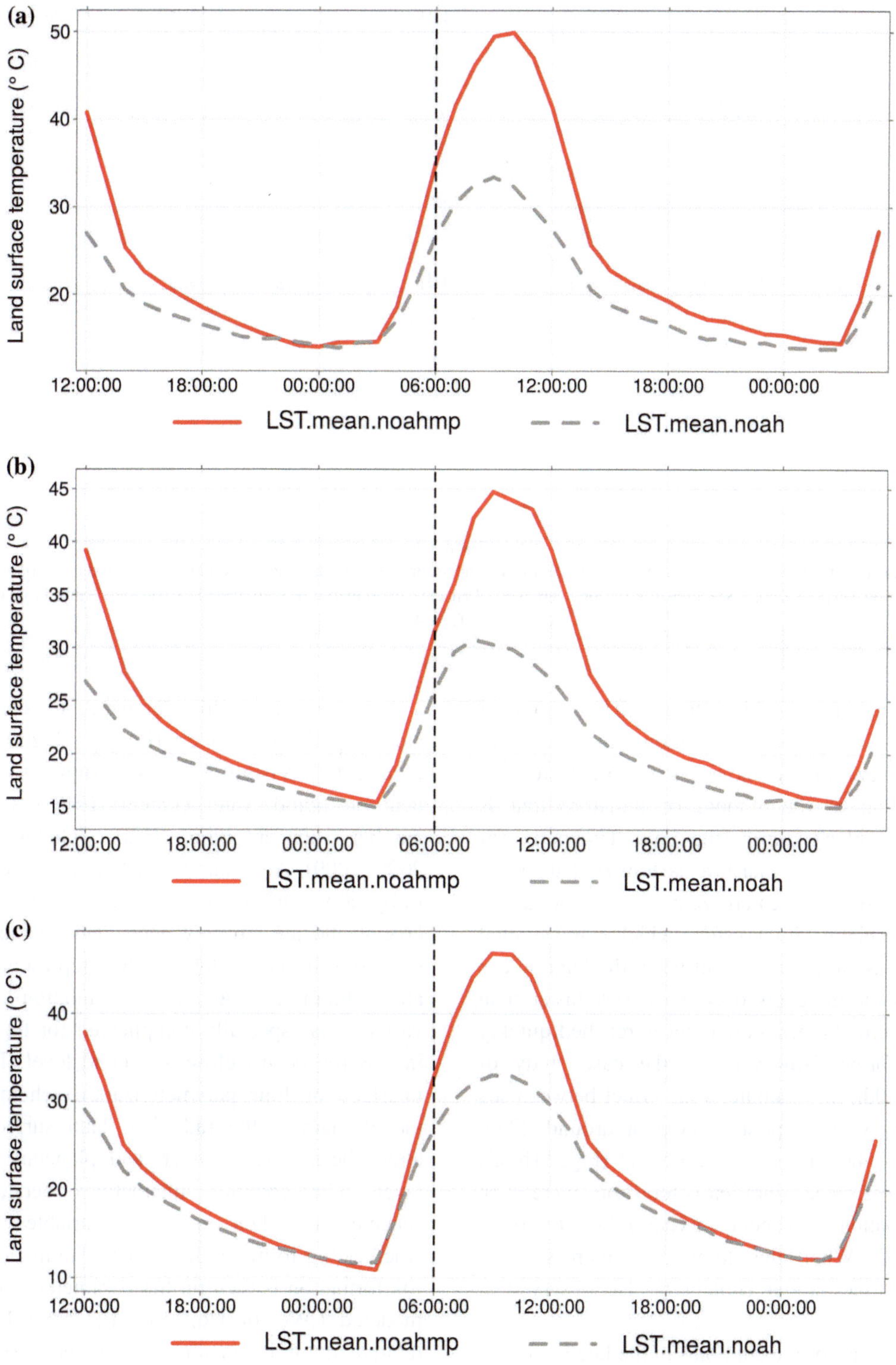

Figure 10
Mean diurnal land surface temperature for all case studies at **a** Abu Dhabi, **b** Dubai, and **c** Al Ain. The black line represents the end of the 24 h forecast and start of the 48 h forecast. Time is in UTC. Site is UTC + 4

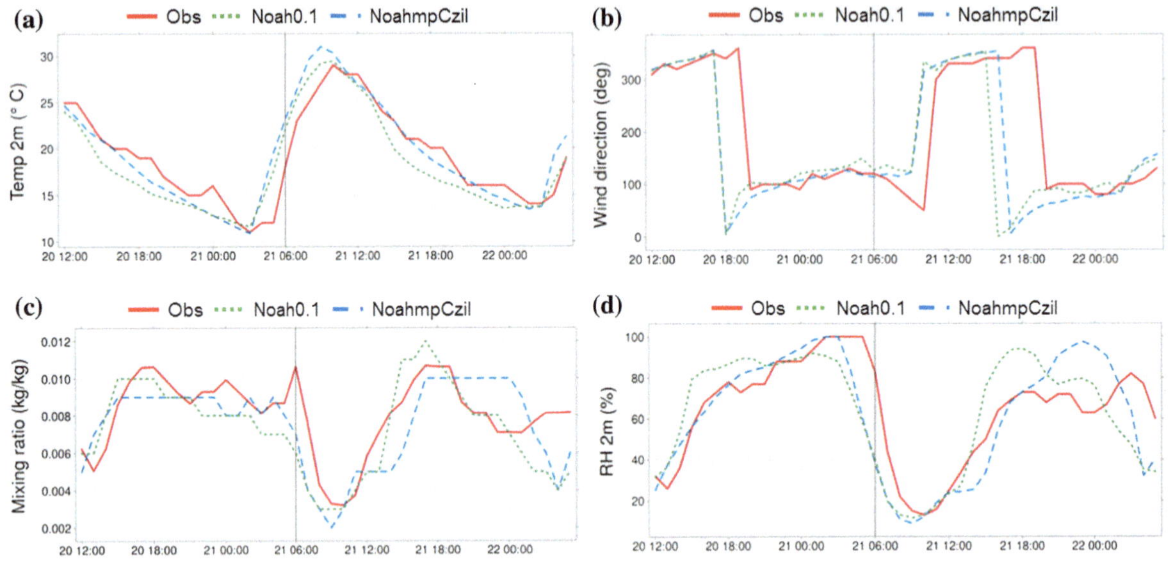

Figure 11

Meteorological parameters for the case study 2016-02-20 for **a** temperature, **b** wind direction, **c** water vapour mixing ratio, and **d** relative humidity at Abu Dhabi Airport. Sea breeze duration increases when the air temperature over land is improved. Time is in UTC. Site is UTC + 4

surface temperature is the other half), our model simulations form part of the solution. The duration of the sea breeze in the model is important for the determination of the amount of moisture that is transported inland during the day. Typically, for observed fog events, moisture is transported inland during the day as the sea breeze blows. This loads the lower atmosphere with moisture, which is transported back towards the coast at night when the land breeze blows. As the air cools over the desert faster than most land-cover types, dew point is reached quickly and fog forms. However, for the case study of February 20th, for example, a lag effect between sea breeze and water vapour content at around 22:00 UTC in the 48 h forecast is observed (Fig. 11b, c). When the sea breeze duration is too short, there is an underestimation in vapour mixing ratio at around 22:00 UTC. When the duration is increased, an increase of the vapour content is also observed at night.

The decrease of the cold bias in the late afternoon and the evening resulted in an improvement in simulating the relative humidity and, therefore, potentially an improvement in the visibility forecast. Visibility is often estimated from relative humidity or a proxy parameter like temperature depression or cloud liquid water content (Bang et al. 2008; Gultepe et al. 2007). Visibility has been shown to be dependent on liquid water content (LWC) and droplet number concentration (Nd) (Gultepe et al. 2006, 2009). An accurate prediction of these meteorological variables that are associated with fog is then crucial and are strongly dependent on the moisture saturation in the model, which is represented by relative humidity. Accurate simulation of these variables is especially complicated for fog forecasting, as fog occurs close to ground level. This is the interface of four parameterization schemes in the model, namely the radiation, land surface model, atmospheric surface layer, and planetary boundary layer. The complex interactions between these schemes and their numerous variables determine whether fog forms in the model. The intensity of the fog formation is then further parameterised based on modeled LWC or Nd. With the control runs, the relative humidity is too high in the early evening which may result in visibility forecasts that are too low for the same period. These high values of the relative humidity and low values of the temperature may result in erroneously low visibility estimates from the model. Thus, the improvement in the timing of the daily maximum relative humidity with the new NoahmpCzil configuration should reduce this error and obtain better visibility estimates.

Finally, the adaptive increase in the Czil value according to land-cover and land-use conditions improved the WRF cold bias during late afternoon and the evening. However, there was no improvement in the early increase of the simulated air temperature and early decrease of the relative humidity with the NoahmpCzil runs. Future work will address the analysis of the diurnal variability of land surface emissivity and its impact on heat exchange. The previous studies showed that desert surface emissivity has an impact of GFS simulations (Yan and Weng, 2011) and showed a significant difference between day and night time surface emissivity estimates (Norouzi et al. 2015). An analysis of the impact of soil moisture on land surface emissivity should be carefully addressed. Reliable estimates of soil moisture in desert are available from several satellite missions (Fares et al. 2013) and could be used to study the impact on the performance of WRF.

5. *Conclusions*

This study focuses on the evaluation of the Noah land surface model within WRF V3.7.1 in the UAE to predict different meteorological variables. Eight dates characterized mostly by fog conditions are simulated to assess the performance of both the default Noah LSM and the Noah-MP with variable thermal roughness length in predicting the air temperature and the relative humidity at 2 m at Abu Dhabi, Dubai, and Al Ain airports.

The findings were as follows:

(a) The modified Noah-MP LSM improved the diurnal temperature profile by decreasing the cold bias. Cold bias is an issue over arid regions and needs to be improved upon in the WRF model. The application of Czil per land-cover type, in combination with the Noah-MP land surface model in WRF, has decreased the mean gross error in 2 m air temperature by up to 1.54 °C. This reduces the cold bias observed in the control run. The improvement is achieved after the daily maximum temperature in the afternoon and early evening for both forecast horizons. The configuration does not solve the rapid heating after sunrise, which suggests that this artifact is controlled by other physical processes in the model.

(b) The modified Noah-MP LSM improved the 2 m relative humidity forecast. The improvement in 2 m temperature mentioned above resulted in a better performance in predicting the relative humidity at 2 m above the ground.

(c) In some cases, the correction to the air temperature above the land resulted in improving the duration of the sea breeze in the model. This is expected as the sea breeze is controlled by the temperature differential of air over the gulf and land.

(d) These results, which represent eight case studies from winter to spring, are encouraging and promote further testing of this configuration over a longer period to determine its validity and application across seasons.

References

Abbs, D. J., & Physick, W. (1992). Sea-breeze observations and modelling: A review. *Australian Meteorological Magazine, 41*, 7–19.

Ajjaji, R., Al-katheri, A. A., & Khaled, A. L. (2008). *Evaluation of United Arab Emirates WRF two-way nested model on a set of thick coastal fog situations*. United Nations: United Nations Development Programme, Air Force and Air Defense Meteorological Department.

Bang, C.-H., Lee, J.-W., & Hong, S.-Y. (2008). Predictability experiments of fog and visibility in local airports over Korea using the WRF model. *Journal of Korean Society for Atmospheric Environment, 24*(E2), 92–101.

Bartok, J., Bott, A., & Gera, M. (2012). Fog prediction for road traffic safety in a coastal desert region. *Boundary Layer Meteorology, 145*(3), 485–506.

Bartoková, I., Bott, A., Bartok, J., & Gera, M. (2015). Fog prediction for road traffic safety in a coastal desert region: Improvement of nowcasting skills by the machine-learning approach. *Boundary Layer Meteorology, 157*(3), 501–516.

Bastidas, L. A., Hogue, T. S., Sorooshian, S., Gupta, H. V., & Shuttleworth, W. J. (2006). Parameter sensitivity analysis for different complexity land surface models using multicriteria methods. *Journal of Geophysical Research Atmospheres, 111*, D20.

Braud, I., Noilhan, J., Bessemoulin, P., Mascart, P., Haverkamp, R., & Vauclin, M. (1993). Bare-ground surface heat and water exchanges under dry conditions: Observations and parameterization. *Boundary Layer Meteorology, 66*(1), 173–200.

Carslaw, D. C., & Ropkins, K. (2012). Openair—an R package for air quality data analysis. *Environmental Modelling and Software, 27–28,* 52–61. https://doi.org/10.1016/j.envsoft.2011.09.008.

Chaouch, N., Temimi, M., Weston, M., & Ghedira, H. (2017). Sensitivity of the meteorological model WRF-ARW to planetary boundary layer schemes during fog conditions in a coastal arid region. *Atmospheric Research, 187,* 106–127.

Chen, F., & Dudhia, J. (2001). Coupling an advanced land surface hydrology model with the Penn State NCAR MM5 modeling system. Part I: Model implementation and sensitivity. *Monthly Weather Review, 129*(4), 569–585.

Chen, F., Janjić, Z., & Mitchell, K. (1997). Impact of atmospheric surface-layer parameterizations in the new land-surface Scheme of the NCEP mesoscale eta model. *Boundary Layer Meteorology, 85,* 391–421.

Chen, Y., Yang, K., He, J., Qin, J., Shi, J., Du, J., et al. (2011). Improving land surface temperature modeling for dry land of China. *Journal of Geophysical Research Atmospheres, 116,* D20.

Chen, Y., Yang, K., Zhou, D., Qin, J., & Guo, X. (2010). Improving the Noah land surface model in arid regions with an appropriate parameterization of the thermal roughness length. *Journal of Hydrometeorology, 11*(4), 995–1006.

Chen, F., & Zhang, Y. (2009). On the coupling strength between the land surface and the atmosphere: From viewpoint of surface exchange coefficients. *Geophysical Research Letters, 36,* 10.

De Villiers, M. P., & Van Heerden, J. (2007). Fog at Abu Dhabi international airport. *Weather, 62*(8), 209–214.

Ek, M. B., Mitchell, K. E., Lin, Y., Rogers, E., Grunmann, P., Koren, V., et al. (2003). Implementation of Noah land surface model advances in the National Centers for Environmental Prediction operational mesoscale Eta model. *Journal of Geophysical Research Atmospheres, 108,* D22.

Fares, A., Temimi, M., Morgan, K., & Kelleners, T. J. (2013). In-situ and remote soil moisture sensing technologies for vadose zone hydrology. *Vadose Zone Journal, 12*(2), vzj2013.03.0058. https://doi.org/10.2136/vzj2013.03.0058.

Gultepe, I., Müller, M. D., & Boybeyi, Z. (2006). A new visibility parameterization for warm-fog applications in numerical weather prediction models. *Journal of Applied Meteorology and Climatology, 45*(11), 1469–1480. https://doi.org/10.1175/jam2423.1.

Gultepe, I., Pearson, G., Milbrandt, J. A., Hansen, B., Platnick, S., Taylor, P., et al. (2009). The fog remote sensing and modeling field project. *Bulletin of the American Meteorological Society, 90*(3), 341–360. https://doi.org/10.1175/2008bams2354.1.

Gultepe, I., Tardif, R., Michaelides, S. C., Cermak, J., Bott, A., Bendix, J., et al. (2007). Fog research: A review of past achievements and future perspectives. *Pure and Applied Geophysics, 164*(6), 1121–1159. https://doi.org/10.1007/s00024-007-0211-x.

Harris, I., Jones, P. D., Osborn, T. J., & Lister, D. H. (2014). Updated high-resolution grids of monthly climatic observations—the CRU TS3.10 Dataset. *International Journal of Climatology, 34*(3), 623–642. https://doi.org/10.1002/joc.3711.

Hogue, T. S., Bastidas, L. A., Gupta, H. V., & Sorooshian, S. (2006). Evaluating model performance and parameter behavior for varying levels of land surface model complexity. *Water Resources Research, 42,* 8.

Hogue, T. S., Bastidas, L., Gupta, H., Sorooshian, S., Mitchell, K., & Emmerich, W. (2005). Evaluation and transferability of the Noah land surface model in semiarid environments. *Journal of Hydrometeorology, 6*(1), 68–84.

Hong, S.-Y., Dudhia, J., & Chen, S.-H. (2004). A revised approach to ice microphysical processes for the bulk parameterization of clouds and precipitation. *Monthly Weather Review, 132*(1), 103–120.

Huang, X., Wang, W., & Powers, J. (2008). A description of the advanced research WRF version 3. *NCAR technical note, NCAR, BoulderSpeer M, Wiles P, Pepler A (2009).*

Iacono, M. J., Delamere, J. S., Mlawer, E. J., Shephard, M. W., Clough, S. A., & Collins, W. D. (2008). Radiative forcing by long-lived greenhouse gases: Calculations with the AER radiative transfer models. *J. Geophys. Res, 113,* D13.

Kain, J. S. (2004). The Kain–Fritsch convective parameterization: An update. *Journal of Applied Meteorology, 43*(1), 170–181.

Kain, J. S., & Fritsch, J. M. (1990). A one-dimensional entraining/detraining plume model and its application in convective parameterization. *Journal of the Atmospheric Sciences, 47*(23), 2784–2802.

LeMone, M. A., Tewari, M., Chen, F., Alfieri, J. G., & Niyogi, D. (2008). Evaluation of the Noah land surface model using data from a fair-weather IHOP_2002 day with heterogeneous surface fluxes. *Monthly Weather Review, 136*(12), 4915–4941.

Liang, X., Wood, E. F., Lettenmaier, D. P., Lohmann, D., Boone, A., Chang, S., et al. (1998). The Project for Intercomparison of Land-surface Parameterization Schemes (PILPS) phase 2 (c) Red-Arkansas River basin experiment: 2. Spatial and temporal analysis of energy fluxes. *Global and Planetary Change, 19*(1), 137–159.

Mitchell, K., Ek, M., Wong, V., Lohmann, D., Koren, V., Schaake, J., et al. (2005) 'Noah Land Surface Model (LSM) User's Guide' *NCAR Research Application Laboratory (RAL).* pp. 1–26.

Mlawer, E. J., Taubman, S. J., Brown, P. D., Iacono, M. J., & Clough, S. A. (1997). Radiative transfer for inhomogeneous atmospheres: RRTM, a validated correlated-k model for the longwave. *Journal of Geophysical Research Atmospheres, 102*(D14), 16663–16682.

Niu, G. Y., Yang, Z. L., Mitchell, K. E., Chen, F., Ek, M. B., Barlage, M., et al. (2011). The community Noah land surface model with multiparameterization options (Noah-MP): 1. Model description and evaluation with local-scale measurements. *Journal of Geophysical Research Atmospheres, 116*(12), 1–19.

Norouzi, H., Temimi, M., Prigent, C., Turk, J., Khanbilvardi, R., Tian, Y., et al. (2015). Assessment of the consistency among global microwave land surface emissivity products. *Atmospheric Measurement Techniques, 8*(3), 1197–1205. https://doi.org/10.5194/amt-8-1197-2015.

Pitman, A. J. (2003). The evolution of, and revolution in, land surface schemes designed for climate models. *International Journal of Climatology, 23*(5), 479–510.

Pitman, A. J., Henderson-Sellers, A., Desborough, C. E., Yang, Z.-L., Abramopoulos, F., Boone, A., et al. (1999). Key results and implications from phase 1 (c) of the project for intercomparison of land-surface parametrization schemes. *Climate Dynamics, 15*(9), 673–684.

Skamarock, W. C., Klemp, J. B., Dudhia, J., Gill, D. O., Barker, D. M., Duda, M. G., et al. (2008). *A description of the advanced research WRF version 3*. Boulder: NCAR.

Sukoriansky, S., Galperin, B., & Perov, V. (2005). Application of a new spectral theory of stably stratified turbulence to the atmospheric boundary layer over sea ice. *Boundary Layer Meteorology, 117*(2), 231–257.

Tewari, M., Chen, F., Wang, W., Dudhia, J., LeMone, M. A., Mitchell, K., et al. (2004). Implementation and verification of the unified NOAH land surface model in the WRF model. *20th conference on weather analysis and forecasting/16th conference on numerical weather prediction.*

Timbal, B., & Henderson-Sellers, A. (1998). Intercomparisons of land-surface parameterizations coupled to a limited area forecast model. *Global and Planetary Change, 19*(1), 247–260.

Wang, W., Bruyère, C., Duda, M., Dudhia, J., Gill, D., Lin, H. C., et al. (2009). Advanced research WRF (ARW) version 3 modeling users guide, mesoscale & microscale meteorology division. *National Center for Atmospheric Research (NCAR), USA*. http://www.mmm.ucar.edu.ar/wrf/users/docs/arw_v3.pdf.

Wei, H., Xia, Y., Mitchell, K. E., & Ek, M. B. (2013). Improvement of the Noah land surface model for warm season processes: Evaluation of water and energy flux simulation. *Hydrological Processes, 27*(2), 297–303.

Wood, E. F., Lettenmaier, D. P., Liang, X., Lohmann, D., Boone, A., Chang, S., et al. (1998). The project for intercomparison of land-surface parameterization schemes (PILPS) Phase 2 (c) Red–Arkansas River basin experiment: 1. Experiment description and summary intercomparisons. *Global and Planetary Change, 19*(1), 115–135.

Yan, B., & Weng, F. (2011). Effects of microwave desert surface emissivity on AMSU-A data assimilation. *IEEE Transactions on Geoscience and Remote Sensing, 49*(4), 1263–1276. https://doi.org/10.1109/TGRS.2010.2091508.

Yang, K., Koike, T., Ishikawa, H., Kim, J., Li, X., Liu, H., et al. (2008). Turbulent flux transfer over bare-soil surfaces: Characteristics and parameterization. *Journal of Applied Meteorology and Climatology, 47*(1), 276–290.

Zheng, W., Wei, H., Wang, Z., Zeng, X., Meng, J., Ek, M., et al. (2012). *Journal of Geophysical Research Atmospheres, 117,* D6.

Zilitinkevich, S. S. (1995). *Air Pollution III—volume I Air pollution theory and simulation: Pollution dispersion aspects of coherent structure of convective flows*. Boston: Computational Mechanics Publ.

(Received November 29, 2017, revised April 4, 2018, accepted May 9, 2018, Published online June 5, 2018)

Pure Appl. Geophys. 176 (2019), 2139–2164

https://doi.org/10.1007/s00024-018-1770-8

Pure and Applied Geophysics

Numerical Simulations of an Inversion Fog Event in the Salt Lake Valley during the MATERHORN-Fog Field Campaign

CATHERINE N. CHACHERE[1] and ZHAOXIA PU[1]

Abstract—An advanced research version of the Weather Research and Forecasting (WRF) Model is employed to simulate a wintertime inversion fog event in the Salt Lake Valley during the Mountain Terrain Atmospheric Modeling and Observations Program (MATERHORN) field campaign during January 2015. Simulation results are compared to observations obtained from the field program. The sensitivity of numerical simulations to available cloud microphysical (CM), planetary boundary layer (PBL), radiation, and land surface models (LSMs) is evaluated. The influence of differing visibility algorithms and initialization times on simulation results is also examined. Results indicate that the numerical simulations of the fog event are sensitive to the choice of CM, PBL, radiation, and LSM as well as the visibility algorithm and initialization time. Although the majority of experiments accurately captured the synoptic setup environment, errors were found in most experiments within the boundary layer, specifically a 3° warm bias in simulated surface temperatures compared to observations. Accurate representation of surface and boundary layer variables are vital in correctly predicting fog in the numerical model.

Key words: Fog, a wintertime inversion, WRF model, numerical simulations, Salt Lake City.

1. *Introduction*

Inversion fog is a wintertime fog that forms when polar air stagnates in a valley under the dominance of a persistent high-pressure system (Taylor 1941). It is different from typical valley radiation fog as warm air advection associated with the high-pressure system creates a subsidence inversion that acts as a lid to keep the fog in a limited vertical expanse. This type of fog is common in the western United States and central Europe during the winter (Byers 1959). The topography of a region is important in the development and persistence of this phenomenon. A basin with limited drainage creates the ideal situation for the formation of inversion fog, as the cold air mass near the surface can remain in place.

Located in the Salt Lake Valley (Fig. 1), Salt Lake City is an ideal basin for inversion fog to form. It is surrounded by mountains except on the northwestern side where the Great Salt Lake is located. Under the dominance of a high-pressure system, the prevailing wind direction is from the northwest, channeling polar air over the Great Salt Lake and into the Salt Lake Valley. During the cold season, cold air pools and persistent inversions are often observed in Salt Lake City and its surrounding valleys (Lareau et al. 2013). According to Chachere and Pu (2016), there exists a statistically significant connection between cold air pools and inversion fog events. Therefore, the highest frequency of fog seen in Salt Lake City is in the cold season, especially during the December and January months (Hodges and Pu 2015).

Due to the large social, economic, and aviation impacts fog can cause, accurate prediction of the fog is very important. It is especially essential for Salt Lake City because it is the home of the Salt Lake International Airport as well as a populous of over one million residents. However, thus far, there have not been many studies to evaluate whether numerical weather prediction can accurately predict fog events (Zhou et al. 2012; Pu et al. 2016). Very few studies emphasize inversion fog events, especially over the Salt Lake Valley. Meanwhile, challenges of fog forecasting with the aid of numerical models have been documented for over 40 years (Zdunkowski and Nielsen 1969; Zdunkowski and Barr 1972; Brown

Submitted as an article contributed to a special issue on *"Aviation Meteorology: Observations and Models"*.

[1] Department of Atmospheric Sciences, University of Utah, 135 S 1460 E, Rm. 819, Salt Lake City, Utah 84112, USA. E-mail: Zhaoxia.Pu@utah.edu

Figure 1
A satellite image of the Salt Lake Valley taken 09 April 2013. The Great Salt Lake is visible in the upper left, the Wasatch range along the right center, Oquirrh range along the left, and the Traverse are seen separating the Salt Lake Valley in the center from the Utah Valley at the bottom center. The MATERHORN-Fog sites are also marked by four-point stars, yellow for the Salt Lake Valley site and orange for the Heber Valley site

and Roach 1976; Turton and Brown 1987; Dyunkerke 1990; Van der Velde et al. 2010; Steeneveld et al. 2014).

Unlike most meteorological processes, fog is classified as a "threshold phenomenon" (Zhou and Ferrier 2008). This means that a certain threshold must be met in order for the phenomenon to be represented. In this case, it is the proper near-surface atmospheric conditions that result in surface visibilities less than 1 km, the definition of a fog event according to the American Meteorological Society (2012). Therefore, fog forecasting requires and adequate representation of the boundary layer and near-surface variables, which in turn requires the ability of numerical results to represent radiative cooling, turbulence, stratified layers, land surface processes, and advection (Bergot and Guédalia 1994). However, these near-surface atmospheric variables have notoriously been incorrectly forecast by numerical weather prediction (e.g., Liu et al. 2008; Zhang et al. 2013). Due to a lack of intensive observations in the boundary layer, few studies have examined how these errors in near-surface temperature, relative humidity, etc., can affect fog prediction.

To improve the understanding of complex-terrain meteorology over a wide range of scales, topographic features, and driving mechanisms, the Mountain Terrain Atmospheric Modeling and Observations Program (MATERHORN) (Fernando et al. 2015) was conducted in northern Utah. As part of the MATERHORN program, the MATERHORN-Fog field project was conducted during 07 January–01 February 2015 (Gultepe et al. 2016). The overall objective of the MATERHORN-Fog campaign is to improve the understanding of mechanisms associated with the formation, evolution, and dissipation of fog over complex-terrain. Over the course of the campaign, 11 intensive observational periods (IOPs) were performed, 10 in January and one in February of 2015 over 4 major fog events. These IOPs were performed in two northern Utah valleys: the Salt Lake Valley and the Heber Valley. Two sites were chosen to better understand the effects of urbanization and localized topography on the formation and dissipation of fog. The locations of these two sites are marked on Fig. 1.

During the MATERHORN-Fog field observations, real-time forecasts conducted with an advanced research version of the Weather Research and Forecasting Model (WRF) (Skamarock et al. 2008) were employed (Pu et al. 2016) to aid planning operations, as well as the evaluation of the performance of the WRF model in capturing these inversion fog events. Early evaluation by Pu et al. (2016) showed an inability of the WRF model to produce correct forecasts of near-surface atmospheric variables. Specifically, it was found that the WRF could capture the elevated subsidence inversion commonly seen with inversion fog events, but errors in temperature, dew point, and wind speed were seen in the boundary layer, especially in the near-surface variables. This data was not evaluated on a case-by-case basis in Pu et al. (2016). In addition, although Pu et al. (2016) speculated that uncertainties in model physical parameterizations (e.g., cloud microphysics, land surface processes, and boundary layer processes, etc.) could be reasons for inaccurate prediction of fog events, sensitivity studies had not yet been conducted to investigate which processes played key roles in fog prediction. As a continuation of Pu et al. (2016), in this paper we emphasize a 2-day inversion fog case over Salt Lake City during the MATERHORN-Fog field project with detailed sensitivity studies using different model configurations and physical parameterization schemes. Leveraged by the intense observations obtained during the field program and the sensitivity studies made with the WRF model, we attempt to explore the factors and processes that could contribute to or limit our ability to construct an accurate numerical simulation of inversion fog.

Section 2 describes the synoptic conditions which contributed to the 2-day fog event seen in IOPs 1 and 2 during the MATERHORN-Fog field program. The numerical model and experiment design are discussed in Sect. 3. Results from the examination of sensitivity to cloud microphysical (CM), planetary boundary layer (PBL), radiation, land surface models (LSM), as well as visibility algorithms and initialization times, are presented and discussed in Sect. 4. A summary and concluding remarks comprise Sect. 5.

2. *An Overview of Inversion Fogs during IOPs 1 and 2*

The first and second IOP (IOPs 1 and 2) from the MATERHORN-Fog campaign took place between 1500 mountain standard time (MST) 08 January 2015 and 1200 MST 09 January 2015, and 1400 MST 09 January 2015 and 1200 MST 10 January 2015, respectively. The field campaign implemented radiosondes, SODAR, and LIDAR instruments at the Salt Lake site, which is close enough (15 km) to the Salt Lake International Airport (KSLC) for METAR observations and NWS soundings to be sufficient proxy for current conditions in addition to MATERHORN-specific observations. Surface mesonet observing stations from the University of Utah's MesoWest network (Horel et al. 2002) provided data real-time from the area. Figure 2 presents a time series of visibility, temperature, and pressure at KSLC for the entirety of IOPs 1 and 2. Two periods of fog occur during these IOPs, both with short durations (0300 MST 09 January–0400 MST 09 January and 1030 MST 10 January–1130 MST 10 January). Considering the challenges associated with predicting fog events with a numerical model and to offer additional time windows to verify fog onset and dissipation, this paper focuses on "near fog" conditions instead of the rigid definition set forth by the AMS glossary. The concept of "Near fog" was

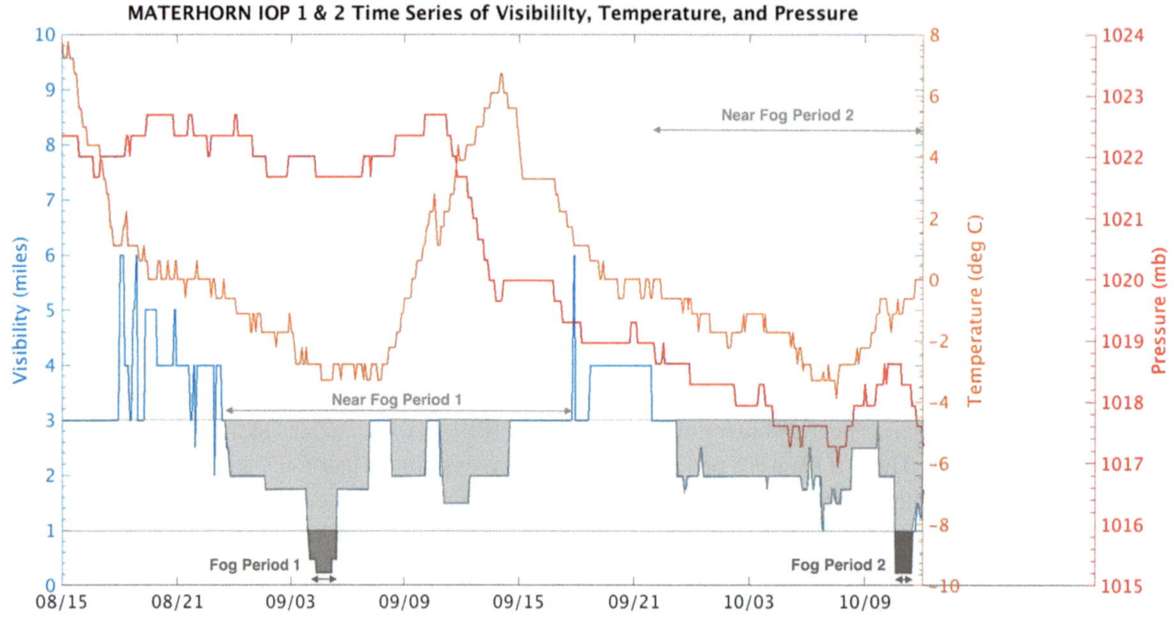

Figure 2
A time series of visibility (blue), temperature (orange), and pressure (red) for Salt Lake City for the entirety of MATERHORN-Fog IOPs 1 and 2 (1500 MST 08 January to 1200 MST 09 January, and 1400 MST 09 January to 1200 MST 10 January, respectively). Two periods of near fog are indicated with light gray arrows and shading, Near Fog Period 1 from 0000 MST to 1800 MST 09 January and Near Fog Period 2 from 2200 MST 09 January to 1200 MST 10 January. Two periods of fog are indicated by dark gray arrows and shading, Fog Period 1 from 0400 MST to 0530 MST 09 January and Fog Period 2 from 1030 MST to 1130 MST 10 January

mentioned in previous studies such as Bergot et al. (2007), Ryerson and Hacker (2014) and Hodges and Pu (2015). It was defined by Hodges and Pu (2015) as a period of visibility below 3 miles coincident with relative humidity values of 70% which lasts for 2 or more hours. Based on this definition, "near fog" occurred during these IOPs with Near Fog Period 2 from 0000 MST to 1800 MST 09 January and Near Fog Period 2 from 2200 MST 09 January to 1200 MST 10 January 2015. IOPs 1 and 2 were conducted in a 2-day fog event, beginning 08 January 2015. Since AMS defines the fog condition as the surface visibility less than 1 km, we also use 1 mile (comparing to 3 miles as it is closer to 1 km) as a reference "near fog" condition in visibility plots presented in this paper.

Detailed synoptic conditions for IOP 1 and 2 can be found in Pu et al. (2016). Outstanding large-scale conditions that are favorable for fog formation during IOP 1 include propagation of a longwave ridge at 500 hPa from the western United States and the advection of cooler temperature from a relative temperature minimum in southeastern Oregon at 850 hPa. IOP 2 began with a 500 hPa ridge aligned in much the same fashion. At 250 hPa, the strongest winds were downstream of the ridge, allowing it to remain relatively stationary and for the central pressure to build. The center of anticyclonic rotation was over Nevada at 1200 UTC 09 January at 500 hPa, leaving advection into this level unperturbed for 24 h. The sounding (Fig. 3a) shows a saturated surface as fog had formed in the valley within the previous diurnal cycle. This can be seen as part of Near Fog Period 1 in Fig. 2. More importantly, it shows a

Figure 3 ▶
Sounding from Salt Lake City International Airport (KSLC) at **a** 1200 UTC 09 January 2015 and **b** 0000 UTC 10 January 2015. Temperature (F) along *x*-axis and pressure (hPa) along *y*-axis. Temperature (°C) (straight beige lines), dry adiabats (curved beige lines), moist adiabats (curved green lines), saturation mixing ratio lines dotted straight green lines, and wind barbs (speed in knots) analyzed. Sounding temperature is represented as the solid black line and sounding dew point temperature is represented as the dotted black line

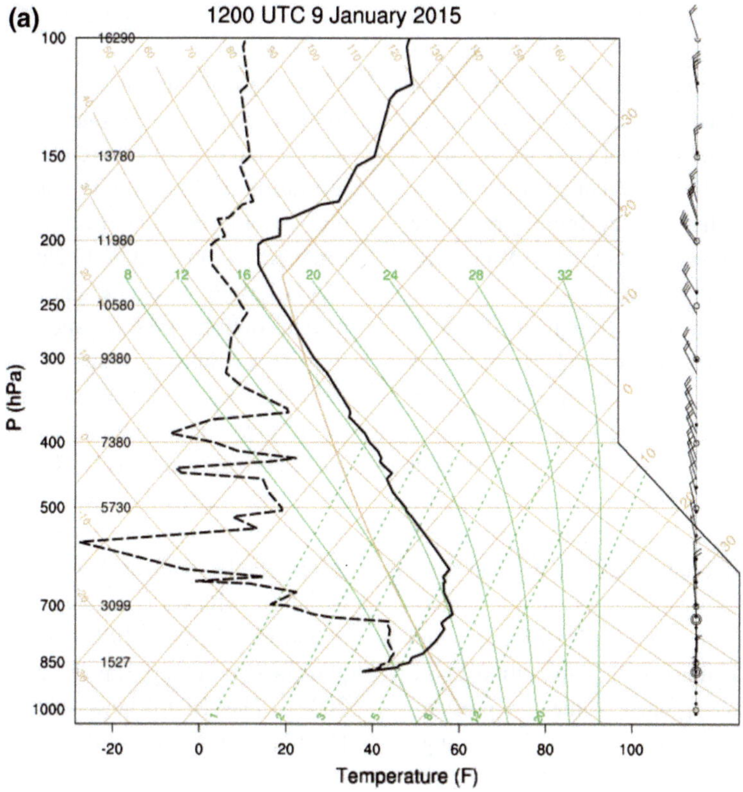
(a)
1200 UTC 9 January 2015
P (hPa)
100
150
200
250
300
400
500
700
850
1000
16290
13780
11960
10580
9380
7380
5730
3099
1527
8
12
16
20
24
28
32
-20
0
20
40
60
80
100
Temperature (F)

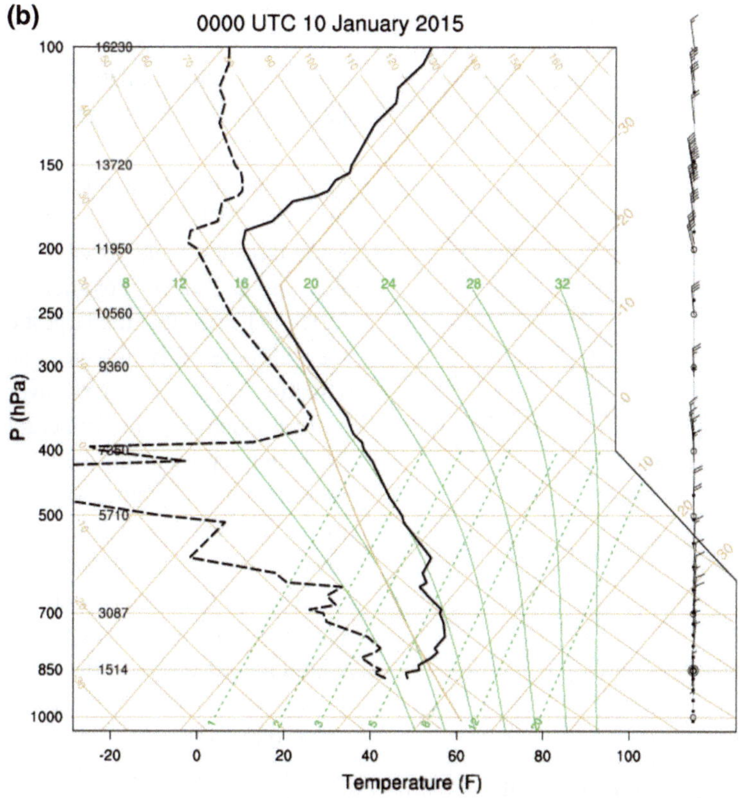
(b)
0000 UTC 10 January 2015
P (hPa)
100
150
200
250
300
400
500
700
850
1000
16230
13720
11950
10560
9360
5710
3087
1514
8
12
16
20
24
28
32
-20
0
20
40
60
80
100
Temperature (F)

decoupled wind profile between the surface air mass and that above terrain height (~ 2200 m). A decoupling can be identified by low-level winds that are independent of speed and direction for the air above them, a feature that is often coincident with a temperature inversion. This is indicative of a cold air pool, an environment that is very important to the evolution of fog (Wolyn and McKee 1989; Whiteman et al. 2001; Chachere and Pu 2016).

By 0000 UTC 10 January, a trough upstream of the dominating longwave ridge had reached the Pacific coast at 500 hPa. Valley cold pools are often broken by strong trough passages, making this a strong indicator that the existing fog environment could soon be destroyed. However, with dropping temperatures associated with sunset and the inversion still present in the sounding (Fig. 3b), fog can once again be expected overnight throughout the end of IOP 2. The next 24 h, from 1200 UTC 10 January and into 0000 UTC 11 January 2015, the atmosphere began mixing with the approach of the low-pressure trough at 500 hPa (figure not shown). The trough passed over northern Utah between 1800 UTC and 2300 UTC, but it was difficult to pinpoint an exact time of passage because there was never a complete dissolution of the stable air mass in the valley.

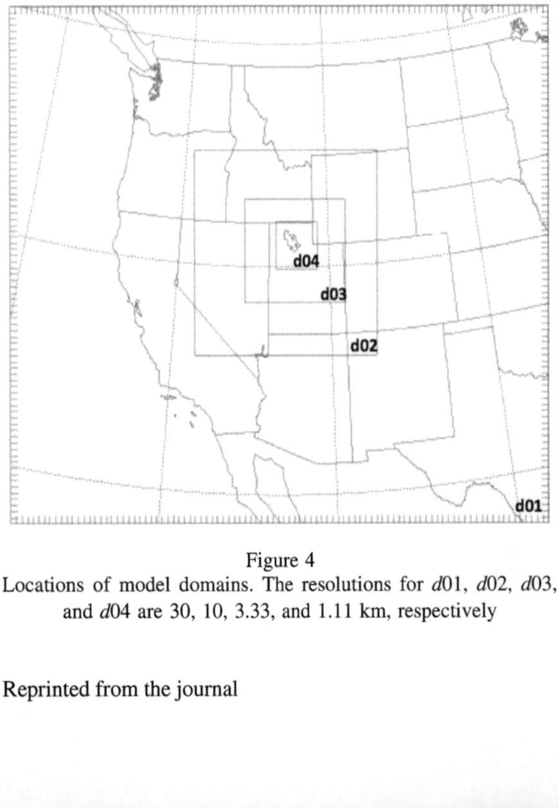

Figure 4
Locations of model domains. The resolutions for *d*01, *d*02, *d*03, and *d*04 are 30, 10, 3.33, and 1.11 km, respectively

3. *Simulation Design*

Numerical simulations were performed using WRF Version 3.7 (Skamarock et al. 2008). Four one-way nested domains were used with a horizontal grid resolution of 30, 10, 3.33, and 1.11 km (Fig. 4). The high-resolution domain (*d*04, 1.11 km resolution) is centered around the Great Salt Lake and the MATERHORN-Fog observation sites. The model physics options are the same for every domain except for the cumulus scheme, where the Kain–Fritsch parameterization (Kain and Fritsch 1990) is only used for the two outer domains *d*01 and *d*02, at 30 and 10 km grids, respectively. The control run, following Pu et al. (2016), is set up with the Thompson microphysics scheme (Thompson et al. 2008), the Rapid Radiative Transfer Model (RRTM) longwave radiation (Mlawer et al. 1997) and Dudhia shortwave radiation (Dudhia 1989) schemes, and the YSU planetary boundary layer scheme (Hong and Lim 2006). The Noah land surface model (Chen and Dudhia 2001) was used, as it is an advanced land surface model that can predict surface temperature, soil moisture, and temperature in multiple soil layers. The NCEP North American Model (NAM) analysis was used as the initial and boundary conditions. A 48-h simulation was conducted and was initialized at 1200 UTC 08 January.

A real-time WRF forecast was conducted during the MATERHORN-Fog field campaign, with the same model settings as described above and 41 vertical sigma levels. Post-field evaluation (Pu et al. 2016) indicated that the largest errors that occurred in the real-time WRF forecasts are due to a misrepresentation of the boundary layer, especially near the surface (such as 2-m temperature and 10-m winds). To better represent the vertical scale of valley phenomena, in this study we changed the number of model vertical levels from 41 to 60. This included lowering the height of the bottom level of the model (~ 18 m in the current control run versus ~ 30 m in the real-time WRF simulations) and placing the additional 19 levels in the planetary boundary layer. Figure 5 compares the near-surface temperature (a), relative humidity (b), and wind speed (c) from model forecasts with tethersonde balloon measurements taken during MATERHORN-Fog IOP 2 at the Heber

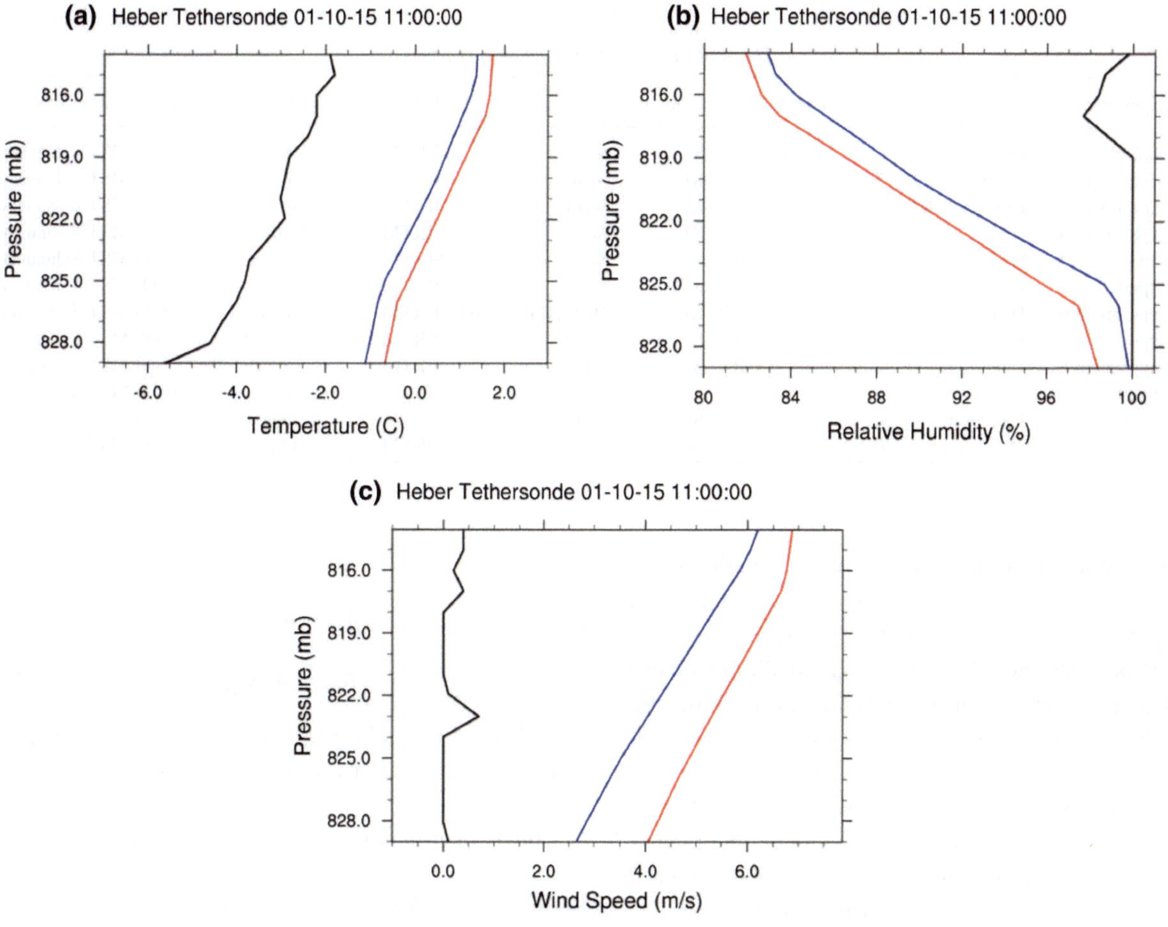

Figure 5
Comparison of tethersonde balloon profiles taken during MATERHORN-Fog IOP 2 (black line) of near-surface **a** temperature, **b** relative humidity, and **c** wind speed with model forecasts made using 41 vertical levels (red line) and 60 vertical levels (blue line)

site (the only tethersonde balloon site) at 0900 UTC 10 January 2015. It clearly illustrates that the simulation with 60 vertical level resolution more accurately represents the lowest levels of the model domain. The results motivate us to set the vertical resolution for all experiments in this study to 50 vertical levels.

In addition, Pu et al. (2016) speculated that uncertainties in model physical parameterizations could be responsible for inaccurate prediction of MATERHORN-Fog events. Therefore, in addition to the control simulation, sensitivity of numerical simulations to various physical parameterization will be conducted for the fog events during IOP 1 and 2. Since the WRF model has multiple options for each type of physical parameterization scheme, to investigate the sensitivity of numerical prediction of inversion fog to various physical processes and to understand which processes might contribute to the fog development and evolution, a total of twelve (including control; see Table 1) sensitivity experiments were conducted to compare results against the control described above. For each sensitivity experiment, only one type of parameterization varies from the control simulation in the WRF model. Specifically, sensitivity to CM, PBL, radiation, and LSM schemes was tested.

The real-time WRF forecasts also show that the accurate simulation of fog events is sensitive to the forecast lead time (Pu et al. 2016). Varying forecast

Table 1

The configuration of sensitivity experiments

Experiment	Microphysics	Climatology	PBL	Cumulus	Land surface	LW radiation	SW radiation	Initialization time
Control	Thompson	–	YSU	KF	Noah	RRTM	Dudhia	0500 MST 8 January
Experiment 1	Thompson AA	True	YSU	KF	Noah	RRTM	Dudhia	0500 MST 8 January
Experiment 2	Thompson AA	False	YSU	KF	Noah	RRTM	Dudhia	0500 MST 8 January
Experiment 3	Lin	–	YSU	KF	Noah	RRTM	Dudhia	0500 MST 8 January
Experiment 4	Morrison	–	YSU	KF	Noah	RRTM	Dudhia	0500 MST 8 January
Experiment 5	Thompson	–	MYJ	KF	Noah	RRTM	Dudhia	0500 MST 8 January
Experiment 6	Thompson	–	YSU	KF	Thermal diffusion	RRTM	Dudhia	0500 MST 8 January
Experiment 7	Thompson	–	YSU	KF	Noah	RRTM	Goddard	0500 MST 8 January
Experiment 8	Thompson	–	YSU	KF	Noah	RRTM	RRTMG	0500 MST 8 January
Experiment 9	Thompson	–	YSU	KF	Noah	Goddard	Dudhia	0500 MST 8 January
Experiment 10	Thompson	–	YSU	KF	Noah	GFDL	Dudhia	0500 MST 8 January
Experiment 11	Thompson	–	YSU	KF	Noah	RRTM	Dudhia	1700 MST 8 January

lead times could produce different simulation results for the fog event regarding its onset and duration. Recent study by Lin et al. (2017) suggested a tread-off between more accurate initial conditions (a short lead time) and a proper (long) forecast lead time that allows model physical processes to spin up adequately during the fog simulation. Therefore, we performed experiments with two different forecast lead times. Table 1 summarizes all configurations of sensitivity experiments. Details are also given in Sect. 4 with the discussion of the results.

To predict fog, a visibility algorithm is necessary. Two such algorithms are used and compared in this study. The first algorithm is based on Stoelinga and Warner (1999). This algorithm is based on the extinction of five types of hydrometeors: cloud water (β_{cw}), rain water (β_{rw}), cloud ice (β_{ci}), snow (β_{sn}), and graupel (β_{gp}). Henceforth, this will be called the GSD parameterization as the algorithm was further developed by the Global Systems Division (GSD), ESRL/NOAA. It is presented as follows:

$$\beta = \beta_{cw} + \beta_{rw} + \beta_{ci} + \beta_{sn} + \beta_{gp} + 10^{-10}, \quad (1)$$

where β is the extinction coefficient for the five water species, calculated:

$$\beta_{cw} = 144.7C_{cw}^{0.88}, \quad (2)$$

$$\beta_{rw} = 2.24C_{rw}^{0.75}, \quad (3)$$

$$\beta_{ci} = 327.8C_{ci}, \quad (4)$$

$$\beta_{sn} = \alpha_{sn} \times C_{sn}, \quad (5)$$

$$\beta_{gp} = 4.0C_{gp}^{0.75}, \quad (6)$$

with C_{cw}, C_{rw}, C_{ci}, C_{sn}, and C_{gp} are the associated mass concentrations (g/m^3) for cloud water, rain water, cloud ice, snow, and graupel, respectively. This is then inputted into a visibility format:

$$\mathrm{VIS_{hydro}} = \min\left(90., \frac{\alpha_{vis}}{\beta}\right), \quad (7)$$

where

$$\alpha_{vis} = -\ln(0.02) \quad (8)$$

by Kunkel (1984). The extinction coefficient of snow (α_{sn}) varies with temperature as:

$$\alpha_{sn} = 10. \times (1 - \alpha_{temp}) + 6.0\alpha_{temp}, \quad (9)$$

where

$$0 < \alpha_{temp} < 1, \quad \text{if } 0 < \text{Temp} < 1, \quad (10)$$

$$\alpha_{temp} = 0, \quad \text{if Temp} < 1, \quad (11)$$

$$\alpha_{temp} = 1, \quad \text{if Temp} \geq 1. \quad (12)$$

The impact of night and day is also taken into effect, and the visibility is modified as:

$$\mathrm{VIS_{hydro\_n}} = 1.69 \times 1.609 \times \left(\frac{\mathrm{VIS_{hydro}}}{1.609}\right)^{0.86}, \quad (13)$$

$$\mathrm{VIS}_{\mathrm{hydro\_}dn} = \alpha_{dn} \times \mathrm{VIS}_{\mathrm{hydro}} + (1 - \alpha_{dn}) \times \mathrm{VIS}_{\mathrm{hydro\_}n}, \quad (14)$$

where α_{dn} is 0 for night hours and 1 for day hours. $\mathrm{VIS}_{\mathrm{hydro}}$ is then compared to a relative humidity visibility parameterization:

$$\mathrm{VIS}_{\mathrm{rh}} = 60.0 \times \exp(-2.5 \times q_{\mathrm{rh}}), \quad (15)$$

where

$$q_{\mathrm{rh}} = \min\left(0.8, \left(\frac{\mathrm{rh}}{100. - 0.15}\right)\right) \quad (16)$$

with rh interpolated from the highest values of the lowest four model levels. When $\mathrm{VIS}_{\mathrm{rh}}$ is less than 10 km, the effect of low-level wind shear between the lowest four levels increasing from 4–6 m/s is taken into account. This modification of visibility is:

$$\mathrm{VIS}_{\mathrm{rh}}(\mathrm{ws}) = \mathrm{VIS}_{\mathrm{rh}} + \alpha_{\mathrm{ws}} \times (10 - \mathrm{VIS}_{\mathrm{rh}}), \quad (17)$$

where α_{ws} is the impact factor of wind shear ranging from 0 to 1, 0 for wind shear equaling 4 m/s and 1 for wind shear over 6 m/s. After this computation is complete, the simplified form of the parameterization can be computed:

$$\mathrm{VIS}_{\mathrm{GSD}} = \min(\mathrm{VIS}_{\mathrm{rh}}, \mathrm{VIS}_{\mathrm{hydro\_}dn}). \quad (18)$$

Since the model commonly does not predict low clouds during the winter over Salt Lake Valley, the GSD algorithm may not be best for all cases. To test the sensitivity of fog prediction to different visibility algorithms, another method was used and compared. This algorithm is based on Gultepe and Milbrandt (2009), referred to as the RH visibility algorithm, and the parameterization is as follows:

$$\mathrm{VIS}_{\mathrm{RH}_{5\%}} = -0.000114\mathrm{RH}^{2.70} + 27.45. \quad (19)$$

Relative humidity values used for this parameterization were corrected, meaning they are calculated differently based on relative humidity to water at temperature above 0.5 °C and relative humidity with respect to ice at temperatures below this value. Instead of choosing 0 °C as the threshold, this study chooses 0.5 °C due to the nature of the temperature in this time period to hover around freezing. This calculation will include all borderline values.

Results from these two visibility algorithms are analyzed and compared to observations during MATERHORN-Fog IOPs 1 and 2.

4. *Evaluation of Simulation Results*

4.1. *Control Simulation Versus Observations*

4.1.1 *Synoptic Conditions*

Figure 6a–f compares synoptic maps valid during the MATERHORN-Fog experiment with the WRF-generated control experiment solution valid at 1200 UTC 09 January 2015 at 250 hPa (a and d), 500 hPa (b and e), and 700 hPa (c and f) pressure levels. As discussed in the synoptic overview, the most important features to be captured are the longwave ridges seen at each pressure level along with their associated high height features. Dashed orange lines and blue "H"s have been drawn on each figure to better see their placement. After a close comparison, it is noted that the WRF model captures the synoptic setup quite well. At each level, agreement is present in height, ridge placement, wind speed and direction, temperature, and relative humidity values are seen. These results are consistent regardless of sensitivity experiment or lead time, indicating that the WRF model has sufficiently represented the synoptic background for IOPs 1 and 2.

The only discrepancies can be found at the 700 hPa level, often at the top or near the top of the boundary layer throughout the entirety of the experiment. Although the errors are slight (temperatures about 2° warmer than observed), they should be associated with simulation errors in the boundary layer. Pu et al. (2016) indicated apparent errors in these layers with the real-time WRF forecasts. In the following sections, further evaluation in the boundary layer is conducted by comparing simulation results from various experiments with observations during MATERHORN-Fog.

4.1.2 *Boundary Layer Conditions*

Figures 7, 8 and 9 illustrate the results of the control simulation for the Salt Lake Valley site. Figure 7 compares the sounding profile from the control simulation and the National Weather Service sounding released at the Salt Lake City International

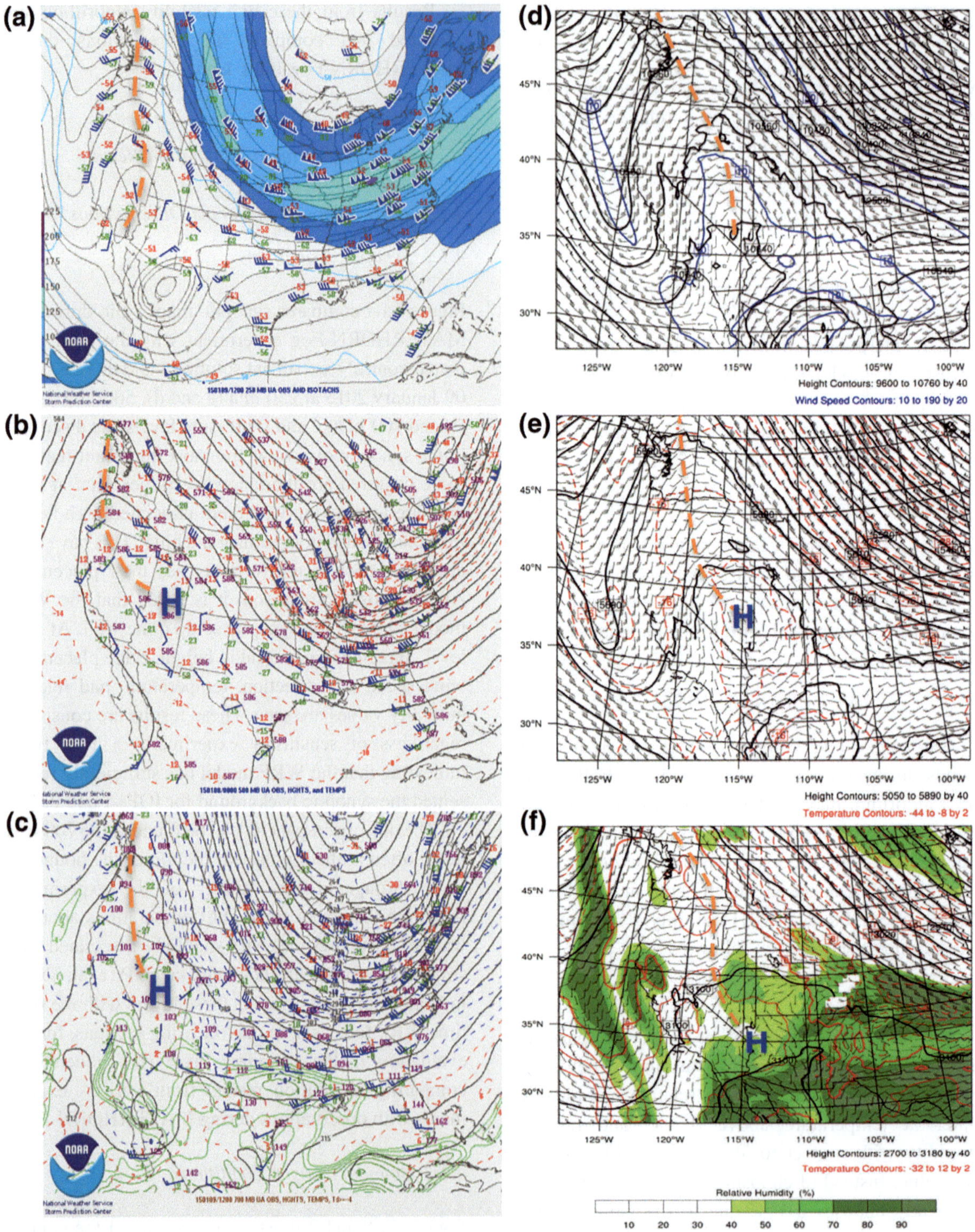

Airport, less than 15 km from the Salt Lake Valley site. The sounding reveals what the synoptic maps alluded to in the previous section: errors are almost entirely confined to the boundary layer. While there is evidence of a decoupling in the wind profile, coincident with the temperature inversion, the model

◀Figure 6
Synoptic comparison of observed **a** 250 hPa, **b** 500 hPa, **c** 700 hPa conditions with WRF-generated **d** 250 hPa, **e** 500 hPa, and **f** 700 hPa solution from the control experiment initialized at 1200 UTC 08 January. In **a–c**, geopotential height (solid black lines), temperature (red and blue dashed lines), dew point depression (solid green lines), and wind speed and direction (blue wind barbs) analyzed. In **d–f**, geopotential height (solid black lines), temperature (red dashed lines), wind speed (blue solid lines), relative humidity (green shading), and wind speed and direction (black wind barbs) analyzed. Ridges are analyzed by dashed orange lines and high-pressure centers by a blue capital H

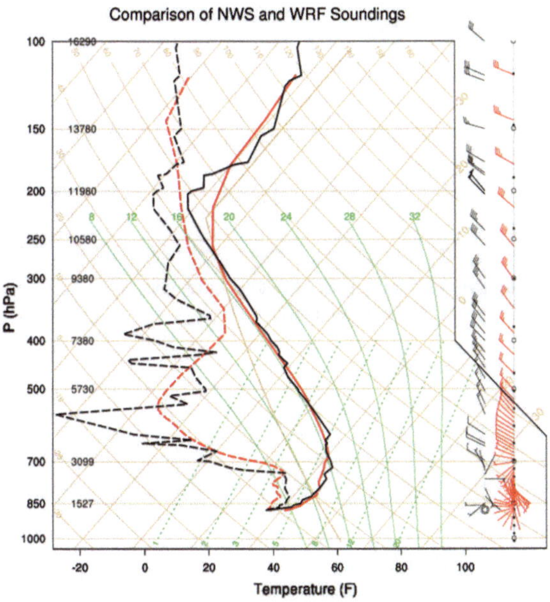

Figure 7
Sounding comparison at Salt Lake City International Airport (KSLC) from 1200 UTC 09 January 2015 between observations and control model solution. Temperature (F) along *x*-axis and pressure (hPa) along *y*-axis. Temperature (°C) (straight beige lines), dry adiabats (curved beige lines), moist adiabats (curved green lines), saturation mixing ratio lines (dotted green lines), and wind barbs (direction and speed in knots) are analyzed. Observed temperature represented by the solid black line, and observed dew point temperature represent by the dotted black line. Modeled temperature for this sounding represented by the solid red line, and modeled dew point temperature represented by the dotted red line

is not producing fog due to errors in temperature and dew point temperature. Temperatures are at most 3° higher than observations throughout the boundary layer, and dew points are 2° lower.

Figure 8a–e displays the time series of WRF control simulations of temperature, dew point temperature, relative humidity, absolute humidity, and wind speed with ASOS station data. Figure 9a, b illustrates how each of the two visibility algorithms performed in predicting the fog events. They show that small errors in the predictions of near-surface variables in the WRF can lead to big errors in a fog forecast. Figure 8a, b shows that the model accurately captures the maximum temperatures and dew point temperatures seen in each diurnal cycle. However, between 2300 MST 08 January and 0800 MST 09 January, temperatures are over-forecasted by as many as 3° and dew point temperatures are over-forecasted by as many as 2°.

Looking at the absolute humidity time series (Fig. 8c), the WRF has produced more moisture during these hours than what is observed (a 0.08 g/m^3 moist bias). This is echoed in the relative humidity time series (Fig. 8d), where the WRF predicted much higher relative humidity values than observed. The wind speed time series (Fig. 8e), however, shows that the WRF solution has values within 1 m/s of those observed. An accurate forecast of this variable is vital in the prediction of fog, as winds that are too high can mechanically mix the column and advect what fog does form away from the observation site.

Figure 9a shows the visibility forecast from the relative humidity visibility algorithm. The relative humidity used in this algorithm is the corrected relative humidity as described above, where ice is taken into effect. In the observations, the "near fog" was observed from 0000 MST to 1500 MST 09 January and 2200 MST 09 January to 0500 MST 10 January, the model captures both "near fog" and fog events couple of time before 0800 MST 09 January but does not predict "near fog" or fog past 0800 MST 09 January. The second period of fog is missed altogether in Fig. 9a. This is mainly due to a 6° error in forecasted temperature by the WRF (Fig. 8a). When combined with relatively little error in dew point temperature, the relative humidity is too low, as seen in Fig. 8d. Erroneously low relative humidity values affect the way the RH visibility algorithm performs.

Figure 9b shows the GSD visibility time series. However, using this visibility algorithm, the WRF does not produce the necessary conditions to predict fog, indicating fog prediction is sensitivity to the choice or design the proper visibility algorithm (e.g., Gultepe and Milbrandt 2009).

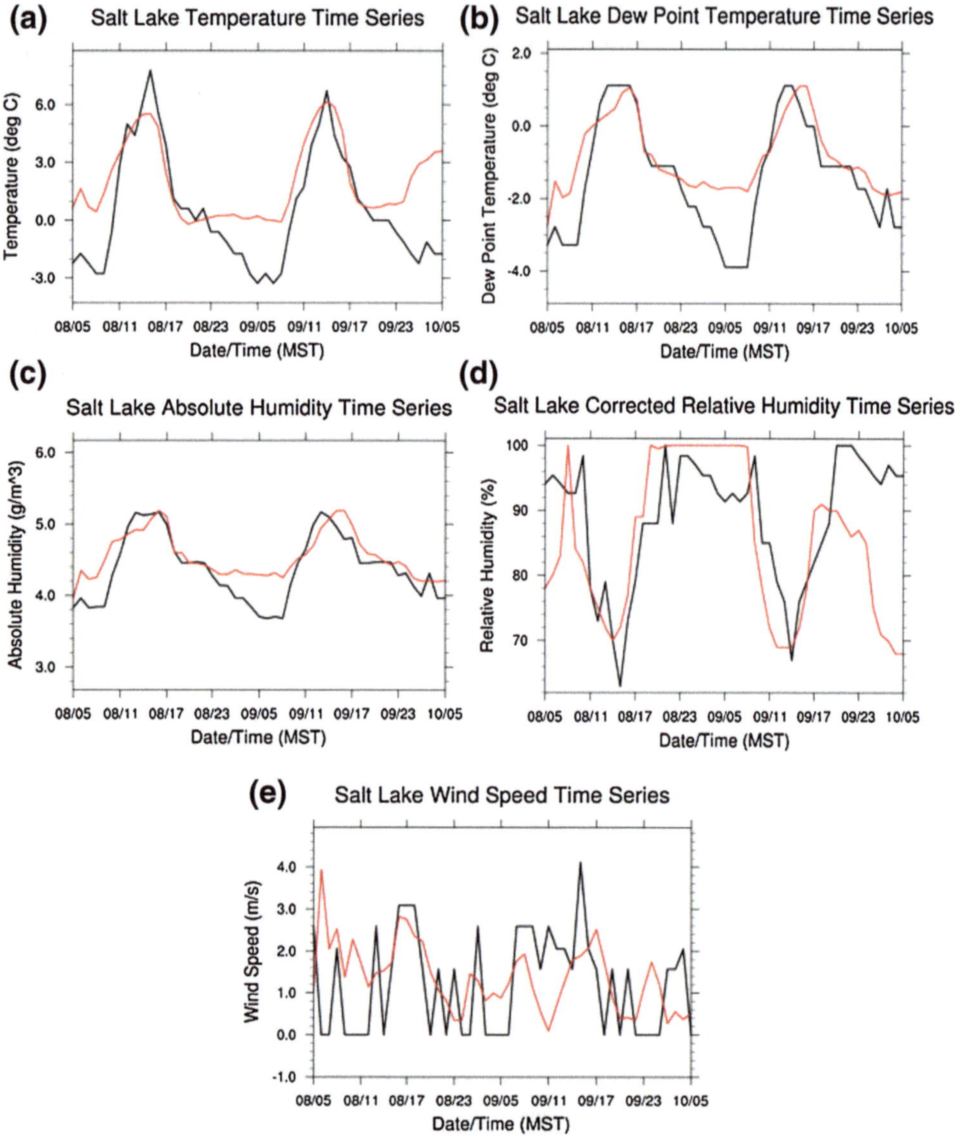

Figure 8

WRF control simulation (solid red lines) compared to MATERHORN-Fog observations. **a** Temperature (°C) time series, **b** dew point temperature (°C) time series, **c** absolute humidity (g/m^3) time series, **d** corrected relative humidity time series calculated using RH with respect to water for temperatures above freezing and RH with respect to ice for temperatures below freezing, **e** wind speed (m/s) time series. Observed variables in black, model solution in red

4.2. Sensitivity of Fog Simulations to WRF Physical Parameterization Schemes

4.2.1 Microphysical Schemes

Four CM sensitivity experiments were performed with the Thompson (control), Thompson aerosol aware, Lin et al. (1983), and Morrison et al. (2009) microphysics schemes. The Thompson Aerosol Aware CM scheme was used to examine the effects of the aerosol concentration and how it acts as cloud condensation nuclei, an important component of fog. This selection is based on Salt Lake City's particularly high concentrations of anthropogenic aerosols in the wintertime. One experiment was run using a logarithmic decay of aerosol concentration with

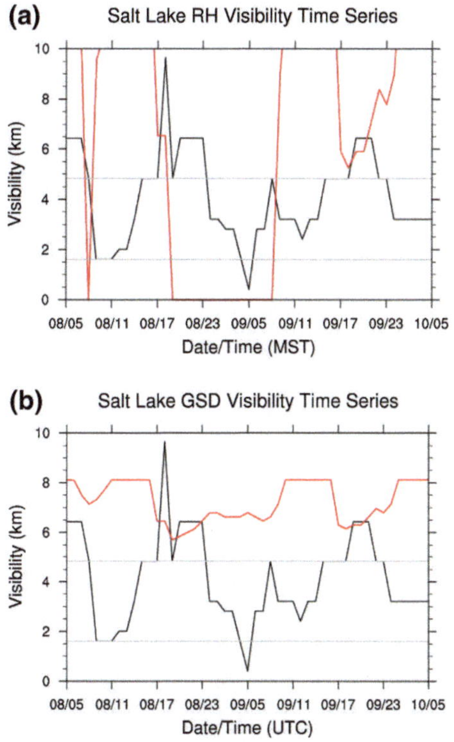

Figure 9
Visibility time series comparison calculated with respect to **a** corrected relative humidity and **b** GSD visibility algorithm for the control simulation. Observed visibility in black, model solution in red. Two gray lines indicate near fog conditions (3-mile visibility and 1-mile visibility), while the fog condition is identified as the visibility is less than 1 km. When either line crosses below these threshold, we consider these conditions met

height (a sufficient approximation), and the other was run using a climatology of aerosols based on 10 years of observed aerosol data that was collected and manually injected into the model (Colarco et al. 2010). The Lin microphysics scheme uses a five-class hydrometeor system that includes ice sedimentation, which was employed to test ice fog conditions. Finally, the Morrison scheme was chosen to test the sensitivity of single-moment versus double-moment parameterizations of cloud microphysics.

Figures 10a and 11a–d show results from the Thompson Aerosol Aware with climatology simulation and the Lin simulation is shown in Figs. 10b and 11e–h. The Thompson Aerosol Awareness without climatology and Morrison simulation results is not shown due to lack of significant different between the Thompson Aerosol Aware with and without climatology, nor between the Morrison and control simulations. The lack of difference between the aerosol schemes is surprising because Salt Lake City suffers from the trapping anthropogenic particles in cold air pools during the winter, but because there are sufficient cloud condensation nuclei, this did not make much of a difference in the simulations. Both Aerosol Aware schemes did not produce a valuable fog forecast mainly due to temperature error similar to those seen in the control.

Using the Lin microphysical scheme, the WRF model predicts near-surface atmospheric variables much more effectively than any of the other schemes, namely a sharp reduction in temperature error after 0500 MST 09 January. Instead of the 3° warm bias seen in the control and Thompson Aerosol Aware simulations, we see a cold bias in the Lin microphysics scheme after the above time. At any given point in time after 0500 MST 09 January, the modeled temperatures are about 2° cooler than observed. This is due to the generation of clouds in the WRF (Fig. 12a). They manifest at 0500 MST 09 January and remain over the Salt Lake Valley MATERHORN observation station through the end of the model simulation.

Due to these changes in temperature, one might expect the WRF to handle absolute humidity differently, and this is the case. Looking at the absolute humidity time series (Fig. 12b), the maximum error seen throughout the simulations is 0.04 g/m^3 more than what is observed. This is half the error in the control simulation. However, because the amount of moisture is closer to what was observed, the cold bias in temperature produces excess relative humidity after 0500 MST 09 January (Fig. 11f). The relative humidity values exceed the observations by as much as 20%. This implies a bias when using the relative humidity to produce a visibility forecast, and this is seen in Fig. 11g. The Lin cloud microphysics scheme, however, does produce a second period of fog where every other sensitivity test thus far has not. There are periods of false positives (between 2000 MST 08 January and 0000 MST 09 January and between 1800 MST and 2200 MST 09 January). The period between fog events is also forecast incorrectly, falling between 1200 MST 09 January and 1600 MST 09 January, but the duration between fog events is 4 h in both observation and simulation. The GSD

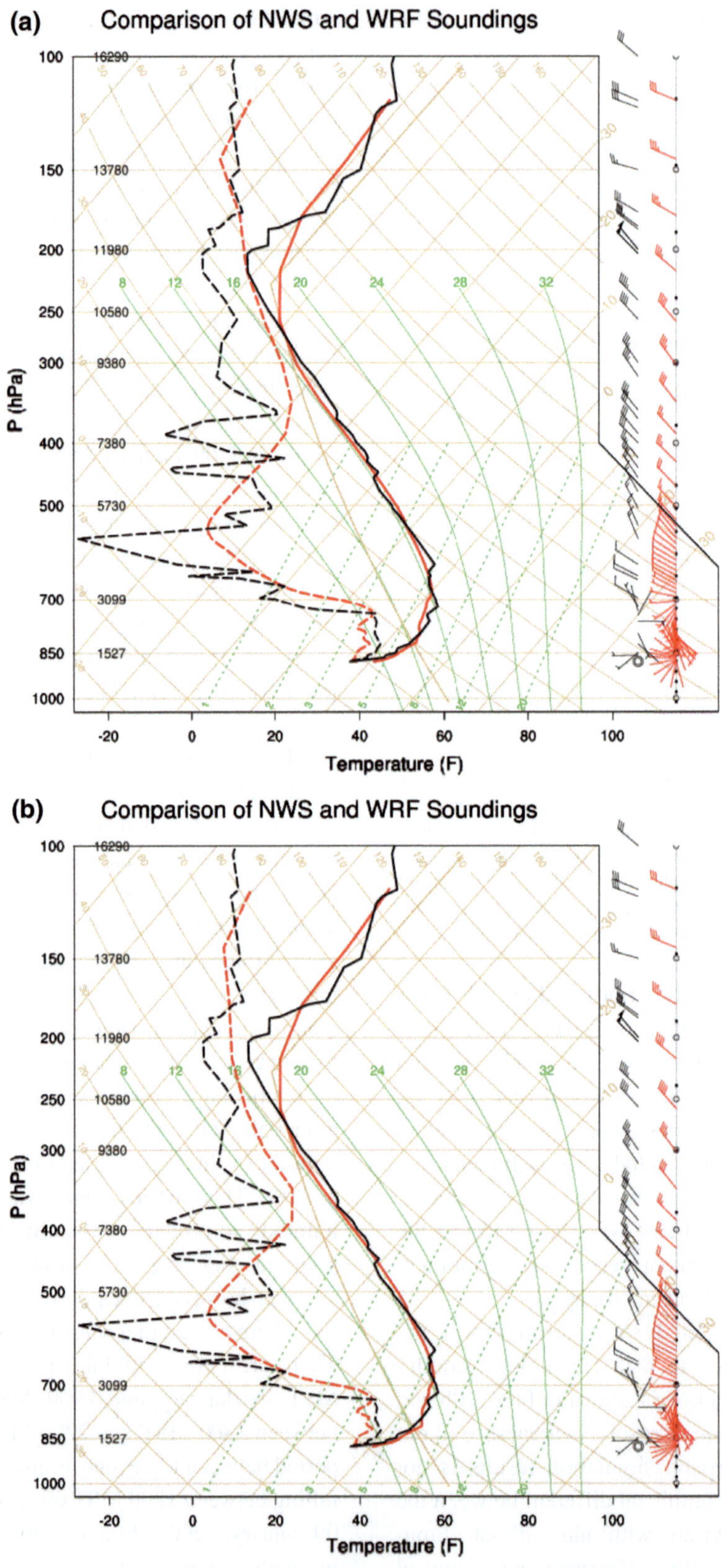
(a)
Comparison of NWS and WRF Soundings
P (hPa)
Temperature (F)
(b)
Comparison of NWS and WRF Soundings
P (hPa)
Temperature (F)

◀Figure 10
Sounding comparison at Salt Lake City International Airport (KSLC) from 1200 UTC 09 January 2015 between observations and **a** Thompson Aerosol Aware cloud microphysics sensitivity test and **b** Lin microphysics sensitivity test. Temperature (F) along *x*-axis and pressure (hPa) along *y*-axis. Temperature (°C) (straight beige lines), dry adiabats (curved beige lines), moist adiabats (curved green lines), saturation mixing ratio lines (dotted green lines) and wind barbs (direction and speed in knots) are analyzed. Observed temperature represented by solid black line and observed dew point temperature by dotted black like. Simulated temperature represented by solid red line and simulated dew point temperature by dotted red line

visibility algorithm was also tested in this experiment, and was able to represent two distinct periods of fog. The first period was much shorter than observed, only lasting between 0600 and 1200 MST 09 January, and the second was longer than observed, spanning from 1600 MST 09 January to 0500 MST 10 January. As with the relative humidity visibility algorithm, although the timing was wrong, the amount of time between each fog period was forecast correctly.

The Morrison cloud microphysics scheme (figures not shown) echoed the errors seen in the control simulation. Although the absolute humidity produced by the simulation was within 1 g/m^3 of what was observed, the warm temperature bias produced relative humidity values that were too low to represent fog. This problem was once again not remedied using the GSD visibility parameterization as no fog was represented in the simulation.

4.2.2 Planetary Boundary Layer Schemes

To examine the effect of the planetary boundary layer (PBL) schemes, the MYJ scheme (Janjic 2002) was chosen to compare against the control YSU PBL scheme. This choice examined the influence of first-order versus 1.5-order parameterization of vertical mixing. The simulation results with the MYJ scheme are consistent with the control, the warm bias in temperature skewing the relative humidity toward drier than observed values. Of particular interest in this iteration is the wind profile seen in the sounding at Salt Lake City (Fig. 13a). The winds simulated throughout the boundary layer, specifically at the top of the temperature inversion, are much higher than observations. The model-produced winds are nearly 5 m/s in excess of the observations at terrain height.

At the surface, the relative humidity values predicted by the model were also much too low in comparison to those observed, arising from errors in the forecast temperature. Like every simulation except for the Lin sensitivity experiment, fog was under-predicted using the relative humidity visibility algorithm (Fig. 14c), but it did not last long enough nor was the second fog period predicted. The GSD visibility algorithm also did not produce visibilities lower than 6 km throughout the duration of the simulation (Fig. 14d). This sensitivity test is among the weakest performers in capturing fog.

4.2.3 Land Surface Schemes

Since the MATERHORN-Fog field campaign was conducted during the winter, it is important that the winter characteristics of the land surface be captured. To accurately represent the present snow and ice cover, a sensitivity test was conducted using the thermal diffusion land surface scheme (Ek and Mahrt 1991) and another was run with forced initial albedo values in the Noah land surface scheme. The land surface was manually modified to put snow over the entire Salt Lake Valley at the initialization time. This was done as present snow and ice cover affects the radiation budget of the surface, as turbulent eddies do not form in the daylight hours that can dissipate fog.

Temperatures produced by the thermal diffusion scheme in the sounding presented in Fig. 13b follow the observations much more closely than any of the sensitivity experiments in this study. Figure 14e shows that the simulation has successfully reproduced temperatures with the most consistency (within 0.5 degree of observations) until 1000 MST 09 January. This scheme then produces similar temperature errors throughout the remainder of the simulation. The simulated relative humidity values (Fig. 14f) also capture most of the signal seen in the observations until 1000 MST 09 January. After this time, temperatures begin to exceed those observed, and the relative humidity in turn becomes up to 20% less than witnessed. Like other simulations, the relative humidity scheme captured the fog at the beginning of the first "near fog" period, but misses

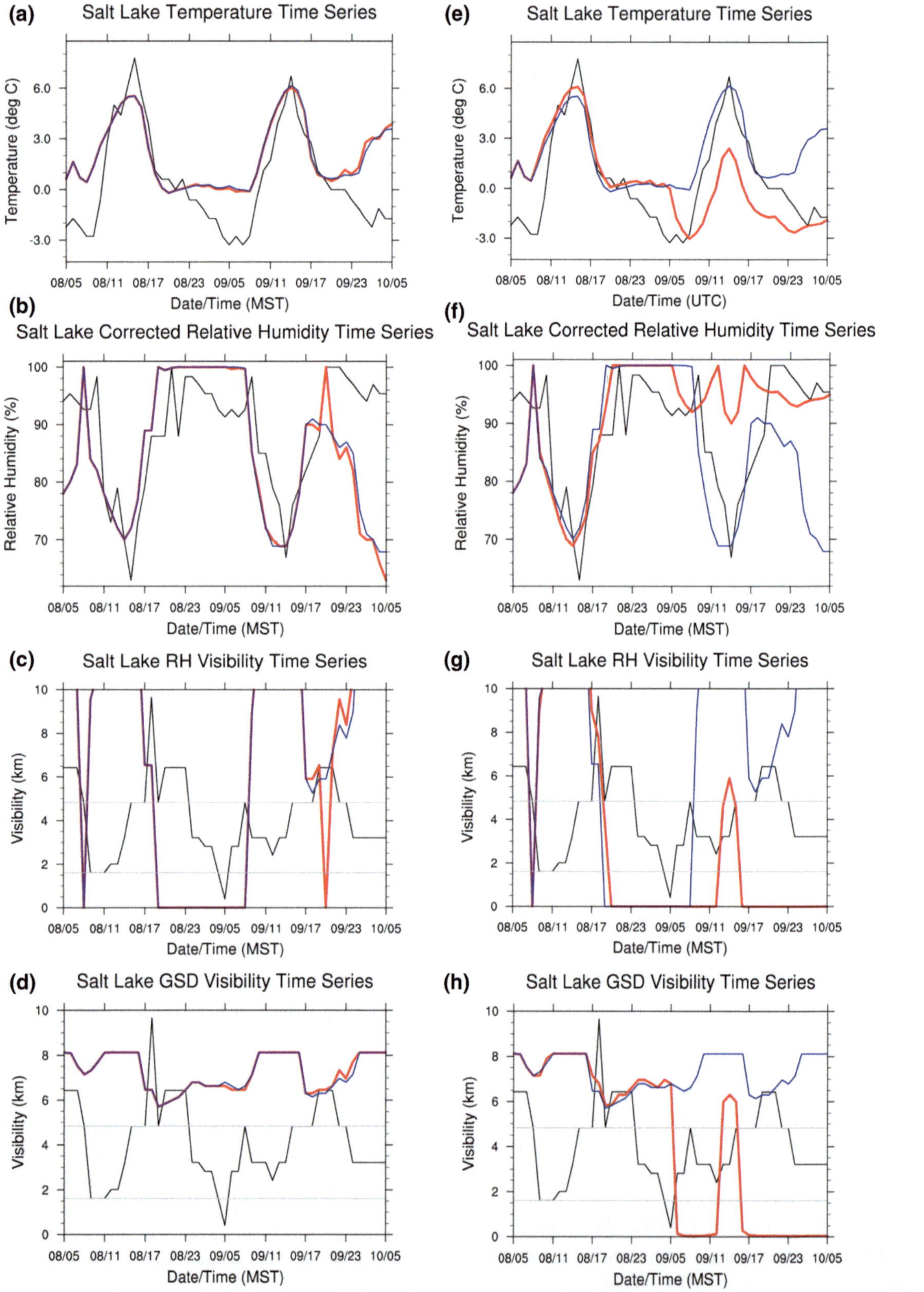
(a)
Salt Lake Temperature Time Series
Temperature (deg C)
6.0
3.0
0.0
-3.0
08/05 08/11 08/17 08/23 09/05 09/11 09/17 09/23 10/05
Date/Time (MST)
(b)
Salt Lake Corrected Relative Humidity Time Series
Relative Humidity (%)
100
90
80
70
08/05 08/11 08/17 08/23 09/05 09/11 09/17 09/23 10/05
Date/Time (MST)
(c)
Salt Lake RH Visibility Time Series
Visibility (km)
10
8
6
4
2
0
08/05 08/11 08/17 08/23 09/05 09/11 09/17 09/23 10/05
Date/Time (MST)
(d)
Salt Lake GSD Visibility Time Series
Visibility (km)
10
8
6
4
2
0
08/05 08/11 08/17 08/23 09/05 09/11 09/17 09/23 10/05
Date/Time (MST)
(e)
Salt Lake Temperature Time Series
Temperature (deg C)
6.0
3.0
0.0
-3.0
08/05 08/11 08/17 08/23 09/05 09/11 09/17 09/23 10/05
Date/Time (UTC)
(f)
Salt Lake Corrected Relative Humidity Time Series
Relative Humidity (%)
100
90
80
70
08/05 08/11 08/17 08/23 09/05 09/11 09/17 09/23 10/05
Date/Time (MST)
(g)
Salt Lake RH Visibility Time Series
Visibility (km)
10
8
6
4
2
0
08/05 08/11 08/17 08/23 09/05 09/11 09/17 09/23 10/05
Date/Time (MST)
(h)
Salt Lake GSD Visibility Time Series
Visibility (km)
10
8
6
4
2
0
08/05 08/11 08/17 08/23 09/05 09/11 09/17 09/23 10/05
Date/Time (MST)

◀Figure 11
Results from WRF Thompson Aerosol Aware cloud microphysics sensitivity test **a–d** and results from WRF Lin cloud microphysics sensitivity test **e–h** (solid red lines) compared to MATERHORN-Fog observations (solid black lines) and WRF control (solid blue lines). **a** and **e** Temperature time series, **b** and **f** corrected relative humidity time series calculated using RH with respect to water for temperatures above freezing and RH with respect to ice for temperatures below freezing, **c** and **g** visibility time series calculated with respect to corrected RH, and **d** and **h** visibility time series calculated with respect to the GSD visibility algorithm. Two gray lines indicate near fog conditions (3-mile visibility and 1-mile visibility), while the fog condition is identified as the visibility is less than 1 km. When either line crosses below the threshold, we consider these conditions met

the second period entirely (Fig. 14g). The GSD visibility scheme did not represent any fog (Fig. 14h).

Forcing observed albedo values at the beginning of the simulation did not improve our forecast from the control simulation. Warm temperature biases of the same magnitude, about 3°, remain in the evening hours, leading to erroneously low relative humidity values (Fig. 14j). This affect the relative humidity visibility algorithm (Fig. 14k), but despite the similarity in temperature between this simulation and control, the results show less fog represented in this simulation. One period of fog was forecast and it was before either period of fog was observed. There is a slightly warmer temperature profile in this run than the control, with the difference in the two runs amounting to less than 0.5°. However, this slight warming allows the temperatures to remain out of the threshold needed to calculate relative humidity with respect to ice instead of water. Therefore, although this run is no drier than the control, it is manifested differently due to how relative humidity is calculated. The GSD visibility algorithm once again fails to show fog for this simulation (Fig. 14l).

4.2.4 Radiation Schemes

Radiation budgets have an influential impact on the formation and evolution of fog. The emission of longwave radiation is closely linked to the formation of daily radiation fog. This process is a contributing factor to the fog seen at the forecast sites for the MATERHORN-Fog campaign. Four sensitivity tests were run to test these influences, two with shortwave parameterizations and two with longwave.

Two tests were used to study the differences that shortwave radiation schemes produce on the results, and the Goddard (Chou 1990) and RRTMG shortwave radiation schemes were chosen to compare to the control. The Goddard shortwave scheme includes an ozone profile that is not present in the control, and the RRTMG can interact with cloud faction instead of only fully resolved clouds. A separate set of experiments were performed to test the influence of longwave radiation parameterizations, conducted with the Goddard (Chou and Suarez 1994) and the GFDL (Eta) (Fels and Schwartzkopf 1975) longwave schemes. The Goddard scheme was chosen for the same reason as above, but the GFDL scheme was chosen because it uses a spectral calculation instead of a discrete calculation system.

For the shortwave sensitivity experiments, the Goddard shortwave and RRTMG shortwave schemes were chosen to test against the RRTM scheme used in the control simulation. The two schemes showed very similar results, so only the Goddard shortwave scheme is shown (Figs. 13d, 15a–d). Persistent ridging across the forecast area during the entirety of MATERHORN-Fog allowed for clear skies throughout the period. This means all incoming solar radiation reached the surface, creating little variability in the shortwave radiation. Both schemes show high temperatures than observations, a staple at this point in the analysis, and relative humidity values too low to account for fog formation. Neither test accurately captured the extent of the first fog period nor forecasted the second.

For the longwave sensitivity experiments, the Goddard was once again chosen, but due to lack of variability between the Goddard and the RRTMG, the GFDL (Eta) longwave scheme was chosen as the second option. However, this did not make much of a difference as the results do not have startling differences. Therefore, the only scheme represented is the Goddard longwave scheme (Figs. 13e, 15e–h). Temperature and relative humidity results are consistent with the other experiments, too warm and thusly too low, respectively. Overall, the results do not seem to be affected by testing the shortwave versus longwave radiation schemes.

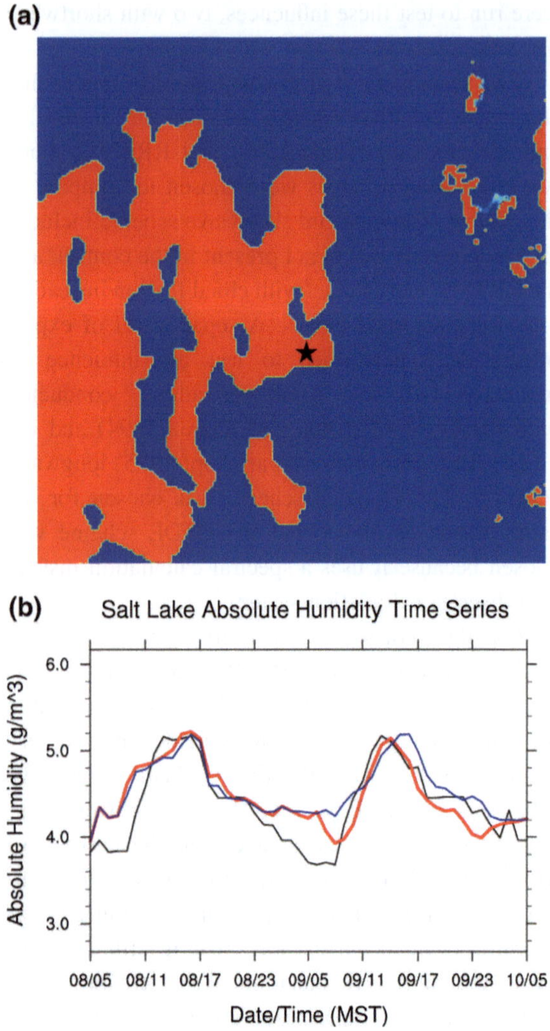

Figure 12

a Map of cloud fraction for *d*04 for Lin cloud microphysics sensitivity test, valid 0500 MST 09 January. Red indicates a cloud fraction of 1, blue indicates a cloud fraction of 0. Black star indicates MATERHORN-Fog Salt Lake Valley site. **b** Absolute humidity time series comparison for duration of Lin cloud microphysics sensitivity test. Observed absolute humidity in black, control model solution in blue, Lin model solution in red

An investigation into the errors caused by incorrect representation of radiation at the surface was conducted. The results are shown in Fig. 16a–d, for the control, Goddard shortwave, Goddard longwave, and control with forced initial albedo. Figure 16a shows that all experiments are capturing correct downward shortwave flux except for the Goddard longwave parameterization (green) which overestimates this value. Downward longwave radiation shows much more spread between the simulations, especially prevalent after 0500 MST 09 January (Fig. 16c). However, none have more than a 20 W/m^2 prolonged deviation from observations. Upward shortwave radiation (Fig. 16b) is a different case, as all experiments overestimate observed values by at least 20 W/m^2, with the Goddard longwave scheme showing most error with values double of what was observed. These errors are also seen in the upward longwave radiation, but with underestimation (Fig. 16d). Most errors show a − 40 W/m^2 bias overnight, a major contribution to the warm bias in temperature seen in these hours in nearly every experiment.

4.3. Sensitivity of Fog Simulations to Lead Time

As seen in nearly every simulation, forecast errors were exacerbated after 30 h of the simulation. The smallest errors introduced in the beginning of the simulation can grow to serious proportions the longer the simulation is run. This is a problem because it occurred in the middle of our fog forecast period, causing most of the simulations to miss the second fog period. To examine the effect of lead time on the accurate representation of inversion fog in the numerical simulation, a sensitivity test was run with an initialization time of 0000 UTC 09 January. This simulation was set with the same configuration as the control to see if forecasted variables were significantly improved.

Figure 17a–f show the results of the experiment initialized at 0000 UTC 09 January. Immediately obvious is the dramatically improved representation of relative humidity (Fig. 17b) compared to the control (Fig. 8b). However, this “better” representation is due to errors in variables that were handled more deftly in the control simulation. A look at the temperature time series (Fig. 17a) reveals that the errors that were present in the control simulation are not mitigated. There still exists a 3° warm bias in temperature overnight and errors are expounded late in the simulation (after 0500 MST 10 January). A result of this simulation that was not present in the control is the warm bias in dew point temperatures. Throughout the last 42 h of the simulation (2300 MST 08 January–1700 MST 10 January), there exists

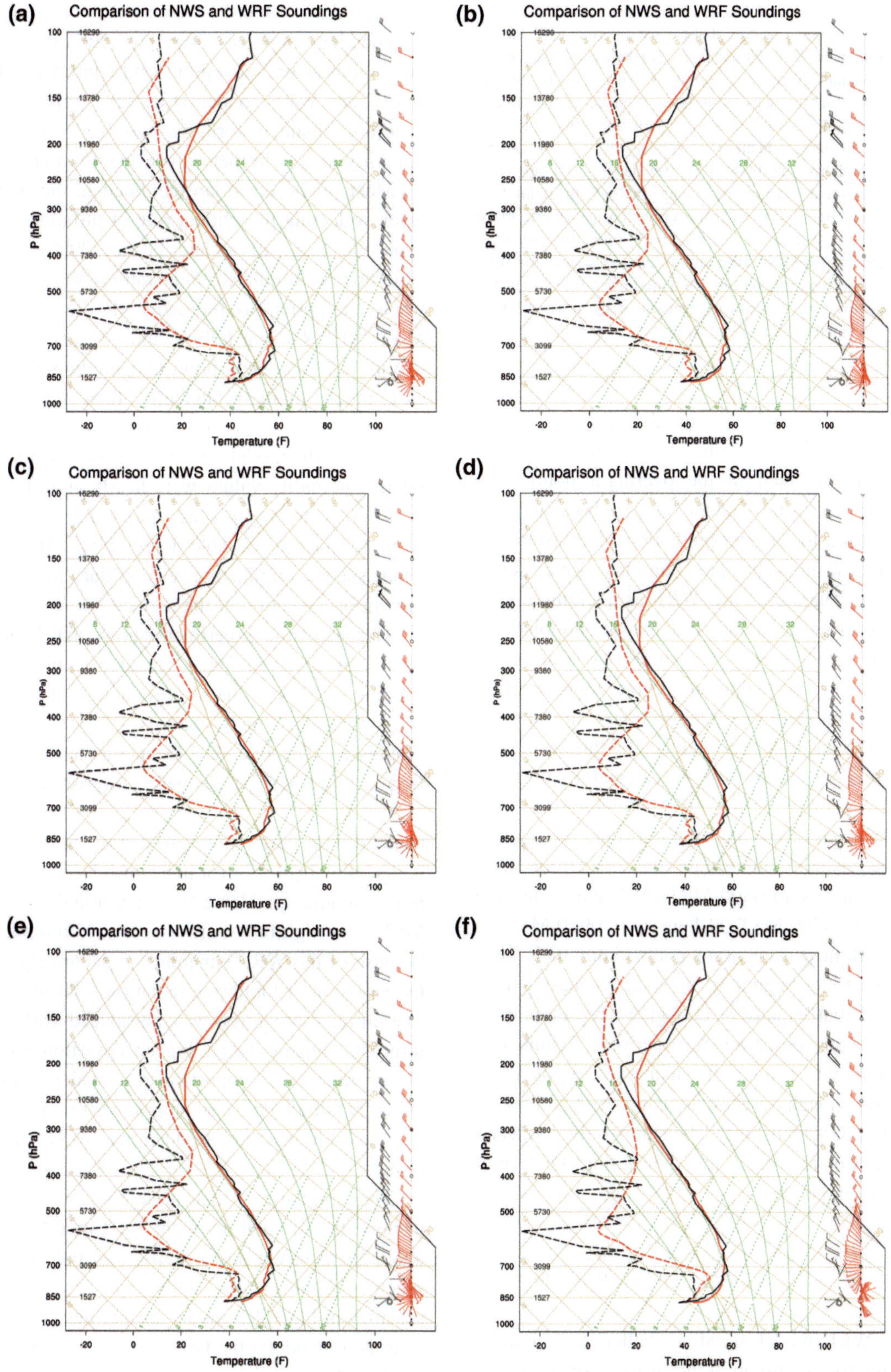
(a)
Comparison of NWS and WRF Soundings
(b)
Comparison of NWS and WRF Soundings
(c)
Comparison of NWS and WRF Soundings
(d)
Comparison of NWS and WRF Soundings
(e)
Comparison of NWS and WRF Soundings
(f)
Comparison of NWS and WRF Soundings
P (hPa)
Temperature (F)

◀Figure 13
Sounding comparisons at Salt Lake City International Airport (KSLC) from 1200 UTC 09 January 2015 between observations **a** MYJ planetary boundary layer sensitivity test, **b** thermal diffusion land surface sensitivity test, **c** forced initial albedo sensitivity test, **d** Goddard shortwave radiation sensitivity test, **e** Goddard longwave radiation sensitivity test, and **f** control simulation initialized at 0000 UTC 09 January. Temperature (F) along *x*-axis and pressure (hPa) along *y*-axis. Temperature (°C) (straight beige lines), dry adiabats (curved beige lines), moist adiabats (curved green lines), and wind barbs (direction and speed in knots) are analyzed. Observed temperature represented by solid black line, observed dew point by the dotted black line. Simulated temperature for each experiment represented by solid red line, simulated dew point by dotted red line

at least a 1° warm bias, with dew point up to 5° warmer than observed. Unlike the control, where the errors were confined to the overnight hours, this warm bias persists through the daylight hours as well.

This leads to an over-prediction of moisture, as evidence by the absolute humidity time series (Fig. 17d). Therefore, the WRF is picking up on more moisture than is observed, causing a moist bias for this simulation. However, these biases are coincided in a way that makes an accurate representation of relative humidity possible. Although the forecast in Figs. 17b and e is best produced by this sensitivity study, it is only due to a miscalculation of other surface variables. Using this configuration, two distinct periods of fog are captured suing the relative humidity visibility algorithm, the first from 2300 MST 08 January to 0800 MST 09 January, and the second from 1700 MST 09 January to 0900 MST 10 January. Both periods were not without their false positives at the beginning of each forecast period (Fig. 17e), but expansive enough to represent portions of both observed fog periods. Using this configuration, we are still unable to produce an expansive enough forecast to capture the observations. The GSD visibility algorithm only produced one period of forecasted fog, 2 h during the second observed fog period.

5. Discussion and Concluding Remarks

The WRF model faced a challenge accurately predicting fog conditions at the surface. Since fog is a "threshold phenomenon", the accuracy of its forecast is dependent on the accurate prediction of boundary layer temperature, dew point, and wind fields. In this study, the WRF model struggled to produce fog conditions using algorithms based on both relative humidity and hydrometeor concentration criteria.

Most of the errors are confined in the boundary layer, reflecting the critical role of turbulence in fog evolution (e.g., Zhou and Ferrier 2008). In most experiments, simulated temperatures were too warm when attempting to reproduce the fog present in MATERHORN-Fog IOPs 1 and 2, especially in the overnight hours. Dew point temperatures were forecast much more proficiently than temperatures, leading to accurate representations of absolute moisture in the model. This implies that errors seen in relative humidity values were nearly entirely dependent on errors in forecast temperature. The most accurate surface variable forecast was the surface wind speed, eliminating errors due to excessive mixing or advection at the surface.

The lead time experiment shows that better prediction of observed fog is sensitive to model initial conditions. This is due to introduction of error early in a forecast that can be exacerbated as the integration progresses. The first set of experiments was initialized at 0500 MST, right before sunrise. As seen in every experiment, the near-surface variables were handled with much more proficiency in the daylight hours, allowing less error to be introduced early in the model simulation. However, when initialized at 1700 MST, near nightfall, more error was introduced early because of the WRF's inability to accurately capture the emission of longwave radiation from the surface. Therefore, more error was introduced early in the forecast, causing for a bigger deviation from the observations. It is important to consider when your model is initialized based on past performance so that you may mitigate the introduction of early simulation error.

The simulations that produced too much fog, namely the simulation with the Lin CM scheme, although incorrect, are useful. Note that all of CM schemes used in this study, and almost all CM schemes in the WRF model are not designed for fog near the ground, particularly not for radiation fog, but for high-level clouds. The microphysical feature of

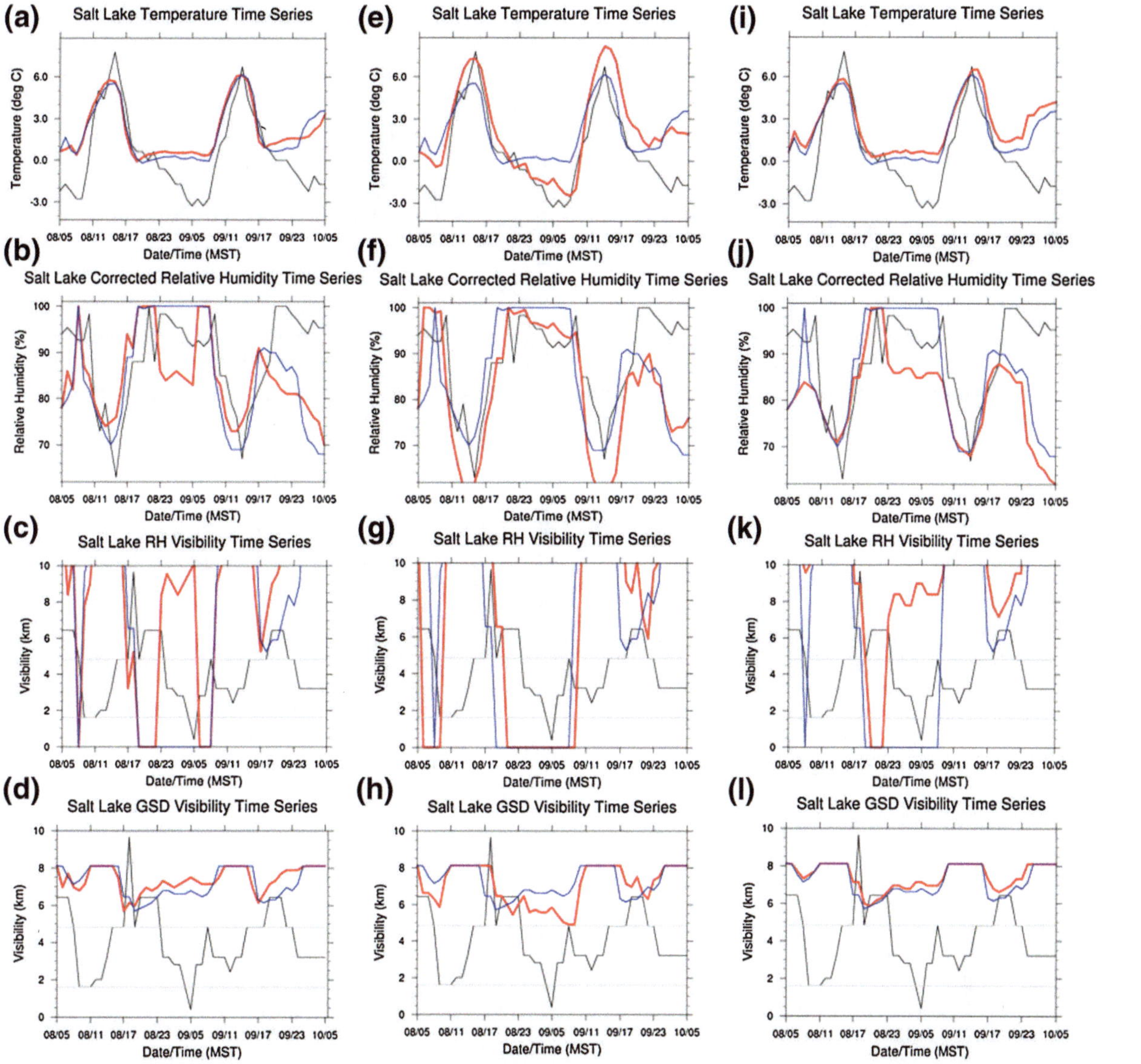

Figure 14

Results from MYJ planetary boundary layer sensitivity test **a–d**, results from thermal diffusion land surface sensitivity test **e–h**, and results from forced initial albedo sensitivity test **i–l** (solid red lines) compared to MATERHORN-Fog observations (solid black lines) and WRF control (solid blue lines). **a**, **e**, and **i** Temperature time series; **b**, **f**, and **j** corrected relative humidity time series calculated using RH with respect to water for temperatures above freezing and RH with respect to ice for temperatures below freezing; **c**, **g**, and **k** visibility time series calculated with respect to corrected RH; and **d**, **h**, and **l** visibility time series calculated with respect to the GSD visibility algorithm. Two gray lines indicated near fog conditions (3-mile visibility and 1-mile visibility), while the fog condition is identified as the visibility is less than 1 km. When either line crosses below the threshold, we consider these conditions met

radiation fog is quite different from that of high-level clouds (e.g., droplet spectrums, etc.). This fact can at least explain part of the failure of fog simulation from the WRF model. In addition, results from this study also indicate that fog prediction is sensitivity to the choice or design the proper visibility algorithm (e.g., Fig. 9). Therefore, the improvement and design of visibility algorithm that accounts the limitation of current CM schemes should also be useful.

In addition, sensitivity experiments from this study indicate that the different combination of various physical schemes could result in diverse fog

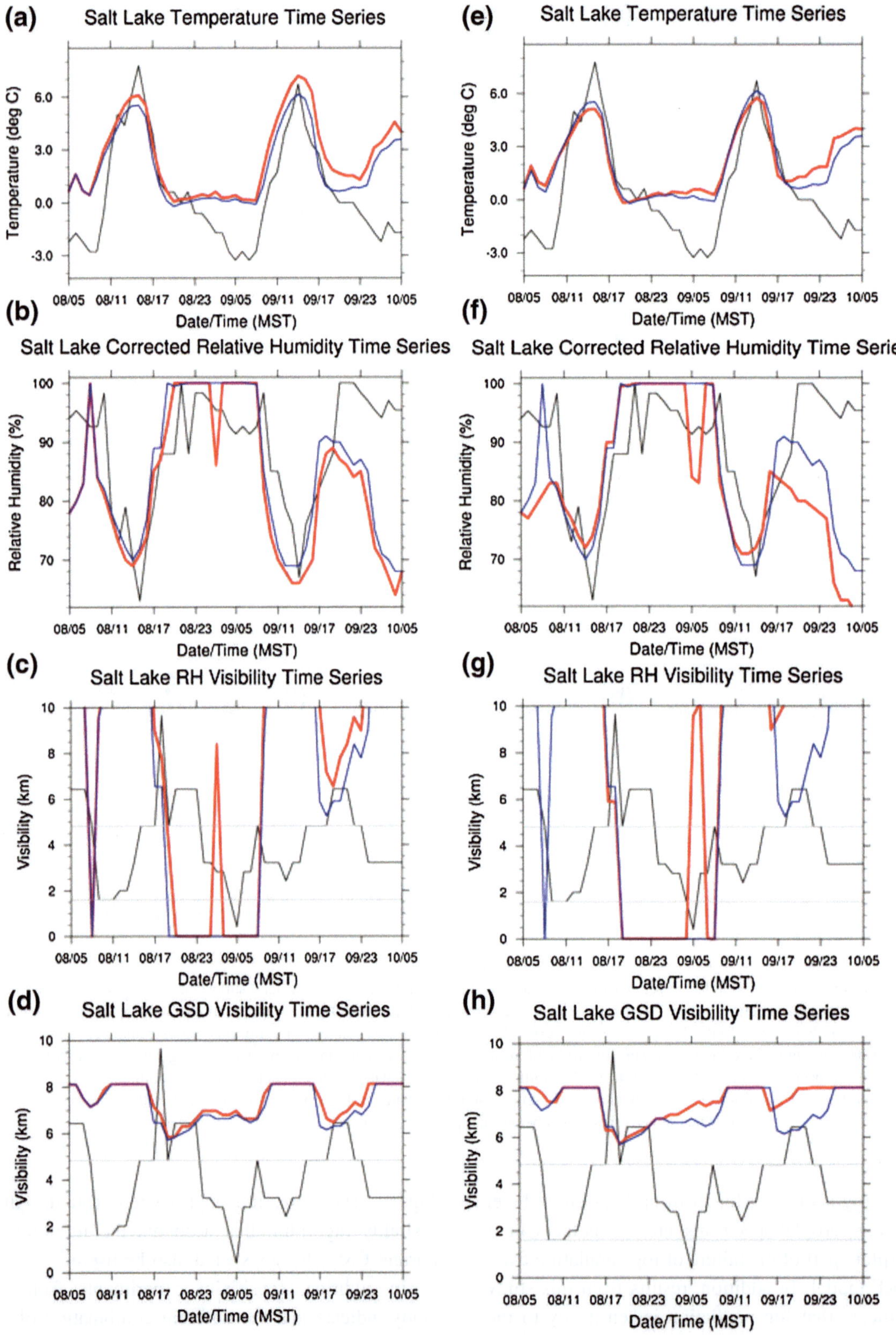
(a)
Salt Lake Temperature Time Series
Temperature (deg C)
6.0
3.0
0.0
-3.0
08/05 08/11 08/17 08/23 09/05 09/11 09/17 09/23 10/05
Date/Time (MST)
(b)
Salt Lake Corrected Relative Humidity Time Series
Relative Humidity (%)
100
90
80
70
08/05 08/11 08/17 08/23 09/05 09/11 09/17 09/23 10/05
Date/Time (MST)
(c)
Salt Lake RH Visibility Time Series
Visibility (km)
10
8
6
4
2
0
08/05 08/11 08/17 08/23 09/05 09/11 09/17 09/23 10/05
Date/Time (MST)
(d)
Salt Lake GSD Visibility Time Series
Visibility (km)
10
8
6
4
2
0
08/05 08/11 08/17 08/23 09/05 09/11 09/17 09/23 10/05
Date/Time (MST)
(e)
Salt Lake Temperature Time Series
Temperature (deg C)
6.0
3.0
0.0
-3.0
08/05 08/11 08/17 08/23 09/05 09/11 09/17 09/23 10/05
Date/Time (MST)
(f)
Salt Lake Corrected Relative Humidity Time Series
Relative Humidity (%)
100
90
80
70
08/05 08/11 08/17 08/23 09/05 09/11 09/17 09/23 10/05
Date/Time (MST)
(g)
Salt Lake RH Visibility Time Series
Visibility (km)
10
8
6
4
2
0
08/05 08/11 08/17 08/23 09/05 09/11 09/17 09/23 10/05
Date/Time (MST)
(h)
Salt Lake GSD Visibility Time Series
Visibility (km)
10
8
6
4
2
0
08/05 08/11 08/17 08/23 09/05 09/11 09/17 09/23 10/05
Date/Time (MST)

◀Figure 15
Results from WRF Goddard shortwave radiation sensitivity test **a–d** and results from WRF Goddard longwave radiation sensitivity test **e–h** (solid red lines) compared to MATERHORN-Fog observations (solid black lines) and WRF control (solid blue lines). **a** and **e** Temperature time series, **b** and **f** corrected relative humidity time series calculated using RH with respect to water for temperatures above freezing and RH with respect to ice for temperatures below freezing, **c** and **g** visibility time series calculated with respect to corrected RH, and **d** and **h** visibility time series calculated with respect to the GSD visibility algorithm. Two gray lines indicate near fog conditions (3-mile visibility and 1-mile visibility), while the fog condition is identified as the visibility is less than 1 km. When either line crosses below the threshold, we consider these conditions met

forecasts. These results may shed light on the optimal selection of the physical schemes to achieve the best possible simulation. However, considering the complicated interaction among different physical processes in various local conditions, model improvement should still be a way to ultimately enhance the fog prediction. Moreover, results from difference sensitivity experiments may also prove that ensemble forecasts with multiple-model and multiple initial conditions could be a promising way to improve the fog prediction (Zhou and Du 2010).

Overall, fog poses a danger to commuters of all types: drivers, bikers, and jet passengers. Therefore,

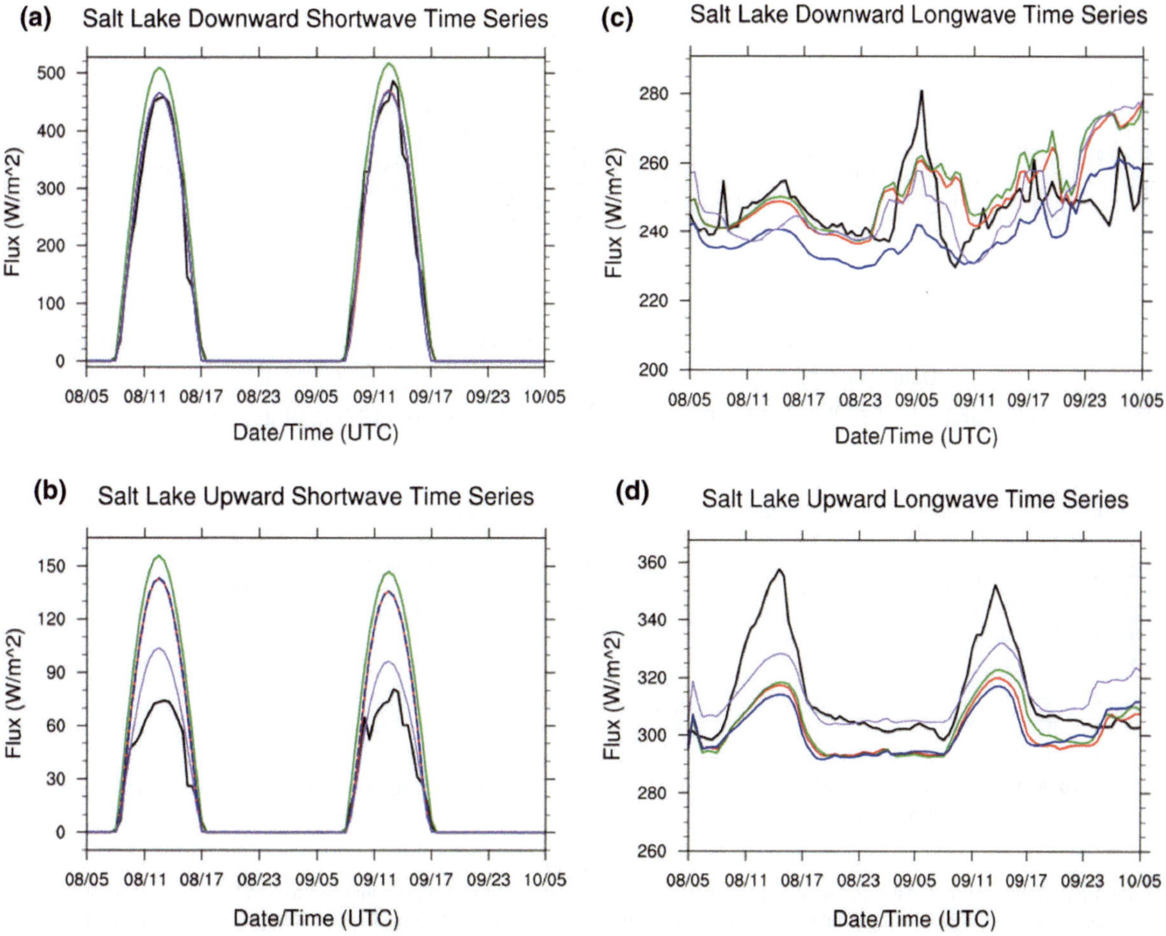

Figure 16
a Downward shortwave radiation time series, **b** downward longwave radiation time series, **c** upward shortwave time series, and **d** upward longwave radiation time series comparison between observations (black), control (red), Goddard shortwave experiment (green), Goddard longwave experiment (blue), and control with forced initial albedo (purple)

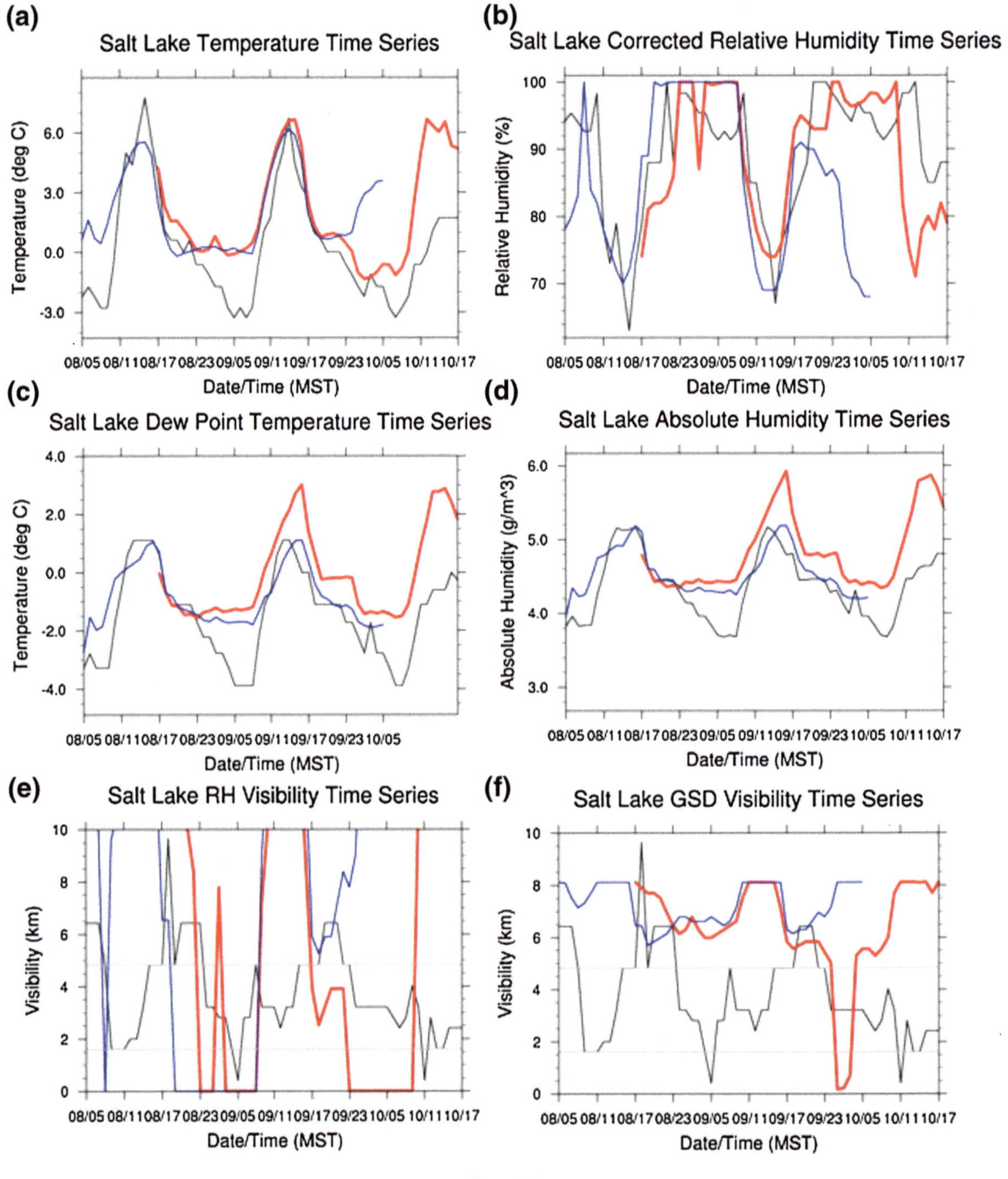

Figure 17

Results from WRF control simulation initialized at 0000 UTC 09 January (solid red lines) compared to MATERHORN-Fog observations (solid black lines) and WRF control simulation initialized at 1200 UTC 08 January (solid blue lines). **a** Temperature time series, **b** corrected relative humidity time series calculated using RH with respect to water for temperatures above freezing and RH with respect to ice for temperatures below freezing, **c** dew point time series, **d** absolute humidity time series, **e** visibility time series calculated with respect to corrected RH, **f** visibility time series calculated with respect to the GSD visibility algorithm. Two gray lines indicate near fog conditions (3-mile visibility and 1-mile visibility), while the fog condition is identified as the visibility is less than 1 km. When either line crosses below the threshold, we consider these conditions met accurate forecast of fog can aid National Weather Service forecasters in the issuance of fog advisories. This information can be invaluable to the citizens of the northern Utah valleys and can potentially lessen the risk of dangerous travel during these foggy periods. It can also aid in the rescheduling of flights ahead of time so that commuters are not delayed, and airports are not at a risk of lost revenue due to last-minute rescheduling. However, as indicated in this study, accurate prediction of fog is still an area that needs further exploration in numerical weather prediction.

Acknowledgements

This study was supported by The MATERHORN Program funded by the Office of Naval Research Award # N00014-11-1-0709. Computer resources and supports from NCAR CISL and University of Utah CHPC are greatly appreciated. We also thank Dr. Ismail Gultepe and two anonymous reviewers for their comments that are helpful for improving the manuscript.

References

American Meteorological Society. (2012). *Glossary of meteorology*. http://glossary.ametsoc.org. Web. 1 Apr 2016.

Bergot, T., & Guédalia, D. (1994). Numerical forecasting of radiation fog. Part I: Numerical model and sensitivity tests. *Monthly Weather Review, 122,* 1218–1230.

Bergot, T., Terradellas, E., Cuxart, J., Mira, A., Liechti, O., Mueller, M., et al. (2007). Intercomparison of single-column numerical models for the prediction of radiation fog. *Journal of Applied Meteorology and Climatology, 46,* 504–521.

Brown, R., & Roach, W. T. (1976). The physics of radiation fog: II—A numerical study. *Quarterly Journal of the Royal Meteorological Society, 102,* 335–354.

Byers, H. R. (1959). *General meteorology*. New York City: McGraw-Hill.

Chachere, C. N., & Pu, Z. (2016). Connections between cold air pools and mountain valley fog events in Salt Lake City. *Pure and Applied Geophysics*. https://doi.org/10.1007/s00024-016-1316-x.

Chen, F., & Dudhia, J. (2001). Coupling an advanced land surface-hydrology model with the Penn State-NCAR MM5 modeling system. Part I: Model implementation and sensitivity. *Monthly Weather Review, 129,* 569–585.

Chou, M.-D. (1990). Parameterization for the absorption of solar radiation by O_2 and CO_2 with application to climate studies. *Journal of Climate, 3,* 209–217.

Chou, M.-D., & Suarez, M. J. (1994). *An efficient thermal infrared radiation parameterization for use in general circulation models*. NASA Tech. Memo. 104606.

Colarco, P., da Silva, A., Chin, M. & Diehl, T. (2010) Online simulations of global aerosol distributions in the NASA GEOS-4 model and comparisons to satellite and ground-based aerosol optical depth. *Journal of Geophysical Research, 115,* D14207. https://doi.org/10.1029/2009JD012820.

Dudhia, J. (1989). Numerical study of convection observed during the winter monsoon experiment using a mesoscale two-dimensional model. *Journal of Atmospheric Science, 46,* 3077–3107.

Ek, M., & Mahrt, L. (1991). *OSU 1-D PBL model user's guide, version 1.04*. Corvallis: Department of Atmospheric Sciences, Oregon State University.

Fels, S. B., & Schwartzkopf, D. (1975). The simplified exchange approximation: a new method for radiative transfer calculations. *Journal of Atmospheric Science, 32,* 1475–1488.

Fernando, H. J. S., Pardyjak, E. R., Di Sabatino, S., Chow, F. K., De Wekker, S. F. J., Hoch, S. W., et al. (2015). The MATERHORN—Unraveling the intricacies of mountain weather. *Bulletin of the American Meteorological Society, 96,* 1945–1967.

Gultepe, I., & Milbrandt, J. (2009). Probabilistic parameterizations of visibility using observations of rain precipitation rate, relative humidity, and visibility. *Journal of Applied Meteorology and Climatology, 49,* 36–46.

Gultepe, I., Fernando, H. J. S., Pardyjak, E. R., Hoch, S. W., Silver, Z., Creegan, E., Leo, L. S., Pu, Z., De Weker, S., & Hang, C. (2016) Mountain ice fog: observations and predictability. *Pure and Applied Geophy*. https://doi.org/10.1007/s00024-016-1374-0.

Hodges, D. G., & Pu, Z. (2015). The climatology, frequency, and distribution of cold season fog events in northern Utah. *Pure and Applied Geophysics*. https://doi.org/10.1007/s00024-015-1187-6.

Hong, S.-Y., & Lim, J.-O. (2006). A new vertical diffusion package with an explicit treatment of entrainment processes. *Monthly Weather Review, 132,* 103–120.

Horel, J., et al. (2002). MesoWest: Cooperative mesonets in the western United States. *Bulletin of the American Meteorological Society, 83,* 211–225.

Janjic, Z. I. (2002). *Nonsingular implementation of the Mellor–Yamada level 2.5 scheme in the NCEP Meso model*. NCEP Office Note 437.

Kain, J. S., & Fritsch, J. M. (1990). A one-dimensional entraining/detraining plume model and its application in convective parameterization. *Journal of Atmospheric Science, 47,* 2784–2802.

Kunkel, B. A. (1984). Parameterization of droplet terminal velocity and extinction coefficient in fog models. *Journal of Climate and Applied Meteorology, 23,* 34–41.

Lareau, N. P., Crosman, E., Whiteman, C. D., Horel, J. D., Hoch, S. W., Brown, W. O. J., et al. (2013). The persistent cold-air pool study. *Bulletin of the American Meteorological Society, 94,* 51–63.

Lin, Y.-L., Farley, R. D., & Orville, H. D. (1983). Bulk parameterization of the now field in a cloud model. *Journal of Climate and Applied Meteorology, 22,* 1065–1092.

Lin, C., Zhang, Z., Pu, Z., & Wang, Y. (2017). Numerical simulations of an advection fog event over Shanghai Pudong International Airport with the WRF model. *Journal of Meteorological Research, 31,* 874–889.

Liu, Y., et al. (2008). The operational mesogamma-scale analysis and forecast system of the U.S. Army test and evaluation command. Part II: Interrange comparison of the accuracy of model analyses and forecasts. *Journal of Applied Meteorology and Climatology, 47,* 1093–1104.

Mlawer, E. J., Taubman, S. J., Brown, P. D., Iacono, M. J., & Clough, S. A. (1997). Radiative transfer for inhomogeneous atmospheres: RRTM, a validated correlated-k model for the longwave. *Journal of Geophysical Research: Atmospheres, 102*(D14), 16663–16682.

Morrison, H., Thompson, G., & Tatarskii, V. (2009). Impact of cloud microphysics on the development of trailing stratiform precipitation in a simulated squall line: Comparison of one- and two-moment schemes. *Monthly Weather Review, 137,* 991–1007.

Pu, Z., Chachere, C., Hoch, S., Pardyjak, E., & Gultepe, I. (2016). Numerical prediction of cold season fog events over complex terrain: The performance of the WRF model during MATERHORN-Fog and early evaluation. *Pure and Applied Geophysics.* https://doi.org/10.1007/s00024-016-1375-z.

Ryerson, W. R., & Hacker, J. P. (2014). The potential for mesoscale visibility prediction with a multimodel ensemble. *Weather and Forecasting, 29,* 543–562.

Skamarock, W. C., Klemp, J. B., Dudhia, J., Gill, D. O., Barker, M., Duda, K. G., Huang, K. G., Huang, Y., Wang, W., Powers, J. G. (2008). *A description of the advanced research WRF version 3.* NCAR Tech. Note, NCAR/TN-475 + STR.

Steeneveld, G. J., Ronda, R. J., & Holtslag, A. A. M. (2014). The challenge of forecasting the onset and development of radiation using mesoscale atmospheric models. *Boundary-Layer Meteorology, 154*(2), 265–289.

Stoelinga, M. T., & Warner, T. T. (1999). Nonhydrostatic, mesobeta-scale model simulations of cloud ceiling and visibility for an east coast winter precipitation event. *Journal of Applied Meteorology, 38,* 385–404.

Taylor, G. F. (1941). *Aeronautical meteorology.* London: Pitman.

Thompson, G., Field, P. R., Hall, W. R., & Rasmussen, R. M. (2008). Explicit forecasts of winter precipitation using an improved bulk microphysics scheme. Part II: Implementation of a new snow parameterization. *Monthly Weather Review, 136,* 5095–5115.

Turton, J. D., & Brown, R. (1987). A comparison of a numerical model of radiation fog at night using channels 3 and 4 on the Advanced Very High Resolution Radiometer (AVHRR). *Meteor Magazine, 115,* 285–290.

Van der Velde, I. R., Steeneveld, G. J., Wichers Schreur, B. G. J., & Holtslag, A. A. M. (2010). Modeling and forecasting the onset and duration of severe radiation fog under frost conditions. *Monthly Weather Review, 138,* 4237–4253.

Whiteman, C. D., Zhong, S., Shaw, W. J., Hubbe, J. M., Bian, X., & Mittelstadt, J. (2001). Cold pools in the Columbia basin. *Weather and Forecasting, 16,* 432–447.

Wolyn, P. G., & McKee, T. B. (1989). Deep stable layers in the intermountain western United States. *Monthly Weather Review, 117,* 461–472.

Zdunkowski, W., & Barr, A. (1972). A radiative-conductive model for the prediction of radiation fog. *Boundary-Layer Meteorology, 3,* 152–157.

Zdunkowski, W., & Nielsen, B. (1969). A preliminary prediction analysis of radiation fog. *Pure and Applied Geophysics, 19,* 45–66.

Zhang, H., Pu, Z., & Zhang, X. (2013). Examination of errors in near-surface temperature and wind from WRF numerical simulations in regions of complex terrain. *Weather and Forecasting, 28,* 893–914.

Zhou, B., & Du, J. (2010). Fog prediction from a multimodel mesoscale ensemble prediction system. *Weather and Forecasting, 25,* 303–322.

Zhou, B., Du, J., Gultepe, I., & Dimego, G. (2012). Forecast of low visibility and fog from NCEP: Current status and efforts. *Pure and Applied Geophysics, 169,* 895–909.

Zhou, B., & Ferrier, B. S. (2008). Asymptotic analysis of equilibrium in radiation fog. *Journal of Applied Meteorology and Climatology, 47,* 1704–1722.

(Received June 27, 2017, revised December 28, 2017, accepted January 4, 2018, Published online January 20, 2018)

Pure Appl. Geophys. 176 (2019), 2165–2177

https://doi.org/10.1007/s00024-018-1863-4

Pure and Applied Geophysics

Probabilistic Nowcasting of Low-Visibility Procedure States at Vienna International Airport During Cold Season

PHILIPP KNERINGER,[1] SEBASTIAN J. DIETZ,[1] GEORG J. MAYR,[1] and ACHIM ZEILEIS[2]

Abstract—Airport operations are sensitive to visibility conditions. Low-visibility events may lead to capacity reduction, delays and economic losses. Different levels of low-visibility procedures (*lvp*) are enacted to ensure aviation safety. A nowcast of the probabilities for each of the *lvp* categories helps decision makers to optimally schedule their operations. An ordered logistic regression (OLR) model is used to forecast these probabilities directly. It is applied to cold season forecasts at Vienna International Airport for lead times of 30-min out to 2 h. Model inputs are standard meteorological measurements. The skill of the forecasts is accessed by the ranked probability score. OLR outperforms persistence, which is a strong contender at the shortest lead times. The ranked probability score of the OLR is even better than the one of nowcasts from human forecasters. The OLR-based nowcasting system is computationally fast and can be updated instantaneously when new data become available.

Key words: Aviation meteorology, low visibility, probabilistic nowcasting, statistical forecasts, ordered logistic regression.

1. Introduction

Low-visibility conditions at airports impact air traffic regarding aviation safety and economic efficiency of airports and airlines. The low-visibility procedures (*lvp*) come into force when horizontal and/or vertical visibility fall below airport-specific thresholds. Additional measures, e.g., increasing spacing between approaching and taxiing aircraft, ensure safe operations but also reduce the capacity of the airport. Consequently, planes might be delayed, diverted to alternative airports or prevented from taking off. Hence, reliable low-visibility forecasts are needed for tactical planning of the aircraft movements within the next few hours.

The two major weather phenomena producing low-visibility conditions are fog and low ceiling. Fog development and dissipation generally depend on temperature, the humidity and the available condensation nuclei of an air mass. Radiative cooling, change of the total water amount due to precipitation or mixing of air masses are only a few processes changing these important parameters (Gultepe et al. 2007). However, despite various field experiments (e.g., Gultepe et al. 2009; Haeffelin et al. 2009; Bergot 2016) the overall effect of some physical processes is still unknown. The complexity of the processes driving low visibility makes predictions challenging.

Physical modeling with numerical weather prediction models and statistical modeling are two general low-visibility forecasting approaches (Gultepe et al. 2007). The statistical approaches are data-driven and computationally faster. Model parameters are estimated on an archive data set and then applied to new data to forecast. The choice of the statistical model depends on the desired form of the forecast variable (continuous or categorical) and the type of the forecast output (deterministic or probabilistic). Regression models were among the first statistical forecasting approaches applied to continuous variables with linear regression (Bocchieri and Glahn 1972) and were extended to binary and multinomial categorical variables (e.g. Hilliker and Fritsch 1999; Herman and Schumacher 2016). Some machine-learning methods have also been used for low-visibility forecasts; for example, tree-based methods (Dutta and Chaudhuri 2015; Bartokov et al. 2015;

[1] Department of Atmospheric and Cryospheric Sciences, University of Innsbruck, Innrain 52f, 6020 Innsbruck, Austria. E-mail: philipp.kneringer@gmail.com

[2] Department of Statistics, University of Innsbruck, Universitätsstr. 15, 6020 Innsbruck, Austria.

Dietz et al. 2017) and artificial neural networks (Bremnes and Michaelides 2007; Marzban et al. 2007; Fabbian et al. 2007).

Since risk management and decision making depend heavily on the probabilistic information (Murphy and Winkler 1984). Probabilistic forecasts are essential to make safe and economic decisions especially for an application such as air traffic regulation. One option for airport visibility forecasts is to predict the horizontal and vertical visibility separately and determine *lvp* afterwards. However, since horizontal and vertical visibility are not statistically independent, there is no obvious way to obtain the combined probability. Instead, the variable of interest to aviation end-users, the *lvp* categories, should be forecast directly.

In this paper, we present a new way to generate probabilistic *lvp* state forecasts for the next 2 h with a statistical regression method. The method of choice is ordered logistic regression (OLR) to capture the categorical and ordered nature of the end-user forecast variable, which is based on fine-grained visibility and ceiling thresholds. Due to the interest in short lead times, the nowcasting system is exclusively based on point measurements (see Vislocky and Fritsch 1997). The performance of the nowcasting system is compared to climatology, persistence, and human forecasts. The methodology to develop the nowcasting system is shown in Sect. 2. Section 3 examines the area of investigation and the data used. Section 4 presents the results of the nowcasting system which are discussed in the final section.

2. Methods

2.1. Ordered Logistic Regression

The method used to develop a probabilistic *lvp* nowcasting system is OLR. This method allows prediction of the probabilities for all categories of an ordered response, such as *lvp*, within one consistent model. Additionally, OLR has the benefit of providing a fast update cycle with low computational costs.

The OLR model describes an ordered categorical variable by assuming an underlying continuous variable mapped with thresholds to the categories. The threshold coefficients and predictor coefficients are determined during model estimation. An arbitrary number of predictors is possible, similar to multiple linear regression. The occurrence probability of the individual categories can be determined by evaluating the chosen error distribution function at the lower and upper thresholds of a category. In mathematical notation the OLR model is described as follows:

Each observation of the response falls into a ordered category $j = (1, 2, \ldots, J)$. The ordering is from no impact to highest impact on airport operations in this case. These ordinal response y_i is modeled by assuming a continuous auxiliary variable y_i^* capturing visibility. For this variable a linear model

$$y_i^* = \mathbf{x}_i^\top \beta + \varepsilon_i \tag{1}$$

holds, where $i = (1, 2, \ldots, n)$ is the index over the observations. But y_i^* is not observed directly, therefore, the observation i is modeled to category j by the thresholds

$$\alpha_{j-1} > y_i^* \geq \alpha_j \tag{2}$$

Equation 1 shows the deterministic component as $\mathbf{x}_i^\top \beta$ and the random term ε_i which is assumed to be i.i.d. with zero mean. The vector $\mathbf{x}_i = (x_{i,1}, \ldots, x_{i,m})$ includes all the m predictors and $\beta = (\beta_1, \ldots, \beta_m)$ includes their coefficients. The thresholds $\alpha = (\alpha_0, \ldots, \alpha_J)$ are determined together with the predictor coefficients β when estimating the OLR model. The lowest and highest threshold values are fixed at the values $\alpha_0 = -\infty$ and $\alpha_J = \infty$. To access the probabilities of the categories an error distribution needs to be selected. Typical distributions are the standard normal and the logistic distribution. The logistic distribution accounts better for observations in the tails of the distribution (Winkelmann and Boes 2006). We select the logistic distribution with its cumulative distribution function:

$$H_{\text{logit}}(\cdot) := \frac{\exp(\cdot)}{1 + \exp(\cdot)}, \tag{3}$$

with $H(-\infty) = 0$ and $H(\infty) = 1$. The probabilities for the categories are derived with the cumulative probability model:

$$\sum_{s=1}^{j} \pi_{is} = H_{\text{logit}}(\alpha_j - \mathbf{x}_i^\top \beta) = \frac{\exp(\alpha_j - \mathbf{x}_i^\top \beta)}{1 + \exp(\alpha_j - \mathbf{x}_i^\top \beta)}, \tag{4}$$

where $\sum_{s=1}^{j} \pi_{is}$ is the cumulative probability that an observation y_i falls into category j or lower. The probability for the individual category $\pi_{ij} = P(y_i = j|x_i) = H(\alpha_j - \mathbf{x}_i^\top \beta) - H(\alpha_{j-1} - \mathbf{x}_i^\top \beta)$ is the difference between the cumulative probabilities at the associated thresholds (Fig. 1).

A notable advantage of this method is the computational speed of the model estimation. It is almost as fast as a linear regression and can be performed instantaneously using standard software. We use the function clm() from the R package ordinal (Christensen 2015), which implements the OLR model with maximum likelihood optimization.

2.2. *Predictor Selection*

The first task in estimating the models is to decide which of the input variables to select and which to omit. We use stepwise selection and verify the forecasts with the ranked probability score (RPS see Sect. 2.3). The initial step of the algorithm estimates the climatology as a first guess. In the next step the variable that most improves the RPS of the model is added. Subsequently this model is used as the new best guess and all remaining variables are tested again. This procedure is repeated until either the model skill does not improve anymore, or no additional variable is left. The variable configuration of the final best-guess model is used to produce the low-visibility forecasts.

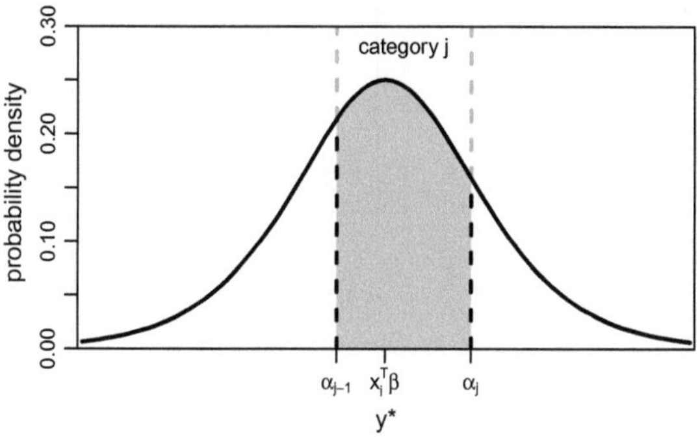

Figure 1
Probability density function of an OLR model. The shaded area highlights the probability for the category j

2.3. *Verification*

To test the model and to cover the uncertainty within the model estimation, we perform tenfold cross validation. Therefore, we split the data set into ten parts, use nine parts for training, and do out-of-sample predictions on the remaining test data set. The test data set is exchanged with one part of the previous training data set and again the model estimation and out-of-sample predictions are performed. This is repeated until we have out-of-sample forecasts for all ten parts of the previously split data set, based on ten different models with slightly different training data sets.

To determine the skill of the probabilistic categorical ordered forecasts a proper scoring rule is required. The ranked probability score (RPS) is such a metric (Wilks 2011). It compares the cumulative distribution function of the forecasts and the observations. The RPS of a forecast i is given by:

$$\text{RPS}_i = \frac{1}{J-1} \sum_{s=1}^{J} \left[\sum_{j=1}^{s} (y_{ij} - o_{ij}) \right]^2, \tag{5}$$

with y_{ij} the predicted probabilities and o_{ij} the observations for each category $j = 1, 2, \ldots, J$. While the predicted probabilities can have continuous values between 0 and 1, the observation is either 0 or 1. The RPS can be interpreted as the normalized shift in categories between the forecast and the observation. In addition, the RPS is normalized by the number of categories $J - 1$ to obtain scaled values within the interval [0, 1]. A perfect forecast has an RPS of zero, the worst forecast has an RPS of 1.

The quality of the prediction is determined by calculating the RPS of all individual observation-prediction pairs within the out-of-sample prediction data set, and averaging them. Bootstrapping is used to estimate the uncertainty due to the limited sample size. Hence, we take 500 random samples of the RPS values from the out-of-sample prediction data set. Each of these samples is taken with replacement and has the size of the full data set. Now we calculate the mean RPS of each random sample. These 500 RPS values are the basis of the results shown in Sect. 4 including the model uncertainty.

2.4. *Reference Forecasts*

The OLR models are compared to three reference forecasts. The first one is the climatology, which uses the climatological occurrence probability of each category (Fig. 2b) as forecast. The second reference is the persistence forecast, which assumes that the *lvp* state at the forecast initialization remains. This state is predicted with a probability of 100% and all other states with 0%. It needs to be mentioned that persistence is already a benchmark at short lead times (Vislocky and Fritsch 1997). As a third reference, we compare the OLR to the operational human forecasts at Vienna International Airport (VIE). Human forecasters use all available information from observations and numerical weather predictions and produce operational forecasts at most airports.

3. *Data*

VIE is selected to develop the statistical low-visibility nowcasting tool. The airport with its two runways is located in the Vienna Basin 20 km southeast of downtown Vienna. The basin is bounded by the Alps to the west and by the Carpathian Mountains to the east. The Airport is surrounded by many humidity sources. Moisture advection from the southeast (Lake Neusiedl and wetlands) and the north (Danube River) favor low-visibility conditions.

3.1. *Definition of the Low-Visibility Procedure States*

Low-visibility events occur when horizontal and/or vertical visibility drop below a set of thresholds. The runway visual range (*rvr*) is used as horizontal visibility measure. It is defined as the range over which the pilot of an aircraft on the center line of a runway can see the runway surface markings or the lights delineating the runway or identifying its center line (World Meteorological Organization 2006). The *rvr* is closely correlated to the horizontal visibility but is truncated with an upper limit of 2000 m for higher visibilities. Vertical visibility is determined by the ceiling height. It is defined as the height of the lowest cloud layer covering more than half of the sky. The ceiling height is measured by ceilometers but finally determined by human observers while the *rvr* is measured automatically by transmissiometers.

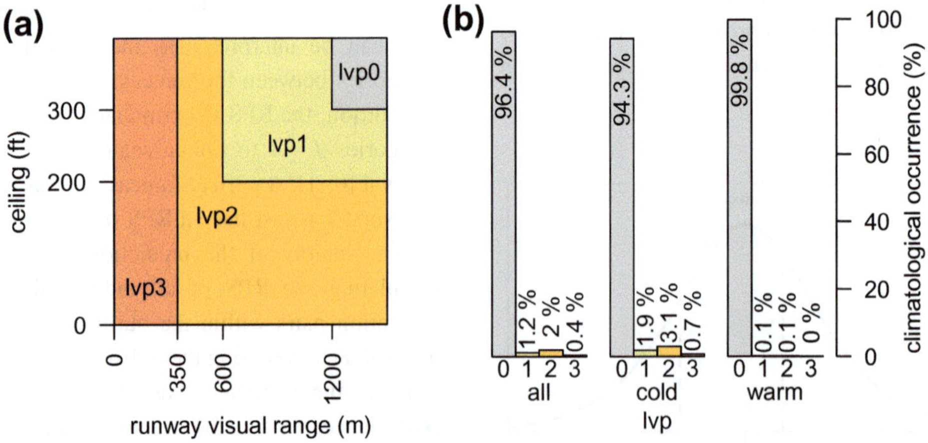

Figure 2

a Ceiling and runway visual range (*rvr*) thresholds for the *lvp* states. **b** Climatological occurrence of the three *lvp* and non-*lvp* (*lvp* 0) states at VIE for the whole year (all), and for the cold (September to March) and the warm season, respectively, over the period July 2008 until March 2017

Table 1

Definition of lvp at VIE and associated capacities relative to a maximum of 40 aircraft per hour

lvp state	*rvr*		Ceiling	Capacity (%)
0				100
1	< 1200 m	or	< 300 ft	75
2	< 600 m	or	< 200 ft	60
3	< 350 m			40

At VIE, *lvp* states from 1 to 3 depend nonlinearly on *rvr* and ceiling (see Table 1). The categories are ordered but not equidistant. The state *lvp* 0 indicates good visibility and ceiling and no restrictions for aviation with a maximum capacity of 40 aircraft per hour. At *lvp* 3 the airport operates at 40% of its maximum capacity (Table 1).

3.2. *Climatology of the Low-Visibility Procedure States*

A climatology of the *lvp* states was compiled to be used as one of the reference models. Figure 2b shows the small proportion (about 4.5%) of *lvp* states 1 or higher for the whole year. Basically no low-visibility events occur during warm season. This seasonal pattern is typical for continental climates with topographical properties similar to the Vienna Basin (Egli et al. 2017). The challenge is to capture these relatively rare *lvp* events with a statistical approach.

The significant annual cycle is also visible in Fig. 3 showing the occurrence probability of any *lvp* 1–3. The annual maximum is during December. Furthermore, the diurnal maximum is during the morning hours, which coincide with one of the airports rush hours. Therefore these *lvp* state events have a high impact on airport operations. The diurnal minimum occurs during afternoon. The correlation of the *lvp* states with wind and thus advection from humidity sources is shown in Fig. 4. The two primary wind directions are from the southeast and the northwest (Fig. 4a). Low visibility states are associated predominantly with winds from the southeast, with a small peak in the northern sector, both of which are known moisture source regions (Fig. 4b). There is also a shift from higher to lower wind speeds (color shading) between the full data set (Fig. 4a) and the *lvp* data set (Fig. 4b), as lower wind speed generally favors the development of radiation fog. As a result of this climatology, we focus only on the cold season beginning in September until the end of March.

3.3. *Measurements and Model Configurations*

Modeling and forecasting the *lvp* state with statistical methods require information related to low-visibility formation and dissipating processes. Since we are interested in lead times up to few hours only, we exclusively use standard meteorological measurements available at all larger airports as basis of this nowcasting system. The data come from the observation system of VIE in a 30-min resolution and are available from September 2008 to Mach 2017.

All potential predictors for the use in this statistical nowcasting system are shown in Table 2. The variable setup includes visibility measures, humidity measures, vertical temperature differences, wind information and climatological information. Most of the predictors are continuous variables except the *lvp* state, the precipitation variable, and the wind direction sector variables. These three variables are categorical variables with two or more levels (see Table 2). Since no soil moisture measurements are available, we used a precipitation sum of at least 1 mm over the previous 12 h as factor variable and proxy. The wind direction is split into two factor variables to capture the main fog wind directions based on the *lvp* climatology (Fig. 4b). Three visibility measures are used. The vertical visibility is captured by the ceiling and the horizontal visibility by the *rvr* and the meteorological visibility. The ceiling variable is set to 25,000 ft for the cases where no ceiling was detected due to the absence of clouds.

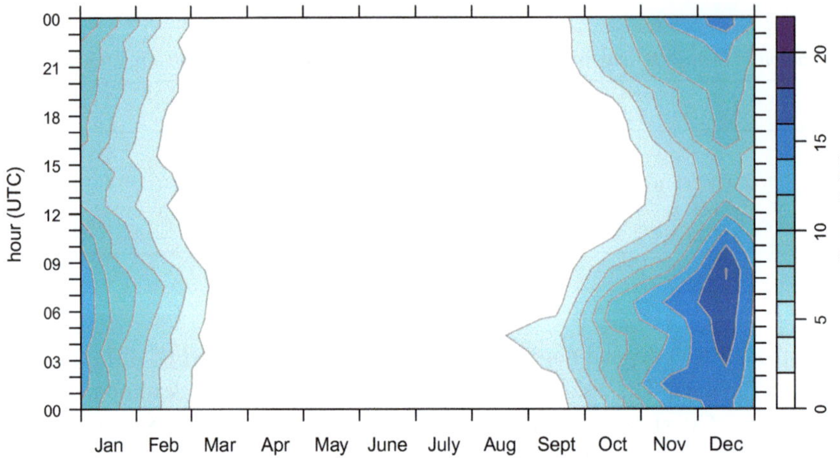

Figure 3
Contour plot of the annual and diurnal cycle of the climatological probability for poor visibility (*lvp* states $\geq$ 1) during the period July 2008 until March 2017

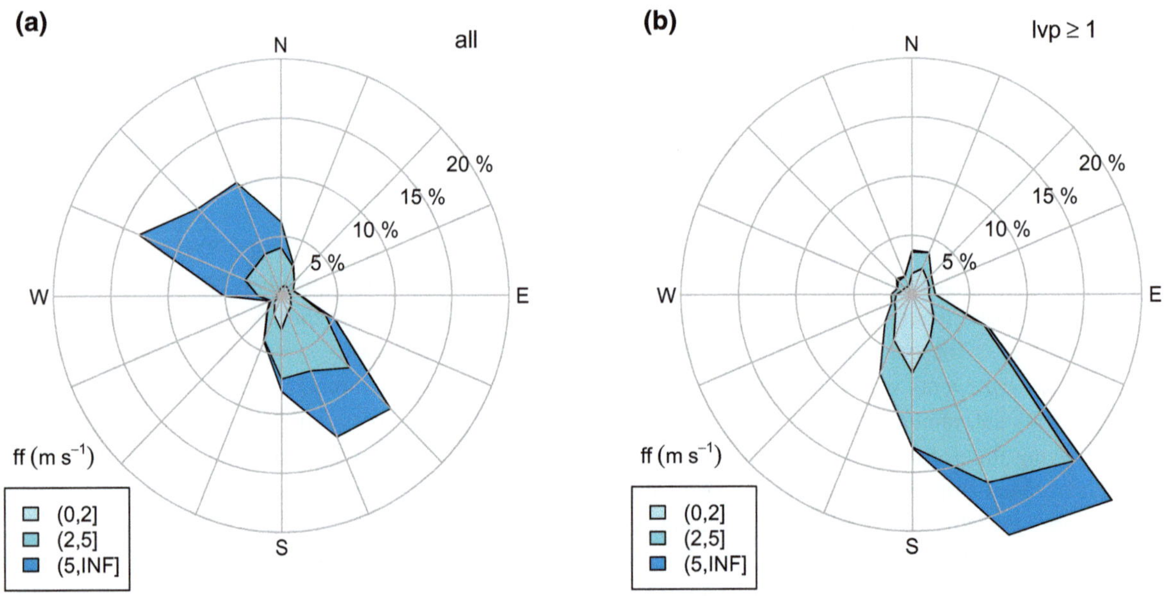

Figure 4
Wind rose plots of VIE for September to March **a** all cases and **b** *lvp* states $\geq$ *lvp* 1. The percentage of the different wind directions (color shadings) is divided into three wind speed clusters (see legend)

The *rvr* and the meteorological visibility provide similar information but *rvr* is truncated at 2000 m, whereas the meteorological visibility is truncated at 20000 m. Two further variables are the relative humidity and the dew point depression which are non-linearly-related humidity measures. Information about the vertical stratification of the lowest air layers is provided by two vertical temperature differences. One is the difference between the 2 m and surface temperature. The other is the difference between 2 m and the tower temperature roughly 100 m above ground. The only climatological variable used is the solar zenith angle, which represents the diurnal cycle of the *lvp* events. Dependent on the desired lead time, we use the associated lag of the above-shown predictors. For instance, if the lead time of + 30-

Table 2

Predictors used within the statistical nowcasting approach

Name	Description	Unit
lvp	Low-visibility procedure state	Ordered [0, 1, 2, 3]
rvr	Runway visual range	m
cei	Ceiling height	ft
vis	Horizontal visibility	m
rh	Relative humidity	%
dpd	Dew point depression	°C
rr12	Precipitation in the last 12 h	Factor [yes, no]
dt.tow	Temperature difference 2 m: tower	°C
dt.surf	Temperature difference 2 m: surface	°C
ff	Wind speed	m s^{-1}
dd.se	Wind direction from sector southeast	Factor [yes, no]
dd.n	Wind direction from sector north	Factor [yes, no]
sza	Solar zenith angle	°

min is of interest the 30-min lags of the measurements are used as predictors, respectively, for other lead times.

With measurements from nine cold seasons in 30-min resolution, we obtain a data set with roughly 85,000 observations. Thus, we have a data set which is sufficient to estimate and verify the statistical models despite the rare occurrence of the low-visibility events (about 5000 observations). The forecasts are produced over the whole day with four lead times from + 30 to + 120-min. Four models, one for each lead time, provide the forecasts. The model skill is based on out-of-sample verification using tenfold cross-validation and estimates the model variance with bootstrapping (Sect. 2.3).

4. Results

4.1. Predictor Selection and Effects

The results of the stepwise model selection (Sect. 2.2) are shown in Fig. 5. The box plots are computed by the bootstrapped RPS mean values from the (a) + 30-min and (b) the + 120-min forecasts. Each box plot shows one forecast model. The model size increases from left to right by adding the annotated variable. The initial best guess is the intercept (= climatological) model (not shown in the graphic) which has the same skill for all lead times with an RPS $\sim$ 0.032. The stepwise algorithm indicates the best input variable combination and also shows the importance of the individual variables. The variable *lvp* is selected first within all models and is the most important variable. This confirms the findings from (Vislocky and Fritsch 1997) that persistence is already a good benchmark for lvp forecasts. Horizontal visibility and humidity measures are selected next. The ceiling has a minor importance and is only selected in the + 90 and + 120-min model. The wind direction (*dd.se*) is selected in all models but improves the models only marginally. Of note is the increasing importance of solar zenith angle (*sza*) with lead time. The vertical temperature stratification (*dt.tow*) and wind speed (*ff*) were never selected as predictors.

Looking at the physical significance of the predictors selected by the objective selection method, the first choice of persistence via *lvp* and the horizontal visibility is explained by the high auto-correlation over such short lead times out to 2 h. Humidity variables are next as clouds and fog require saturation. Moist soil with the proxy of previous precipitation (*rr12*) and near-surface temperature gradient (*dt.surf*) influence radiative fog. A third important part within the models is the climatology captured by the solar zenith angle. Least important are the inputs modeling advection fog like the wind direction sectors. The variable selection implies that the information about upcoming low visibility is lead-time dependent and generally shifts from persistence predictors to others for longer lead times.

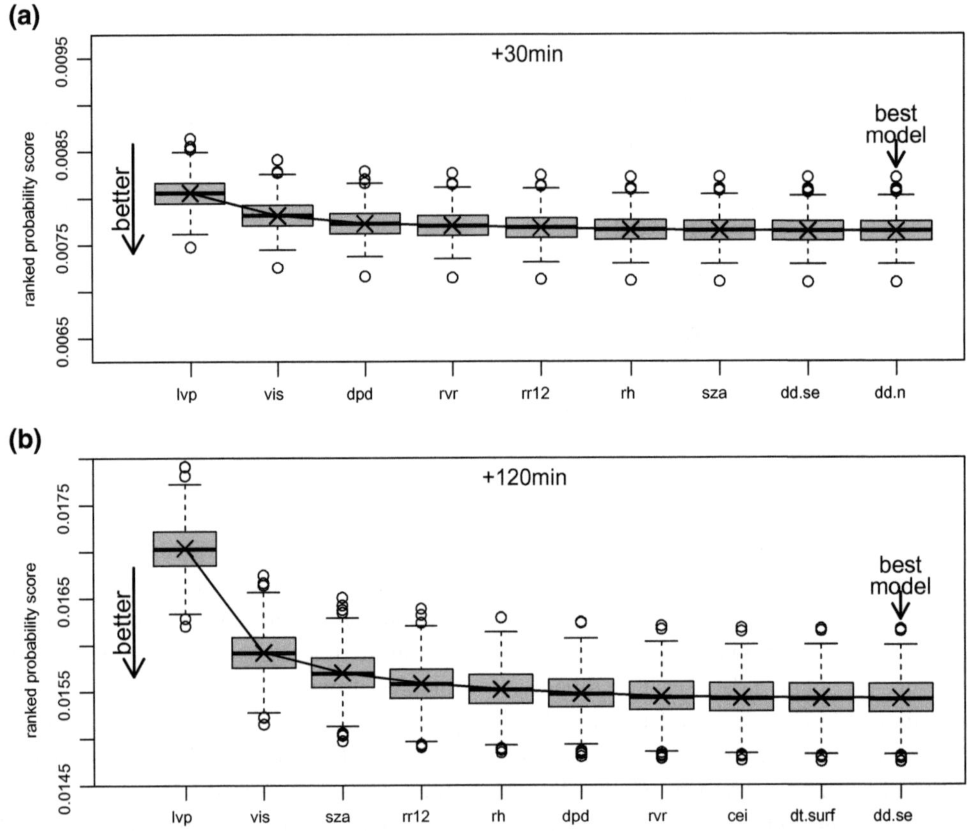

Figure 5
RPS model skill due to parameter selection with the forward stepwise selection method for the two lead times **a** + 30-min and **b** + 120-min. The model size increases from left to right adding the predictor (see Table 2) which improves the model most. The black line shows the median of the RPS. The last model indicates the best model where the stepwise algorithm stops

Table 3

Coefficients β of the OLR models for the four different lead times

	+ 30-min	+ 60-min	+ 90-min	+ 120-min
α_1	10.177	11.006	11.800	12.436
α_2	12.379	12.479	12.972	13.440
α_3	16.778	15.972	16.022	16.200
$lvp_{1\|0}$	2.773	1.870	1.507	1.252
$lvp_{2\|1}$	1.930	1.134	0.846	0.698
$lvp_{3\|2}$	2.684	1.749	1.290	0.907
rvr	− 0.668	− 0.523	− 0.426	− 0.372
cei	–	–	0.010	0.012
vis	− 0.316	− 0.338	− 0.313	− 0.281
rh	0.134	0.131	0.131	0.132
dpd	− 0.721	− 0.731	− 0.714	− 0.649
rr12	− 0.451	− 0.511	− 0.543	− 0.573
dt.surf	–	0.063	0.071	0.059
dd.se	0.195	0.065	0.060	0.046
dd.n	0.099	–	–	–
sza	0.003	0.005	0.007	0.009

For the predictor names see Table 2, α_{1-3} indicate the thresholds between the *lvp* classes. Contrary from Table 2 units from *rvr* and *vis* in km, unit from *cei* in kft

The model coefficients β from the best models using the step wise algorithm are shown in Table 3. The sign of the coefficients indicates the sign of the effects, i.e. positive coefficients indicate higher probabilities of high *lvp* states when the predictor values increase. The first three parameters are the threshold coefficients α (see Eq. 2) between the four different *lvp* states. They can be interpreted as intercepts for the three thresholds. A further special feature of these models is the predictor *lvp*, which is an ordered variable with four levels. Within the model, *lvp* is replaced by three categorical variables which indicate if the latest *lvp* is higher *lvp* 0, higher *lvp* 1 and higher *lvp* 2. The predictors for the different lead times vary just slightly and remain the same for the lead times + 90-min and longer. As already shown in Fig. 5, the model for the shortest lead time performs best with fewer predictors compared to longer lead times.

Figure 6 shows the variation in *lvp* forecasts due to a individual predictor when all other predictors are kept constant at representative values. Representative values (red marks in Fig. 6) are values typical before low-visibility events (see Table 4). Figure 6a shows that high *lvp* categories become more likely when *rvr*, visibility (*vis*) and dew point depression (*dpd*) decrease or when relative humidity (*rh*) or solar zenith angle (*sza*) increase. Decreasing values of *sza* indicate stronger radiative cooling and, therefore, favor low visibility. The predictor *lvp* has the strongest effect, which again confirms the importance of the persistence for short lead times. The effects for *lvp*, visibility and humidity parameters are nonlinear while all the other predictors produce linear effects. The factor variable *dd.se* with winds from the southeast sector has only a minor effect. However, the probability for high *lvp* states increases with winds from southeast compared to other directions. Surprisingly, the probability for high *lvp* states increases with higher ceiling height (Fig. 6b). An interpretation of this negative effect is that a low ceiling increases the long-wave downward radiation and prevents strong radiative cooling of the ground. Thus, it is less likely for a fog layer to develop. The negative effect of the precipitation within the preceding 12 h (*rr12*) is also unexpected, but again a possible explanation is that the remaining precipitation clouds reduce radiative cooling of the ground.

Comparing the effect plots from the + 30-min (Fig. 6a) and + 120-min (Fig. 6b), forecasts illustrate in general lower probabilities for the higher *lvp* states at the + 120-min forecasts. This indicates that the OLR model forecasts tend to converge to climatology with increasing lead times. Furthermore, the large effect of the predictor *lvp* (see Fig. 6a) indicates that the forecast information is concentrated on this predictor at the lead time + 30-min. In contrast at the lead time + 120-min, the effect of *lvp* becomes weaker while the other effects remain similar. This behavior is consistent with the model coefficients shown in Table 3, where the coefficients for *lvp* decrease with increasing lead times while the coefficients of several other predictors, e.g. *sza*, *dpd*, or *dt.surf*, increase or at least remain approximately unchanged.

4.2. Model Stability

To estimate the sensitivity of the OLR models to the training data set length, Fig. 7 shows RPS of + 60-min forecasts based on one up to six cold seasons training data. The forecast are evaluated out-of-sample for three cold seasons. The box plots for the OLR models indicate a slight model improvement with an increasing training data set from one cold season to two cold seasons. The OLR models based on two or more cold seasons have similar skill but the model based on four cold seasons has the best RPS. The shortest training data set length of one cold season includes about 5000 observations which are already an appropriate data set to estimate an OLR model with the applied model configuration. Taken together, forecast accuracy improves with increasing training data set length up to two cold seasons and remains similar for longer training data sets.

4.3. Model Performance

The results of the final *lvp* forecasts are shown in Fig. 8. The models are based on data of nine cold seasons using the best model configuration resulting from the stepwise predictor selection. Figure 8a

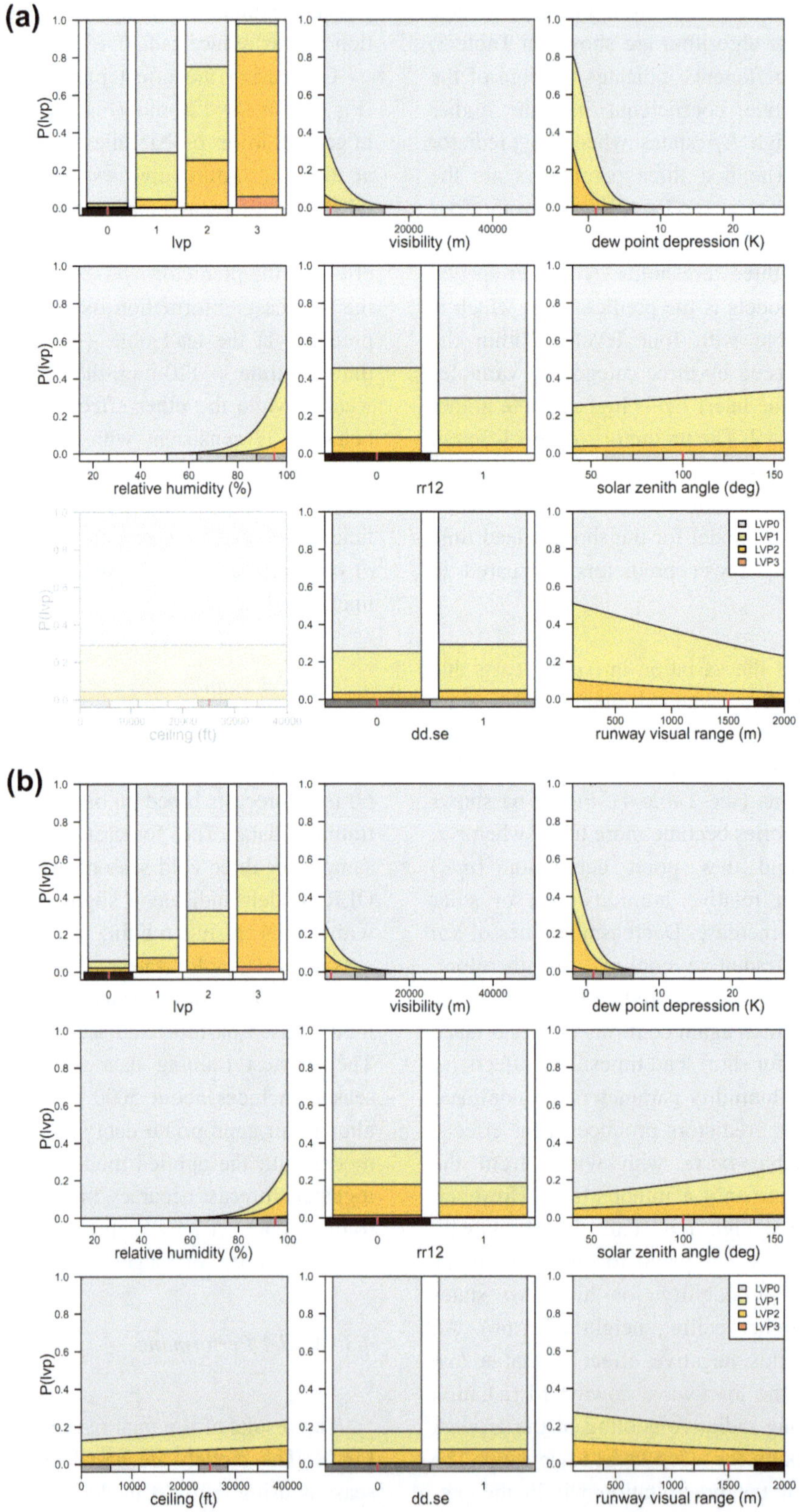
(a)
(b)
P(lvp)
lvp
visibility (m)
dew point depression (K)
relative humidity (%)
rr12
solar zenith angle (deg)
ceiling (ft)
dd.se
runway visual range (m)
LVP0
LVP1
LVP2
LVP3

◀Figure 6
Effect of the predictors for the **a** + 30-min and **b** + 120-min forecast. The four colors indicate the probability for each *lvp* state dependent on the particular predictors. The observed distribution of the particular predictors is shown as bar with gray shading below each effect plot. Typical conditions before an *lvp* event (red markers see Table 4) are chosen as representative values for the marginal effects. The effect plot with the fade out indicate that these predictors have not been selected in the particular model and therefore have constant effects

illustrates the verification for OLR compared to the climatology and the persistence forecasts. The OLR outperforms climatology and persistence for each lead time. The skill of all forecasts decreases with increasing lead time but the benefit of OLR over persistence increases. OLR and persistence converge to climatology but within the lead times considered here, the performance of climatology as a forecast of *lvp* is clearly worse. The spread of the RPS indicates the uncertainty due to the model estimation determined by bootstrapping (see Sect. 2.3). The uncertainty is smallest for the OLR model compared to persistence and climatology, and this uncertainty does not vary substantially with lead time.

When the nowcasting performance is evaluated only for two (un/affected by low visibility conditions)

Table 4

Fixed values for the marginal effects shown in Fig. 6

Parameter	Value	Parameter	Value
lvp	0	*rr12*	Yes
rvr	1500	*dt.surf*	1
cei	25,000	*dd.se*	Yes
vis	1500	*dd.n*	No
rh	95	*sza*	100
dpd	1		

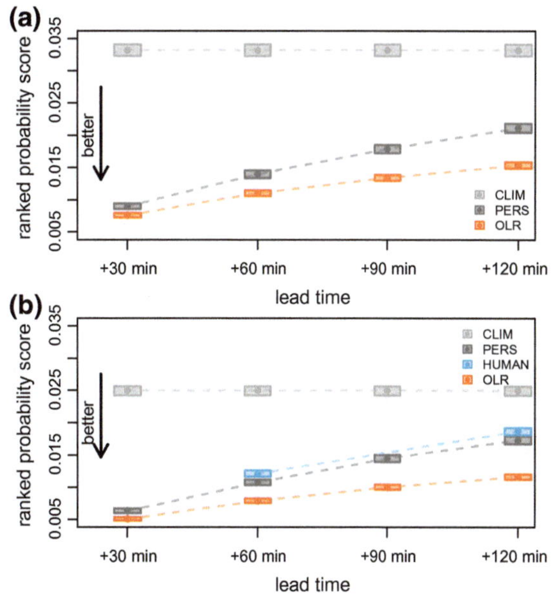

Figure 8
Model skill based on RPS as function of the lead time for **a** the full data set and **b** for the subset where the human observations are available. The dots represent the 50% quantile, the shadings in the lighter color show the 95% confidence interval. The model acronyms stand for climatology (CLIM), persistence (PERS), ordered logistic regression (OLR), and the human forecaster (HUMAN). The human forecasts are available from September 2012 to March 2017 for the lead times + 60-min and + 120-min between 5 and 20 UTC only

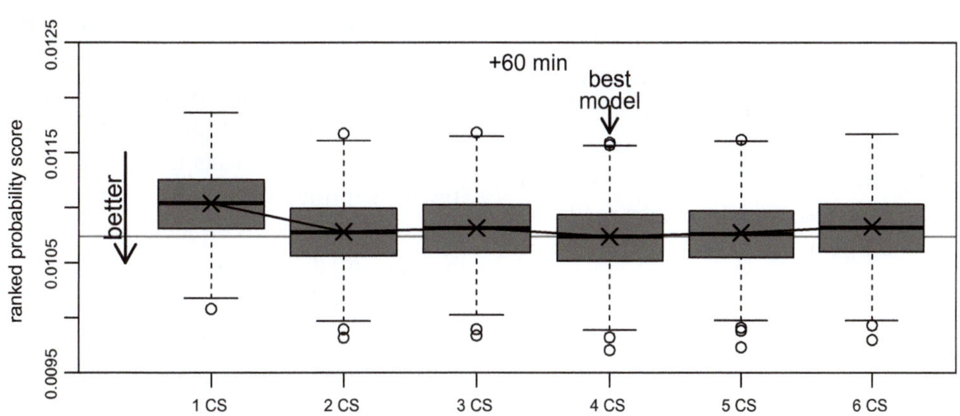

Figure 7
Model stability of the OLR due to variation of training data set length from one to six cold seasons (CS). RPS values are out-of-sample on a three-cold season test data set for the + 60-min forecast. Persistence is shown as reference model

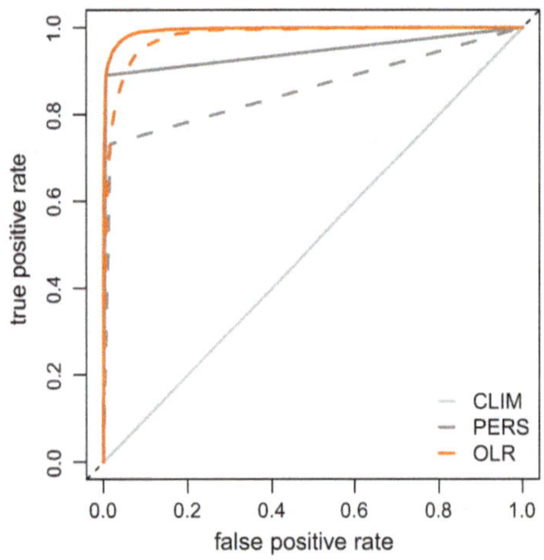

Figure 9
Model performance evaluated with ROC curves for the two (instead of four) category nowcasts for operations unaffected (*lvp* 0) or affected (*lvp* 1–*lvp* 3) by low-visibility conditions. Models are climatology (CLIM), persistence (PERS) and the ordered logistic regression (OLR) for the lead times + 30-min (solid line) and + 120-min (dashed line)

instead of four categories, the ROC curves in Fig. 9 also show the OLR to outperform persistence already at the shortest lead time of + 30-min and even more so for + 120-min. A climatological forecast lies on the diagonal of the ROC diagram, which (per definition) represents the forecasting skill of randomly drawing from the climatological distribution.

Human forecasts are available as third reference (Fig. 8b), albeit only over a shorter period (September 2012 to March 2017), part of the day (5–20 UTC) and 60 min lead-time intervals instead of 30 min. The human forecast is comparable to persistence; therefore, OLR also outperforms the human forecasts. Due to the rare occurrence of *lvp* events, and the larger number of *lvp* 0 during the day, when considering only the daytime OLR, persistence and climatology perform better compared to the results on the full data set in Fig. 8a.

5. Discussion and Conclusion

Predicting *lvp* (low-visibility procedure) states is very challenging because small changes in the physical drivers of low-visibility formation and dissipation processes have a large impact on *lvp*. The OLR (ordered logistic regression) model has been used to develop a probabilistic nowcasting system producing direct airport-specific, low-visibility forecasts tested at Vienna International Airport.

The OLR model leads to promising forecasts and outperforms persistence and climatology, although persistence offers predictions to be reckoned with especially at the very short lead times. This nowcasting approach is also competitive with human forecasts. A benefit of this method is to produce probabilistic forecasts of a variable relevant for to the end user within one consistent model. The probabilistic forecasts of such a model allows different end users, e.g., air traffic controllers or air traffic managers, to extract the information regarding their requirements.

The variable influencing the forecasts most is *lvp*. Additional important variables are the horizontal visibility and the humidity measures. Contrary to our expectations, wind speed and the temperature stratification of the first 100 m above the ground did not contribute information for future *lvp* states at Vienna International Airport.

The presented data-driven nowcasting system is transferable to other airports but it needs to consider site-specific predictors. For instance, considering the *lvp* climatology guides us to focus on the cold season only and justifies omission of the warm season, during which low-visibility events almost never occur. The nowcasting system uses the available measurements on the site and its vicinity and objectively selects the important variables to form the best model. Dependent on the regime of the site (e.g., coastal or continental) different physical effects drive the development and dissipation of low-visibility conditions. These effects need to be adequately represented in the model variables to achieve accurate forecasts. The usage of additional model inputs and increase of the forecast temporal resolution would have an additional gain for the operational use of the nowcasting system (cf. Dietz et al. 2017).

In summary, these findings suggest that the OLR model is a suitable tool for computationally fast probabilistic visibility forecasts which can provide useful guidance to aviation decision makers.

Acknowledgements

Open access funding provided by University of Innsbruck and Medical University of Innsbruck. This study has been supported by the Austrian Research Promotion Agency FFG 843457 and benefits greatly from Markus Kerschbaum, Andreas Lanzinger and the operational personnel of Vienna International Airport. The authors thank the Austrian Aviation Agency ACG for providing access to the data.

References

Bartokov, I., Bott, A., Bartok, J., & Gera, M. (2015). Fog prediction for road traffic safety in a coastal desert region: Improvement of nowcasting skills by the machine-learning approach. *Boundary-Layer Meteorology*, *157*(3), 501–516. https://doi.org/10.1007/s10546-015-0069-x.

Bergot, T. (2016). Large-Eddy simulation study of the dissipation of radiation fog. *Quarterly Journal of the Royal Meteorological Society*, *142*(695), 1029–1040. https://doi.org/10.1002/qj.2706.

Bocchieri, J. R., & Glahn, H. R. (1972). Use of model output statistics for predicting ceiling height. *Monthly Weather Review*, *100*(12), 869–879. https://doi.org/10.1175/1520-0493(1972)100%3c0869:UOMOSF%3e2.3.CO;2.

Bremnes, J. B., & Michaelides, S. C. (2007). Probabilistic visibility forecasting using neural networks. *Pure and Applied Geophysics*, *164*(6–7), 1365–1381. https://doi.org/10.1007/s00024-007-0223-6.

Christensen. R.H.B. (2015). Ordinal—Regression Models for Ordinal Data. R package version 2015.6-28. http://www.cran.r-project.org/package=ordinal/

Dietz, S. J., Kneringer, P., Mayr, G. J., & Zeileis, A. (2017). Forecasting low-visibilit procedure states with tree-based statistical methods. Working papers, Faculty of Economics and Statistics, University of Innsbruck. https://eeecon.uibk.ac.at/wopec2/repec/inn/wpaper/2017-22.pdf. Accessed 15 Sept 2017.

Dutta, D., & Chaudhuri, S. (2015). Nowcasting visibility during wintertime fog over the airport of a metropolis of India: Decision tree algorithm and artificial neural network approach. *Natural Hazards*, *75*(2), 1349–1368. https://doi.org/10.1007/s11069-014-1388-9.

Egli, S., Thies, B., Drnner, J., Cermak, J., & Bendix, J. (2017). A 10 year fog and low stratus climatology for europe based on meteosat second generation data. *Quarterly Journal of the Royal Meteorological Society*, *143*(702), 530–541. https://doi.org/10.1002/qj.2941.

Fabbian, D., de Dear, R., & Lellyett, S. (2007). Application of artificial neural network forecasts to predict fog at Canberra International Airport. *Weather and Forecasting*, *22*(2), 372–381. https://doi.org/10.1175/WAF980.1.

Gultepe, I., Hansen, B., Cober, S. G., Pearson, G., Milbrandt, J. A., Platnick, S., et al. (2009). The fog remote sensing and modeling field project. *Bulletin of the American Meteorological Society*, *90*(3), 341–359. https://doi.org/10.1175/2008BAMS2354.1.

Gultepe, I., Tardif, R., Michaelides, S. C., Cermak, J., Bott, A., Bendix, J., et al. (2007). Fog research: A review of past achievements and future perspectives. *Pure and Applied Geophysics*, *164*(6–7), 1121–1159. https://doi.org/10.1007/s00024-007-0211-x.

Haeffelin, M., Bergot, T., Elias, T., Tardif, R., Carrer, D., Chazette, P., et al. (2009). PARISFOG: Shedding new light on fog physical processes. *Bulletin of the American Meteorological Society*, *91*(6), 767–783. https://doi.org/10.1175/2009BAMS2671.1.

Herman, G. R., & Schumacher, R. S. (2016). Using reforecasts to improve forecasting of fog and visibility for aviation. *Weather and Forecasting*, *31*(2), 467–482. https://doi.org/10.1175/WAF-D-15-0108.1.

Hilliker, J. L., & Fritsch, J. M. (1999). An observations-based statistical system for warm-season hourly probabilistic forecasts of low ceiling at the San Francisco International Airport. *Journal of Applied Meteorology*, *38*(12), 1692–1705. https://doi.org/10.1175/1520-0450(1999)038%3c1692:AOBSSF%3e2.0.CO;2.

Marzban, C., Leyton, S., & Colman, B. (2007). Ceiling and visibility forecasts via neural networks. *Weather and Forecasting*, *22*(3), 466–479. https://doi.org/10.1175/WAF994.1.

Murphy, A. H., & Winkler, R. L. (1984). Probability forecasting in meteorology. *Journal of the American Statistical Association*, *79*(387), 489–500. https://doi.org/10.1080/01621459.1984.10478075.

Vislocky, R. L., & Fritsch, J. M. (1997). An Automated, observations-based system for short-term prediction of ceiling and visibility. *Weather and Forecasting*, *12*(1), 31–43. https://doi.org/10.1175/1520-0434(1997)012%3c0031:AAOBSF%3e42.0.CO;2.

Wilks, D. S. (2011). *Statistical methods in the atmospheric sciences*. Oxford: Academic Press.

Winkelmann, R., & Boes, S. (2006). *Analysis of microdata*. Berlin: Springer.

World Meteorological Organization. (2006). Guide to meteorological instruments and methods of observation. Tech. rep, Genf, Schweiz: Secretariat of the WMO

(Received September 26, 2017, revised February 9, 2018, accepted April 6, 2018, Published online April 17, 2018)

Pure Appl. Geophys. 176 (2019), 2179–2190

https://doi.org/10.1007/s00024-018-1882-1

Pure and Applied Geophysics

Fog Occurrence and Associated Meteorological Factors Over Kempegowda International Airport, India

Saumya G. Kutty,[1] G. Agnihotri,[2] A. P. Dimri,[1] and I. Gultepe[2,3,4]

Abstract—The increase in fog frequency over the past few decades is a major cause of concern for the aviation and transportation sectors. Accurate forecasting of the spatio-temporal extent of fog is crucial for minimizing socioeconomic losses. The present study attempts to characterize the fog frequency and associated meteorological factors over Kempegowda International Airport, Bengaluru (KIAB), in Karnataka, India. Maximum fog occurrence is observed during the month of December, followed by January. The time of onset of fog lies usually between 1800 and 0300 UTC. No fog is formed between 0400 and 1700 UTC indicating the role of radiation fog. The predominant wind direction during fog events is east or southeasterly. There is significant positive correlation between the fog frequency and both the northeast monsoon, October–November (0.72), as well as December–January–February (DJF) rainfall (0.80). Soil moisture conditions during the DJF period also play a key role in fog occurrence and its climatology, which is evident from the correlation coefficient of order 0.68. These suggest that further research is needed for understanding the extent of impact on aviation at KIAB.

Key words: Fog, radiative cooling, visibility, variability, forecast.

1. Introduction

Fog is characterized by suspension of very small droplets and ice crystals of water in air that result in a visibility of less than 1 km (Gultepe et al. 2017; WMO 1992). Haze and mist are terms used intermittently when the visibility is poor, but greater than 1 km. The effects of fog are manifold. Low visibility owing to fog events has an adverse influence on human life and the economy, in particular the aviation and transportation sectors (Gultepe and Heymsfield 2016). The location for the present study, Kempegowda International Airport, Bengaluru (KIAB), in Karnataka, India, has witnessed huge losses due to delays and cancelation of flights owing to fog occurrence as reported in popular media (e.g., The Hindu 2015; Times News Network 2016, 2017). Whiffen (2004) and Gultepe et al. (2007) termed the losses due to fog comparable to other weather events like tornadoes and hurricanes. Thus, the characterization of fog, especially in the changing climate scenario becomes imperative. However, its complexity by virtue of occurrence in the boundary layer of the atmosphere and spatio-temporal variability combined with our lack of understanding of the physical mechanisms, such as nucleation processes involved, makes real-time fog prediction a challenge.

2. Background

The fog can be classified based on thermodynamic conditions using temperature, RH, and their related phase as described in Gultepe and Heymsfield (2016) or on environmental conditions such as upslope, frontal, radiation fog, marine, or advection fog (Tardif and Rasmussen 2007; Gultepe et al. 2007; Saurabh and Dimri 2015). Out of these, the most widely studied fog type is radiative cooling fog over land (Willet 1928; Gultepe et al. 2007; Liu et al.

Electronic supplementary material The online version of this article (https://doi.org/10.1007/s00024-018-1882-1) contains supplementary material, which is available to authorized users.

[1] School of Environmental Sciences, Jawaharlal Nehru University, New Delhi 110067, India. E-mail: apdimri@hotmail.com

[2] Flood Meteorological Office, India Meteorological Department, Bangalore, India.

[3] Cloud Physics and Severe Weather Research Section, Meteorological Research Division, Environment Canada, Toronto, ON, Canada.

[4] Faculty of Engineering, UOIT, Oshawa, ON, Canada.

2011; Cuxart and Jiménez 2012; Román-Cascón et al. 2016). This is partly due to the poor logistics involved in measurements at oceans as compared to the land-based observations. This type includes further radiation fog, high inversion fog and advection–radiation fog. Radiation fog is formed by radiative cooling of the surface during conducive meteorological conditions including low wind speed, high relative humidity, clear sky, stable boundary layer and micro as well as mesoscale interactions (Roach 1995; Bergot and Guedalia 1994). With the advent of industrialization resulting in increasing pollution, fog frequency and variability have been altered (Witiw and LaDochy 2008; Niu et al. 2010; Li et al. 2012; Williams et al. 2015). Therefore, it has become a major source of concern for the expanding aviation sector. Consequently, a number of studies focused on understanding the influence of localized factors as well as large scale forcings on fog dynamics have been undertaken as follows:

The Fog Remote Sensing and Modeling project (FRAM) was initiated with the objective of developing microphysical parameterizations, remote sensing methods for fog forecasting in continent (FRAM-C), marine (FRAM-L) environment and ice fog (FRAM-B) among others (Gultepe et al. 2008, 2009, 2014). Syed et al. (2012) studied the fog variability over South Asia using both ground observations and reanalysis datasets and found two distinct regime shifts associated with a gradual increase in relative humidity. They attributed this variability to the wave train originating from the North Atlantic in the upper troposphere. Zhao et al. (2016) tied the decadal variability in haze over China with the Pacific Decadal Oscillations. Niu et al. (2010) linked the increase in fog frequency over east-central China to the weakening East Asian monsoon circulation, which in turn is due to enhanced urban aerosol loading.

In India, fog is a common phenomenon in the winter season, especially in the Indo-Gangetic plains (IGP). The trough-like topography in conjunction with the moisture advected by the eastward moving low pressure systems is associated with the mid latitude Subtropical westerly jet (SWJ) called Western Disturbances (WDs) (Dimri et al. 2015) favor fog formation. Radiation fog is formed in the rear sector, while advection fog is formed in the forward sector of WDs (Sawaisarje et al. 2014). Srivastava et al. (2016) looked into the geo-statistical aspects of fog in the IGP and found more than 66% of fog days in the winter season (December–January) alone during 1971–2010. The fog frequency over this area has increased at an alarming rate in recent years. Jenamani (2007), Bhowmik et al. (2004), Singh (2011) and various other studies based on ground observations have attributed this to enhanced pollution (Bisht et al. 2016; Fosu et al. 2017; Ghude et al. 2017). Furthermore, the spatial extent of fog and associated radiative impacts were studied using satellite-based observations. Increase in spatial extent (Choudhury et al. 2007) and significantly higher aerosol optical depth (AOD), especially in the eastern part (Gautam 2014), were observed. Negative temperature anomalies over regions of fog-induced radiative cooling suggest colder foggy days in comparison to non-foggy days (Sathiyamoorthy et al. 2015). Despite considerable existing literature, fog variability under the changing climate scenarios and its spatio-temporal distribution over different terrains and climate regimes is not very well understood. Also, even though fog is a common phenomenon in India, there is a scarcity of literature regarding fog formation, its dynamics and prediction in southern peninsular India. This lack of understanding of the climatology and associated fog characteristics acts as a constraint on the choice of optimal parameterizations and thereby accurate fog forecasting.

Attempts at operational forecasts of fog have had limited success due to the inability of models to correctly represent the physical processes involved in its formation. Inadequate representation of the nocturnal boundary layer, surface heterogeneities and atmospheric interactions with soil and vegetation acts as a constraint. Low-resolution models fail to capture the local-scale feedbacks. Presently, two major approaches for predicting fog include using one-dimensional (1-D) numerical models with and without mesoscale terms. Duynkerke (1991) used a simple 1-D model and obtained a three-layer structure within the boundary layer before fog formation. These layers describe the cooling mechanisms that lead to fog formation. The surface layer was characterized by a small difference between longwave radiative cooling and turbulent warming, while cooling in the midregion was due to turbulent heat flux divergence.

Radiative flux divergence led to cooling in the topmost layer. It was also found that resistance of grass to heat transport gave way to higher temperature difference between vegetation and upper soil layer, leading to a subsequent drop in air temperature. This emphasized the role of vegetation in radiative cooling and its inadequate representation in models. The models forced with mesoscale features incorporate measures such as advection, pressure gradients and diffusion. Bergot and Guedalia (1994) used one such model to investigate, among others, the role of surface homogeneity and found four external factors, namely geostrophic wind, soil conductivity, cloud cover and advection that affect the atmospheric cooling rate. The time of fog formation depends highly on dew deposition and low cloud height. Attempts at fog prediction in India are limited and include statistically derived (Saurabh and Dimri 2015; Madan et al. 2000; Pasricha et al. 2003), dynamically derived (Goswami and Tyagi 2007), satellite derived (Choudhury et al. 2007; Ahmed et al. 2014) and diagnostic approaches (Payra and Mohan 2014).

In this work, with the aim of attempting fog forecast accurately in future, and owing to limited literature pertaining to fog occurrence in southern India (Mohapatra and Thulsidas 1998; Suresh et al. 2007), an analysis of fog over the Kempegowda International Airport in Bengaluru, Karnataka, India, is carried out. An understanding of fog climatology over the region of study will provide a foundation for future studies aiming to improve and make accurate predictions of fog and its impact on aviation applications. The specific objectives of this work are (1) to analyze the trend in fog frequency at the study area using ground-based observations and (2) to understand the extent of influence, and correlation between different meteorological factors e.g., precipitation and soil moisture, and fog intensity/occurrence.

3. *Study Area and Observations*

3.1. *Project Location*

Kempegowda International Airport (13.18°N, 77.70°E), originally named as the Bengaluru International Airport, is located in Bengaluru District of Karnataka, India. It opened in 2008 and is the third busiest airport in India. It lies near the town of Devanahalli, in the rural Bengaluru District about 40 km northeast of urban Bengaluru and is situated at an altitude of 920 m (Fig. 1). Catering to a population of 8.5 million in 2011, Bengaluru is classified as an urban agglomeration and is the fifth most populous city in India (Census of India 2011). Geographically, its northern part consists of a level plateau lined with a prominent ridge along the NE–SW direction, while a complex terrain lies to the south (Karnataka Gazette 1990). The ridge along with the dominant rock type gneiss and granite have a considerable influence on the drainage pattern and plays a part in soil moisture variation. The climate is very agreeable with temperature ranging from 33 °C in April to a minimum of 14 °C in January. Fog is a common phenomenon from October to February. The district receives rainfall from both summer as well as northeast monsoons. The average annual rainfall is about 880 mm (Sudhira et al. 2007). Soils include fine loamy to clay types and vegetation is largely deciduous. The seasonal variation in vegetation is observed from the true color Google Earth images in Fig. S1a–f. These composites taken from Google Earth Pro have been derived from Landsat/Copernicus satellite program (30 m spatial resolution). The time of acquisition for Landsat 7 and 8 are 10:00 am ± 15 min and 10:11 am ± 15 min' local time, respectively. As the airport began operation in

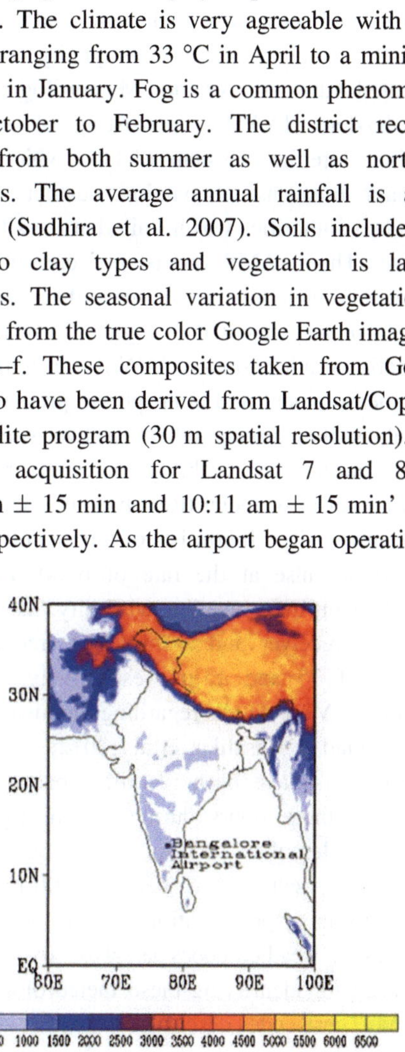

Figure 1
Elevation (m) and location of study area within India. The airport lies at an elevation of 915 m

2008, fog data corresponding to 03rd February 2004 (Fig. S1a) are not available. Fog occurred on 08 December 2009 (Fig. S1b) during the early morning hours of 0000–0300 UTC and therefore is not captured in the imagery. No fog occurred on 09 June 2015 (Fig. S1c).

3.2. Observations

Hourly surface observations of dry bulb and dew point temperature, wind speed and direction as well as visibility from 2008 to 2017 were obtained from India Meteorological Department (IMD) station at the site for months October to March (ONDJFM). IMD archives and preserves meteorological data after thorough scrutiny which ensures quality control. An Integrated Automatic Aviation Meteorological Instrument System (IAAMS), consisting of Automatic Terminal Interface System (ATIS) which ensures automatic broadcast of aviation weather reports to concerned authorities was installed at the site for this purpose. The IAAMS consists of current weather instruments including wind speed/direction sensor (Vaisala make), 16-bit data logger (Telvent make), automatic weather station (AWS) for temperature, wind measurement and atmospheric pressure sensor (Vaisala make) among others at two sites on the airport. Further, for visibility measurement, a dual baseline transmissometer (Revolver) is installed at all three runway sites. It consists of a transmitter, which emits light pulse at the rate of 6–60 flashes/min depending on the prevailing visibility, as well as a long base receiver and a short base receiver at a distance of 75 and 15 m, respectively, from the transmitter. More details regarding the measurements can be found in Hosalikar et al. (2012).

Fog hours have been identified based on IMD criteria as those hours during which fog reduces visibility to less than 1000 m. Those hours in which visibility is reduced to below 1000 m due to thunderstorms or precipitation, etc. are not included. The present weather codes as given by IMD have been used for identifying these meteorological phenomena. These codes describe the weather during the period of 1 h immediately preceding the time of observation. For example, an observation at 0100 UTC describes the weather between 0000 and 0100 UTC. Thus, fog at any particular time indicates that fog occurred either in patches (i.e., duration in minutes) or prevailed throughout the preceding hour. Any day on which fog occurred is considered a foggy day. Even if the visibility drop was observed for half hour or less, the day is termed as a foggy day. For determination of fog onset time and duration, fog events have been defined as follows:

Fog observed in consecutive observations, i.e., for one or more hours, is termed as a fog event. A fog hour has been termed as an event provided there was no fog prior to or following the fog hour. For example, if fog occurred at 0500 UTC but no fog was reported at 0400 and 0600 UTC, then it will be termed as a fog event of duration 0–60 min. If fog occurs again after a break of 1 h or more, it will be termed as another fog event. As an example, on 8 December 2015, fog was observed between 0100 and 0300 UTC (Event 1) and thereafter from 2200 to 2300 UTC (Event 2). These two are considered separate fog events, as there was a break from 0400 to 2100 UTC (no fog was reported during this period). Thus, multiple fog events can occur during a day. For event 1, as fog could have occurred in patches for any duration between 0100 and 0300 UTC, the duration of event 1 has been taken as 0–120 min. Also, as fog was not reported at 0000 UTC observation, it is assumed that fog onset took place between 0000 and 0100 UTC and, therefore, time of onset for event 1 was taken as 0100 UTC. Thus, while a fog event can constitute a single hour or multiple, consecutive hours of observation, a foggy day can constitute multiple fog events. IMD classification is used to define the extent of fog intensities (IMD 1982; Laskar 2013). This classification defines four types of fog based on the extent of visibility. These are namely:

1. light (< 1000 and ≥ 500 m),
2. moderate (< 500 and ≥ 200 m),
3. thick (< 200 and ≥ 50 m) and
4. very thick (< 50 m).

4. Method

To understand the temporal variation in fog occurrence over KIAB, yearly and monthly averaged distribution of fog hours was plotted. Fog hours were classified based on the extent of visibility into light, moderate, thick and very thick fog. Time of onset and duration of fog events were observed for each month averaged over 2008–2017. The sky conditions were plotted 1–2 h prior to the onset of fog. The state of sky is derived from IMD classification (IMD 2008) which divides cloud cover (octa) into five categories, namely clear sky (0 octa), mainly clear (1–2 octa), partly clear (3–4 octa), generally cloudy (5–7 octa) and cloudy (> 7 octa). Wind rose diagrams were used to observe the prevailing wind speeds and directions during both ONDJFM and ON. While fog occurrence/non-occurrence, meteorological variables including dry bulb temperature, dew point temperature, wind speed and direction as well as visibility were obtained from IAAMS installed at the site, and other variables including rainfall, soil moisture and specific humidity were obtained from gridded datasets. For estimating the seasonal variability and influence of rainfall on fog occurrence; IMD precipitation daily data set in gridded format at 0.25° spatial resolution for the given duration was used. Reanalysis datasets for soil moisture at surface and specific humidity at 1000 hPa, both having 0.25° spatial resolution with 6 h temporal resolution, were obtained at time 0000 UTC from the European Centre for Medium Range Weather Forecasts (ECMWF) Reanalysis, hereafter ERA Interim (Dee et al. 2011). Interannual variation in the meteorological fields was observed by way of anomalies calculated from 2008 to 2017 climatology (ONDJFM). To further assess the nature of relationship between the above-mentioned variables and fog frequency, statistical measures of correlation (both parametric and non-parametric) and regression analysis were used. While parametric tests are more accurate, their tendency to make assumptions on a fixed set of parameters renders them less robust in comparison to non-parametric tests. Pearson correlation coefficient, r (parametric), is used to measure the degree of relationship between two variables (Hall 2015). Its value varies between + 1 and − 1. It gives both the strength of relationship (0 indicates poor correlation, while 1 indicates strong correlation) as well as the direction of relationship (+ indicates positive correlation, while − indicates negative correlation). While Pearson's r assumes an inherent linear relationship between variables, Spearman and Kendall tests (non-parametric) do not assume any such linearity. As the strength and direction of relationship between the variables could be seasonally dependent, the correlation of fog frequency with soil moisture as well as rainfall, both averaged over DJF, ON and ONDJFM, is analyzed. To further analyze the extent of influence of these variables on fog, regression analysis was used where the coefficient of determination (R^2) gives the percentage of the explained variance.

5. *Results*

The average number of fog days per year in November–February (NDJF) over Kempegowda International Airport is found to be 30.11 during 2008–2017. Mohapatra and Thulsidas (1998) conducted a study on Hindustan Airport, also located in Bengaluru, and found a significant rise in fog frequency from 15.83 days per year during 1983–1989 to 24.75 days per year during 1990–1993. This implies that there is steady rise in fog frequency in the region. Although the ratio of foggy days to non-foggy days over KIAB for ONDJFM in the duration 2008–2017 is ~ 0.25, the intensity and persistence of these events adversely affect the aviation sector. Figure 2a, b shows both the yearly and monthly percentage frequency of fog hours, respectively. The highest incidence of fog is observed during the month of December, followed by January and November. There is interannual variation in the total number of fog hours when averaged over ONDJFM, with year to year variation evident for different months (Fig. 2c–h). In December 2010–2011 and 2015–2016, the higher incidences of fog are attributed to soil moisture and relative humidity as is explained in the following paragraphs.

Ahmed et al. (2014) investigated aerosol fog interaction and found that the minimum requirement for drop in visibility below 1 km includes fog droplet number concentration of 3.23/cm^3 and aerosols at 95% relative humidity. Fog intensity based on IMD classification (IMD 1982; Laskar 2013) for each month is shown in Fig. 3. Binned histogram for fog hours based on the extent of visibility is plotted in

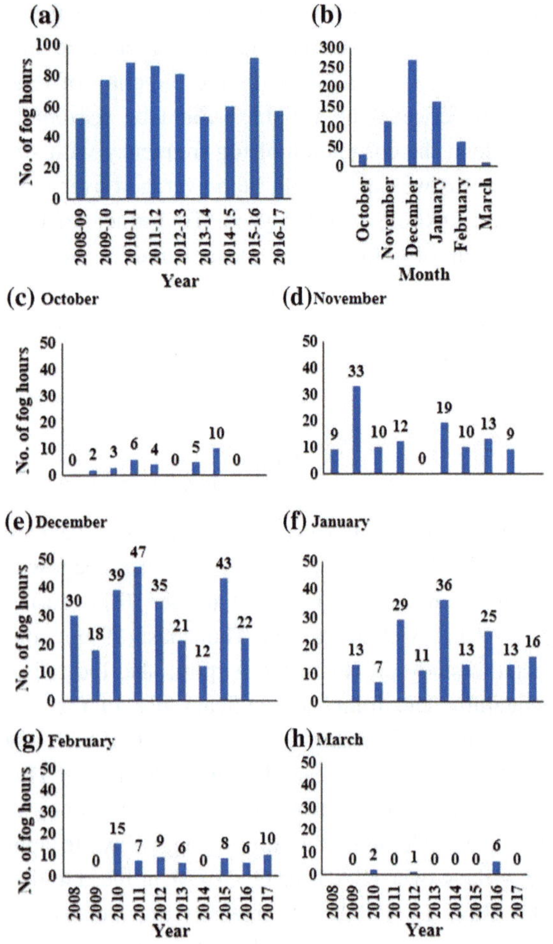

Figure 2
Distribution of fog hours; **a** yearly average over ONDJFM, **b** monthly average over 2008–2017 and **c–h** monthly averages for each year over the ONDJFM period, respectively. The maximum number of fog hours occur in December, followed by January

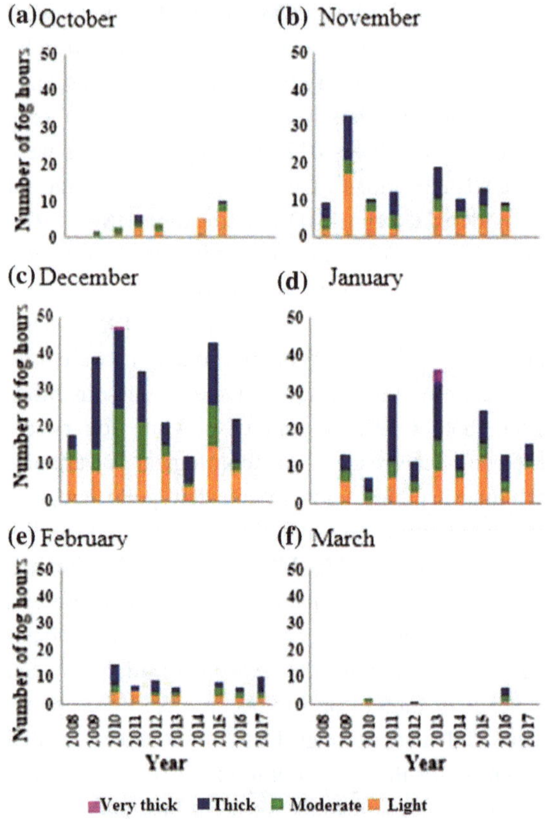

Figure 3
Monthly distribution of fog hours based on the extent of visibility: very thick (< 50 m), thick (≥ 50 m and < 200 m), moderate (≥ 200 m and < 500 m) and light (≥ 500 m and < 1000 m)

Fig. S2. Light fog is predominant in almost all months, particularly in the months of October–November. Thick fog is seen mostly during the months of November, December and January. Very thick fog is formed only during December and January indicating the possible role of low temperature and radiative cooling at night. During winter season, lower sun angle and the reduced number of daylight hours result in cooling as well as inversion conditions. This in turn traps more pollutants resulting in the formation of thick fog.

Fog onset in model predictions is associated with dew point depression in the order of 1 °C (Meyer and Rao 1999) and visibility < 1 km. Figure 4a–c shows the temperature, visibility and wind speed anomalies for fog hours over the period of the study. A dew point depression of almost 0 °C along with dry bulb temperature ≤ 20 °C and negative wind speed anomalies seem to favor fog formation. Wind speed anomaly indicates deviation of monthly mean from climatology, i.e., positive (negative) anomalies correspond to higher (lower) than the average wind speed. For significant sample sizes, visibility is lower during December and January, and a possible reason for this could be the role of aerosol loading in winters, similar to that observed over IGP in Fosu et al. (2017) could be present. Sreekanth (2013) found mean aerosol optical depth (AOD) values of 0.16 and 0.14 nm derived from MODIS Terra and Aqua satellites during the month of December with mean fine mode fraction (FMF) of 0.34 nm in winter. The

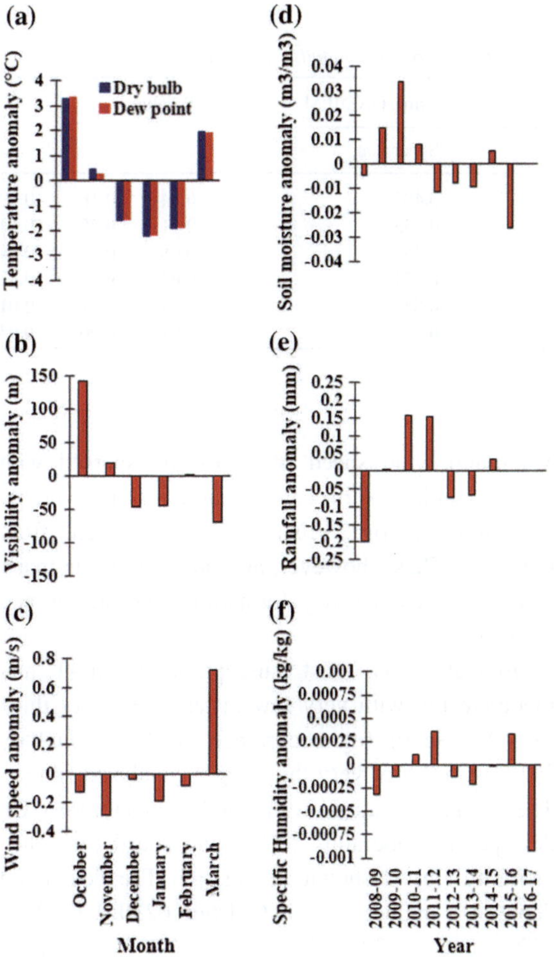

Figure 4

Anomaly diagrams of meteorological conditions corresponding to foggy days. **a–c** ONDJFM averaged over 2008–2017 and **d–f** year-wise distribution averaged over ONDJFM. **a** Dry bulb and dew point temperature (°C), **b** visibility (m), **c** wind speed (m/s), **d** soil moisture (m3/m3), **e** rainfall (mm) and **f** specific humidity (kg/kg)

tendency of long range transport of aerosols toward southern peninsula was also stated in agreement with Badarinath et al. (2009). The extent of fog occurrence in the project location and IGP differs in terms of duration and monthly distribution. While fog occurs between October and March in Bengaluru, fog in the IGP is mostly limited to the months of November–December–January. The duration of fog, however, is comparatively higher in the IGP (Srivastava et al. 2016). This is possibly due to higher aerosol loading (Krishna Moorthy et al. 2007; Sreekanth 2013) and conducive meteorological conditions including lower temperature during the winter season.

Rainfall is another critical factor, as it provides moisture for fog formation. The rainfall anomaly time series in Fig. 4e shows positive (negative) anomaly in the year 2015 (2013 and 2014), indicating good (poor) northeast monsoon. Concurrently, the fog frequency is higher in 2015–2016 as compared to 2013–2014 and 2014–2015 indicating possible positive correlation. The correlation of percentage frequency of fog hours with soil moisture and rainfall (Table 1) indicates a significant positive association of fog occurrence with ON ($R^2 = 59.6\%$) and DJF rainfall as well as with soil moisture in DJF ($R^2 = 48.4\%$). Drastic seasonal variation in vegetation can be inferred from Fig. S1a–f. Dense vegetation during December (Fig. S1b), followed by scarce vegetation in the months of February (Fig. S1a) and June (Fig. S1c) indicate the role of the northeast monsoon, soil moisture available during winter months and that of cropping patterns. Dense vegetation in December is also evident from the normalized difference vegetation index (NDVI) images (Fig. S1d–f) where higher values of NDVI (> 0.425) are seen (e) as opposed to February (d) and June (f). Seasonal as well as yearly changes in vegetation could possibly enhance (deteriorate) the frequency of future fog events in case of wet (dry) northeast monsoon and winter season. Further, both soil moisture anomalies and specific humidity anomalies during fog days have been plotted in Fig. 4d, f, respectively. Distinct peaks in 2010–2011 and 2015–2016, in agreement with the fog frequency in Fig. 2a, indicate a strong correlation that is also interpreted from Table 1. Positive anomalies for rainfall and specific humidity are observed in 2010–2011. Low temperatures combined with sufficient moisture and low wind speeds may result in an inversion, provided conducible clear sky conditions and turbulence are present.

The role of radiative cooling is evident from Fig. 5a, where the time of onset for fog events lies between 1800 and 0300 UTC. Further, 67.7% fog events occur between 0000 and 0300 UTC, while the rest occurs between 1800 and 2300 UTC. No fog is formed between 0400 and 1700 UTC. Radiation fog forms due to radiative cooling at night and dissipates after sunrise. The duration of fog (Fig. 5b) varies within less than 1 h to around 6 h. Only one

Table 1

Correlation of percentage frequency of fog hours with soil moisture and rainfall, respectively

Variable	Season	Correlation with percentage frequency of fog hours during ONDJFM			*P* value		
		Pearson	Kendall	Spearman			
Soil moisture	DJF	**0.684**	0.388	0.483	**0.04**	0.18	0.19
	ON	0.568	0.333	0.533	0.11	0.25	0.14
	ONDJFM	0.598	0.277	0.433	0.08	0.35	0.24
Rainfall	DJF	**0.802**	**0.714**	**0.821**	**0.02**	**0.03**	**0.03**
	ON	**0.726**	**0.714**	**0.833**	**0.04**	**0.01**	**0.01**
	ONDJFM	0.709	0.619	**0.785**	0.07	0.06	**0.04**

Highlighted values correspond to $P < 0.05$ indicating significance at 95% confidence level

exceptional event on 11–12 December 2012 lasted for around 8 h. The cloud cover indicative of the sky conditions 1–2 h prior to onset of fog events is shown in Fig. 5c. It is observed that 54.8% (68.9%) of fog events were associated with sky cover between 0 and 4 octa (i.e., clear, mainly clear and partly clear) during 1 (2) h prior to fog onset. This reasserts the presence of clear sky condition and its role in fog formation. One fog event, however, was found to have cloudy condition (8 octa) 1 h before fog onset. The percentage clear sky (0 octa) during 1 and 2 h prior to fog onset is 4.5 and 9.0%, respectively. Mist (90.7%) and haze (83.0%) were found to be prominent among weather phenomena 1–2 h prior to fog onset. While the frequency of events which were associated with haze decreased from 50.2%, 2 h prior to onset to 32.8%, 1 h prior to onset, the opposite (i.e., increase in frequency) was seen in case of mist. This indicates that for a majority of events, haze and mist could act as a precursor to fog formation. However, more observations in this regard would be required to reach any conclusions.

Sawaisarje et al. (2014) attributed the fog formation over the region to the westward moving low pressure waves over peninsular India (including south interior part of Karnataka, Tamil Nadu and Telangana). These waves advect moisture which is a prerequisite to fog formation. Its strategic geographical location enables Bengaluru to benefit from both the summer as well as northeast monsoon. The contribution of northeast monsoon during October–November can be seen in Fig. 5e, f from the northeasterly winds. However, the predominant wind direction during ONDJFM is east–southeasterly. Maximum wind speed of 5 m/s interspersed with 14.2% incidences of calm weather is a favorable condition for fog formation and is evident from Fig. 5e, f. These, however, are only few of the many factors influencing fog formation, persistence and dissipation.

Bengaluru is most susceptible to light and moderate fog with very few instances of very thick fog. A case of fog occurrence on 16–17 January 2013, with 2 h of very thick fog, is analyzed further. The diurnal variation in dry bulb temperature (°C), dew point temperature (°C), wind speed (m/s) and visibility (m) is shown in Fig. S3. The dew point depression reduces to 0 °C from 0230 LT (2100 UTC) until 0830 LT (0300 UTC). Wind speed remains < 3 m/s from 1830 LT (1300 UTC) on 16 January to 0830 LT (0300 UTC) on 17 January after which it increases gradually to a maximum value of 7.2 m/s. A dip in visibility is seen from 0130 LT (2000 UTC) to 0530 LT (0000 UTC), followed by 2 h of zero visibility. Increased wind and incoming solar radiation in the morning from around 0730 LT (0200 UTC) onward give way to fog dissipation and enhance the visibility.

6. Discussions

Most studies on fog, owing to its highly localized nature, rely to a great extent on ground-based observations. However, data availability is a limitation in southern India where fog occurrence is less in comparison to the Indo-Gangetic plains in northern

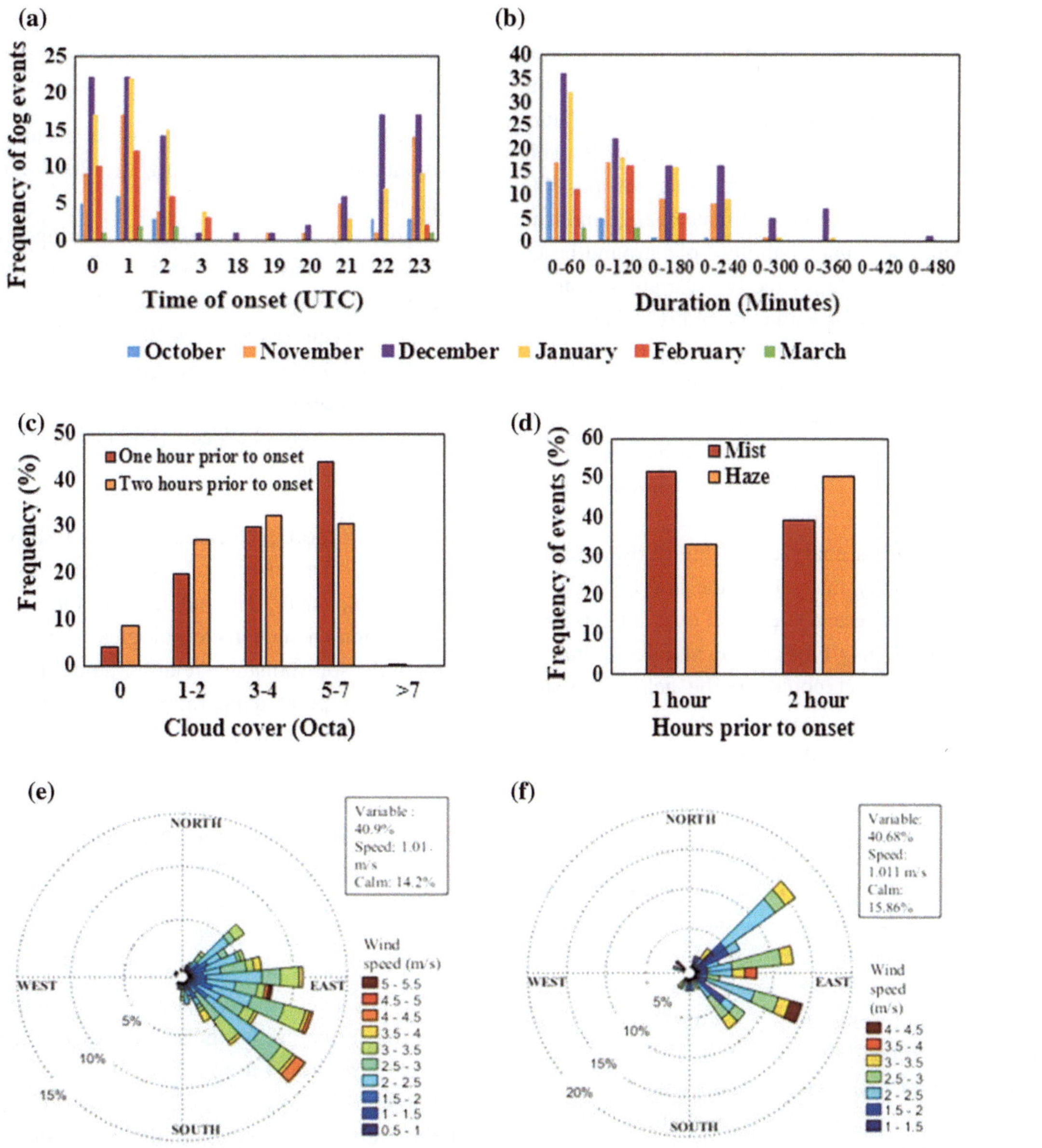

Figure 5

a Time of fog onset in UTC, **b** duration of fog events in minutes, **c** cloud cover (octa) 1–2 h prior to onset of fog events, **d** prominent weather phenomenon 1–2 h prior to onset of fog events and **e**, **f** wind rose for ONDJFM and ON, respectively, obtained from sensors installed by IMD

India. This contributes to the lack of understanding of the underlying physical mechanisms during fog formation and its persistence. Consequently, the impact of surface roughness, turbulence, vegetation and anthropogenic factors including urbanization and the effect of pollutants could not be adequately quantified. However, based on online imagery (Figs. S4, S5) for years 2011–2012, derived from Bhuvan-Thematic Services portal (Bhuvan-Thematic Services 2018) maintained by the National Remote Sensing Centre (NRSC) and Indian Space Research Organization (ISRO), it is seen that the built-up area around KIAB is minimal and, as fog is a localized phenomenon, therefore the effect of urbanization on fog over KIAB is almost negligible. Vegetation including agriculture, forests and grasslands constituted an area of 5002.67 km^2 (86.03%) (Fig. S5). However, as the airport is functional since 2008, no prior literature regarding fog in the study area was available. This limited any attempts at understanding the possible decadal changes or long-term climate change-induced effects. Due to unavailability of ground-based observation from IMD for the two variables, namely soil moisture and specific humidity, they were obtained from gridded reanalysis data at 0.25° spatial resolution. The coarser resolution of gridded data in comparison to the ground-based observations adds uncertainty and limits the distinction between foggy and non-foggy days. However, the general agreement in the interannual variation of fog days for all meteorological variables analyzed in this study indicates that the uncertainty may affect the hourly/daily fog occurrence, but is greatly reduced over longer timescales. Distinguishing types of fog and determination of the time of onset could not be achieved using ground-based observations in the absence of a dynamical forecast model. The drop in dew point depression in conjunction with fog onset can be captured using a one-dimensional model. Even though fog is a localized phenomenon requiring expertise of the observer's familiarity with the area of study, most models incorporate detailed observations and data that a routine observer may not possess. Meyer and Rao (1999) developed a simple 1-D model which requires local, routinely available observations and to be run on desktop computers. The model gives 24 h prediction, and input parameters include dry bulb temperature, dew point temperature and wind speed at 1800 LT, rainfall in the past 24 h and location-specific details such as soil type. With an objective of providing effective prediction of fog onset over the Kempegowda International Airport in Bengaluru, this one-dimensional model is being amended to suit the localized weather as well as topography and will form future scope of work.

7. Conclusions

The study aimed at characterizing fog occurrence based on associated meteorological conditions over KIAB, Karnataka, India, during the study period of 2008–2017. The following conclusions are drawn from this work:

- The highest number of fog events occurred in the month of December during the 2008–2017 time period.
- Light fog is most common during October–November and thick fog is seen mostly during November–December–January. Most fog events fall between duration 0–60, with less common extreme 0–360 min duration events occurring during December.
- Fog onset occurs between 1800 and 0300 UTC and its formation seems to cease sometime after sunrise.
- As the airport is located in the rural Bengaluru District, the effect of urbanization is almost negligible and the major contributors to fog formation include meteorological factors such as rainfall, soil moisture, relative humidity, temperature, dew point temperature and the predominant east south east wind.
- A higher number of fog events are observed during 2010–2011 and 2015–2016, consistent with the inter-annual variation in the meteorological variables studied.
- There is significant correlation between fog occurrence with rainfall (0.80 for DJF and 0.78 for ON) and soil moisture (0.68 for DJF).
- The lower temperatures and soil moisture may aid in fog formation by means of radiative cooling at night, resulting in fog onset between 1800 and 0300 UTC. Thus, we conclude that radiation fog is the predominant fog type in the project location.
- The seasonal variation in vegetation and the presence of dense vegetation during December indicate its plausible role in fog formation.

Overall, it is concluded that, the day to day variability in the meteorological and environmental conditions, fog physical parameters need to be further looked into. The roles of boundary layer, radiative fluxes, surface roughness, pollutants as well as possible advective mechanisms among others are to be understood to reach conclusive decisions for development/application of a model for fog prediction. A one-dimensional model developed by Meyer and Rao

(1999) is being modified according to the localized meteorology of the region and will form the future scope of work. This will provide for prediction of fog over the airport in future.

Acknowledgements

The first author wishes to acknowledge the University Grants Commission funding for Junior Research Fellowship and Jawaharlal Nehru University for providing the requisite facilities and assistance as part of the UPOE scheme.

References

Ahmed, R., Dey, S., & Mohan, M. (2014). A study to improve night time fog detection in the Indo-Gangetic Basin using satellite data and to investigate the connection to aerosols. *Meteorological Applications*. https://doi.org/10.1002/met.1468.

Badarinath, K. V. S., Kharol, S. K., Sharma, A. R., & Roy, P. S. (2009). Fog over Indo-Gangetic Plains—a study using multisatellite data and ground observations. *IEEE Journal of Selected Topics in Applied Earth Observations and Remote Sensing, 2*(3), 185–1995.

Bergot, T., & Guedalia, D. (1994). Numerical forecasting of radiation fog. Part I: Numerical model and sensitivity tests. *Monthly Weather Review, 122*(6), 1218–1230.

Bhowmik, S. K. R., Sud, A. M., & Singh, C. (2004). Forecasting fog over Delhi—an objective method. *Mausam, 55*(2), 313–322.

Bhuvan-OCM2/NDVI/NRC, NRSC/ISRO-India. (2018). http://bhuvan.nrsc.gov.in/data/download/index.php?c=t&s=LV&p=ndvi. Accessed 8 Apr 2018.

Bhuvan-Thematic Services. (2018). *LULC-50 K map/NRC, NRSC/ISRO-India.* http://bhuvan.nrsc.gov.in/gis/thematic/index.php. Accessed 8 Apr 2018.

Bisht, D., Tiwari, S., Dumka, U., Srivastava, A., Safai, P., Ghude, S., et al. (2016). Tethered balloon-born and ground-based measurements of black carbon and particulate profiles within the lower troposphere during the foggy period in Delhi, India. *The Science of the Total Environment, 573,* 894–905.

Census of India. (2011). *Urban Agglomerations/Cities having population 1 lakh and above"* (PDF). http://www.censusindia.gov.in/2011. Accessed 20 Sept 2017.

Choudhury, S., Rajpal, H., Saraf, A., & Panda, S. (2007). Technical note: Mapping and forecasting of North Indian winter fog: An application of spatial technologies. *International Journal of Remote Sensing*. https://doi.org/10.1080/01431160600993470.

Cuxart, J., & Jiménez, M. A. (2012). Deep radiation fog in a wide closed valley: Study by numerical modeling and remote sensing. *Pure and Applied Geophysics, 169*(5–6), 911–926.

Dee, D. P., Uppala, S. M., Simmons, A. J., Berrisford, P., Poli, P., Kobayashi, S., et al. (2011). The ERA-Interim reanalysis: Configuration and performance of the data assimilation system. *Quarterly Journal of the Royal Meteorological Society, 137,* 553–597.

Dimri, A. P., Niyogi, D., Barros, A. P., Ridley, J., Mohanty, U. C., Yasunari, T., et al. (2015). Western Disturbances: A review. *Reviews of Geophysics*. https://doi.org/10.1002/2014rg000460.

Duynkerke, P. G. (1991). Radiation fog: A comparison of model simulation with detailed observations. *Monthly Weather Review, 119,* 324–341.

Fosu, B. O., Wang, S. Y. S., Wang, S. H., Gillies, R. R., & Zhao, L. (2017). Greenhouse gases stabilizing winter atmosphere in the Indo-Gangetic plains may increase aerosol loading. *Atmospheric Science Letters*. https://doi.org/10.1002/asl.739.

Gautam, R. (2014). Challenges in early warning of the persistent and widespread winter fog over the Indo-Gangetic plains: A satellite perspective. *Reducing disaster: Early warning systems for climate change* (pp. 51–61). Berlin: Springer.

Ghude, S. D., Bhat, G. S., Prabhakaran, T., Jenamani, R. K., Chate, D. M., Safai, P. D., et al. (2017). Winter fog experiment over the Indo-Gangetic plains of India. *Current Science, 112*(4), 767–784.

Goswami, P., & Tyagi, A. (2007). *Advance forecasting of onset, duration and hourly fog intensity over Delhi. Research Report No. RR CM 0714, CSIR Centre for Mathematical Modelling and Computer Simulation, Bangalore, India.* http://citeseerx.ist.psu.edu/viewdoc/download?doi=10.1.1.599.2641&rep=rep1&type=pdf. Accessed 26 Sept 2017.

Gultepe, I. B., Hansen, S. G., Cober, G., Pearson, J. A., Milbrandt, S., Platnick, P., et al. (2009). The fog remote sensing and modeling field project. *Bulletin of American Meteorological Society, 90,* 341–359.

Gultepe, I., & Heymsfield, A. J. (2016). Ice fog, ice clouds, and remote sensing; introduction. *Pure and Applied Geophysics, 173,* N.9. https://doi.org/10.1007/s00024-016-1380-2.2977-2982.

Gultepe, I., Heymsfield, A. J., Gallagher, M., Ickes, L., & Baumgardner, D. (2017). Ice fog: The current state of knowledge and future challenges. *Meteorological Monographs, 58,* 4.1–4.24.

Gultepe, I., Kuhn, T., Pavolonis, M., Calvert, C., Gurka, J., Isaac, G. A., et al. (2014). Ice fog in Arctic during FRAM-IF project: Aviation and nowcasting applications. *Bulletin of American Meteorological Society., 95,* 211–226.

Gultepe, I., Minnis, P., Milbrandt, J., Cober, S. G., Nguyen, L., Flynn, C., et al. (2008). The fog remote sensing and modeling (FRAM) field project: Visibility analysis and remote sensing of fog. *Remote Sensing Applications for Aviation Weather Hazard Detection and Decision Support, 7088*(12), 708–803.

Gultepe, I., Tardif, R., Michaelides, S., et al. (2007). Fog research: A review of past achievements and future perspectives. *Pure and Applied Geophysics, 164,* 1121–1159.

Hall, G. (2015). *Pearson's correlation coefficient.* http://www.hep.ph.ic.ac.uk/~hallg/UG_2015/Pearsons.pdf. Accessed 7 Apr 2018.

Hosalikar, K. S., Mohan, K. N., Vashishta, R. D., & Tyagi, A. (2012). An integrated automatic aviation meteorological instrument system at CSI Airport, Mumbai. *Mausam, 63*(2), 247–260.

India Meteorological Department. (1982). *Weather codes*. Pune: IMD.

India Meteorological Department. (2008). *Forecaster's Guide, IMD, Pune*. http://imdpune.gov.in/Weather/Reports/forecaster_guide.pdf. Accessed 7 Apr 2018.

Jenamani, R. (2007). Alarming rise in fog and pollution causing a fall in maximum temperature over Delhi. *Current Science, 93*(3), 314–322.

Karnataka Gazette. (1990). *Government of Karnataka* http://gazetteer.kar.nic.in/gazetteer/distGazetteer.html. Accessed 25 Sept 2017.

Krishna Moorthy, K., Babu, S. S., Badarinath, K. V., Sunilkumar, S. V., Kiranchand, T. R., & Ahmed, Y. N. (2007). Latitudinal distribution of aerosol black carbon and its mass fraction to composite aerosols over peninsular India during winter season. *Geophysical Research Letters, 34*(8), Article ID L08802.

Laskar, S. (2013). Some statistical characteristics of occurrence of fog over Patna airport. *Mausam, 64,* 345–350.

Li, Z. H., Yang, J., Shi, C. E., & Pu, M. J. (2012). Urbanization effects on fog in China: Field research and modeling. *Pure and Applied Geophysics, 169*(5–6), 927–939.

Liu, D., Yang, J., Niu, S., & Li, Z. (2011). On the evolution and structure of a radiation fog event in Nanjing. *Advances in Atmospheric Sciences, 28*(1), 223–237.

Madan, O. P., Ravi, N., & Mohanty, U. C. (2000). A method for forecasting of visibility at Hindon. *Mausam, 51*(1), 47–56.

Meyer, W., & Rao, G. (1999). *Pure and Applied Geophysics.* https://doi.org/10.1007/s000240050254.

Mohapatra, M., & Thulsidas, A. (1998). Analysis and forecasting of fog over Bangalore airport. *Mausam, 49*(1), 135–142.

Niu, F., Li, Z., Li, C., Lee, K. H., & Wang, M. (2010). Increase of wintertime fog in China: Potential impacts of weakening of the Eastern Asian monsoon circulation and increasing aerosol loading. *Journal of Geophysical Resources.* https://doi.org/10.1029/2009JD013484.

Pasricha, P. K., Gera, B. S., Shastri, S., Maini, H. K., John, T., Ghosh, A. B., et al. (2003). Role of the water vapour greenhouse effect in the forecasting of fog occurrence. *Boundary Layer Meteorology, 107*(2), 469–482.

Payra, S., & Mohan, M. (2014). Multirule based diagnostic approach for the fog predictions using WRF modelling tool. *Advances in Meteorology.* https://doi.org/10.1155/2014/456065.

Roach, W. T. (1995). Back to basics: Fog: Part 2—the formation and dissipation of land fog. *Weather, 50,* 7–11.

Román-Cascón, C., Steeneveld, G. J., Yagüe, C., Sastre, M., Arrillaga, J. A., & Maqueda, G. (2016). Forecasting radiation fog at climatologically contrasting sites: Evaluation of statistical methods and WRF. *Quarterly Journal of the Royal Meteorological Society, 142*(695), 1048–1063.

Sathiyamoorthy, V., Arya, R., & Kishtawal, C. (2015). Radiative characteristics of fog over the Indo-Gangetic Plains during northern winter. *Climate Dynamics.* https://doi.org/10.1007/s00382-015-2933-2.

Saurabh, K., & Dimri, A. P. (2015). Non-linearity explanation in artificial neural network application with a case study of fog forecast over Delhi region. *Pure and Applied Geophysics.* https://doi.org/10.1007/s00024-015-1205-8.

Sawaisarje, G., Khare, P., Shirke, S., Deepakumar, S., & Narkhede, N. M. (2014). Study of winter fog over Indian subcontinent: Climatological perspectives. *Mausam, 65,* 19–28.

Singh, C. (2011). Unusual long and short spell of fog conditions over Delhi and northern plains of India during December–January 2009–2010. *Mausam, 62*(1), 41–50.

Sreekanth, V. (2013). Satellite derived aerosol optical depth climatology over Bangalore, India. *Advances in Space Research, 51*(12), 2297–2308.

Srivastava, S. K., Sharma, A. R., & Sachdeva, K. (2016). A ground observation based climatology of winter fog: Study over the Indo-Gangetic Plains, India. *International Journal of Environmental Chemical Ecological Geological and Geophysical Engineering, 10*(7), 705–716.

Sudhira, H. S., Ramachandra, T. V., & Subrahmanya, M. H. B. (2007). City profile Bangalore. *Cities, 24,* 379–390.

Suresh, R., Janakiramayya, M. V., & Sukumar, E. R. (2007). An account of fog over Chennai. *Mausam, 58,* 501–512.

Syed, F., Körnich, H., & Tjernström, M. (2012). On the fog variability over South Asia. *Climate Dynamics.* https://doi.org/10.1007/s00382-012-1414-0.

Tardif, R., & Rasmussen, R. M. (2007). Event-based climatology and typology of fog in the New York City region. *Journal of Applied Meteorology and Climatology, 46*(8), 1141–1168.

The Hindu. (2015). Fog hits flight operations in Bengaluru, Published November 28, 2015. http://www.thehindu.com/news/cities/bangalore/fog-hits-flight-operations-at-kempegowda-international-airport-in-bengaluru/article7927436.ece. Accessed Nov 28 2017.

Times News Network. (2016). *Fog disrupts 127 flights at Kempegowda International Airport, Published December 30, 2016.* https://timesofindia.indiatimes.com/city/bengaluru/Fog-disrupts-127-flights-at-KIA/articleshow/56244624.cms. Accessed Nov 28 2017.

Times News Network. (2017). *Fog grounds 89 flights at Kempegowda International Airport; none cancelled, Published January 2, 2017.* https://timesofindia.indiatimes.com/city/bengaluru/fog-grounds-89-flights-at-kempegowda-international-airport-none-cancelled/articleshow/56284573.cms. Accessed Nov 28 2017.

Whiffen, B., Delannoy, P., & Siok, S. (2004). Fog: Impact on road transportation and mitigation options. National Highway Visibility Conference, Madison, Wisconsin, 18–19 May 2004.

Willet, H. C. (1928). Fog and haze, their causes, distribution, and forecasting. *Monthly Weather Review, 56,* 435–468.

Williams, A. P., Schwartz, R. E., Iacobellis, S., Seager, R., Cook, B. I., Still, C. J., et al. (2015). Urbanization causes increased cloud base height and decreased fog in coastal Southern California. *Geophysical Research Letters, 42*(5), 1527–1536.

Witiw, M. R., & LaDochy, S. (2008). Trends in fog frequencies in the Los Angeles Basin. *Atmospheric Research, 87*(3–4), 293–300.

World Meteorological Organization (WMO). (1992). *International Meteorological Vocabulary* (2nd ed.). Geneva: WMO. **(ISBN 978-92-630-2182-3).**

Zhao, S., Li, J., & Sun, C. (2016). Decadal variability in the occurrence of wintertime haze in central eastern China tied to the Pacific Decadal Oscillation. *Scientific Reports.* https://doi.org/10.1038/srep27424.

(Received November 30, 2017, revised April 13, 2018, accepted April 25, 2018, Published online May 3, 2018)

Pure Appl. Geophys. 176 (2019), 2191–2202

https://doi.org/10.1007/s00024-018-1781-5

Pure and Applied Geophysics

Fog at the Guarulhos International Airport from 1951 to 2015

Gutemberg Borges França,[1] Luiz Felipe Rodrigues do Carmo,[1] Manoel Valdonel de Almeida,[1] and Francisco Leite Albuquerque Neto[1]

Abstract—This paper presents and discusses the fog occurrences before and after the construction of the Guarulhos International Airport, using data from 1951 to 2015. The analysis showed the following: (1) a total of 19,816 h of fog were registered. (2) The minimum average, mean and maximum average of the fog temperature had significantly increased after the airport was constructed from 1.2 to 6.9, 12.1 to 14.5 and 20.2 to 20.7 °C, respectively, due to the urban development around the airport during the study period. (3) The average fog hours per year decreased by approximately 73.1%, i.e., from 492 ± 84.45 to 132 ± 54.51 h per year. (4) Most of the fog events occurred due to longwave cooling on clear nights with relatively low wind speeds (characterizing radiation fog), with over 65% having duration of 2 h and occurring in the early hours of the day during March–September period. (5) The maximum probability of fog occurrence dropped about 10% from before to after the construction of the airport. Finally, two fog events are investigated using data collected during the fog evolution using atmospheric sounding profiles (from an acoustic sounder) and automatic meteorological stations and preliminary results showed that the values of cooling rate and turbulent kinetic energy play key roles in the onset and growth-dissipation phases of the fog, respectively.

Key words: Fog, meteorology, airport, wind, visibility, SODAR.

1. Introduction

Fog is the major restrictive physical phenomenon that influences the landing and take-off operations at Guarulhos International Airport, São Paulo, Brazil. Restricted visibility for take-off and landing means that flights will be diverted or delayed; thereby causing substantial financial losses in the aviation and its related sectors. According WMO (1966), fog is defined as microscopic water droplets or ice crystals suspended in air that reduce horizontal visibility to less than 1000 m. It is usual to classify fog event based on its geneses as radiation fog (caused by cooling thermal radiation at the land surface, predominantly cloud-free atmosphere) and advection fog (caused by the blowing of warm-humid air on a cooler land/water surface). Studies about fog formation and forecasting have been conducted over time, for example, Emmons and Montgomery (1947) described the process of fog formation; Kunkel (1984) studied the parameterization for extinction coefficients and mean terminal velocities based on 11 fogs drop size measurements; Bott et al. (1990) and Bott (1991) had investigated the microphysics and physico-chemical properties of aerosols of radiation fog; Bergot and Guedalia (1994), Teixeira (1999) and Bergot et al. (2005) carried out studies on fog prediction using deterministic models; Tardif (2007) studied radiation fog and the impacts on vertical resolution in the numerical prediction model; Gultepe and Milbrandt (2007) and Gultepe et al. (2009) had been carried out studies about fog forecasting based on field campaigns, which had been investigated marine and continental fog conditions. In particular, Silva Dias and Jaschke Machado (1997) used a regional-scale numerical model to investigate the local circulation evolution and its impacts on convection and nocturnal fog occurrence in São Paulo. Oliveira (2002) proposed an alternative stochastic model for fog forecasting for the Guarulhos International Airport. Fedorova et al. (2008) investigated the physical and synoptic processes of fog and stratus cloud formation over the northern and southern coasts of Brazil, and Fedorova et al. (2013) investigated the visibility and its forecasting at the Maceio International Airport in the Brazilian Northeast. Da Rocha

[1] Laboratório de Meteorologia Aplicada-LMA, Departamento de Meteorologia, Universidade Federal do Rio de Janeiro-UFRJ, Av. Athos da Silveira Ramos, 274-Cidade, Universitária-Ilha do Fundão, Rio de Janeiro, RJ 21044-020, Brazil. E-mail: valdonel@lma.ufrj.br

et al. (2015) evaluated the impact of the horizontal resolution of a regional climate model on reproducing local climate features caused by foggy conditions at the Metropolitan Area of São Paulo. Colabone et al. (2015) proposed a neural network-based fog nowcasting model for the Brazilian Air Force airfield, called Campo Fontenelle, in Pirassununga city in the State of São Paulo, Brazil. Silva et al. (2016) proposed a conceptual model for runway changes based on acoustic sound data for the Guarulhos International Airport. The latter is the largest airport in Brazil based on the number of passengers served and is also the hub for most national/international aviation companies operating in Brazil. The airport is often affected by fog events every year bringing major problems to landing and take-off operations, and thus was recently equipped with an acoustic sounder, or SODAR, aiming to study fog evolution by observing the top fog layer. Therefore, the objective of this work is to present a unique diagnostic of the variations and daily probability fog occurrences before and after construction of the Guarulhos International Airport, São Paulo, Brazil, based on data collection from 1951 to 2015, and also, show the preliminary results of the characterizations of fog events using high temporal resolution data from a SODAR and automatic meteorological stations.

2. *Study Region and Data*

The Guarulhos International Airport was inaugurated on 20 January 1985 and originated as a Brazilian Air Force airfield called Cumbica, which, in the Tupi-Guarani indigenous language, means low cloud or fog. In fact, the Guarulhos airport location favors the formation of fog since its runways are located near the Forest of Cantareira Mountain and permanent marsh region due to waters from the Baquirivu-Guaçu River. Figure 1 gives an overview of the study area. Over time, the data collection facilities at the Guarulhos International Airport have been improved. Table 1 provides details about the dataset used to characterize the fog events in this work. Most of the fog data used here was generated from human observations, except for the period starting in 2006 when an Automatic Meteorological Station (AMS) was installed. In addition, the visibilities on the two runways were automatically registered by Runway Visual Range (RVR-MITRAS Transmissometer-Vaisala) starting in 2005. Table 1 specifies the meteorological variables used, which are represented by W, u, v w, BLH, TKE, PRP and RH, which correspond to surface wind (i.e., direction in degrees and speed in m/s or knots), zonal wind component, meridional wind component, vertical wind component, boundary layer height (km), turbulent kinetic energy (m^2/s^2), precipitation (mm) and relative humidity (%), respectively. To standardize the differences between the automatically collected information, including the instantaneous visibility registered by the RVR, and that collected by human beings, only the fog events that lasted a full hour (i.e., 1, 2 h, etc.) were included in the study; therefore, the fog climatology presented in the results sections may be slightly less than that observed.

3. *Discussion of Results*

Figure 2a shows the average of T_{air} and the variations of fog hours per year during the period from 1951 to 2015 corresponding to the total of 19,816 h of fog. There are three distinct periods of numbers of fog hours, i.e., (1) from 1951 to 1979, during which the fog hours (FH) per year and STandard Deviation (STD) are equal to 491.47 and 84.45, respectively; (2) from 1980 to 1985 (FH = 266.2 and STD = 118.37); and (3) from 1986 to 2015 (FH = 132 and STD = 54.51). Overall, the average T_{air} and FH have opposite behaviors, i.e., when T_{air} was gradually growing, the FH declined. Period (2) coincides with the building period of the Guarulhos International Airport. It is also noted that between period (1) and period (3) there were a significant increase of the mean T_{air} of 1.4 °C, from 18.4 to 19.8 °C, and decrease of the average of number of fog hours of approximately 73.1%, from 491.47 to 132 h. The latter may be attributed to the development of the area surrounding the Guarulhos International Airport caused by the construction of new roads, houses, buildings and population increase. Figure 2b shows the variation minimum average

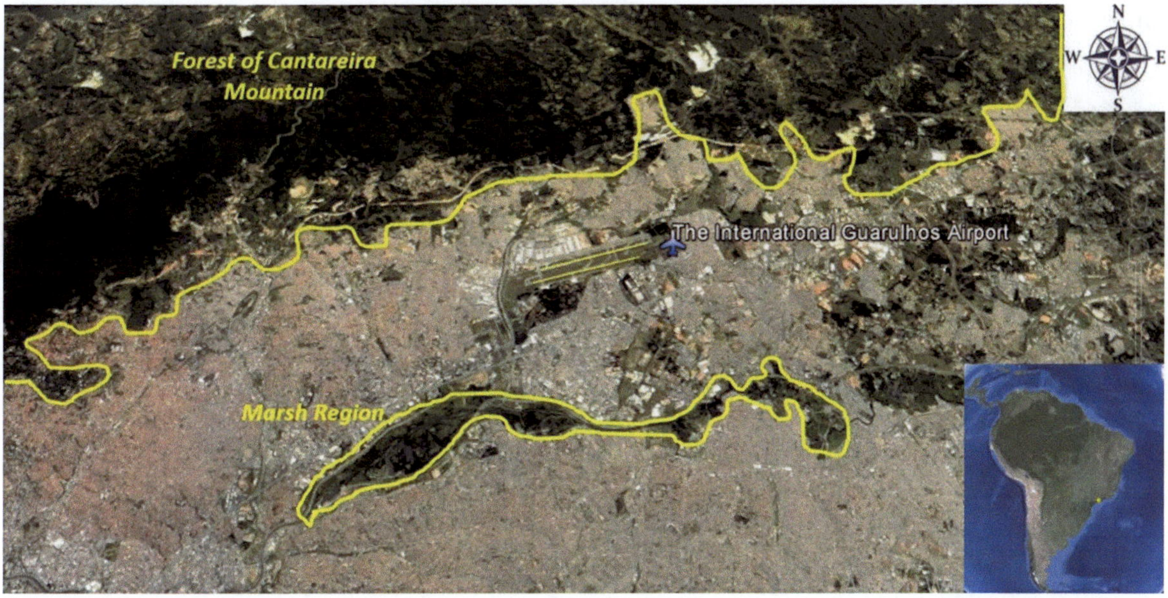

Figure 1
Characteristics of the region of the Guarulhos International Airport. Source: Adapted from http://www.google.com.br/maps

Table 1

Meteorological variables used and their data sources

Data source	Meteorological variables	Sampling frequency	Period
SODAR	u, v, w, BLH and TKE	15-min	2011–2015
Conventional meteorological station	W, T_{air}, RH, PRP and AP	1-h	1951–2006
Automated meteorological station	W, T_{air}, RH, PRP and atmospheric pressure	1-min	2006–2015
VAISALA-RVR	Visibility	1-s	2005–2015

(dotted line), mean (solid line) and maximum average (dashed line) of the fog T_{air} per year during the data period and the gray rectangle represents data recorded in period (2), as in Fig. 2a. And it can be observed that there were significant increases in the mean values of minimum average, mean and maximum average of the fog T_{air}, between period (1) and period (3), from 1.2 to 6.9 °C (this is one more expressive of 5.65 times), 12.1 to 14.5 and 20.2 to 20.7 °C, respectively.

Figure 3 shows the percentage variations of FH versus the hour of the day (local time) corresponding to 19.816 events (or hours) with visibilities of less than 1 km at the Guarulhos International Airport during the period from 1951 to 2015. The highest frequencies of fog occurrence (representing 91.1% of fog population) are distributed in early hours of the day, i.e., from 00 to 09 h (local time), with peak frequency of 15.6% at the time of sunrise at 06 h, and the main reason is that the most fog events in the study area are generated due to longwave cooling on predominantly cloud-free atmosphere with relatively low wind speeds (which typifies a radiation fog), as discussed here in Sect. 4. Figure 4a depicts the monthly frequency of FH. The highest fog frequencies occurred from March to September that is the coldest period of the year—corresponding to approximately 80.6% of fog events—with peaks of 16.3 and 16.0% in May and June, respectively. Figure 4b shows the percentage of fog occurrences in relation to fog duration in hours, which shows that the highest occurrence was 2 h, corresponding to 66.9%

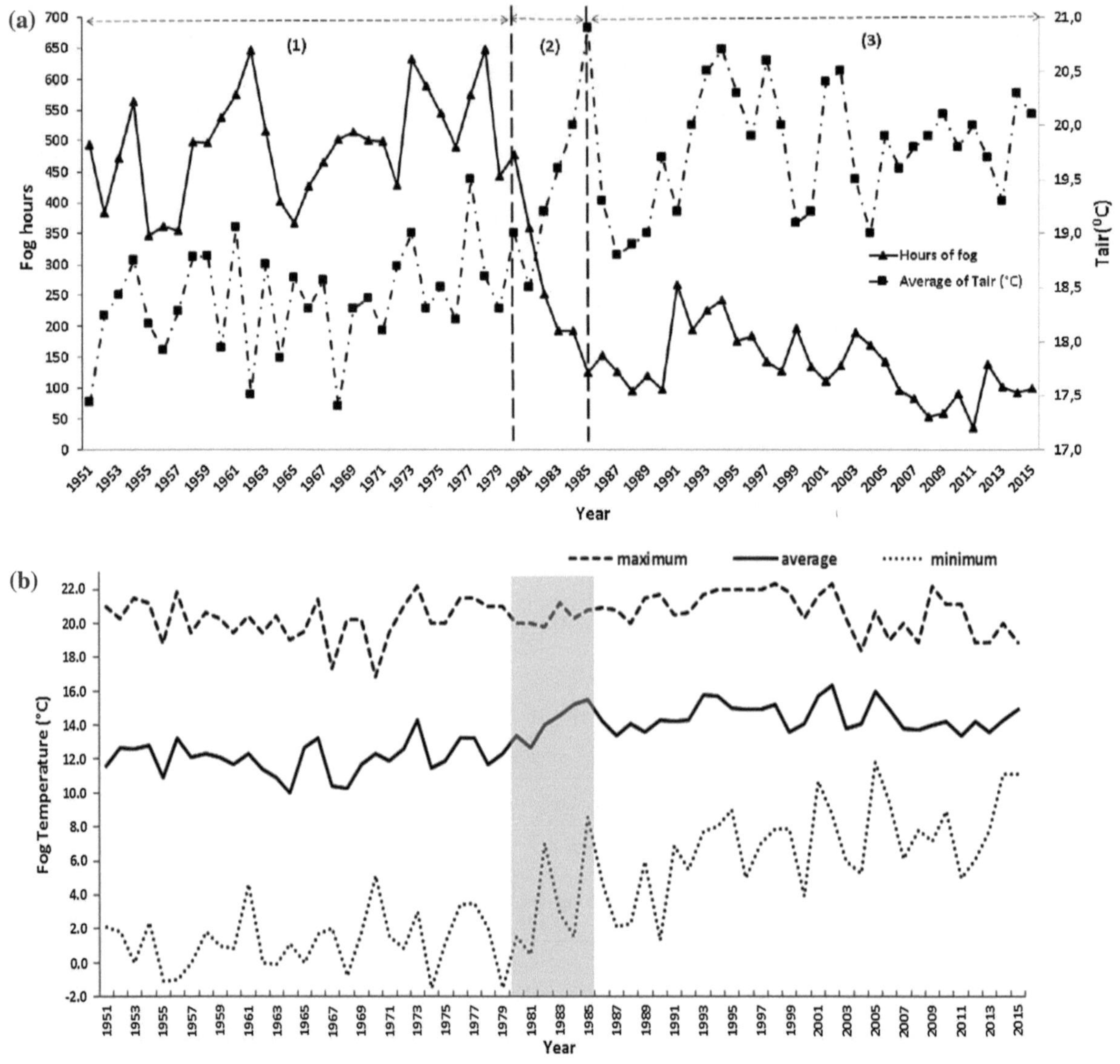

Figure 2
a Variations of fog hours and average T_{air} in the degree Celsius per year and where (1), (2) and (3) represent, in the dataset, the periods before, during and after the installation of the Guarulhos International Airport, respectively. **b** The maximum average (dashed line), mean (solid line) and minimum average (dotted line) values of the fog temperature per year during the study period

of the population. The longest recorded fogs were 12 h long. The fog periods and their percentages of occurrence are, respectively, written and specified in parentheses, i.e., 1 h (47.40%), 2 h (18.50%), 3 h (9.25%), 4 h (7.51%), 5 h (5.78%), 6 h (3.47%), 7 h (2.31%), 8 h (2.31%), 9 h (1.73%), 10 h (1.16%), 11 h (1.16%) and 12 h (0.58%). The dataset reveals that, during fog events, the mean values of visibility, the mean fog T_{air}, RH versus fog duration up to 1 h; greater than 1 h and less than or equal to 3 h and greater than 3 h are approximately equal to 798 ± 128 m, 16.6 ± 0.6 °C; 89.6 ± 8%; 592 ± 324 m; 16.3 ± 2.0 °C, 99.0 ± 1.0%; and 322 ± 173 m, 15.1 ± 1.8 °C, 99.1 ± 0.8%, respectively. As expected, the latter shows that longer fog events, i.e., lasting more than 3 h, have their average visibility and temperature lower than the shorter fog events.

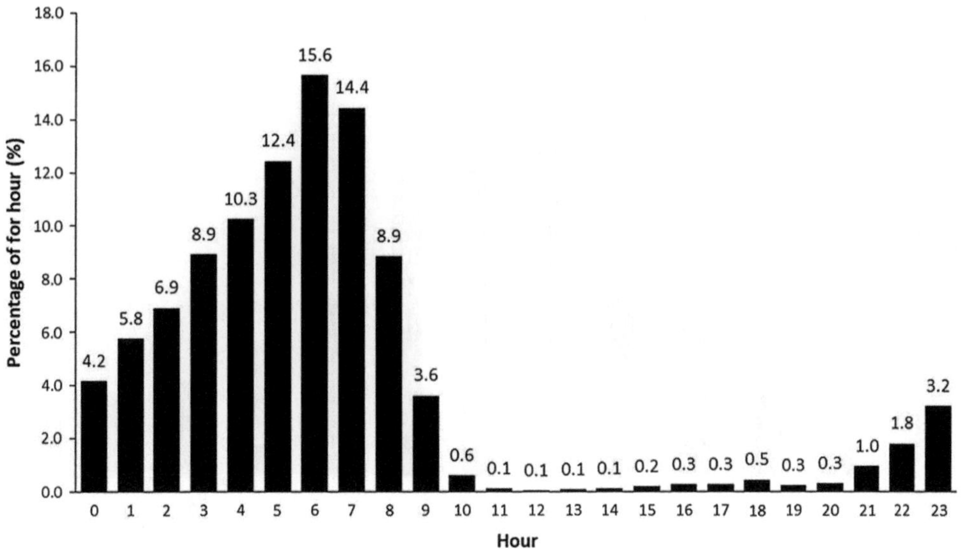

Figure 3
Percentage of fog occurrence versus local time during the period from 1951 to 2015

With the aim of studying foggy conditions before and after the construction of the Guarulhos International Airport, the dataset was subdivided into two subsets, corresponding to 1961–1990 and 1991–2015, and referred to herein as A and B. The period A is classified as the fog climatology, following the World Meteorological Organization (2011) definition (when fog conditions were recorded over a period of 30 years). On the other hand, although the B period is an incomplete climatology spanning 25-year, this period is used here to analyze the post-construction airport period. Figure 5a shows the variation fog hours versus the month of each period A and B represented by the dotted and solid lines, respectively. The peaks of the curves occurred in different months, i.e., in May for period A and in June for period B. In addition, a significant decrease in the amplitude variation of fog hours per month was observed during the B period. Possibly it can affirm that the decrease in fog hours, as previously discussed, is due to the increase in mean T_{air} that occurred between 70 and 80 s (related to the development of the area surrounding the airport), as observed clearly in Fig. 2a. In summary, the dataset revealed that there were significant changes, indicated in parentheses, of the physics quantities between period A and B, respectively, i.e., RH (from 82.9 to 81.7%), mean fog T_{air} (from 12.8 to 14.6 °C), mean minimum fog T_{air} (from 2.2 to 7.7 °C), mean maximum fog T_{air} (from 20.3 to 20.7 °C) and fog hours (from 298 to 152 h).

Figure 6 shows the percentage of fog occurrence versus fog duration in hours for period A and B and observes that the occurrence of fog persists with the same durations of the period before the construction of the airport. However, fogs lasting 1 and 2 h increased, respectively, their occurrence frequencies of approximately 6.9 and 0.4% between period A and B, while the longer fogs decreased their frequencies.

Additionally, Fig. 7a, b show the probabilities of fog occurrences per month and hour of the day, i.e., when the horizontal visibility is less than 1 km, for periods A and B, respectively. Generally, the maximum probability of fog occurrences drops from over 50%, in Fig. 7a in period A to less than 40%, in Fig. 7b, in period B, which is likely due to the increase of T_{air} during period B.

4. Synoptic Analysis and the Use of SODAR and AMS During Two Fog Events

In the study of this airport, the fog typically occurs because of longwave cooling, which reduces

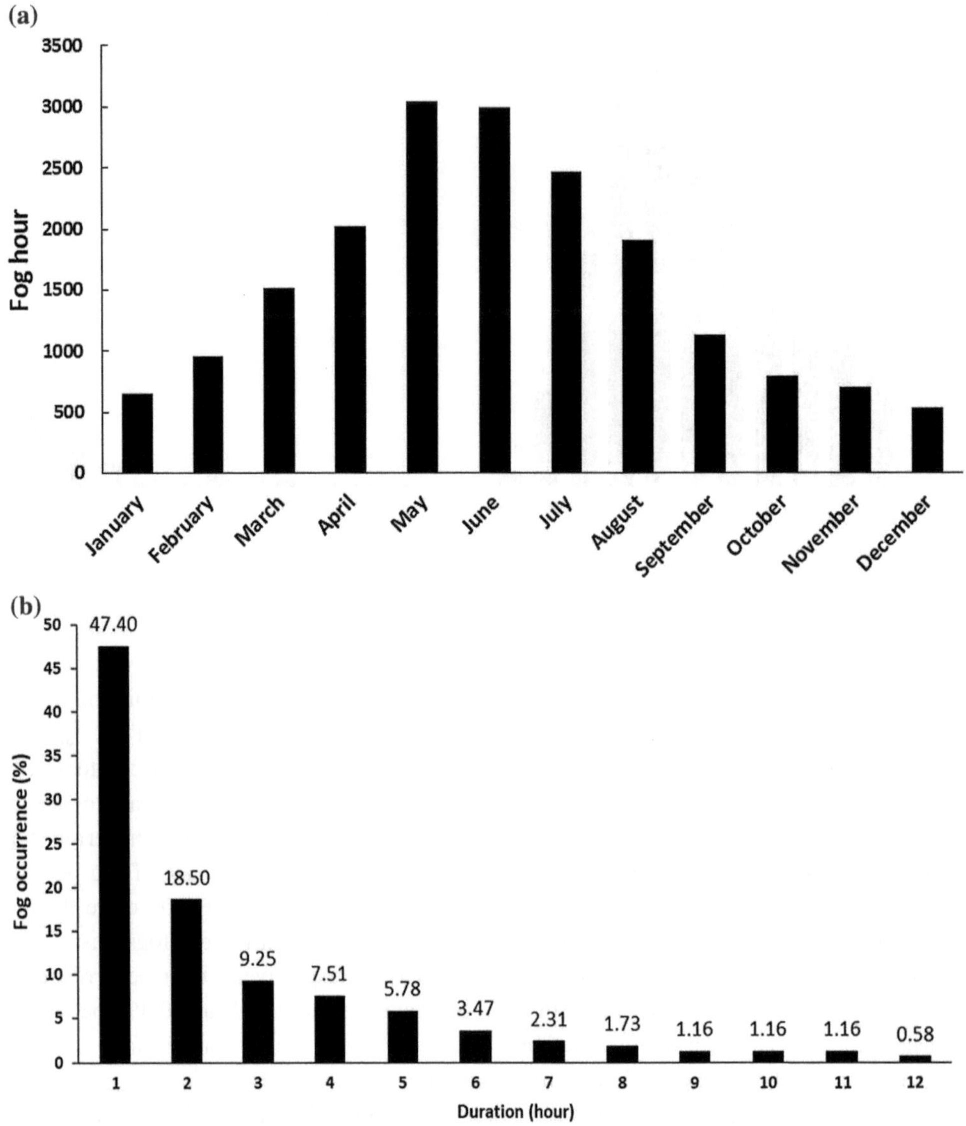

Figure 4

a, **b** The monthly variations of fog hours and percentages of fog occurrence versus its duration (h) during the period from 1951 to 2015, respectively

T_{air} to the dew point temperature (T_d), on clear nights with low wind speeds. After the fog onset, the liquid water in the fog increases due to the cooling rate, which will be higher with a dryer atmosphere (Teixeira and Miranda 2001). After sunrise, T_{air} begins to gradually increase at the surface, working in conjunction with the longwave cooling rate at the top layer of the fog, which causes an increase in the turbulent kinetic energy (TKE), and thus, the fog is intensified for some time until the fog dissipates. To analyze the importance of each measured and calculated physical quantity in the fog evolution (i.e., its onset, growth, and dissipation), two typical radiation fog events of 3:45 h that occurred on 08 September 2011 and 05 April 2012 (referred to herein as events 1 and 2) and are here characterized and discussed.

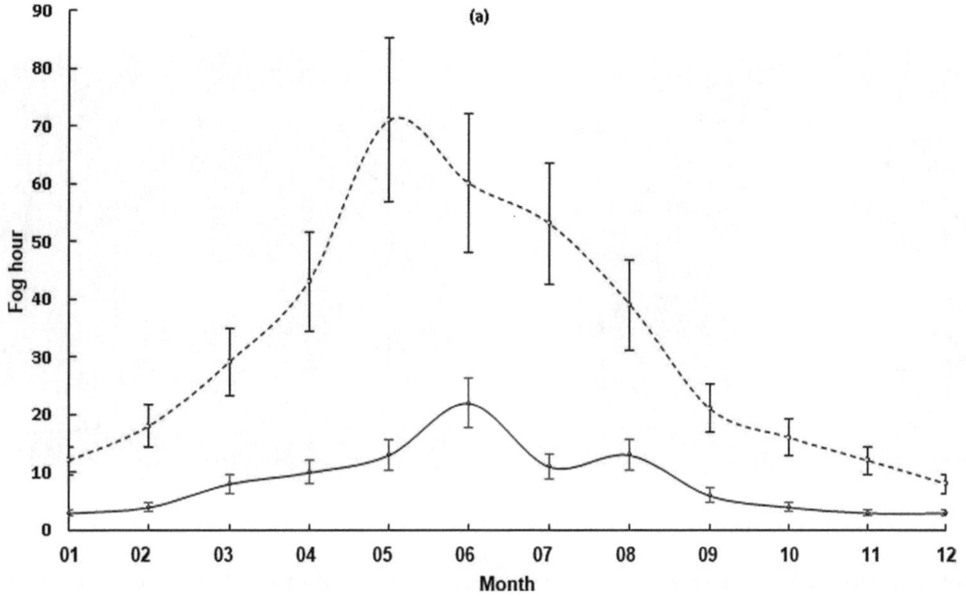

Figure 5
The average, minimum, maximum monthly fog hours for the periods corresponding to 1961–1990 (dash line) and 1991–2015 (solid line)

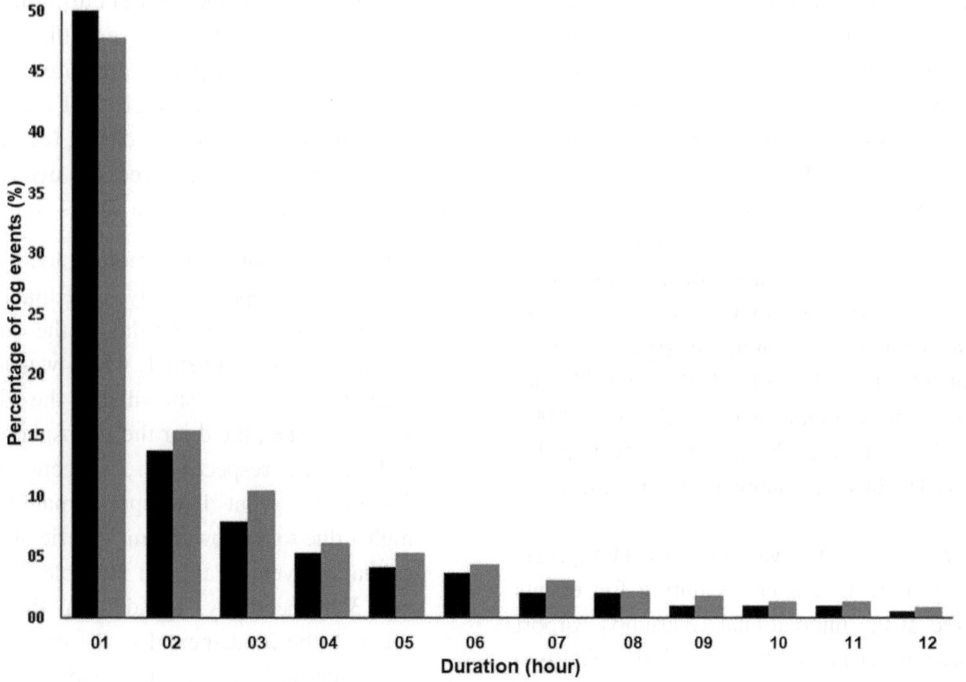

Figure 6
Percentages of fog versus its duration (h) during from 1951 to 1980 (gray bar) and 1991 to 2015 (black bar), respectively

The synoptic surface analysis event 1, represented by Fig. 8a, b, shows that the study area, represented roughly by the black rectangle, had been influenced by the proximity of South Atlantic Subtropical High, whose point core of 1027 hPa is approximately located at 26°S, 27°W, which brought divergence and

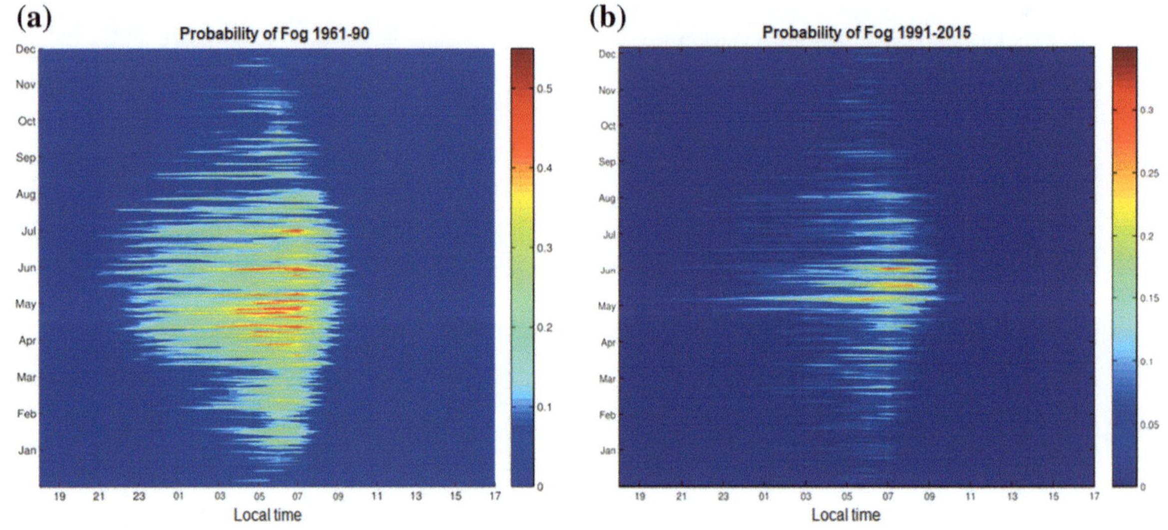

Figure 7

a, **b** The probabilities of the occurrence of fog (represented by the color bar) for the period before and after the construction of the Guarulhos International Airport for the periods of 1961–1990 and 1991–2015, respectively

clear sky, and thus, conditions for formation of radiation fog. It noted also that there was a very low intensity trough positioned over the area with NW–SE orientation. Similarly, to event 1, the synoptic surface analysis for event 2 (Fig. 8c, d) shows the positioning of South Atlantic Subtropical High, whose point core of 1026 hPa is approximately located at 38°S, 30°W influencing also the circulation in the study area bringing typical conditions, as aforementioned, for the formation of radiation fog.

Figure 8a, b and b, d are GOES 12 brightness temperature images of Channel 4 (spectral band centered at 11 μm) collected at 1000 (GMT) and 0900 (GMT) and synoptic analysis charts of 0000 (GMT) on 08 September 2011 and 05 April 2012, respectively. The black rectangle represents the study regions.

Figure 9a, b show the variations of TKE, BLH and visibility at 15-min intervals during fog events that occurred at the International Guarulhos Airport on 08 September 2011 and 05 April 2012. The two vertical lines limit the periods of the fog in the aforementioned figures, i.e., the lines delineate the period when the visibility is less than 1 km. Figure 9c, d show the Skew-*T* profile plots of T_{air} (solid black line on the right) and Td (solid black line on the left), collected via a sounding launched at Marte receiving station, located approximately 18 km from the airport, at 09:00 (local time) during fog events 1 and 2, respectively. Table 2 presents the averages and variations of the atmospheric pressure, T_{air}, cooling rate (CR), TKE and wind registered during the 7 h before (columns 2 and 3) and during (columns 4 and 5) the fog evolution registered for event 1 and 2, respectively. The investigation shows the following:

(a) The thickness of the fog represented by the BLH of event 1 is approximately twice that of event 2; a reasonable reason for this is the atmospheric conditions during event 1, which was much drier than event 2, as shown by the differences T_{air} − Td (i.e., the drier the atmosphere), as seen in Fig. 9b, d, respectively. As a consequence, the CR of the event 1 is approximately twice as high—due to the lower atmospheric absorption—as that of event 2, as seen in Table 2, columns 2 and 3, respectively. In addition, the CR value seems to be fundamental in the onset process of the radiation fog, as event 1 and 2;
(b) Despite a T_{air} difference of approximately 5 °C between events 1 and 2, T_{air} had no apparent impact on the duration of the fog;
(c) The wind blew from the northeast influenced by the proximity of South Atlantic Subtropical High,

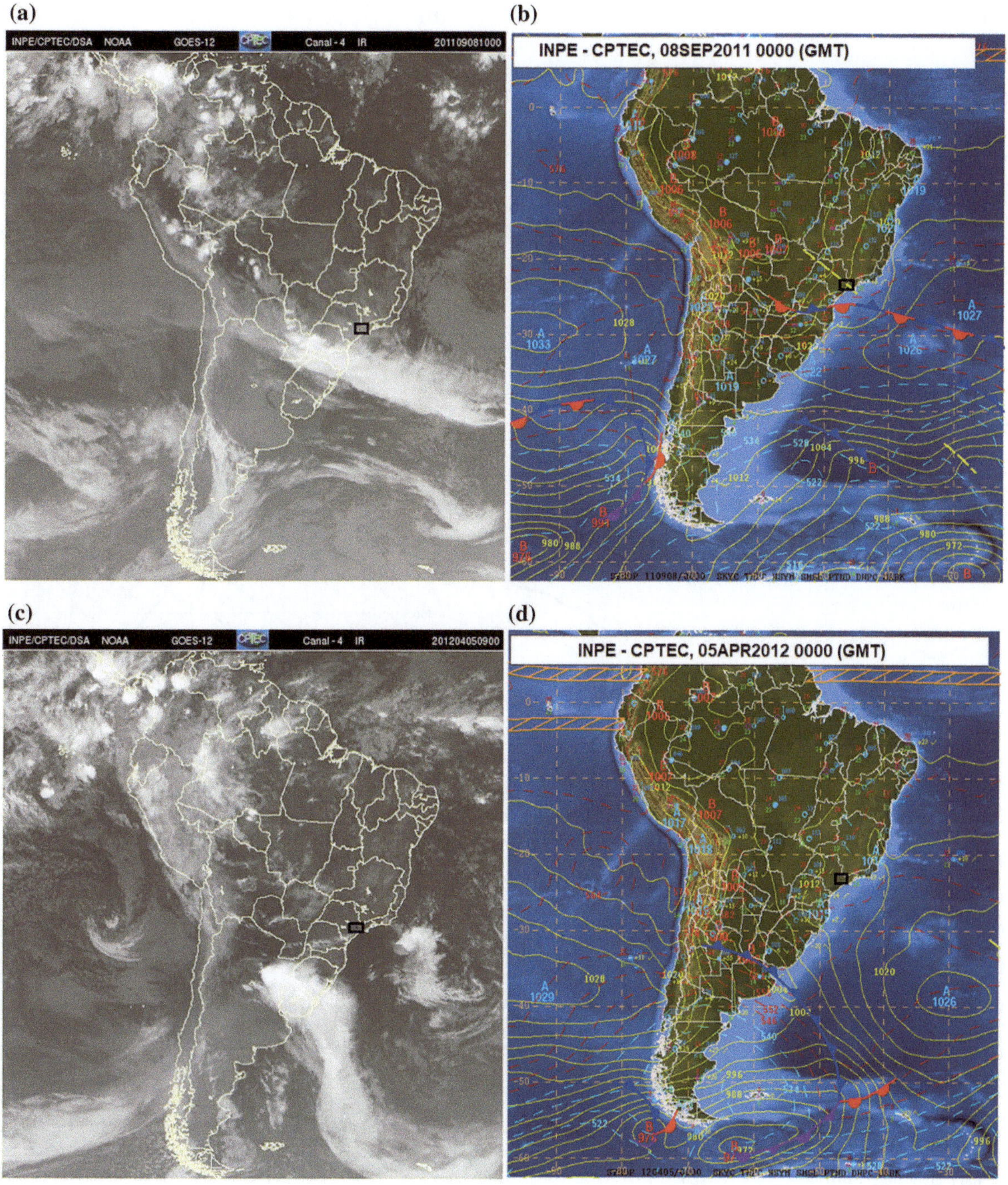

Figure 8
a–c and **b–d** depict images of brightness temperature of channel 4 registered by GOES-12 satellite at 1000 (GMT) and 0900 (GMT), which roughly represents the onset times of events 1 and 2 (as in Fig. 9a, c), and synoptic charts of 0000 (GMT) on 08 September 2011 and 05 April 2012, respectively, which are regularly made available by the National Institute for Space Research (INPE) on http://satelite.cptec.inpe.br

with wind speeds varying between approximately 1.5 and 4.0 m/s. The average speed is approximately three times greater in event 1 than in event 2 before the fog onset and is almost the same magnitude during the event maturations;

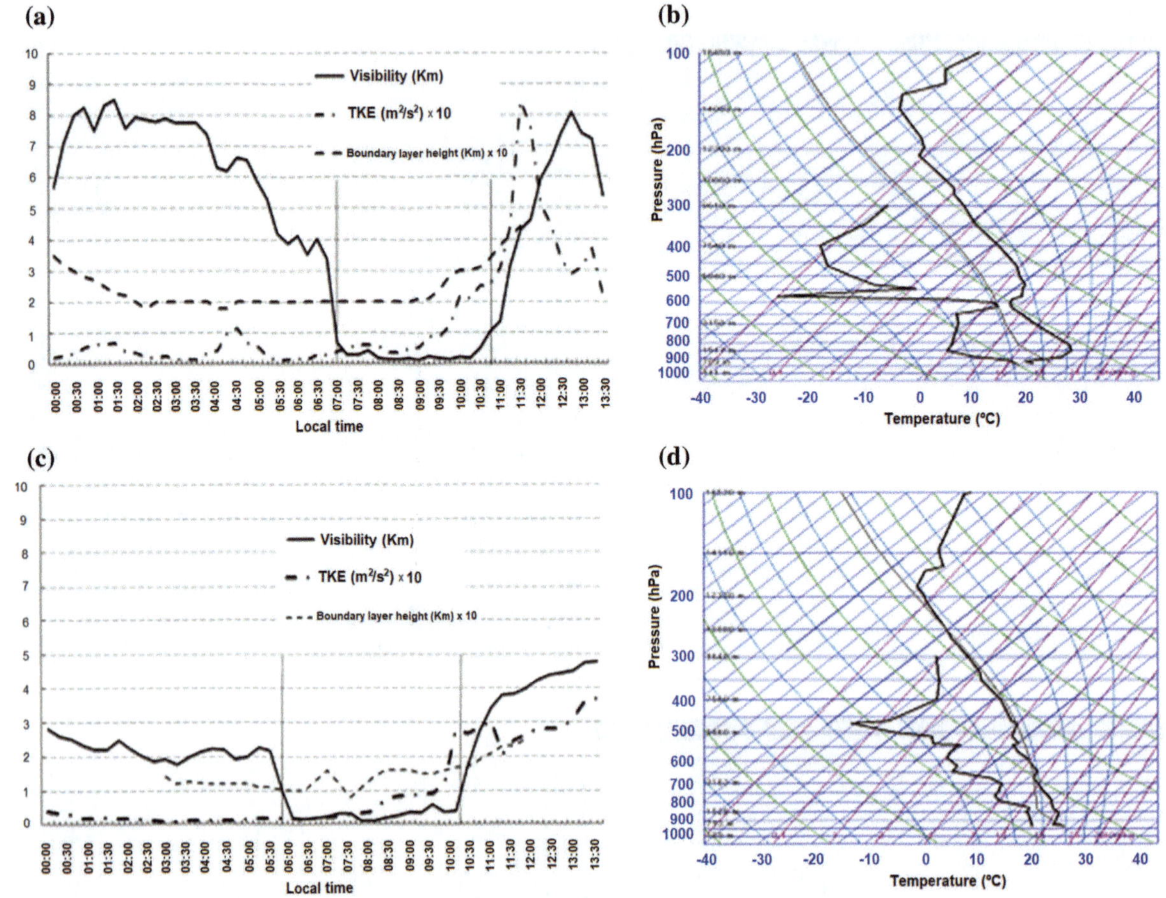

Figure 9

a–c The two time series of visibility (solid line), turbulent kinetic energy (dashed line with dot) and boundary layer height (dashed line with dot) at 15-min sampling intervals, and **b–d** show the Skew-*T* plots from the Marte soundings launched at 0900 h (local time) on 08 September 2011 and 05 April 2012, respectively

Table 2

Averages and variations of atmospheric pressure, T_{air}, cooling rate, TKE and the wind before and during the two fog events registered on 8 September 2011 and 5 April 2012

Physical quantity	7 h before fog onset		Fog period: 3:45 h	
	08-SET-2011	05-APR-2012	08-SET-11	05-APR-2012
Pressure (hPa)	1017.0 ± 1.1	1017.0 ± 1.1	1016.0 ± 0.5	1017.2 ± 0.8
Boundary layer height (m)	220 ± 1	140 ± 2	211 ± 1	134 ± 2
T_{air} (°C)	15.3 ± 0.8	22.7 ± 2.3	14.2 ± 0.3	19.5 ± 0.9
Cooling rate °C per h	− 0.41 ± 0.01	− 0.26 ± 0.78	− 0.25 ± 0.01	− 1.01 ± 0.21
TKE (m^2/s^2)	0.56 ± 0.22	0.031 ± 0.024	0.65 ± 0.31	0.039 ± 0.021
Wind speed (m/s)	2.9 ± 1.0	1.17 ± 0.48	2.1 ± 0.6	1.69 ± 0.35
Wind direction (°)	58.9 ± 15	68.5 ± 30.5	69.8 ± 41	60.89 ± 13

(d) As far as the fog dissipation is concerned, the TKE seems—as shown in Fig. 9a, c—to play a pivotal role in fog dissipation, giving a clear advance signal of the dissipation. In both figures, TKE (dashed line with dot) and BLH (dashed line) are positively correlated since their values tend to follow each other due to the turbulence process noted above, which is responsible for the intensification of the fog, and a few minutes later, triggers the fog dissipation, in concordance with the results presented by Dabas et al. (2011), which also used SODAR data to study fog evolution. On the other hand, the trigger for fog onset is not so clear, and thus, further investigations will be needed. In summary, the preliminary results presented show that SODAR is able to identify the BLH or the thickness of a fog layer; therefore, its data will be important for the future parameterization of fog onset, growth, and dissipation.

5. Conclusion

This paper mainly presents the climatological study of the fog occurrences before and after the construction of the International Guarulhos Airport for the period from 1951 to 2016. In addition, fog events were characterized using acoustic sounder and automatic meteorological station data with 15-min sampling intervals. The principal conclusions of this work are given below.

- The construction of the studied airport and local urban development promoted a significant increase in the fog temperature, which caused a significant decrease of fog hours (to approximately one fifth the initial value), which has lowered the restrictions on the landing and take-off operations at the airport.
- More fog events occur in the early hours of the day, with over 72% of them occurring during the fall–spring period. The majority of fog events (over 65% of fog events) had durations about 2 h.
- Most fogs were caused by longwave cooling on clear nights with low wind speeds and the values of the cooling rate and TKE play important roles their onset and growth-dissipation phases, respectively;
- The maximum probability of fog occurrence fell significantly from before to after the construction of the International Guarulhos Airport;
- The SODAR provided reliable measurements of the thickness of the fog, in concordance with the results obtained by Dabas et al. (2011). The preliminary results have revealed that the fog evolution is positively correlated with turbulent kinetic energy and is potentially an important physical quantity for the parameterization of fog onset, growth and dissipation.

Currently, researchers from the Applied Meteorological Laboratory of the Federal University of Rio de Janeiro and the operational community are involved in testing and validating models based on computational intelligence or deterministic modeling for short-term fog forecasting; the results will be published in future works.

Acknowledgements

The authors thank the Brazilian Aeronautic Command for supplying the meteorological data. This study was funded by the Organização Brasileira para o Desenvolvimento Cientifico e Tecnológico do Controle do Espaço Aéreo (CTCEA) GRANT: TC005-(C002/COPPETEC_CTCEA/2007).

References

Bergot, T., Carrer, D. J., & Noilhan, P. Bougeault. (2005). Improved site-specific numerical prediction of fog and low clouds: A feasibility study. *Weather Forecasting, 20,* 627–646.

Bergot, T., & Guedalia, D. (1994). Numerical forecasting of radiation fog. Part I: Numerical model and sensitivity tests. *Monthly Weather Review, 122,* 1218–1230.

Bott, A. (1991). On the influence of the physico-chemical properties of aerosols on the life cycle of radiation fogs. *Boundary-Layer Meteorology, 56,* 1–31.

Bott, A., Sievers, U., & Zdunkowski, W. (1990). A radiation fog model with a detailed treatment of the interaction between radiative transfer and fog microphysics. *Journal of the Atmospheric Sciences, 47,* 2153–2166.

Colabone, R. O., Ferrari, A. L., Vecchia, F. A. S., & Tech, A. R. B. (2015). Application of artificial neural networks for fog forecast. *Journal of Aerospace Technology and Management.* https://doi.org/10.5028/jatm.v7i2.446.

Da Rocha, R. P., Gonçalves, F. L. T., & Segalin, B. (2015). Fog events and local atmospheric features simulated by regional climate model for the metropolitan area of São Paulo, Brazil.

Atmospheric Research, 151, 176–188. https://doi.org/10.1016/j.atmosres.2014.06.010.

Dabas, J., Chauhan, A., Kumar, M. (2011). Existence of the mild solutions for impulsive fractional equations with infinite delay. *International Journal of Differential Equations, 2011.* https://doi.org/10.1155/2011/793023.

Emmons, G., & Montgomery, R. B. (1947). Note on the physics of fog formation. *Journal of Meteorology, 4,* 206.

Fedorova, N., Levit, V., & Fedorov, D. (2008). Fog and stratus formation on the coast of Brazil. *Atmospheric Research, 87,* 268–287. https://doi.org/10.1016/j.atmosres.2007.11.008.

Fedorova, N., Levit, V., Silva, A. O., & Santos, D. M. B. (2013). Low visibility formation and forecasting on the northern coast of brazil. *Pure and Applied Geophysics, 170*(4), 689–709. https://doi.org/10.1007/s00024-012-0565-6.

Gultepe, I., & Milbrandt, J. A. (2007). Microphysical observations and mesoscale model simulation of a warm fog case during FRAM project. *Journal of Pure and Applied Geophysics, 164,* 1161–1178.

Gultepe, I., Pearson, G., Milbrandt, J. A., Hansen, B., Platnick, S., Taylor, P., et al. (2009). The fog remote sensing and modeling (FRAM) field project. *Bulletin of the American Meteorological Society, 90,* 341–359.

Kunkel, B. A. (1984). Parameterization of droplet terminal velocity and extinction coefficient in fog models. *Journal of Climate and Applied Meteorology, 23,* 34–41.

Oliveira, G. A. (2002). Método estatístico no auxílio à previsão de nevoeiro para o aeroporto de Guarulhos, Master Thesis, Federal University of Santa Catarina, p. 104. https://repositorio.ufsc.br/xmlui/bitstream/handle/123456789/83415/188292.pdf?sequence=1&isAllowed=y.

Silva Dias, M. A. F., & Jaschke Machado, A. (1997). The role of local circulations in summertime convective development and nocturnal fog in São Paulo, Brazil. *Boundary-Layer Meteorology, 82,* 135–157. https://doi.org/10.1023/A:1000241602661.

Silva, W. L., Neto, F. A., França, G. B., & Matschinske, M. R. (2016). Conceptual model for runway change procedure in Guarulhos International Airport based on SODAR data. *The Aeronautical Journal, 120*(1227), 725–734.

Tardif, R. (2007). The impact of vertical resolution in the explicit numerical forecasting of radiation fog: A case study. *Journal Pure and Applied Geophysics, 164*(Special issue), 1221–1240.

Teixeira, J. (1999). Simulation of fog with the ECMWF prognostic cloud scheme. *Quarterly Journal of the Royal Meteorological Society, 125,* 529–553.

Teixeira, J., & Miranda, P. M. A. (2001). Fog prediction at Lisbon Airport using a one-dimensional boundary layer model. *Meteorological Applications, 8,* 497–505.

WMO. (1966). *International Meteorological Vocabulary, World Meteorological Organization, (WMO-No 1182)* (p. 276). Geneva: WMO. https://doi.org/10.1002/qj.49709339524.

WMO. (2011). *Guide to Climatological Practices, World Meteorological Organization, (WMO-No 100).* Geneva: WMO.

(Received September 14, 2017, revised January 15, 2018, accepted January 17, 2018, Published online February 1, 2018)

MIX
Papier aus verantwortungsvollen Quellen
Paper from responsible sources
FSC® C105338

If you have any concerns about our products,
you can contact us on
ProductSafety@springernature.com

In case Publisher is established outside the EU,
the EU authorized representative is:
Springer Nature Customer Service Center GmbH
Europaplatz 3, 69115 Heidelberg, Germany

Printed by Libri Plureos GmbH
in Hamburg, Germany